EmQM13: Emergent Quantum Mechanics 2013

Journal of Physics: Conference Series Volume 504

Vienna, Austria
3 – 6 October 2013

ISBN: 978-1-63439-284-6
ISSN: 1742-6588

Printed from e-media with permission by:

Curran Associates, Inc.
57 Morehouse Lane
Red Hook, NY 12571

Printed by Curran Associates, Inc. (2014)

For permission requests, please contact the Institute of Physics
at the address below.

Institute of Physics
Dirac House, Temple Back
Bristol BS1 6BE UK

Phone: 44 1 17 929 7481
Fax: 44 1 17 920 0979

techtracking@iop.org

Additional copies of this publication are available from:

Curran Associates, Inc.
57 Morehouse Lane
Red Hook, NY 12571 USA
Phone: 845-758-0400
Fax: 845-758-2634
Email: curran@proceedings.com
Web: www.proceedings.com

EmQM13: Emergent Quantum Mechanics 2013

Journal of Physics: Conference Series Volume 504

Vienna, Austria
3 – 6 October 2013

ISBN: 978-1-63439-284-6
ISSN: 1742-6588

IOP Institute of Physics

TABLE OF CONTENTS

INVITED LECTURES

THE NON-SIGNALLING THEOREM IN GENERALIZATIONS OF BELL'S THEOREM 1
J Walleczek, G Grössing
WHERE IS QUANTUM THEORY HEADED? 11
S L Adler
PHYSICS ON THE BOUNDARY BETWEEN CLASSICAL AND QUANTUM MECHANICS 17
G 'T Hooft
THE LINEARITY OF QUANTUM MECHANICS FROM THE PERSPECTIVE OF HAMILTONIAN CELLULAR AUTOMATA 31
H-T Elze
NONLINEAR RICCATI EQUATIONS AS A UNIFYING LINK BETWEEN LINEAR QUANTUM MECHANICS AND OTHER FIELDS OF PHYSICS 40
D Schuch
RELATIONAL CAUSALITY AND CLASSICAL PROBABILITY: GROUNDING QUANTUM PHENOMENOLOGY IN A SUPERCLASSICAL THEORY 51
G Grössing, S Fussy, J Mesa Pascasio, H Schwabl
EMERGENCE OF QUANTIZATION: THE SPIN OF THE ELECTRON 66
A M Cetto, L De La Peña, A Valdés-Hernández
A SUBQUANTUM ARROW OF TIME 78
T M Nieuwenhuizen
ENTROPIC DYNAMICS: AN INFERENCE APPROACH TO QUANTUM THEORY, TIME AND MEASUREMENT 88
A Caticha
SPIN AND CHARGE FROM SPACE AND TIME 100
M Faber
WHICH CAME FIRST, SPACETIME OR CLOCKS? 115
G N Ord
COOPERATIVE DYNAMICAL PROCESSES: THE EMERGENCE OF RELATIVISTIC QUANTUM THEORY 127
P Jizba, F Scardigli
STOCHASTIC MECHANICS OF RELATIVISTIC FIELDS 136
E Nelson
ELEMENTS OF PHYSICS FOR THE 21ST CENTURY 140
W A Hofer
CAUSALITY AND LOCAL DETERMINISM VERSUS QUANTUM NONLOCALITY 156
M Kupczynski
WEAK MEASUREMENT AND ITS EXPERIMENTAL REALISATION 167
R Flack, B J Hiley
NON-LOCALITY AND DESTRUCTIVE INTERFERENCE OF MATTER WAVES 180
H Rauch
TESTING SUPERDETERMINISTIC CONSPIRACY 189
S Hossenfelder
CLASSICAL PROBABILITY MODEL FOR BELL INEQUALITY 193
A Khrennikov
NEWTON FORCE FROM WAVE FUNCTION COLLAPSE: SPECULATION AND TEST 203
L Diósi
GRAVITATIONAL DECOHERENCE, ALTERNATIVE QUANTUM THEORIES AND SEMICLASSICAL GRAVITY 210
B L Hu
COLLAPSE MODELS AND PERCEPTUAL PROCESSES 225
G C Ghirardi, R Romano
COLLAPSE MODELS: FROM THEORETICAL FOUNDATIONS TO EXPERIMENTAL VERIFICATIONS 239
A Bassi, H Ulbricht

HEISENBERG'S UNCERTAINTY RELATION: VIOLATION AND REFORMULATION 249
M Ozawa
INVESTIGATIONS OF FUNDAMENTAL PHENOMENA IN QUANTUM MECHANICS WITH NEUTRONS 261
Y Hasegawa
EVENT-BASED SIMULATION OF NEUTRON EXPERIMENTS: INTERFERENCE, ENTANGLEMENT AND UNCERTAINTY RELATIONS 274
K Michielsen, H De Raedt
HAMILTONIAN FLOWS, SHORT-TIME PROPAGATORS AND THE QUANTUM ZENO EFFECT 289
M A De Gosson, B J Hiley
PARTICLES, WAVES AND TRAJECTORIES: 210 YEARS AFTER YOUNG'S EXPERIMENT 302
Á S Sanz
NEW INSIGHTS ON EMERGENCE FROM THE PERSPECTIVE OF WEAK VALUES AND DYNAMICAL NON-LOCALITY 316
J Tollaksen
Author Index

Preface

These proceedings comprise the invited lectures of the second international symposium on Emergent Quantum Mechanics (EmQM13), which was held at the premises of the Austrian Academy of Sciences in Vienna, Austria, 3-6 October 2013.

The symposium was held at the "Theatersaal" of the Academy of Sciences, and was devoted to the open exploration of emergent quantum mechanics, a possible "deeper level theory" that interconnects three fields of knowledge: emergence, the quantum, and information. Could there appear a revised image of physical reality from recognizing new links between emergence, the quantum, and information? Could a novel synthesis pave the way towards a 21st century, "superclassical" physics? The symposium provided a forum for discussing (i) important obstacles which need to be overcome as well as (ii) promising developments and research opportunities on the way towards emergent quantum mechanics. Contributions were invited that presented current advances in both standard as well as unconventional approaches to quantum mechanics.

The EmQM13 symposium was co-organized by Gerhard Grössing (Austrian Institute for Nonlinear Studies (AINS), Vienna), and by Jan Walleczek (Fetzer Franklin Fund, USA, and Phenoscience Laboratories, Berlin). After a very successful first conference on the same topic in 2011, the new partnership between AINS and the Fetzer Franklin Fund in producing the EmQM13 symposium was able to further expand interest in the promise of emergent quantum mechanics.

The symposium consisted of two parts, an opening evening addressing the general public, and the scientific program of the conference proper. The opening evening took place at the Great Ceremonial Hall (Grosser Festsaal) of the Austrian Academy of Sciences, and it presented talks and a panel discussion on "The Future of Quantum Mechanics" with three distinguished speakers: Stephen Adler (Princeton), Gerard 't Hooft (Utrecht) and Masanao Ozawa (Nagoya).

The articles contained in these proceedings represent the talks of the invited speakers as written immediately after the symposium. The volume starts with a contribution by organizers Jan Walleczek and Gerhard Grössing, essentially explaining why emergent quantum mechanics, and other deterministic approaches to quantum theory, must be considered viable approaches in quantum foundations today. This is followed by the exposition of Stephen Adler's talk who introduced to a general audience key questions at the current frontiers of quantum mechanics during the opening evening (with the contents of his conference talk appearing elsewhere). The conference proceedings then continues with the presentations as given in their chronological order i.e. starting with the opening talk of the scientific program by Gerard 't Hooft. While the page number was restricted for all invited speakers, the paper by Jeff Tollaksen was given more space, as his invited collaborator Yakir Aharonov was unable to deliver a separate talk, in order to represent both contributions in one paper. Note that the talks of all speakers, including the talks of those who could not be represented in this volume (M. Arndt, B. Braverman, C. Brukner, S. Colin, Y. Couder, B. Poirier, A. Steinberg, G. Weihs and H. Wiseman) are freely available on the conference website as video presentations (`http://www.emqm13.org`).

The organizers wish to express their gratitude to Siegfried Fussy and Herbert Schwabl from AINS for the organizational support. The organizers also wish to thank Bruce Fetzer, President and CEO, John E. Fetzer Memorial Trust, and the Members of the Board of Trustees, for their strong support and for funding this symposium.

We also wish to thank the Austrian Academy of Sciences for allowing the symposium to be held on their premises, and Anton Zeilinger, President of the Austrian Academy of Sciences, for his welcome address. The expertise of the Members of the Scientific Advisory Board of the EmQM13 symposium, Ana Maria Cetto (Mexico), Lajos Diósi (Budapest), Maurice de Gosson (Vienna), Edward Nelson (Princeton), Theo Nieuwenhuizen (Amsterdam) and Helmut Rauch (Vienna), is also gratefully acknowledged.

Finally, it is a pleasure to again thank Sarah Toms and her team at IOP Publishing (Bristol) for their friendly advice and help during the preparation of these proceedings.

Vienna, Pisa, Berlin, February 2014

Gerhard Grössing,
Hans-Thomas Elze,
Johannes Mesa Pascasio,
Jan Walleczek

The front cover image shows two bouncing oil droplets on an oscillating oil surface, as they are employed by Couder, Fort, Bush, and others to show macroscopic analogues of wave-particle complementarity (courtesy of Dan Harris and John Bush, MIT).

Journal of Physics: Conference Series **504** (2014) 011002 doi:10.1088/1742-6596/504/1/011002

Peer review statement

All papers published in this volume of *Journal of Physics: Conference Series* have been peer reviewed through processes administered by the proceedings Editors. Reviews were conducted by expert referees to the professional and scientific standards expected of a proceedings journal published by IOP Publishing.

Journal of Physics: Conference Series **504** (2014) 012001 doi:10.1088/1742-6596/504/1/012001

The Non-Signalling theorem in generalizations of Bell's theorem

J Walleczek[1] **and G Grössing**[2]

[1] Phenoscience Laboratories, Chausseestrasse 8, 10115 Berlin, Germany

[2] Austrian Institute for Nonlinear Studies, Akademiehof, Friedrichstr. 10, 1010 Vienna, Austria

E-mail: `walleczek@phenoscience.com`, `ains@chello.at`

Abstract. Does "epistemic non-signalling" ensure the peaceful coexistence of special relativity and quantum nonlocality? The possibility of an affirmative answer is of great importance to deterministic approaches to quantum mechanics given recent developments towards generalizations of Bell's theorem. By generalizations of Bell's theorem we here mean efforts that seek to demonstrate the impossibility of any deterministic theories to obey the predictions of Bell's theorem, including not only local hidden-variables theories (LHVTs) but, critically, of nonlocal hidden-variables theories (NHVTs) also, such as de Broglie-Bohm theory. Naturally, in light of the well-established experimental findings from quantum physics, whether or not a deterministic approach to quantum mechanics, including an emergent quantum mechanics, is logically possible, depends on compatibility with the predictions of Bell's theorem. With respect to deterministic NHVTs, recent attempts to generalize Bell's theorem have claimed the impossibility of any such approaches to quantum mechanics. The present work offers arguments showing why such efforts towards generalization may fall short of their stated goal. In particular, we challenge the validity of the use of the non-signalling theorem as a conclusive argument in favor of the existence of free randomness, and therefore reject the use of the non-signalling theorem as an argument against the logical possibility of deterministic approaches. We here offer two distinct counter-arguments in support of the possibility of deterministic NHVTs: one argument exposes the circularity of the reasoning which is employed in recent claims, and a second argument is based on the inconclusive metaphysical status of the non-signalling theorem itself. We proceed by presenting an entirely informal treatment of key physical and metaphysical assumptions, and of their interrelationship, in attempts seeking to generalize Bell's theorem on the basis of an ontic, foundational interpretation of the non-signalling theorem. We here argue that the non-signalling theorem must instead be viewed as an epistemic, operational theorem i.e. one that refers exclusively to what epistemic agents can, or rather cannot, do. That is, we emphasize that the non-signalling theorem is a theorem about the operational inability of epistemic agents to signal information. In other words, as a proper principle, the non-signalling theorem may only be employed as an **epistemic**, **phenomenological**, or **operational** principle. Critically, our argument emphasizes that the non-signalling principle must not be used as an ontic principle about physical reality **as such**, i.e. as a theorem about the nature of physical reality independently of epistemic agents e.g. human observers. One major reason in favor of our conclusion is that any definition of signalling or of non-signalling invariably requires a reference to epistemic agents, and what these agents can actually measure and report. Otherwise, the non-signalling theorem would equal a general "no-influence" theorem. In conclusion, under the assumption that the non-signalling theorem is epistemic (i.e. "epistemic non-signalling"), the search for deterministic approaches to quantum mechanics, including NHVTs and an emergent quantum mechanics, continues to be a viable research program towards disclosing the foundations of physical reality at its smallest dimensions.

Introduction

Historically, the first successful attempt in fashioning a deterministic approach to quantum mechanics was the one now known as the de Broglie-Bohm theory, and that theory introduced the notion of "hidden variables" (Bohm [ ]). In turn, Bohm's consideration that hidden variables might possibly underlie quantum phenomena, in conjunction with the earlier analysis by Einstein, Podolsky and Rosen [ ], inspired John Bell to develop the theorem named after him (Bell [ ]). 50 years on, it is widely accepted that Bell's groundbreaking theorem has proven beyond doubt that no (deterministic) quantum theory based on local hidden variables could account for the observations of quantum physics. However, while Bell's theorem ruled out as an explanation any local hidden-variables theory (LHVT), Bohm's original proposal of a nonlocal hidden-variables theory (NHVT) remained a viable possibility at least on logical grounds, even after Bell (e.g. Bell [ ]).

In recent years, however, efforts have gained prominence that implied that **all** types of deterministic approaches to quantum mechanics, not only LHVTs, but also all forms of NHVTs such as de Broglie-Bohm theory, must be in violation of Bell's theorem. In the context of the present article, such efforts we will refer to as efforts seeking to generalize Bell's theorem. Recent publications with important implications for generalization, in the here proposed sense, of Bell's theorem are, for example, Colbeck and Renner [,] and Gallego et al. [ ]. (Their work builds on previous work that has similar implications such as Barrett et al. [ ]; Kofler et al. [ ]; Pironio et al. [ ]). Both Colbeck and Renner [,], and Gallego et al. [ ], have argued that even those deterministic quantum theories, which are unconstrained by Bell's locality assumption (i.e. an NHVT such as de Broglie-Bohm theory), must necessarily violate the non-signalling theorem or principle. Gallego et al. [ ] also argued that their own work had important "philosophical and physics-foundational implications". Because of their interest in foundational issues as well and also because this work is the most recent one we included in our analysis, we will use Gallego et al. [ ] as the key reference in our exploration of basic physical and metaphysical assumptions concerning the non-signalling theorem. While in no way do we question the great technical achievements which are presented in the mentioned articles, we do question the finality of assumptions expressed therein concerning the possibility of deterministic approaches to quantum mechanics. Importantly, these authors, like many researchers in quantum physics in general, take for granted a foundational, ontic interpretation of the non-signalling theorem. By contrast, the present work will argue that for deterministic approaches to quantum mechanics an exclusively **epistemic**, **phenomenological**, or **operational**, interpretation of the non-signalling theorem does in fact ensure compatibility between special relativity and quantum nonlocality. Before proceeding with our analysis of the non-signalling theorem, and of possible assumptions underlying the theorem, we will start with a brief summary of quantum phenomenology as apparent from EPR-type experiments.

1. Quantum phenomenology of EPR-type experiments

What is the phenomenon that is in need of explanation by quantum-theoretical approaches whether by standard indeterminist or by unconventional determinist approaches? We here focus on quantum phenomenology, such as on nonlocal correlation phenomena, as apparent from EPR-type experiments (e.g. Aspect et al. [,]; Weihs et al. [ ]; Tittel et al. [ ]; Ursin et al. [ ]; Giustina et al. [ ]). In the following sections (1.1. to 1.3.) we present a brief overview of different metaphysical interpretations of the same quantum-phenomenological observations.

1.1. Nonlocal, instantaneous correlations signify the existence of instantaneous influences

Orthodox quantum theory makes no claim regarding the nature of physical reality that might underlie the formation of nonlocal i.e. instantaneous correlations. Everyone agrees that the term 'correlation' merely represents a descriptive, epistemic term i.e. 'correlation' defines a state of knowledge only. Therefore, the use of the term neither implies underlying causal processes nor the existence of any fundamental physical elements or ontic structures in general. However, at the same time, we consider the fact to be non-controversial that instantaneous correlations manifest **physically** at the level of distant measuring devices. It goes without saying that lacking the power to manifest physically, instantaneous correlations would be beyond the capacity of epistemic agents to observe. Therefore, the fact that instantaneous correlations can indeed be observed as **physical effects** in concrete i.e. ontically-real systems, we take as evidence that some kind of "instantaneous" or "quasi-instantaneous" influence must be at work as part of the formation, between space-like separated quantum detectors, of nonlocal correlations. We here do not engage the difficult question of whether the existence of "nonlocal correlations" is evidence of nonlocal or superluminal **causation** (e.g. see Maudlin [ ]). In fact, it is of no consequence for our subsequent argument that instantaneous influences – at a minimum – represent near-instantaneous superluminal influences, whether or not influences are assumed to be "nonlocally causal" (e.g. as an instantaneous "common cause"), or are characterized only as "a-causal" or "formal".

1.2. Instantaneous influences associated with nonlocal correlations are, at a minimum, near-instantaneous, superluminal influences

The instantaneous influences usually associated with observation in EPR-type experiments of nonlocal correlations would be influences, again whether causal or a-causal (see 1.1.), that operate, by definition, at an infinite velocity. However, as is well-known, the assumption of infinite velocity could never be proven experimentally. In other words, the question appears to be undecidable by any conceivable experiment of whether or not an observation, such as the appearance of nonlocal correlations, is in fact due to an infinite-velocity influence (i.e. an "instantaneous correlation"). By contrast, the generation of positive evidence for finite-velocity, yet superluminal, influences is possible by way of scientific experimentation, at least in principle (e.g. see Salart et al. [ ]; Cocciaro et al. [ ]). In any case, what can, and indeed has been, established by way of EPR-type experiments is that influences exist in nature which, if they are not instantaneous, then they at least must be superluminal (however, they must not be signalling, of course). Consequently, we here take a conservative position and assume that nonlocal correlations could either be instantaneous correlations or near-instantaneous, superluminal correlations; again, the reason is that proof of instantaneity is beyond scientific measurement.

1.3. Nonlocal correlations are unpredictable but not necessarily intrinsically random correlations

It is well-known to physicists that no mathematical proof exists which could decide "whether or not a given series of digits is in fact random or only seems random" (Chaitin [ ]). This does not mean, of course, that powerful statistical tests are unavailable for demonstrating the lack of **certain** patterns in a digital sequence; nevertheless "no finite set of tests can ever be considered complete, as there may be patterns not covered by such tests" (Pironio et al. [ ]). In short, it is undecidable – as a matter of principle again – whether a digital sequence, for example, a data sequence obtained from an EPR-type experiment, is genuinely random or merely pseudorandom. The principal undecidability of whether data sequences derived from quantum experiments are either random or merely pseudorandom has the following consequence: no experiment could refute the possibility that deterministic processes might rule at all levels of physical reality, including at the level of the quantum. Therefore, while indeterminism remains a metaphysical assumption which is greatly preferred over determinism by many quantum physicists, it must be noted also that indeterminism is neither a conclusive experimental fact nor a logical requirement on the basis of existing evidence. Importantly, in relation to EPR-type experimental findings, the two competing metaphysical assumptions – determinism and indeterminism – share an all-important feature: the **statistical unpredictability** of measurement outcomes. Therefore, given the above-described undecidability concerning the existence, or not, of free randomness, we here take again a conservative position, and we characterize EPR-type correlation phenomena as merely "unpredictable" instead of as a sure sign of "intrinsic randomness".

For example, statistical unpredictability may be the result of intrinsic complexity rather than intrinsic randomness. That is, unpredictability could be entirely a function of deterministic processes, i.e. at the level of individual events predictability would be impossible due to lack of precise knowledge. Famously, in deterministic chaos, (long-term) unpredictability of behavior is due to **intrinsic complexity**, and not due to intrinsic randomness. Therefore, in the case of complexly-structured, deterministic systems an assumption of free randomness is not required to account for the appearance of statistical unpredictability of measurement outcomes.

Therefore, it should remain an open question in quantum physics whether, or not, the correlations that are observed in EPR-type experiments do in fact represent **intrinsically random, instantaneous** correlations. From the alternative perspective of deterministic theory, we here argue that the quantum phenomenology as established by EPR-type experiments merely indicates that these correlations are **unpredictable**, not intrinsically random, and that they either may be **near-instantaneous**, **superluminal** or **instantaneous** (1.1. to 1.3.).

2. Evaluating the validity of the non-signalling theorem in generalizations of Bell's theorem

During the last decade, as was already mentioned in the Introduction, an argument has gained traction, especially in work concerning the foundations of quantum cryptography, which uses the validity of the non-signalling theorem, to generalize the predictions of Bell's theorem i.e. to rule out the possibility of NHVTs including de Broglie-Bohm theory. In summary, the argument has three stages: (1) assume the validity of the non-signalling theorem as a foundational principle concerning the nature of physical reality, (2) if so, then the unpredictability of nonlocal correlations as observed in EPR-type experiments is essentially due to intrinsic randomness, and therefore (3) deterministic approaches to quantum mechanics, including even NHVTs, are impossible. A critique of this argument will be the main topic for the remainder of this article.

2.1. On the use of the non-signalling theorem in the foundations of quantum cryptography

There is a wide-spread belief, or basic assumption, especially in the growing literature on quantum cryptography, that the non-signalling theorem guarantees the presence of "free randomness" i.e. of fundamental indeterminism. Conversely, the belief is wide-spread also that a deterministic approach, in combination with nonlocality, necessarily must imply superluminal signalling. For example, one representative statement from that literature reads "any state that is deterministic and nonlocal allows signalling" (e.g. Barrett et al. [8]). However, if that statement were true in an ultimate sense, then Bell's theorem would not only negate the possibility of LHVTs but also of any NHVTs such as de Broglie-Bohm theory. Traditionally, however, compatibility of NHVTs and the predictions of Bell's theorem had always been assumed (e.g. Bell [?]; Bohm and Hiley [21]; Holland [22]; Valentini [23]). Consequently, there now appears to have surfaced a fundamental contradiction between the well-known predictions of Bell's theorem, on the one hand, and the recent claims in the literature on quantum cryptography concerning the use of the non-signalling theorem as a foundational, instead of as an operational, principle. As evidence for the "new" incompatibility we will next discuss a number of statements by the authors who were mentioned in the Introduction already (Gallego et al. [7]). We will point out, in appropriate places, that the presumption of incompatibility results from the assumption that the non-signalling principle serves as an ontic, foundational principle, which, as we will explain, presents a view of the principle that has lost its essential connection to epistemological concerns.

According to Gallego et al. [ ], the non-signalling principle "states that no instantaneous communication is possible, which in turn imposes a local structure on events, as in Einstein's special relativity." In other words, any kind of superluminal signalling or communication is prohibited, whether by instantaneous or quasi-instantaneous, superluminal influences. (For the remainder of this analysis we will only discuss possible consequences of instantaneous influences in the context of non-signalling.) We fully agree with Gallego et al. [ ] that this indeed is the essential meaning of the non-signalling principle. Put differently, the impossibility by epistemic agents to signal each other by way of quantum nonlocality protects against violations of special relativity. However, we question the basic assumption by Gallego et al. [ ] that the non-signalling principle applies also to instantaneous influences that are beyond control for the purposes of communication or signalling. Crucially, the non-signalling principle does not state "no instantaneous influences are possible", but it only states "no instantaneous communication is possible"; otherwise, for example, the formation of nonlocal correlations in EPR-type experiments would be impossible to account for, or even to talk about, in terms of possible explanations in physics (compare 1.1.). Importantly, we point out that as long as control is impossible of instantaneous influences, these influences, even if acting instantaneously, do not violate the non-signalling constraint. In other words, if such influences are informationally inaccessible to epistemic agents, i.e. if influences are genuinely "hidden influences" (e.g. "hidden variables"), then instantaneous influences by way of nonlocal hidden variables do not have to contradict special relativity. This is essentially what is implied by an epistemic, operational, or phenomenological interpretation of the non-signalling theorem.

Valentini [ ], for example, showed that "all deterministic hidden-variable theories, that reproduce quantum theory for a 'quantum equilibrium' distribution of hidden variables, predict the existence of instantaneous signals at the statistical level for hypothetical 'nonequilibrium ensembles'." Crucially, however, the assumption of "instantaneous signals at the statistical level" for quantum theories sharing both determinism and nonlocality does not mean that such theories would have to contradict the non-signalling theorem. For example, after considering the possibility, between two distant members of a correlated pair of spin-$\frac{1}{2}$ particles, "of nonlocal information flow at the hidden-variable level", Valentini [ ] explained: "Of course, in equilibrium this information flow is not visible at the statistical level, because as many outcomes flip from +1 to -1 as from -1 to +1." An epistemic, phenomenological interpretation of the non-signalling theorem is therefore implicit in the argument by Valentini [ ]: "It is as if there is a conspiracy in the laws of physics that prevents us from using nonlocality for signalling. But another way of looking at the matter is to suppose that our universe is in a state of statistical equilibrium at the hidden-variable level, a special state in which nonlocality happens to be hidden. The physics we see is not fundamental; it is merely a phenomenological description of an equilibrium state."

However, it is apparent that Gallego et al. [ ] have excluded, as of course have many others (e.g. Barrett et al. [ ]; Kofler et al. [ ]; Pironio et al. [ ]; Colbeck and Renner [,]), as a possibility an epistemic, phenomenological interpretation of the non-signalling constraint: "In fact, Bohm's theory is both deterministic and able to produce all quantum predictions, but it is **incompatible with no-signalling at the level of hidden variables**." And, they continue: "Thus, we assume throughout the validity of the no-signalling principle." From this last statement it becomes clear that these authors, like many before them, have committed to a view of the non-signalling theorem which has eliminated any vital reference to epistemic agents, and to what epistemic agents can or can't do; they have "ontologized", so to speak, a principle derived solely on the basis of epistemic considerations i.e. signalling or non-signalling. In consequence, the "non-signalling" principle was tacitly turned into a "no hidden-influence"

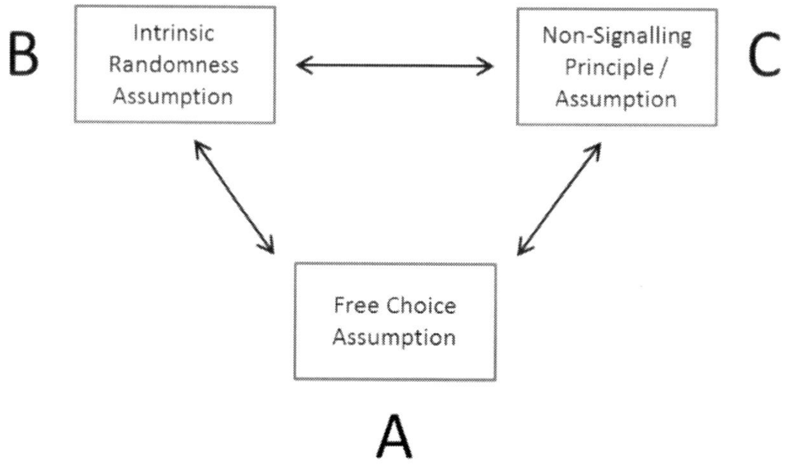

Figure 1. Relational diagram showing the interdependency of three basic assumptions (A–C) which are discussed, for example, in Colbeck and Renner [,], and Gallego et al. [ ]. The validity of each of these assumptions is central to the conclusion that deterministic theories, including those based upon nonlocal hidden variables (i.e. de Broglie-Bohm theory) are impossible because they contradict the non-signalling principle.

principle. The justification for this move and for transforming the very meaning of non-signalling has not yet been revealed. Is this a valid interpretation of the non-signalling principle? Can the notion of "non-signalling" really be reduced to a notion of "non-influencing"?

Regarding the problem of "non-signalling" versus "non-influencing", specifically in relation to NHVTs such as de Broglie-Bohm theory, for example, Holland [ ] commented: "To summarize, the quantum potential implies that a certain kind of 'signalling' does, in fact, take place between the sites of distantly separated... particles in an entangled state, if one of the particles undergoes a local interaction. This transfer of information cannot, however, be extracted by any experiment which obeys the laws of quantum mechanics." This is in agreement with the above-cited position by Valentini [ ] who explained that in the manifestation of nonlocal correlations "information flow is not visible at the statistical level"; therefore control of information flows by epistemic agents for the purposes of signalling is again prohibited. Finally, Holland [ ] concluded that therefore "...the failure of quantum correlations to provide a signalling mechanism at the empirical level is consistent with the requirement of special relativity that no signal be transmitted faster than the speed of light."

2.2. Circular reasoning in the relationship between the non-signalling principle and the assumption of intrinsic randomness

We will next describe another argument against the use of the non-signalling principle as an ontic foundational principle in favor of the existence of intrinsic randomness, in addition to the argument already presented above (2.1.). This second argument is based on the following simple observations: it is exactly because standard quantum theory assumes the unpredictability of nonlocal correlations to be evidence of intrinsic randomness that these correlations are presumed to be "intrinsically non-signalling" in the first place; it is equally valid therefore to say that the non-signalling assumption (C in Fig. 1) is a consequence of the prior free (intrinsic) randomness assumption (B in Fig. 1).

Similarly to the first step of the argument by Gallego et al. [ ] which did correctly recognize

Journal of Physics: Conference Series **504** (2014) 012001 doi:10.1088/1742-6596/504/1/012001

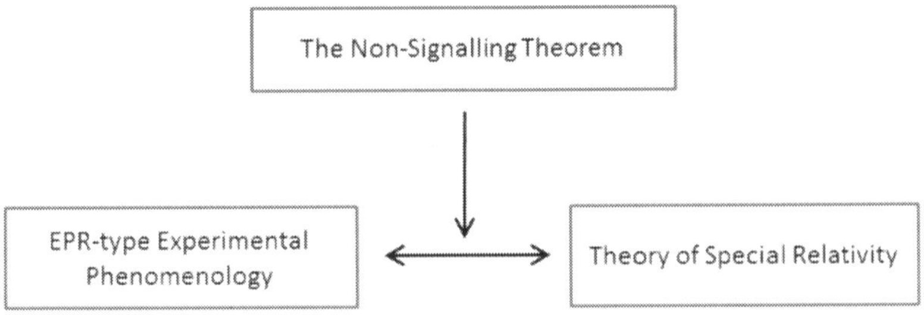

Figure 2. The non-signalling theorem or principle.

the inseparability, or relationality, of assumptions A and B (see bidirectional arrow between B and A in Fig. 1), the second step of their argument also, **as is pointed out in the present work only**, is equally characterized by an inseparability, or relationality, of assumptions B and C (see bidirectional arrow between C and B in Fig. 1). Therefore, it is questionable whether the non-signalling principle can be used to argue for the necessity of free randomness in nature, as only the **prior** assumption of intrinsic randomness could justify any foundational, physical implications of the non-signalling theorem to begin with. In the final section of this article (2.3), we will first provide a short (historical) overview of reasons for introducing the non-signalling principle, and then briefly return to discussion of the incomplete metaphysical status of the non-signalling principle.

2.3. The non-signalling theorem

The non-signalling theorem was introduced, of course, to acknowledge that EPR-type experiments, which indicate "instantaneous correlations" between space-like separated quantum detectors, do not violate special relativity (e.g. Eberhard [[illegible]]). Special relativity imposes a fundamental constraint on all physical phenomena whether quantum or classical. In particular, special relativity imposes a fundamental limit on the maximum speed of any physical influences existing in the vacuum of spacetime – the speed of light. Therefore, special relativity should outright prohibit the appearance in nature of any kind of "superluminal influences" whether infinite-velocity influences (i.e. instantaneous influences) or finite-velocity influences (e.g. near-instantaneous, superluminal influences). However, as was described before, EPR-type experiments have provided ample evidence for the existence of instantaneous, or at least, superluminal influences (see 1.3). Such influences are apparent from the measurement of physical effects which manifest – at velocities far exceeding the speed of light – between distant quantum detectors. Again, how could the existence of faster-than-light influences be compatible with special relativity? In regards to EPR-type experimental phenomenology, the tension which arises as a function of the **apparent** incompatibility between special relativity, on the one hand, and the observation of instantaneous, or at least, superluminal correlations, on the other hand, is relieved through the introduction of the non-signalling theorem or principle (see Fig. 2).

Clearly, the need to invoke non-signalling as a theorem for avoiding conflict with special relativity reveals just how extraordinary the phenomenology is of EPR-type experiments. This introduction is all the more remarkable as the theory of special relativity puts firm constraints already on the nature of allowable micro-causal processes, and yet the non-signalling principle itself does not refer to causal processes at all. The non-signalling principle refers to signals only, or rather their absence, i.e. to epistemic "states of knowledge" rather than to any ontic physical or causal states. The fact that an **epistemological** rather than an ontological argument

Journal of Physics: Conference Series **504** (2014) 012001 doi:10.1088/1742-6596/504/1/012001

	Signalling / Non-signalling Influences	**Instantaneous Influences**
Basic relations:	Sender and Receiver	"Unknown relations" [1]
Operational ability:	Ability, or inability, by epistemic agents to control information transfers	Registering random events
Influence is:	Purposeful; a "message" is, or is not, conveyed	Not purposeful; blind
Available must be:	Epistemic agents **and** physical processes	Physical processes
Conceptual framework:	Epistemology **and** Ontology	Ontology

Table 1. Distinguishing signalling/non-signalling influences from instantaneous influences. [1]The reference in the Table to "unknown relations" only states the obvious namely the lack of scientific understanding of the **physical nature** of "instantaneous influences". For example, an instantaneous influence does neither imply "efficient causation" nor any kind of "action-reaction"-type scenarios unless when considering a wholly **operational** meaning of an "action-reaction" relation. For example, in the context of instantaneous influences, as part of an EPR-type experiment, an "action" may be the defining of a measurement setting in location A, whereas a "reaction" might refer to the physical measurement effect at the quantum detector in location B; again operationally only.

suffices to resolve the apparent incompatibility between special relativity and EPR-type data has astonished many thinkers on quantum foundations. This includes for example John Bell also who asked (Bell [25]):

"Do we then have to fall back on 'no signalling faster than light' as the expression of the fundamental causal structure of contemporary theoretical physics? This is hard for me to accept. For one thing we have lost the idea that correlations can be explained, or at least this idea awaits reformulation. More importantly, the 'no-signaling…' notion rests on concepts which are desperately vague, or vaguely applicable. The assertion that 'we cannot signal faster than light' immediately provokes the question: Who do **we** think we are? **We** who can make 'measurements', **we** who can manipulate 'external fields', **we** who can signal at all, even if not faster than light? Do we include chemists, or only physicists, plants, or only animals, pocket calculators, or only mainframe computers?"

We conclude that the metaphysical status of the non-signalling principle still remains a very much open question even today. For example, as was alluded to above many times already, it is still undecided whether the non-signalling principle is a foundational principle, or an operational, phenomenological one. We conclude that until this all-important question is answered, the argument, for example, by Gallego et al. [7] must be considered, at a minimum, to be incomplete concerning their claim that the non-signalling principle is a conclusive argument in favor of intrinsic randomness in nature (compare Fig. 1, B and C).

Importantly, our analysis maintains that while the non-signalling principle prohibits faster-than-light communication it does not prohibit the possibility of instantaneous influences. Put differently, in combination with the non-signalling principle, the meaning of special relativity is changed from an absolute "no influence can exist which is faster than the speed of light" to the less restrictive statement that "no instantaneous/ superluminal influence can exist **that can be controlled for signalling purposes**". Critically, the conflation in meaning of the notions of "signalling / non-signalling influence" and "instantaneous influence" has led to much confusion in the literature, and worse, to misleading conclusions. The Table summarizes differences for better overview.

The present work sought to argue against proposals that seek to construe the non-signalling theorem as a "non-influence theorem". Specifically, we have argued that the non-signalling theorem serves to eliminate only the capacity of epistemic agents to signal information by way of instantaneous, or near-instantaneous, influences. However, the non-signalling theorem does not equal a non-influence theorem, e.g. the non-signalling theorem does not prohibit the existence of instantaneous influences, for example, in the formation of nonlocal correlations as part of EPR-type experiments (see 1.1. to 1.3.).

Finally, we conclude that "epistemic non-signalling" is likely to be sufficient to ensure the peaceful coexistence of special relativity and quantum nonlocality. A major reason in favor of our conclusion is that any definition of signalling or non-signalling invariably requires a reference to epistemic agents, and what these agents can actually measure and report (see Table 1).

Acknowledgments
Work by Jan Walleczek at Phenoscience Laboratories (Berlin) is partially funded by the Fetzer Franklin Fund of the John E. Fetzer Memorial Trust. Work by Gerhard Grössing at the Austrian Institute for Nonlinear Studies is also partially funded by the Fetzer Franklin Fund of the John E. Fetzer Memorial Trust.

References

[1] Bohm D 1952 A suggested interpretation of the quantum theory in terms of "Hidden" variables. I *Phys. Rev.* **85**(2) 166–179
[2] Einstein A, Podolsky B and Rosen N 1935 Can quantum-mechanical description of physical reality be considered complete? *Phys. Rev.* **47**(10) 777–780
[3] Bell J S 1964 On the Einstein Podolsky Rosen paradox *Physics* **1** 195–200
[4] Bell J S 1987 *Speakable and Unspeakable in Quantum Mechanics* (Cambridge: Cambridge University Press)
[5] Colbeck R and Renner R 2011 No extension of quantum theory can have improved predictive power *Nat. Commun.* **2** 411—415
[6] Colbeck R and Renner R 2012 Free randomness can be amplified *Nat. Phys.* **8**(6) 450–454 (*Preprint* arXiv:1105.3195 [quant-ph])
[7] Gallego R, Masanes L, De La Torre G, Dhara C, Aolita L and Acín A 2013 Full randomness from arbitrarily deterministic events *Nat. Commun.* **4** 2654 (*Preprint* arXiv:1210.6514 [quant-ph])
[8] Barrett J, Hardy L and Kent A 2005 No signaling and quantum key distribution *Phys. Rev. Lett.* **95**(1) 010503
[9] Kofler J, Paterek T and Brukner C 2006 Experimenter's freedom in Bell's theorem and quantum cryptography *Phys. Rev.* A **73**(2) 022104 (*Preprint* arXiv:quant-ph/0510167)
[10] Pironio S, Acin A, Massar S, de la Giroday A B, Matsukevich D N, Maunz P, Olmschenk S, Hayes D, Luo L, Manning T A and Monroe C 2010 Random numbers certified by Bell's theorem *Nature* **464**(7291) 1021–1024 (*Preprint* arXiv:0911.3427 [quant-ph])
[11] Aspect A, Grangier P and Roger G 1982 Experimental realization of Einstein-Podolsky-Rosen-Bohm gedankenexperiment: A new violation of Bell's inequalities *Phys. Rev. Lett.* **49**(2) 91–94
[12] Aspect A, Dalibard J and Roger G 1982 Experimental test of Bell's inequalities using time-varying analyzers *Phys. Rev. Lett.* **49**(25) 1804–1807
[13] Weihs G, Jennewein T, Simon C, Weinfurter H and Zeilinger A 1998 Violation of Bell's inequality under strict Einstein locality conditions *Phys. Rev. Lett.* **81**(23) 5039–5043

[14] Tittel W, Brendel J, Zbinden H and Gisin N 1998 Violation of Bell inequalities by photons more than 10 km apart *Phys. Rev. Lett.* **81**(17) 3563–3566
[15] Ursin R, Tiefenbacher F, Schmitt-Manderbach T, Weier H, Scheidl T, Lindenthal M, Blauensteiner B, Jennewein T, Perdigues J, Trojek P, Oemer B, Fuerst M, Meyenburg M, Rarity J, Sodnik Z, Barbieri C, Weinfurter H and Zeilinger A 2007 Free-space distribution of entanglement and single photons over 144 km *Nat. Phys.* **3**(7) 481–486 (*Preprint* arXiv:quant-ph/0607182)
[16] Giustina M, Mech A, Ramelow S, Wittmann B, Kofler J, Beyer J, Lita A, Calkins B, Gerrits T, Nam S W, Ursin R and Zeilinger A 2013 Bell violation using entangled photons without the fair-sampling assumption *Nature* **497**(7448) 227–230
[17] Maudlin T 2011 *Quantum Non-Locality and Relativity: Metaphysical Intimations of Modern Physics* 3rd ed (Malden, Mass.: Wiley-Blackwell) ISBN 9781444396973
[18] Salart D, Baas A, van Houwelingen J A W, Gisin N and Zbinden H 2008 Spacelike separation in a Bell test assuming gravitationally induced collapses *Phys. Rev. Lett.* **100**(22) 220404
[19] Cocciaro B, Faetti S and Fronzoni L 2011 A lower bound for the velocity of quantum communications in the preferred frame *Phys. Lett.* A **375**(3) 379–384 (*Preprint* arXiv:1006.2697 [quant-ph])
[20] Chaitin G J 1987 *Information, Randomness & Incompleteness: Papers on Algorithmic Information Theory* (*Series in Computer Science* no 8) (Singapore: World Scientific)
[21] Bohm D and Hiley B J 1993 *The Undivided Universe: An Ontological Interpretation of Quantum Theory* (London: Routledge) ISBN 0415065887
[22] Holland P R 1993 *The Quantum Theory of Motion* (Cambridge: Cambridge University Press) ISBN 0-521-35404-8
[23] Valentini A 2002 Signal-locality in hidden-variables theories *Phys. Lett.* A **297**(5-6) 273–278 (*Preprint* arXiv:quant-ph/0106098)
[24] Eberhard P H 1978 Bell's theorem and the different concepts of locality *Nuov. Cim.* B **46**(2) 392–419
[25] Bell J S 2004 La nouvelle cuisine, *Speakable and Unspeakable in Quantum Mechanics* (Cambridge: Cambridge University Press) revised ed

Where is quantum theory headed?

Stephen L Adler
Institute for Advanced Study, Einstein Drive, Princeton, NJ 08540, USA
E-mail: adler@ias.edu

Abstract. Public talk at the EmQM13 conference opening event on "The Future of Quantum Mechanics".

The organizers have asked me to state my views on the direction of the future development of quantum mechanics. Will it evolve within the standard framework, without the addition of new foundational physics? Or will the foundations require modification in an, at least in principle, experimentally detectable way?

First, let's discuss the current status of quantum theory. Quantum mechanics is our most successful physical theory. It underlies our detailed understanding of atomic physics, chemistry, and nuclear physics, and the many technologies based on this knowledge. Additionally, relativistic quantum mechanics is the basis for the very successful standard model of elementary particles.

However, from its beginnings there have been conceptual problems associated with the nature of measurement in quantum mechanics. These can be simply illustrated with the famous Stern-Gerlach experiment (Fig. 1).

Silver atoms boiled off from a furnace are sent through a non-uniform magnetic field, and impinge on a photographic plate. Instead of a continuous distribution of spots, one sees two spots, corresponding to spin up and spin down relative to the magnetic field axis. Each atom

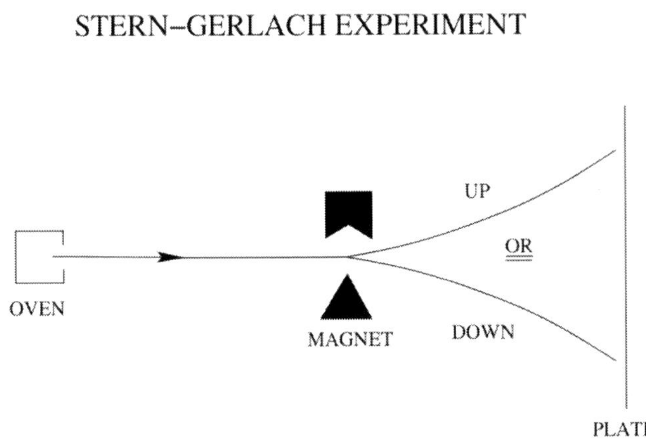

Figure 1. Schematic representation of the Stern–Gerlach experiment

goes up OR down, but one cannot predict which in any given run – the results of the experiment are probabilistic. There is a 50% chance of an atom going up, and a 50% chance that it will go down.

From the point of view of the Schrödinger equation of quantum theory, this result has no explanation. In quantum theory, the state of the particle is described by its wave function, and the Schrödinger equation says that at a post-measurement final time T_f, the wave function is related to that at a pre-measurement initial time T_i, by a deterministic relation

$$\begin{aligned} \Psi(T_f) &= U(T_f, T_i)\Psi(T_i) \quad , \\ U(T_f, T_i) &= e^{iH(T_f - T_i)} \quad , \end{aligned}$$

with the transition operator U completely specified by the Hamiltonian H. To explain what is observed, the Schrödinger equation must be supplemented by the reduction postulate and the Born rule. These state that the wave function only gives a description of probabilities when a measurement is made, with the probabilities for an "up" outcome and a "down" outcome given by the squares of the coefficients of the corresponding components in the initial wave function $\Psi(T_i)$,

$$\begin{aligned} &\text{Born Rule for Probabilities} \\ \Psi(T_i) &= C_{\rm up}\Psi_{\rm up} + C_{\rm down}\Psi_{\rm down} \quad , \\ {\rm prob}_{\rm up} &= |C_{\rm up}|^2 \quad , \\ {\rm prob}_{\rm down} &= |C_{\rm down}|^2 \quad , \\ &|C_{\rm up}|^2 + |C_{\rm down}|^2 = 1 \quad , \end{aligned}$$

with the sum of the up and down probabilities equal to one. The reduction postulate and Born rule are an add-on to the Schrödinger equation. According to the Copenhagen interpretation of quantum mechanics, the Schrödinger equation applies when a microscopic system, the silver atom, is time-evolving in isolation. But when the atom interacts with a macroscopic measuring apparatus, as in the Stern–Gerlach setup, you have to use the reduction postulate and Born rule.

This situation leads to puzzles that have been debated for over eighty years. If quantum mechanics describes the whole universe, then why can't one use the Schrödinger equation to describe the system consisting of the silver atom plus the measuring apparatus? But we never see a superposition state of the atom plus apparatus. This is Schrödinger's famous cat paradox. Arrange the experiment so that an"up" outcome triggers a mechanism that kills the cat, while a "down" outcome keeps the cat alive. Of course we don't do this, but if we were to do it, we would always see a live cat OR a dead one, never a superposition of the two (Fig. 2). So we have the problem of definite outcomes: where does the "either"–"or"dichotomy arise?

A related question is where do the probabilities come from? Quantum mechanics has probabilities without a sample space! An example of a sample space is a population of people, 40% blonde and 60% brunette. If you pick a person at random from the population, there is a probability ${\rm prob}_{\rm blonde} = 0.4$ that you will have a blonde, and ${\rm prob}_{\rm brunette} = 0.6$ that you will have a brunette. But the population (the sample space) is composed of individuals, with definite hair coloring – the probabilities only reflect our ignorance of details if we make a random pick without looking. Another example of a sample space, closer to our Stern–Gerlach experiment, is a coin toss. Consider 1000 coin tosses. If the coin is tossed without bias, you will find close to 500 heads and 500 tails, corresponding to ${\rm prob}_{\rm heads} = 0.5$ and ${\rm prob}_{\rm tails} = 0.5$. Here the sample space consists of the 1000 detailed trajectories of the toss, which your eye cannot follow,

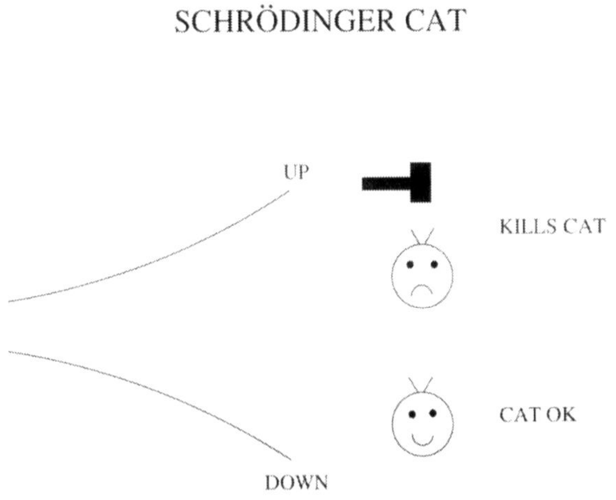

Figure 2. The Stern–Gerlach experiment with a Schrödinger cat as the outcome registration

but which if analyzed by a very fast computer could predict which toss would give a head and which a tail (Fig. 3). Again, the probabilities are just reflections of our ignorance of the details, but the details are there. So we have the questions – are there hidden details underlying the probabilities in quantum mechanics? Is there a hidden sample space?

Now we come to the question with which I began – where is future work to deal with these problems headed? Two routes proceed within quantum theory: (i) The first route within quantum theory is to try to change the interpretational rules. Examples are the so-called "many worlds" interpretation (all possibilities are there, we just only see one), and the so-called "histories" program, which sets up an observer-free generalization of the Copenhagen rules. (ii) The second route within quantum theory is to say there is a sample space, but we don't see

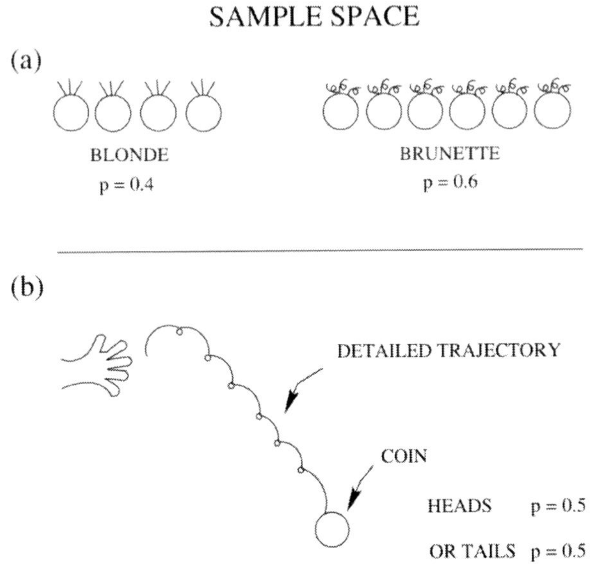

Figure 3. Some sample spaces: (a) Populations of individuals with blonde/brunette hair coloring (b) Trajectories in a coin toss

it. One example is the proposal of Bohmian trajectories as the sample space underlying the Born rule. Another example comprises various statistical interpretations, which attribute the probabilistic outcomes to different internal states of the apparatus and or its environment, with the aim of deriving the Born rule.

None of these routes within orthodox quantum theory has gained general acceptance. Also, none makes experimental predictions at odds with standard theory, so experimentally they are not distinguishable.

The other possibility is to modify the foundations of quantum theory. Specifically, one gets a sample space by postulating additional degrees of freedom – so called "hidden variables".

There are two possibilities for hidden variables – they can be *local*, or *nonlocal*. Local variables have the property that variables $V(x_1, t_1)$ and $W(x_2, t_2)$, with x_1, x_2 the spatial points and t_1, t_2 the times of occurrence, cannot influence one another if the distance between them $|x_1 - x_2|$ is greater than the distance $c|t_1 - t_2|$ that light can travel, at velocity c, in the time interval from t_1 to t_2. Such local variables are called causally separated. In quantum theory, causally separated variables have a commutative multiplication law

$$V(x_1, t_1) \times W(x_2, t_2) = W(x_2, t_2) \times V(x_1, t_1) \quad .$$

Ordinary numbers obey such a commutative law of multiplication, for example $AB = BA$ for $A = 7$ and $B = 11$, whereas in non-commutative multiplication, one would have $AB \neq BA$.

John Bell's theorem asserts that local hidden variables plus the usual rules for probabilities imply certain inequalities that are not satisfied by quantum mechanical systems – and experiment sees that these inequalities are in fact violated. There is much discussion of possible loopholes, but I believe the result is robust, and that local hidden variables are excluded.

The other possibility is that the hidden variables are non-local: the hidden variables can act faster than the speed of light to establish correlations (as long as no faster than light signaling is possible). The hidden variables can also obey a non-commutative multiplication law.

For the rest of the talk I'll focus on the possibility of non-local hidden variables – this is where my research interests lie. At a phenomenological level, there are very interesting models for the emergence of probabilities within the usual wave function formulation of nonrelativistic quantum theory, pioneered by Ghirardi, Rimini, and Weber in Trieste and by also by Pearle at Hamilton College in the U. S., and worked on by many others. These models postulate that space is filled with a very low level noise with a coupling to matter proportional to the imaginary unit i, rather than with a real-valued coupling (more technically, they couple through an anti-Hermitian Hamiltonian term). For example, there could be a small, rapidly fluctuating contribution to the gravitational potential or g_{00} metric component proportional to the imaginary unit i. If such a theory obeys two general properties, (1) the total probability of a particle being present remains one for all times (that is, the wave function normalization is preserved), and (2) there is no faster than light signaling, then the extra terms in the Schrödinger equation equation must have a special structure. This special structure allows one to *prove* definite outcomes obeying the Born rule!

In these models, for each repetition of the Stern–Gerlach experiment, the noise variable takes different values. For a large apparatus, these have a measurable effect, whereas for an atom not interacting with an apparatus, the effect is not measurable. The noise leads to different outcomes for different runs, with probabilities given by the Born rule. The different noises for different runs of the experiment are analogous, in the coin toss example I gave earlier, to different details of the tumbling coin trajectories for the different coin tosses (Fig. 4).

I've worked on phenomenological reduction models, but my main long term interest has been at the foundational level. I am trying to make an analogy between quantum mechanics emerging from a possible pre-quantum theory, and the known fact that thermodynamics emerges from the laws of statistical mechanics. Thermodynamics – the science of heat and work (Fig. 5) – reflects

Journal of Physics: Conference Series **504** (2014) 012002 doi:10.1088/1742-6596/504/1/012002

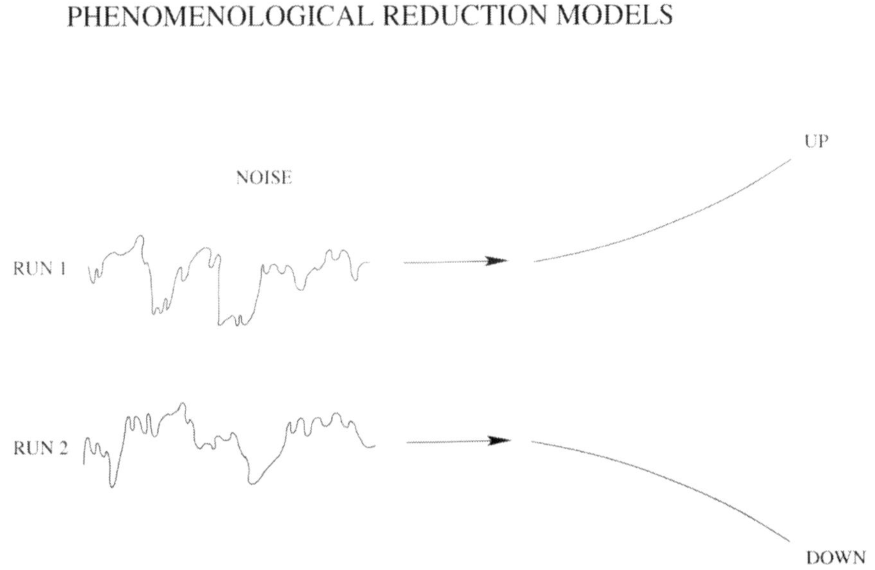

Figure 4. Different noise histories, in objective reduction models, can explain "up" and "down" registrations in the Stern–Gerlach experiment

averaged properties of huge numbers of atoms. It is a complete, consistent system by itself, and remarkably was discovered in the 19th century *before* the existence of atoms was established. But from statistical mechanics – the laws of motion of large systems of atoms, one can deduce the laws of thermodynamics, together with details of fluctuation corrections to thermodynamics, so called Brownian motion corrections. (The figure shows the random walk trajectory of a pollen grain being bombarded by molecules in thermal motion.)

My suggestion, in articles with collaborators and a small book that I wrote in 2004 [1] , is a theory that I call "trace dynamics". It is a classical-like system of non-commutating variables – even distant systems in the universe are interacting instantaneously with us. One can make sense of the mathematics of non-commuting variables by using the cyclic property of a mathematical operation called the Trace: Trace$ABCD..FG$ = Trace$GABCD..F$ = Trace$FGBCD...$ This is a very powerful tool. One can use it to set up a system of equations analogous to classical mechanics, and to do statistical averaging. Getting a little technical now, for those in the audience familiar with quantum theory and statistical physics, what distinguishes trace dynamics from ordinary classical mechanics is the existence of a generic conserved quantity in addition to the energy and momentum. This quantity is operator-valued, and has the form

$$\begin{aligned} &\text{Conserved operator in trace dynamics} = \\ &\textstyle\sum_{\text{bosons}}[q_{\text{boson}}\; p_{\text{boson}} - p_{\text{boson}}\; q_{\text{boson}}] \\ - &\textstyle\sum_{\text{fermions}}[q_{\text{fermion}}\; p_{\text{fermon}} + p_{\text{fermion}}\; q_{\text{fermion}}] \quad , \end{aligned}$$

with the qs the canonical coordinates and the ps the canonical momenta. This is reminiscent in structure to the canonical commutation and anti-commutation relations of quantum theory. Just as energy in statistical mechanics is equally partitioned between the various degrees of freedom, one might expect this conserved operator, in a statistical mechanical treatment of trace dynamics, to also be equi-partitioned, giving the starting point for quantum theory.

My conjectures thus are: statistical averages in trace dynamics give the Schrödinger equation

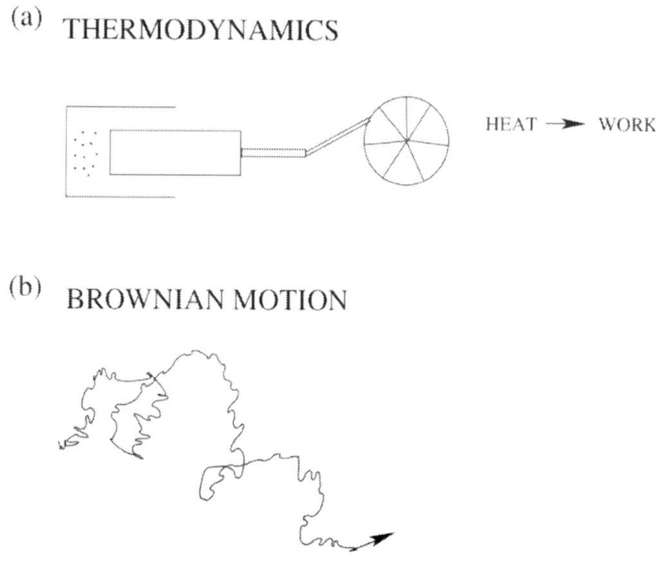

Figure 5. (a) Thermodynamics: the science of heat and work (b) Brownian motion of a pollen grain being bombarded by molecules in the liquid in which it is suspended

and operator algebra of quantum theory, while Brownian motion corrections give the low level noise on which phenomenological reduction models are based. I talked about this program in my keynote address at the Vienna conference two years ago. Currently I am working on incorporating gravity into trace dynamics [2] , and that is what I will talk about tomorrow. My approach to an emergent quantum theory is still a work in progress – there is much yet to be done!

Acknowledgments
I wish to thank Gerhard Grössing for organizing the conference EmQM13 on Emergent Quantum Mechanics, the Fetzer Franklin fund for its financial support, and Susan Higgins for redrawing my figure sketches for publication.

References
[1] Adler S L 2004 *Quantum Theory as an Emergent Phenomenon: The Statistical Mechanics of Matrix Models as the Precursor of Quantum Field Theory* (Cambridge: Cambridge Univ. Press)
[2] Adler S L 2013 Incorporating gravity into trace dynamics: the induced gravitational action *Class. Qu. Grav.* **30** 195015 (*Preprint* arXiv:1306.0482); *do.* 2013 *Class. Qu. Grav.* **30** 239501 (Erratum); Adler S L and Ramazanoğlu F R 2013 Spherically symmetric vacuum solutions arising from trace dynamics modifications to gravitation *Preprint* arXiv:1308.1448

Physics on the boundary between classical and quantum mechanics

Gerard 't Hooft

Institute for Theoretical Physics, Utrecht University
and
Spinoza Institute, Postbox 80.195, 3508 TD Utrecht, the Netherlands

Abstract. Nature's laws in the domain where relativistic effects, gravitational effects and quantum effects are all comparatively strong are far from understood. This domain is called the Planck scale. Conceivably, a theory can be constructed where the quantum nature of phenomena at such scales can be attributed to something fundamentally simpler. However, arguments that quantum mechanics cannot be explained in terms of any classical theory using only classical logic seem to be based on sound mathematical considerations: there can't be physical laws that require "conspiracy". It may therefore be surprising that there are several explicit quantum systems where these considerations apparently do not apply. In the lecture we will show several such counterexamples. These are quantum models that do have a classical origin. The most curious of these models is superstring theory. This theory is often portrayed as to underly the quantum field theory of the subatomic particles, including the "Standard Model". So now the question is asked: how can this model feature "conspiracy", and how bad is that? Is there conspiracy in the vacuum fluctuations?

1. Introduction

When the foundations of quantum mechanics are investigated, this is often done from different perspectives. What is agreed about is that the theory contains mysteries which somehow require explanation, but then the question comes what 'clarification' of such mysteries might mean:

i One might assume that quantum mechanics as we experience it in today's understanding of the physical laws, is an approximation, an idealization, of some deeper underlying theory. The underlying theory is not quantum mechanical, but obeys classical logic, which in some approximation takes the shape of Schrödinger's equation. It is difficult to imagine a theory that is more basic than quantum mechanics, while it would only roughly mimic pure quantum mechanics, but this is the only choice one has if one takes for granted that theories with classical logic cannot reproduce quantum mechanics exactly.

ii Alternatively, one may accept quantum mechanics as it is, but desires to obtain a more precise picture of how exactly quantum mechanics describes physical reality. What is physical reality? Do we have an infinity of different universes that all can be called 'real', as in the Many World Hypothesis? Or should the wave function be interpreted exactly as in the Copenhagen interpretation, as a complex quantity that replaces the standard probability function W in classical statistical mechanics?

iii Search for an explanation as to how quantum mechanics can emerge as an exact statistical description of an underlying theory, which, nevertheless, is suspected to exhibit classical logical equations anyway. In this case, one must also provide an explanation of how this can be reconciled with the usual arguments that such theories are impossible. Since these arguments are usually considered to be impeccable, such approaches are rare, but what can be done is this: let us simply try and investigate various candidate models that seem to be exactly equivalent to quantum mechanical systems, but will usually suffer from some defects. Then investigate these defects. Maybe we hit upon models that, even if they are classical, must exhibit some unusual features as well. These will be worth-while to investigate for their own sake.

This report is about progress made[1][2] in this last approach, number *iii*. It is clear that theories along these lines, which are deterministic but also appear to obey locality, so that they seem to be "local hidden variable theories", must be quite different from what we are used to. Quite probably, their elementary mass- and distance scales are in the Planck domain, about which we have very little direct information. Regardless what kind of theories we will try, there are all sorts of difficulties here: not only do we have the difficulties associated with the unrenormalizable divergences, the black hole information paradox, and questions concerning non-trivial topologies in space-time, but one must also not forget that the group of Lorentz transformations is essentially non-compact, and this creates problems when we wish to consider the possibility of a 'smallest distance' in space and/or time. This may explain why our progress is slow, but we do wish to report progress anyway, and it is interesting.

To set the scene, we have a number of very illuminating 'toy models', which display very well the promising nature of this approach. Three examples will be discussed here. All three are integrable, which makes their discussion very transparent, and all three models share the property that they are classical and quantum mechanical at the same time:

1 The quantum harmonic oscillator. It is mathematically equivalent to a *classical point particle moving around the unit circle at constant speed.* More generally, we claim that the quantum harmonic oscillator is represented by any periodic classical system.
2 The chiral Dirac fermion. It is mathematically equivalent to *an infinite plane moving with the speed of light in one of the two possible directions orthogonal to the plane.*
3 The non-interacting sector of a fully quantized superstring, which is mathematically equivalent to a classical string moving over a target space that is a lattice[4]. The lattice spacing a obeys $a = 2\pi\sqrt{\alpha'}$, where α' is the string slope parameter.

The interactions of strings with themselves still lead to problems that have not been solved. In the following sections we discuss these models. At the end, we discuss their potentials for the general quantum discussion.

2. The harmonic oscillator

Consider the spectrum of the harmonic oscillator. Its energy eigen states $|n\rangle_E$ have energy eigen values

$$E_n = (n + \tfrac{1}{2})\hbar\omega \; . \tag{2.1}$$

Journal of Physics: Conference Series **504** (2014) 012003 doi:10.1088/1742-6596/504/1/012003

First, consider a high-energy cut-off: the spectrum has an upper limit: $n < N$. We now also remove the vacuum energy $\frac{1}{2}\hbar\omega$ as being not very essential for the discussion that follows[1]. Our Hilbert space is N-dimensional, so a basis has always N elements. Because of the equal spacings in the energy levels, we have indeed periodicity with period T:

$$e^{-2\pi i H/\hbar\omega} = 1 , \qquad T = 2\pi/\omega . \tag{2.2}$$

This suggests that we look at a new basis of states, labled with an integer $k = 0, \cdots, N-1$, and defined by the finite Fourier transformation:

$$|k\rangle_{\rm ont} \equiv \frac{1}{\sqrt{N}} \sum_{n=0}^{N-1} e^{-2\pi i\, kn/N} |n\rangle_E , \qquad k = 0, \cdots N-1 . \tag{2.3}$$

This turns our system into a model with classical periodicity. At time steps that are integer multiples of $\delta t = 2\pi T/N$, the system evolves by permuting these "ontological" states:

$$|0\rangle_{\rm ont} \rightarrow |1\rangle_{\rm ont} \rightarrow \cdots \rightarrow |N-1\rangle_{\rm ont} \rightarrow |0\rangle_{\rm ont} . \tag{2.4}$$

This is as classical as a model can be. The quantum number k in Eq. (2.3) will be called *beable*, in honor of J.S. Bell[3].

The *continuum limit* $N \rightarrow \infty$ can be taken, but requires some caution[2]. It is instructive to compute the matrix elements of harmonic oscillator operators such as $\hat{x}$ and $\hat{p}$ in the ontological basis. Since we know the explicit wave functions of the energy eigen states in the x and the p basis, we can calculate these operators. For simplicity we put $\omega = 1$. The period is fixed, $T = 2\pi$, while the fundamental time steps δt become infinitesimally short.

Let the continuum limit then be described by the ontological states

$$|\varphi\rangle = \frac{1}{\sqrt{\mathcal{N}}} |k\rangle_{\rm ont} , \qquad \varphi = 2\pi k/N , \tag{2.5}$$

where $\mathcal{N}$ is the norm that normalizes the $|\varphi\rangle$ states such that $\langle\varphi|\varphi'\rangle = \delta(\varphi - \varphi')$. The matrix transforming these to the x eigen states and back is found to be

$$\langle\varphi|x\rangle = \sum_{n=0}^{\infty} \frac{H_n(x) e^{-\frac{1}{2}x^2}}{2^{(n+1)/2}\pi^{3/4}\sqrt{n!}} e^{in\varphi} \tag{2.6}$$

where $H_n(x)$ are the hermite polynomials.

At all fixed values of x and φ, the sum converges, but very slowly, because we need to include all energy eigen states that contribute at a given x, until the energies are so high that the bulk of the wave function stretches much further, away from x. Mathematically working out the sum is laborious, and it does not lead to simple special functions because of the square root. We did find powerful approximation methods, and arrived at functions that can be plotted, see Figure 1. The most salient features of this matrix are visible here: the function oscillates wildly, and eigen values of the operator $\hat{x}$ jump from predominantly positive to predominantly negative as φ passes the values $\pm\frac{1}{2}\pi$.

[1] The term $\frac{1}{2}.\hbar\omega$ makes the system *anti*periodic with period T. But here we consider a total shift of the energy that removes the $\frac{1}{2}\hbar\omega$ as a harmless procedure, so that the minus sign over an odd number of periods is ignored.
[2] In the case of finite N, the $\hat{x}$ operator connects the lowest energy state $|0\rangle_E$ to the highest energy state $|N-1\rangle_E$; one would be tempted to ignore this term in the limit $N \rightarrow \infty$, in which case unitarity will be violated. The fact that there is a small subset of states where the mapping classical – quantum mechanical fails may be an important feature to keep in mind; it happens in other examples as well.

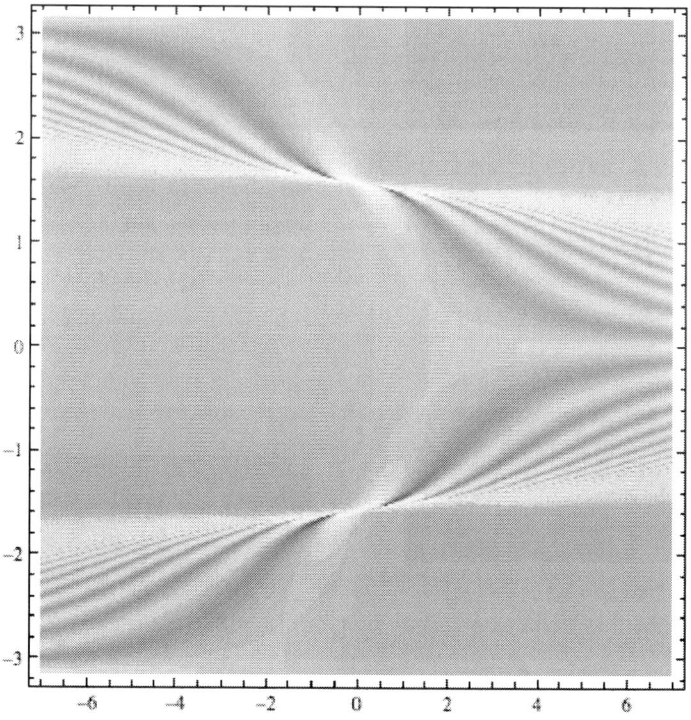

Figure 1. Plot of the real part of the transformation matrix $\langle\varphi|x\rangle$. Horizontally, x runs from -10 to 10, vertically, φ runs from $-\pi$ to π. In the busiest parts of the picture, the rapid oscillations are no longer visible.

The most important conclusion of this chapter is that the quantum harmonic oscillator is closely related to completely classical periodic systems in Nature. In elementary particle physics, we encounter many oscillating modes whose energy packets are the fundamental particles. We now see that these may be linked to periodic structures in a classical underlying theory. Of course this leaves the question of interactions wide open, but one may guess where one could try to go from here: seek how interactions can be understood as violations of exact periodicity in the behavior of oscillating modes in whatever the 'hidden variable' theory might contain.

3. Massless chiral fermions in three spacelike dimensions

In a massless fermion, even when coupled to a gauge field such as the Maxwell or the Yang-Mills field, the helicity, defined as being the spin projected along the direction of motion, is a conserved quantity. This implies that we can postulate that the particle is always left-handed, or only right handed. In that case, the wave function is only a two-component spinor rather than the usual four components.

Let us consider the non-interacting case. We call our particle a "neutrino", between quotation marks because it is the idealized version: no mass and no interaction. In the Majorana representation, the single, first quantized "neutrino" is described by the Hamiltonian

$$H=\vec{\sigma}\cdot\vec{p}\,, \tag{3.1}$$

where $\vec{p}$ is the momentum operator and $\vec{\sigma}$ are the Pauli matrices:

$$\sigma_x = \begin{pmatrix} 0 & 1 \\ 1 & 0 \end{pmatrix} , \qquad \sigma_y = \begin{pmatrix} 0 & -i \\ i & 0 \end{pmatrix} , \qquad \sigma_z = \begin{pmatrix} 1 & 0 \\ 0 & -1 \end{pmatrix} . \tag{3.2}$$

Define

$$\hat{p} = \pm\frac{p}{|p|} , \qquad s = (\hat{p}\cdot\vec{\sigma}) , \qquad r = \tfrac{1}{2}(\hat{p}\cdot\vec{x} + \vec{x}\cdot\hat{p}) . \tag{3.3}$$

The sign here is arbitrary; it could be defined such that $\hat{p}_3 \geq 0$. The equations of motion for these operators are derived as follows:

$$\frac{\mathrm{d}}{\mathrm{d}t}\vec{x} = -i[\vec{x}, H] = \vec{\sigma} , \qquad \frac{\mathrm{d}}{\mathrm{d}t}\vec{p} = 0 , \tag{3.4}$$

$$\frac{\mathrm{d}}{\mathrm{d}t}\sigma_i = 2\varepsilon_{ijk}\,p_j\sigma_k ; \qquad \frac{\mathrm{d}}{\mathrm{d}t}\hat{p} = 0 ; \tag{3.5}$$

$$\frac{\mathrm{d}}{\mathrm{d}t}(\hat{p}\cdot\vec{\sigma}) = \pm 2\varepsilon_{ijk}(p_i/|p|)p_j\sigma_k = 0 , \tag{3.6}$$

$$\frac{\mathrm{d}}{\mathrm{d}t}(\hat{p}\cdot\vec{x}) = \hat{p}\cdot\vec{\sigma} , \tag{3.7}$$

so that we have

$$\frac{\mathrm{d}}{\mathrm{d}t}\hat{p} = 0 , \qquad \frac{\mathrm{d}}{\mathrm{d}t}s = 0 , \quad \text{and} \quad \frac{\mathrm{d}}{\mathrm{d}t}r = s = \pm 1 . \tag{3.8}$$

Furthermore, we have

$$[\hat{p}, s] = 0 , \qquad [\hat{p}, r] = 0 , \qquad \text{and} \qquad [s, r] = 0 . \tag{3.9}$$

The first of these is obvious, and the last follows from the second. The fact that $[\hat{p}, r] = 0$ can be found by explicit calculation, but also understood as a consequence of the fact that, in momentum space, r is the dilation operator divided by $|p|$, while $\hat{p}$ is invariant under dilations (its norm stays one).

Together, Eqs. (3.8) and (3.9) imply that $\hat{p}$, r and s form a set of *beables*[3], which are operators that commute with one another at all times. The best way to describe these observables physically is to say that they represent a *sheet* moving with the speed of light in one of the two possible orthogonal directions, see Fig. 2.

This is why we claim that the ontological theory behind a 'neutrino' is an infinite sheet moving with velocity $v = c$ in one of its two possible orthogonal directions. If interactions or masses would be introduced, we have no obvious ontological theory anymore; one would have to search for the theory replacing the sheet in that case.

The mathematics of the transformation relating infinite sheets to neutrino states is delicate and interesting. The Hilbert space of 'neutrinos' can be assumed to be spanned by the states $|\vec{p}, \alpha\rangle_{\rm neutrino}$ or the states $|\vec{x}, \alpha\rangle_{\rm neutrino}$, where $\alpha = \pm 1$ is the spin variable, or, the eigenvalue of the Pauli spin operator σ_z. The sheet states will be denoted as $|\hat{p}, r, s\rangle_{\rm sheet}$. By first Fourier transforming the r variable to get the states $|\hat{p}, p_r, s\rangle_{\rm sheet}$, the orthogonal matrix relating these two sets of states can be calculated to be

$${}_{\rm neutrino}\langle\vec{p}, \alpha\,|\,\hat{p}, p_r, s\rangle_{\rm sheet} = p_r\,\delta^3(\vec{p} - \hat{p}\,p_r)\chi^s_\alpha(\hat{p}) , \tag{3.10}$$

$${}_{\rm neutrino}\langle\,\vec{x}, \alpha\,|\,\hat{p}, r, s\,\rangle_{\rm sheet} = \frac{i}{2\pi}\delta'(r - \hat{p}\cdot\vec{x})\,\chi^s_\alpha(\hat{p}) , \tag{3.11}$$

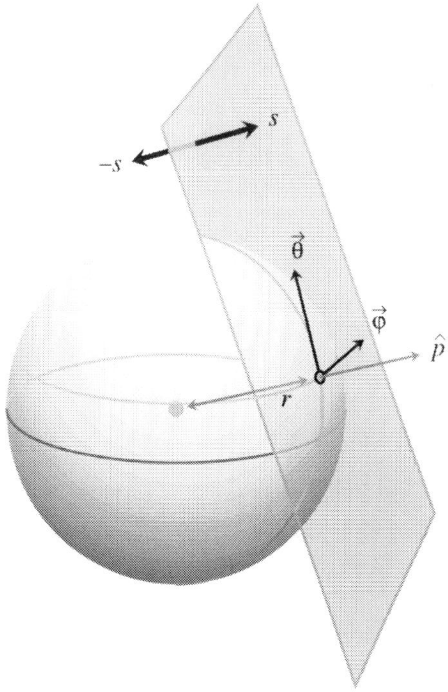

Figure 2. The "neutrino" as an infinite sheet. $\hat p$ is a unit vector, and therefore is a point of a unit sphere (actually half of a sphere, since its sign can be kept free). It defines the orientation of a sheet. The original coordinate variable $\vec x$ sits in this sheet. r is the distance of the sheet to the origin. The vectors $\vec\theta$ and $\vec\varphi$ are used in the text.

where $\chi^s_\alpha(\hat p)$ is the spin eigen state of the operator s with eigen value s, in the basis of σ_z. Furthermore, $\delta'(z)\equiv(\partial/\partial z)\delta(z)$. The derivative stems from the factor p_r, which is necessary for a proper normalization of the sheet states.

To express the neutrino operators $\vec x$ and $\vec p$ in terms of sheet operators is not quite so easy. The *phases* of the sheet states still have to be defined. To do that right, we need an orthonormal set of two unit vectors $\vec\theta$ and $\vec\varphi$ on the sheet. They could be defined as in Figure 2. We then define the rotation operators $L^{\rm ont}_i$ for the sheet as the generators that rotate the sheet. $L^{\rm ont}_i$ do nothing but rotate the unit vectors $\hat p$.

We have

$$p_r=\pm|p|\ ,\qquad \vec p^{\ \rm neutrino}=\hat p\,p_r\ , \tag{3.12}$$

but finding an operator $\vec x$ with the proper commutation rules with $\vec p$ is tedious. We first have to define spin flip operators s_1 and s_2, obeying, just like the Pauli matrices:

$$s_3=s\ ,\qquad s_is_j=\delta_{ij}+i\varepsilon_{ijk}s_k\ . \tag{3.13}$$

We found (no derivation is given here)[1]:

$$x^{\rm neutrino}_i=\hat p_i\,(r-\frac{i}{p_r})+\varepsilon_{ijk}\,\hat p_j\,L^{\rm ont}_k/p_r+\frac{1}{2p_r}\Big(-\varphi_i\,s_1+\theta_i\,s_2+\frac{\hat p_3}{\sqrt{1-\hat p_3^2}}\,\varphi_i\,s_3\Big)\ . \tag{3.14}$$

Thus we arrive at a picture where 'neutrinos' are identified with flat sheets spanning to infinity, able to move with light speed in one of the two directions in a line orthogonal to the sheet, in a totally classical way. This indicates the potential of our theory but also the difficulties. The potential is that systems normally considered as purely quantum mechanical may yet be completely classical, in disguise. One might object that these sheets may transfer signals with infinite speed inside a sheet, so that indeed the model might be considered non-local in a sense. Perhaps this is the kind of non-locality that will be inevitable in "hidden variable" theories.

The difficulty clearly shows up when one wishes to consider mass terms and/or interactions. What can happen if two sheets interact at a point? Most likely one first has to second-quantize the system: have many neutrinos described by many sheets. But even then it is difficult to see how things can interact; a sheet cannot easily be made to change its direction of motion at one point by means of local interactions. One will have to consider the possibility of having imperfect sheets, but more likely models will be found that are more profoundly different.

4. Real numbers and integers

Imagine that, in contrast to appearances, the real world, at its most fundamental level, were not based on real numbers at all. Real numbers are a man-made invention in response to the apparent continuous nature of our world, such as the positions and sizes of things. Our impression of everything being continuous received its first blow when it became quite evident that matter is made by atoms, which are discrete. Subsequently, it was found that energy is quantized. Now the energy quantum is controlled by the frequency of a wave function, and since time still seems to be continuous as far as we know, frequencies can be varied continuously, and in spite of our discovery of *quantum* mechanics, we still use real numbers in most of our calculations.

On the other hand, however, there are indications that, eventually, real numbers might get to be gradually replaced as our understanding of space and time further improves. A curious indication that this might happen is the theory of the information content of black holes. The microscopic regions near the horizon of a black hole seem to harbor only discrete amounts of information, in the form of *black hole microstates*. Information is fundamentally discrete. It is generally believed that space and time cannot be represented as a continuum of real-valued coordinates at distance scales as tiny as the Planck scale.

This is a motivation for searching for procedures to replace real numbers by integers. As it turns out, standard mathematics used in quantum mechanics provides us with the tools to do this. Again we use mappings, where one basis of Hilbert space is replaced by another. Surprisingly, what we shall derive is the following:

> There is a natural way to map a Hilbert space where the basis elements are characterized by a set of two integers, which we shall call P and Q, onto a set of states where the basis elements are characterized by one real number.[5]

The mapping is one-to-one, with only one single state escaping at one side: it generates a singularity. The mapping can be trivially extended to the mapping of $2N$ integers, P_i and Q_i, $i = 1, 2, \cdots, N$, onto basis elements of *either* N real numbers q_i, $i = 1, 2, \cdots, N$ *or* N real numbers p_i. The construction is such that p_i and q_i can be handled as positions and momenta in accordance with canonical theories of mechanics. Because $[q_i, p_j] = i\hbar\,\delta_{ij}$, the $|\vec{p}\rangle$ states and the $|\vec{q}\rangle$ states are not independent.

The integers, which we shall always write as capitals Q_i and P_i, completely commute. For this reason, it will be relatively easy to formulate deterministic theories based on Q_i and P_i, since these can be the beables of our system.

The mapping is easy to formulate. Consider first just one set of states $|Q\rangle$, where Q is an integer. It may run from $-\infty$ to ∞. We can then introduce the unitary step operator U by

$$U|Q\rangle = |Q+1\rangle \ . \tag{4.1}$$

Since U is unitary, and is easily seen to be non-degenerate, we can write it as

$$U = \epsilon^{i\eta} \ , \qquad \epsilon \equiv e^{2\pi} \approx 535.5 \ ; \qquad \epsilon^{Ni\eta}|Q\rangle = |Q+N\rangle \ , \tag{4.2}$$

for any N, where $-\frac{1}{2} < \eta \le \frac{1}{2}$. The quantity η is an operator, and its relation to the Q operator is easy to derive:
Consider the function η in the domain $-\frac{1}{2} < \eta < \frac{1}{2}$. It is easy to Fourier transform this function,

$$\eta = \sum_{N=-\infty}^{\infty} a_N \epsilon^{iN\eta} \ , \qquad a_N = \int_{-\frac{1}{2}}^{\frac{1}{2}} \eta \, \mathrm{d}\eta \, \epsilon^{-iN\eta} = \frac{i(-1)^N}{2\pi N} \quad \text{if } N \ne 0 \ , \tag{4.3}$$

and $a_0 = 0$. Therefore, we can write

$$\eta = \sum_{N\ne 0} \frac{i(-1)^N}{2\pi N} \epsilon^{iN\eta} \ , \tag{4.4}$$

so that

$$\langle Q_1|\eta|Q_2\rangle = \frac{i}{2\pi}(1-\delta_{Q_1,Q_2})\frac{(-1)^{Q_1-Q_2}}{Q_1-Q_2} \ . \tag{4.5}$$

It follows that

$$[\eta, Q] = \frac{i}{2\pi}(\mathbb{I} - |\psi_0\rangle\langle\psi_0|) \ , \text{ with } \langle Q|\psi_0\rangle \equiv (-1)^Q \ . \tag{4.6}$$

Here the single state $|\psi_0\rangle$ describes the one state for which the commutator is not the canonical one. We shall encounter such states more often; it is the exceptional state alluded to earlier, to be referred to as the *edge state*. We observe that this is also the state for which the operator η has the value $\pm\frac{1}{2}$, exactly the edge of the Fourier domain in Eq. (4.3).

The mapping that we wish to use is the mapping from the Q basis to the η basis and back. Just because we prefer to put η on an open interval rather than a closed circle, we encounter the 'problem' of the edge state; it will have to be taken proper care of. The transformation matrix is simply $\epsilon^{iQ\eta}$. Because of our special choice of ϵ as the base of our exponentials, the normalization of the matrix became trivial.

Now, consider a real number q. Let us write

$$q = Q + \eta_P \ , \tag{4.7}$$

where Q is integer and η_P lies between 0 and 1 (as much as possible, we intend to write real numbers as lower case letters, integers as capitals, and numbers *modulo* 1, or fractional numbers between 0 and 1, as Greek letters). Since this splitting is unique, we can write basis elements $|q\rangle$ as

$$|q\rangle = |Q, \eta_P\rangle \ . \tag{4.8}$$

Next, we can Fourier transform

$$|\eta_P\rangle = \sum_{P=-\infty}^{\infty} e^{iP\eta_P}|P\rangle \,. \tag{4.9}$$

where P is integer. Any state $|\psi\rangle$ can thus be written in the real q basis or in the $Q,\,P$ basis:

$$\langle q|\psi\rangle = \langle Q,\,\eta_P|\psi\rangle = \sum_P e^{-iP\eta_P}\langle Q,\,P|\psi\rangle \,. \tag{4.10}$$

This is what we mean when we state that quantum mechanics on a basis described by a real number q can be transformed to quantum mechanics with a basis of a pair of integers, Q and P.

Subsequently, one may transform to momentum space, using the matrix

$$\langle q|p\rangle = e^{ipq} \,. \tag{4.11}$$

Notice however, that this procedure is not quite symmetric under the interchange $Q \leftrightarrow P$, $q \leftrightarrow p$, but this symmetry can be restored. At the same time, one can reduce the effects of edge states. This we do by multiplying the coefficients of a wave function $|\psi\rangle$ in the space of the pairs $\eta_Q,\,\eta_P$, by a complex phase, $e^{i\varphi(\eta_Q,\,\eta_P)}$. This phase can be chosen such that the coefficients become exactly periodic both in η_P and in η_Q. This would remove the edge state completely, except for the fact that the phase function $e^{i\varphi}$ will feature a vortex singularity. This means that, at the very location of this vortex, we still will have an edge state.

The result of this procedure, requiring a calculation that will be published elsewhere[5], is that we obtain a transformation from the basis $\langle q|$ or $\langle p|$ to a basis of states $\langle Q,P|$ that, apart from a few signs, is entirely symmetric. We write for the q and p operators

$$q = Q + a_Q \,, \qquad p = P + a_P \,, \tag{4.12}$$

where the operators a_Q and a_P are not exactly restricted to the interval $(-\frac{1}{2},\,\frac{1}{2})$, but they will stay of order one:

$$\begin{aligned} \langle Q_1,P_1|a_Q|Q_2,P_2\rangle &= \frac{(-1)^{P+Q+1}\,iP}{2\pi(P^2+Q^2)} \,, \\ \langle Q_1,P_1|a_P|Q_2,P_2\rangle &= \frac{(-1)^{P+Q}\,iQ}{2\pi(P^2+Q^2)} \,. \end{aligned} \tag{4.13}$$

From these, one derives that

$$[q,\,p] = \frac{i}{2\pi}(1-|\psi_{\rm edge}\rangle\langle\psi_{\rm edge}|) \,, \quad \text{with} \quad \langle Q,P|\psi_{\rm edge}\rangle = (-1)^{Q+P} \,. \tag{4.14}$$

In (η_Q,η_P) space, the edge state $|\psi_{\rm edge}\rangle$ is the delta peak on the spot $(\eta_Q,\eta_P)=(\pm\frac{1}{2},\,\pm\frac{1}{2})$, exactly where we located the vortex of our phase function $\varphi(\eta_Q,\eta_P)$. The eigen values of both operators q and p in Eq. (4.12) occupy the entire real line without overlappings, but we do have to restrict ourselves to the states that are orthogonal to the edge state $|\psi_{\rm edge}\rangle$.

5. Free massless bosons in 1+1 dimensions

The field operators in a quantized field theory also have their eigenvalues on the real line. We would like to apply the results of the previous chapter to such theories, but in general this would not produce viable deterministic models. The reason is, that the splitting (4.12) does not transform in a simple way when two real numbers are added or subtracted.

However, the $(Q, P) \leftrightarrow (\hat{q}, \hat{p})$ mapping can nevertheless be significant for field theories, if we can arrange the dynamical variables in such a way that the classical field equations involve displacements without additions or subtractions.

The rest of this chapter was written down earlier in Ref.[2]. Take a free, massless quantum field theory in 1+1 dimensions, having field variables $\phi(x,t)$ and a canonical momentum field $p(x,t)$ obeying commutation rules

$$[\phi(x,t),\, p(x',t)] = \frac{i}{2\pi}\delta(x-x')\ , \qquad [\phi(x,t),\, \phi(x',t)] = [p(x,t),\, p(x',t)] = 0\ . \tag{5.1}$$

The Klein-Gordon equation,

$$(\partial_x + \partial_t)(\partial_x - \partial_t)\phi(x,t) = 0\ , \tag{5.2}$$

implies the existence of left-movers $\phi^L(x+t)$ and right-movers $\phi^R(x-t)$:

$$\phi(x,t) = \phi^L(x+t) + \phi^R(x-t)\ , \qquad p(x,t) = \tfrac{1}{2}a^L(x+t) + \tfrac{1}{2}a^R(x-t)\ , \tag{5.3}$$

$$a^L(x+t) = p(x,t) + \partial_x\phi(x,t)\ , \qquad a^R(x-t) = p(x,t) - \partial_x\phi(x,t)\ . \tag{5.4}$$

In fact, up to some coefficients, these are Fourier transforms of the familiar particle creation and annihilation operators. We have the hamiltonian density

$$\mathcal{H} = \tfrac{1}{2}(p^2 + (\partial_x\phi)^2) = \tfrac{1}{4}({a^L}^2 + {a^R}^2)\ . \tag{5.5}$$

In terms of the left- and right-movers, the commutation rules are

$$[a^L,\, a^R] = 0\ , \quad [a^L(x),\, a^L(y)] = \frac{i}{\pi}\partial_x\delta(x-y)\ , \quad [a^R(x),\, a^R(y)] = \frac{-i}{\pi}\partial_x\delta(x-y)\ . \tag{5.6}$$

Replacing the spacetime continuum by a (dense) lattice, one sees that these commutation rules can be replaced by

$$[\phi(x,t),\, p(y,t)] = \frac{i}{2\pi}\delta_{x,y}\ , \quad [a^L(x), a^L(y) = -[a^R(x), a^R(y)] = \pm\frac{i}{2\pi} \text{ if } y = x \pm 1\ . \tag{5.7}$$

Now, suppose we introduce integer valued operators $A^{L,R}(x)$ and their associated momentum operators $\eta^{L,R}(x)$, obeying

$$[\eta^L(x),\, A^L(y)] = [\eta^R(x), A^R(y)] = \frac{i}{2\pi}\delta_{x,y}\ , \quad [A^L, A^R] = [A^L, \eta^R] = 0\ ,\ \text{etc.} \tag{5.8}$$

(ignoring the usual edge states), one finds that we can write

$$a^L(x) = A^L(x) + \eta^L(x+1)\ , \qquad a^R(x) = A^R(x) + \eta^R(x-1)\ , \tag{5.9}$$

so that the commutation rule (5.7) is automatically obeyed.

With some more advanced mathematics, one can reduce the effects of the edge states to a minimum (they cannot be ignored completely), as was briefly explained in the previous section.

The importance of this procedure, and the reason why it *only* works in one space, one time dimension, is that the time evolution of the $a^{L,R}$ fields involves nothing more than shifts in x space, without further linear transformations such as additions or subtractions. Linear transformations in the variables $A^{L,R}$ and $\eta^{L,R}$ do not lead to similar operators where the A are integer and the η stay in the interval $(\frac{1}{2}, \frac{1}{2}]$.

We now observe that, in string theory, the ϕ fields are the left- and right moving coordinates X^μ in a D dimensional space-time. For convenience, we had chosen the world sheet lattice to have lattice length $a = 1$, but it is easy to verify that, if we had chosen any other lattice length, the relation between the quantized variables $A(x)$ and the space-time coordinates $X^\mu(x)$ would remain the same. In other words, just because we wished to relate the commutation rules (5.1) and (5.6) to the commutation rules (5.8), and have the relations (5.9) such that the real numbers are smoothly covered, we find that the classical system has its spacetime coordinates X^μ defined on a grid, and the grid length is fixed. Re-inserting the usual string units, one finds a space-time lattice with lattice length[4]

$$a_{\text{spacetime}} = 2\pi\sqrt{\alpha'} \,, \tag{5.10}$$

a remarkable result. Note, that the classical theory has its string equations simply formulated in terms of the integer valued operators $A^{L,R}(x,t)$ (where x and t are the world sheet variables usually denoted as σ and τ. From the field equations (5.2) – (5.4), one derives that the integer-valued string coordinates obey simple classical equations,

$$X^\mu(\sigma, \tau + a) + X^\mu(\sigma, \tau - a) = X^\mu(\sigma + a, \tau) + X^\mu(\sigma - a, \tau) \,. \tag{5.11}$$

These equations must be assumed to describe the transverse string coordinates only. The longitudinal and timelike coordinates have to be derived from the usual string constraint equations. This part of the quantum theory remains unaffected. It means that, only in 26 dimensions, or in 10 dimensions if we add the fermionic degrees of freedom (see the next section), the quantum theory has an enhanced symmetry: Lorentz invariance.

Curiously, Lorentz invariance here is a *quantum* symmetry, which means that its transformations act non-trivially in the quantum q, p Hilbert space; it is not conspicuously present in the classical Q, P variables.

The quantized *superstring* contains fermionic degrees of freedom in addition to the bosonic coordinates. Remarkably, these fermionic modes can also be rephrased in classical terms: *Fermions are boolean variables* (variables taking only two values, such as ± 1), existing on the fundamental links of the string. The independent fermionic modes are again the ones in the transverse direction; the longitudinal fermionic modes are constrained by (quantum) equations.

The constraint equations that are needed to guarantee Lorentz invariance have no effect on the bulk behavior of the (super)string, but do appear to cause problems when we consider string interactions. Deterministic interactions can be considered, but it is not obvious that these coincide with the standard quantum interactions of the usual quantum string. This may turn out to be a serious limitation of the present considerations, but even if we would have our mappings only for the non-interacting strings, we found it quite worth-while to report this.

6. Bell inequalities and other fundamental quantum features

Not much of this chapter is new, but it had to be added to clarify the author's position concerning J.S. Bells's important observation[6]: local classical theories cannot reproduce the

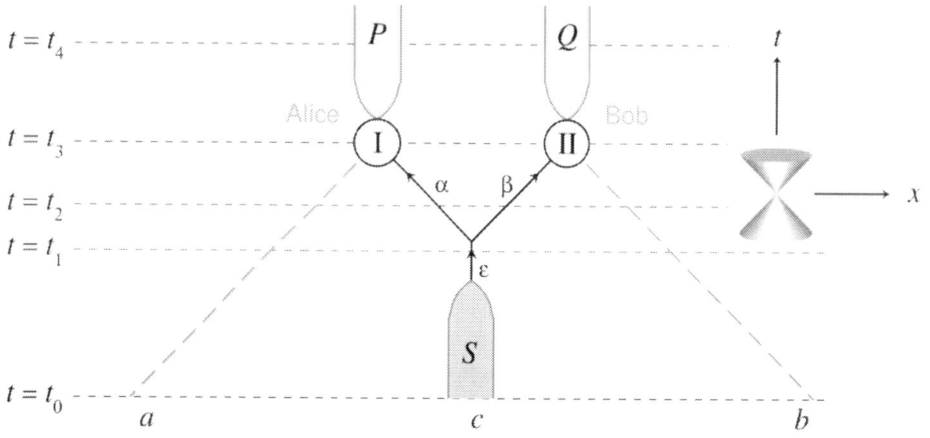

Figure 3. Space-time diagram for a Bell-type experiment, see text.

correlations generated in entangled particles that may typically arise in a quantum experiment. Schematically, we imagine that at $t = t_1$ (see Fig. 3), two entangled particles α and β are produced; think of photons whose spins add up to zero. At $t = t_3$, two widely separated observers, Alice and Bob, check whether their photon passes through a filter. The polarization angles of the filters are given by the normalized vectors $\vec{a}$ and $\vec{b}$, or angles a and b. Alice and Bob each have the freedom to choose their angles.

Using quantum mechanics, we can calculate under all circumstances how Alice's and Bob's measurements are correlated. According to Bell, these correlations can never arise from classical signals travelling with the photons; a direct, "spooky" signal has to be assumed to be transmitted between Alice and Bob *after* they chose the orientations of their filters. Alice and Bob chose their orientation angles both by free will; there was no way these angles could have been known prior to $t = t_3$.

At first sight, this argument can easily be countered when deterministic theories are considered: *Alice and Bob do not have such a free will;* their decisions were already decided at the very beginning of the Universe.

But this is not good enough. Suppose that Alice and Bob both base their decisions on measurements of fluctuations in signals coming from distant sources a and b, We may imagine that these are distant quasars that emitted their photons at $t = t_0$, shortly after the Big Bang. One should be allowed to assume that these signals are uncorrelated. In that case also, the outcome of the experiment then cannot be explained by classical arguments.

The correlations that emerge in simple quantum mechanical descriptions of entangled photons will require correlations in the quasars a and b as well. A simple calculation yields the required correlations as follows. The probability distribution W between the angle a imposed by Alice in her experiment, the angle b that Bob decided upon, and the angle c that one might assume to be carried by both entangled photons (in a classical effective description), is then required to be at least of the form

$$W(\vec{a},\, \vec{b},\, \vec{c}) \propto |\sin(2a + 2b - 4c)| \,. \tag{6.1}$$

Notice that, if we randomize either a, or b or c, the two remaining quantities become uncorrelated, so if we drop our information of the photons c, then Alice and Bob need no exchange of signals. The problem is that in a classical theory, one would have expected a classical observable c, not correlated to a or b at all.

Eq. (6.1) describes what is usually called "conspiracy"[7]. The photons should not "know" what Alice and Bob are going to do. One could imagine exceedingly complex devices that randomize the signals a and b on their way from the quasars a and b to the filters I and II used by Alice and Bob. It does not make sense if the photons could anticipate the outcome of these calculations.

On the other hand there may be reasons to demand caution. Relativistic quantum field theory does explicitly lead one to observe the existence of *vacuum fluctuations*. These are fluctuations that fields undergo in the lowest energy state, the vacuum. Not only do vacuum fluctuations show correlations between data that are spacelike separated (the propagators of the fields do not vanish outside the light cone), but they also have the property that all scattering amplitudes, and therefore also the future behavior, can be obtained from these correlation functions, simply by analytic continuation. In this sense, there is conspiracy, while this is a kind of conspiracy that never worried anyone.

We must conclude that statistical arguments may be invalid in deterministic theories. Also, counterfactual arguments are not acceptable. In fact, Bell's gedanken experiment was useful to set the scene, but the problem in question already arises for single, unentangled photons. As soon as an observer decides to choose the polarisation angle of a filter with the aim of measuring photons, the wave functions of these photons are diagonalized according to the coordinate frame generated by the filter, and extrapolating to the distant past, the photon *always was* in an ontological state defined by the filter that lies in its future. This conceptual difficulty will have to be accepted or ignored in all deterministic theories. What counts is that we have models that seem to exhibit such behavior. Yet at the classical level there is no trace of any conspiracy.

As an intermezzo, consider an example from number theory. Consider an "arbitrary" large prime number P. Consider an even large number Q defined by

$$Q = 2^{P-1} - 1 \; . \tag{6.2}$$

What is the expected distribution of the number

$$P_1 = Q \text{ modulo } P \; ? \tag{6.3}$$

One might have expected a flat distribution, but in fact, P_1 is always zero, a simple result from number theory. This illustrates our statement that statistical arguments in deterministic systems should not be trusted.

Let us emphasize that the peculiarity found by Bell in his considerations for entangled particles indeed leads to "conspirational" correlations in the classical theories we describe. We suspect that statistical arguments, and also counterfactual arguments[8] are illegal for our classical models, but how to explain this apparent 'conspiracy'?

A state considered in some experimental setup may either be a physical state, which we shall call 'ontological', or it is a superposition of ontological states. In our analysis of the experiment, such states are indistinguishable. However, if an 'ontological basis' exists, which we believe to be the case, then there is a *conservation law*: the ontological nature of a state is conserved in time. If, at some late time, a photon is observed to be in a given polarization state, just because it passed through a filter, then *that* is its ontological state, and the photon has been in that ontological state from the moment it was emitted by its source. It seems to be inevitable to demand that the 'ontological basis' is unobservable, that is, indistinguishable from other bases, as it is for instance in our description of string theory. 'Conspiracy' is then unobservable.

Let us also stress the importance of another observation: in quantum field theory, locality means that the Hamiltonian can be written as the integral over space of a Hamiltonian density. The Hamiltonian density is the generator of local time translations, and hence it is part of the algebra of general coordinate transformations. This may mean that general relativity will be an essential ingredient of a complete quantum theory; without general relativity, locality of quantum mechanics may be difficult to understand. Indeed, the same conclusion can be reached by considering the black hole microstates, which quite possibly correspond to the ultimate, classical degrees of freedom of an underlying theory, while they fundamentally arise at the Planck scale only. The problem of "quantizing" curved space-time, quantum gravity for short, is notoriously complex and far from understood. It may well be that a complete understanding of the quantum nature of our world will have to come together with the complete resolution of the quantum gravity problem.

References

[1] G. 't Hooft, *The Cellular Automaton Interpretation of Quantum Mechanics*, to be publ.

[2] G. 't Hooft, *The Fate of the Quantum*, Proceedings of the 4th International Conference on Time and Matter, 4-8 March 2013, Venice, Italy, M. O'Loughlin et al eds., University of Nova Gorica Press, pp. 127 - 143. ITP-UU-13/22; SPIN-13/15.

[3] J.S. Bell, *Subject and Object*, in J. Mehra (ed), The Physicist's Conception of Nature, pp 687-690, D. Reidel Publ. Comp., Dordrecht, Holland.

[4] G. 't Hooft, *Discreteness and Determinism in Superstrings*, ITP-UU-12/25; SPIN-12/23, arXiv:1207.3612v2[hep-th].

[5] G. 't Hooft, *Relating the quantum mechanics of discrete systems to standard canonical quantum mechanics*, ITP-UU-12/14; SPIN-12/12, arXiv:1204.4926

[6] J.S. Bell, *On the Einstein Podolsky Rosen Paradox*, Physica **1** (1964) 195;
J.F. Clauser, M.A. Horne, A. Shimony and R.A. Holt, *Proposed experiment to test local hidden-variable theories*, Phys. Rev. Lett. **23** n. 15 (1969) 880.

[7] B. van Fraassen, *The Charybdis of Realism: Epistemological Implications of Bells Inequality*, in James T. Cushing and Ernan McMullin (eds.), Philosophical Consequences of Quantum Theory. Notre Dame: University of Notre Dame Press, 97113; Synthese **52** (1):25 - 38 (1982)

[8] J. Conway and S. Kochen, *The Free Will Theorem*, quant-ph/0604079 (March 31, 2006)

The linearity of quantum mechanics from the perspective of Hamiltonian cellular automata

Hans-Thomas Elze

Dipartimento di Fisica "Enrico Fermi", Università di Pisa, Largo Pontecorvo 3, I-56127 Pisa, Italia

E-mail: elze@df.unipi.it

Abstract. We discuss the action principle and resulting Hamiltonian equations of motion for a class of integer-valued cellular automata introduced recently [1]. Employing sampling theory, these deterministic finite-difference equations are mapped reversibly on continuum equations describing a set of bandwidth limited harmonic oscillators. They represent the Schrödinger equation. However, modifications reflecting the bandwidth limit are incorporated, *i.e.*, the presence of a time (or length) scale. When this discreteness scale is taken to zero, the usual results are obtained. Thus, the linearity of quantum mechanics can be traced to the postulated action principle of such cellular automata and its conservation laws to discrete ones. The cellular automaton conservation laws are in one-to-one correspondence with those of the related quantum mechanical model, while admissible symmetries are not.

1. Introduction

The linearity of quantum mechanics (QM) is a fundamental feature most notably embodied in the Schrödinger equation. This linearity does not depend on the particular object under study, provided it is sufficiently isolated from anything else. It is naturally reflected in the superposition principle and entails the "quantum essentials" interference and entanglement, without which modern applications of QM, *e.g.* in precision measurement and information technologies, and the ongoing study of the foundations of QM would not be the same.

The linearity of QM has been questioned now and then and various nonlinear modifications have been proposed, in order to test experimentally the robustness of QM against such nonlinear deformations. This has been thoroughly discussed by Jordan presenting a stepwise proof that QM has to be linear based on the separability assumption *"... that the dynamics we are considering can be independent of something else in the universe, that the system we are considering can be described as part of a larger system without interaction with the rest of the larger system"* [2]. It is also worth recalling that most proposals of nonlinearities have been beset with the problem of superluminal signalling or communication between branches of the wave function, see, *e.g.*, Refs. [3, 4, 5, 6]; since then, "no signalling" has become an important criterium that attempted modifications of QM are confronted with.

We have recently provided a different point of view concerning the linearity of QM [1]. We have related QM to the mechanics of a class of Hamiltonian cellular automata. Which shows that the linearity of QM is necessarily required by the consistency of the postulated action principle for such discrete dynamics. This may further the understanding of interference, entanglement, and measurement in QM and lead to new approximation schemes in quantum theory.

Our approach was motivated by explorations of discrete deterministic mechanics by Lee [7], by the study of bandwidth limited fields and their possible role in the physics of gravity and spacetime by Kempf [8], and by the representation of QM in terms of classical notions (observables, phase space, Poisson bracket algebra) by Heslot [9, 10]. Presently, we recover our earlier arguments, which combine these ideas, and add a few details on the way.

2. Discrete Hamiltonian mechanics

Discreteness has many facets in physics, besides quantization, *e.g.*, discrete maps for numerical studies of complex systems, regularized versions of quantum field theories on spacetime lattices, or intrinsically discrete processes. Finite difference equations are expected to play a prominent role here, instead of the usual preponderance of differential calculus.

Lee and collaborators proposed to incorporate fundamental discreteness into all of dynamics [7] (and references therein). This was apparently motivated by the difficulties encountered in trying to formulate a consistent theory of "quantum gravity", or even "the" unified theory. Thus, deterministic discrete mechanics derives from the assumption that *time is a discrete dynamical variable*. This invokes a *fundamental time or length scale* (in natural units), l, and can be rephrased that in a fixed $(d+1)$-dimensional spacetime volume Ω maximally Ω/l^{d+1} measurements can be performed or this number of events take place [7].

We consider the new discreteness scale l in the spirit of *deformations* of Lorentz symmetry, in the form of "doubly special relativity" (DSR) [11], or in its explicit breaking, and in the nonlinear *deformation* of QM [3, 4, 5, 6] mentioned earlier. While such studies necessarily introduce additional parameters, the aim is to probe the stability of the existing theories, the Standard Model and QM in particular, against such deformations, predicting new phenomena that eventually could lead to a deeper theory with a smaller set of fundamental parameters [12]. It is commonly expected that $l \equiv l_{Pl}$, *i.e.*, that discreteness and Planck scale coincide.

Various discrete models have been elaborated, which share desirable symmetries with the corresponding continuum theories while presenting finite degrees of freedom. Different forms and (dis)advantages of a Lagrangian formulation [7] have been discussed, *e.g.*, in Refs. [13, 14, 15]. In the following, we introduce an action principle instead which leads to particularly transparent and symmetric Hamiltonian equations of motion, corresponding to a discrete phase space picture.

3. The action principle for Hamiltonian cellular automata

The state of a classical cellular automaton (CA) with a denumerable set of degrees of freedom will be represented by *integer-valued* "coordinates" x^α_n, τ_n and "conjugated momenta" p^α_n, π_n, where $\alpha \in \mathbf{N}_0$ denote different degrees of freedom and $n \in \mathbf{Z}$ different states.

The x_n and p_n might be higher dimensional vectors, while τ_n and $\mathcal{P}_n$ are assumed one-dimensional. In separating the "coordinate" τ_n from the x^α_n's (and correspondingly π_n from the p^α_n's), we follow Refs. [7, 13, 14]. This degree of freedom represents the *dynamical time* variable here, instead of the external parameter time of Newtonian mechanics or QM.

Finite differences, for all dynamical variables, are defined by

$$\Delta f_n := f_n - f_{n-1} \quad . \tag{1}$$

Furthermore, we define (using henceforth the summation convention for Greek indices, $r^\alpha s^\alpha \equiv \sum_\alpha r^\alpha s^\alpha$):

$$\mathcal{A}_n \quad := \quad \Delta\tau_n(H_n + H_{n-1}) + a_n \quad , \tag{2}$$

$$H_n \quad := \quad \frac{1}{2}S_{\alpha\beta}(p^\alpha_n p^\beta_n + x^\alpha_n x^\beta_n) + A_{\alpha\beta}p^\alpha_n x^\beta_n + R_n \quad , \tag{3}$$

$$a_n \quad := \quad c_n\pi_n \quad , \tag{4}$$

where constants, c_n, and symmetric, $\hat{S} \equiv \{S_{\alpha\beta}\}$, and antisymmetric, $\hat{A} \equiv \{A_{\alpha\beta}\}$, matrices are all integer-valued; R_n stands for higher than second powers in x_n^α or p_n^α. The choice of the right-hand side of Eq. (4) determines the behaviour of the variable τ_n; for our present purposes, a very simple choice suffices and will be further discussed shortly (nontrivial 'potential' terms have been made use of in Ref. [16]).

In terms of the definitions just given, we introduce the integer-valued CA *action*

$$\mathcal{S} := \sum_n [(p_n^\alpha + p_{n-1}^\alpha)\Delta x_n^\alpha + (\pi_n + \pi_{n-1})\Delta\tau_n - \mathcal{A}_n] \; , \tag{5}$$

and postulate that a Hamiltonian CA is described by the following

Action Principle: The CA obeys the discrete updating rules (equations of motion) which are determined by $\delta\mathcal{S} = 0$, referring to arbitrary integer-valued variations of all dynamical variables,

$$\delta g(f_n) := [g(f_n + \delta f_n) - g(f_n - \delta f_n)]/2 \; , \tag{6}$$

where f_n stands for one of the variables on which polynomial g may depend. •

Several remarks are in order here. – We observe that the variations of constant, linear, or quadratic terms yield results that are analogous to the continuum case. – While infinitesimal variations do not conform with integer valuedness, there is no *a priori* constraint on integer ones. However, for arbitrary δf_n, the *remainder of higher powers* in Eq. (3), which enters the action, has to vanish for consistency, $R_n \equiv 0$. Otherwise the number of equations of motion generated by the action principle, generally, would exceed the number of variables. We note that a suitably chosen R_0 or a sufficient small number of such remainder terms could encode the necessary *initial conditions* for the CA evolution.

With the notation $\dot{O}_n := O_{n+1} - O_{n-1}$, the following CA *equations of motion* are obtained:

$$\dot{x}_n^\alpha = \dot{\tau}_n(S_{\alpha\beta}p_n^\beta + A_{\alpha\beta}x_n^\beta) \; , \tag{7}$$

$$\dot{p}_n^\alpha = -\dot{\tau}_n(S_{\alpha\beta}x_n^\beta - A_{\alpha\beta}p_n^\beta) \; , \tag{8}$$

$$\dot{\tau}_n = c_n \; , \tag{9}$$

$$\dot{\pi}_n = \dot{H}_n \; , \tag{10}$$

which are discrete analogues of Hamilton's equations, where all terms are integer-valued. Discreteness of the *automaton time* n is reflected in the finite difference equations here.

Note that the $\dot{\tau}_n$ present *background* parameters for the evolving x, p-variables, as a consequence of Eqs. (4), (9). Generally, $\dot{\tau}$ is a *lapse function* in Eqs. (7)–(8).

The Eqs. (7)–(10) are *time reversal invariant*; the state $n+1$ can be calculated from knowledge of the earlier states n and $n-1$ and the state $n-1$ from the later ones $n+1$ and n.

4. Conservation laws and CA symmetries

Surprisingly, there are conservation laws that are always respected by the discrete equations of motion Eqs. (7) and (8). – Introducing the self-adjoint matrix $\hat{H} := \hat{S} + i\hat{A}$, these equations can be combined into:

$$\dot{x}_n^\alpha + i\dot{p}_n^\alpha = -i\dot{\tau}_n H_{\alpha\beta}(x_n^\beta + ip_n^\beta) \; , \tag{11}$$

and its adjoint. Thus, we recover a *discrete analogue of Schrödinger's equation*, with $\psi_n^\alpha := x_n^\alpha + ip_n^\alpha$ as the amplitude of the "α-component" of "state vector" $|\psi\rangle$ at "time" n. Then, the Eqs. (7)–(8) imply this:

Theorem A: For any matrix $\hat{G}$ that commutes with $\hat{H}$, $[\hat{G}, \hat{H}] = 0$, there is a *discrete conservation law*:

$$\psi_n^{*\alpha} G_{\alpha\beta} \dot{\psi}_n^\beta + \dot{\psi}_n^{*\alpha} G_{\alpha\beta} \psi_n^\beta = 0 \ . \tag{12}$$

For self-adjoint $\hat{G}$, with complex integer elements, this relation concerns real integer quantities. •

Corollary A: For $\hat{G} := \hat{1}$, the Eq. (12) implies a *conserved constraint* on the state variables:

$$\psi_n^{*\alpha} \dot{\psi}_n^\alpha + \dot{\psi}_n^{*\alpha} \psi_n^\alpha = 0 \ . \tag{13}$$

For $\hat{G} := \hat{H}$, an *energy conservation* law follows. •

Note that Eqs. (12) and (13) *cannot* be trivially "integrated", since the *Leibniz rule* is modified. Recalling $\dot{O}_n := O_{n+1} - O_{n-1}$, we have, for example, $O_{n+1}O'_{n+1} - O_{n-1}O'_{n-1} = \frac{1}{2}(\dot{O}_n[O'_{n+1} + O'_{n-1}] + [O_{n+1} + O_{n-1}]\dot{O}'_n)$, instead of the product rule of differentiation.

Furthermore, we cannot obtain a continuum limit simply by letting the discreteness scale $l \to 0$, as for example in Refs. [7, 16]. integer valuedness here conflicts with continuous time and related derivatives.

4.1. Another CA action and conservation laws without admissible unitary symmetries?

The CA action is invariant under suitable unitary transformations. This can be most easily recognized by considering an equivalent form of the action, *i.e.*, which generates the same discrete equations of motion as before. We may replace the definition given in Eq. (5) by:

$$\mathcal{S} := \sum_n [\mathrm{Im}(\psi_n^\alpha {}^* \psi_{n-1}^\alpha) + (\pi_n + \pi_{n-1})\Delta\tau_n - \mathcal{A}_n] \ , \tag{14}$$

with $\psi_n^\alpha := x_n^\alpha + ip_n^\alpha$ and $\mathrm{Im}X := (X - X^*)/2i$, together with the replacement of the definition of H_n (recall $R_n \equiv 0$), Eq. (3), by:

$$H_n := \frac{1}{2} H_{\alpha\beta} \psi_n^\alpha {}^* \psi_n^\beta \ , \tag{15}$$

which enters the action through $\mathcal{A}_n$, Eq. (2). Then, n-independent unitary transformations $\hat{U}$, with $\psi'_n = \hat{U}\psi_n$ and $[\hat{U}, \hat{H}] = 0$, leave the action $\mathcal{S}$ invariant.

The self-adjoint matrices $\hat{G}$ of *Theorem A* generate unitary transformations which leave $\mathcal{S}$ invariant. However, since the CA variables $\psi_n^\alpha := x_n^\alpha + ip_n^\alpha$ are restricted to be *complex integer-valued*, only unitary transformations that preserve this property are *admissible*. In general, we expect that there are very few interesting ones. Thus, we encounter here the situation that there can exist CA conservation laws, according to *Theorem A* or *Corollary A*, which are *not* related to *admissible symmetry transformations*. This differs from from what is usually the case in QM.

5. Sampling theory

It is worth recalling the underlying assumption of discrete mechanics that the density of events and, thus, of information content of spacetime regions is cut off by the scale l [7, 17]. Therefore, despite the observed similarities of our Hamiltonian CA with QM systems, we may wonder whether the discreteness of a deterministic CA can be reconciled with any continuum description at all and, in particular, with QM?

Indeed, we have argued that physical fields, wave functions in particular, could be *simultaneously discrete and continuous*, represented by sufficiently smooth functions containing a finite density of degrees of freedom [1]. This idea has recently been introduced by Kempf and has led to a covariant ultraviolet cut-off suitable for theories including gravity [8]. However, neither *integer-valued CA* nor the *structure of QM* have been addressed in this way.

The fact that information can have simultaneously continuous and discrete character has been pointed out by Shannon in his pioneering work [18]. This is routinely applied in signal processing, converting analog to digital encoding and *vice versa*. Sampling theory demonstrates that a bandlimited signal can be perfectly reconstructed, provided discrete samples of it are taken at the rate of at least twice the band limit (Nyquist rate). For an extensive review and modern developments of the theory, see Refs. [19, 20], respectively.

We consider the *Sampling Theorem* in its simplest form [8, 19]: Consider square integrable *bandlimited functions* f, *i.e.*, which can be represented as $f(t) = (2\pi)^{-1}\int_{-\omega_{max}}^{\omega_{max}} \mathrm{d}\omega\ \mathrm{e}^{-i\omega t}\tilde{f}(\omega)$, with bandwidth ω_{max}. Given the set of amplitudes $\{f(t_n)\}$ for the set $\{t_n\}$ of equidistantly spaced times (spacing π/ω_{max}), the function f is obtained for all t by:

$$f(t) = \sum_n f(t_n)\frac{\sin[\omega_{max}(t-t_n)]}{\omega_{max}(t-t_n)} \quad . \tag{16}$$

Since the CA time is given by the integer n, the corresponding discrete *physical time* is obtained by multiplying with the fundamental scale l, $t_n \equiv nl$, and the bandwidth by $\omega_{max} = \pi/l$.

When attempting to *map invertibly* Eqs. (7)–(8) on reconstructed continuum equations, according to Eq. (16), the nonlinearity on the right-hand sides is problematic: the product of two functions, with bandwidth ω_{max} each, is not a function with the same bandwidth. Therefore, we presently assume that $\dot{\tau}_n$ is a constant and postpone consideration of more general situations.

Let us recall Eq. (11). Inserting $\psi_n^\alpha := x_n^\alpha + ip_n^\alpha$ and applying the *Sampling Theorem*, this discrete time equation is mapped to the *continuous time equation*:

$$\frac{\hat{D}_l - \hat{D}_{-l}}{2}\psi^\alpha(t) = \sinh(l\partial_t)\psi^\alpha(t) = \frac{1}{i}H_{\alpha\beta}\psi^\beta(t) \ , \tag{17}$$

where we employed the translation operator defined by $\hat{D}_T f(t) := f(t+T)$ and set $\dot{\tau}_n \equiv \dot{\tau} = 2$.

Thus, we obtain the *Schrödinger equation*, however, modified in important ways. (We use the QM terminology freely and concentrate on new effects arising here.) The wave function ψ^α has bandwidth ω_{max}, due to reconstruction formula (16). This leads to an *ultraviolet cut-off* of the energy E of stationary states of the generic form $\psi_E(t) := \exp(-iEt)\tilde{\psi}$. Diagonalizing the self-adjoint Hamiltonian, $\hat{H} \rightarrow \mathrm{diag}(\epsilon_0, \epsilon_1, \ldots)$, Eq. (17) yields the eigenvalue equation:

$$\sin(E_\alpha l) = \epsilon_\alpha \ \ , \tag{18}$$

and a *modified dispersion relation*, $E_\alpha = l^{-1}\arcsin(\epsilon_\alpha) = l^{-1}\epsilon_\alpha[1 + \epsilon_\alpha^2/3! + \mathrm{O}(\epsilon_\alpha^4)]$ [21]. The spectrum $\{E_\alpha\}$ is cut off by the condition $|\epsilon_\alpha| \leq 1$, entailing $|E_\alpha| \leq \pi/2l = \omega_{max}/2$, *i.e.* half the bandlimit.

For CA with a finite (or truncated) number of degrees of freedom or states, labeled by α, the constant $\dot{\tau}$ could be determined instead by the largest eigenvalue of H, $\epsilon_{max} := \max\{|\epsilon_\alpha|\}$, choosing $\dot{\tau} = 1/\epsilon_{max}$. In this way, there would be stationary states corresponding to all eigenvectors of $\hat{H}$. Otherwise, states corresponding to larger eigenvalues of $\hat{H}$ are unstable, related to complex solutions of the dispersion relation.

The modified Schrödinger equation (17) incorporates an infinite series of higher-order time derivatives. These are negligible for low-energy wave functions, which vary little with respect to the cut-off scale, *i.e.* $|\partial^k\psi/\partial t^k| \ll l^{-k} = (\omega_{max}/\pi)^k$. However, in general, they would require an infinity of initial data.

Journal of Physics: Conference Series **504** (2014) 012004 doi:10.1088/1742-6596/504/1/012004

5.1. Continuous time conservation laws

The relation between Eq. (11) and Eq. (17), together with the linearity of both equations, suggest that the correct continuous time conservation laws can be obtained by the replacement

$$\dot{\psi}_n := \psi_{n+1} - \psi_{n-1} \quad \longrightarrow \quad \frac{1}{l}\sin(il\partial_t)\psi(t) \ , \tag{19}$$

from Eqs. (12) and (13), respectively. Indeed, by Eq. (17), the following holds:

Theorem B: For any matrix $\hat{G}$ with $[\hat{G}, \hat{H}] = 0$, there is a *continuous time conservation law*:

$$\psi^{*\alpha} G_{\alpha\beta} \sin(il\partial_t)\psi^{\beta} + [\sin(il\partial_t)\psi^{*\alpha}]G_{\alpha\beta}\psi^{\beta} = 0 \ , \tag{20}$$

in particular,

$$\psi^{*\alpha} \sin(il\partial_t)\psi^{\alpha} + [\sin(il\partial_t)\psi^{*\alpha}]\psi^{\alpha} = 0 \ , \tag{21}$$

which modifies the QM wave function *normalization*, referring to a basis labeled by α. •

Only now we can remove the ultraviolet cut-off, with $l \to 0$, and recover familiar QM results from the leading order terms in Eqs. (20)–(21). (If l is a fundamental *constant*, this limit may be interesting for heuristic reasons alone.)

For example, consider the real symmetric *two-time function*,

$$2C_{\hat{G}}(t_1, t_2) := \psi^{*\alpha}(t_1)G_{\alpha\beta}\psi^{\beta}(t_2) \ + \ \text{c.c.} \ , \tag{22}$$

where $X + \text{c.c.} := X + X^*$ and $\hat{G}$ is a self-adjoint matrix, with $[\hat{G}, \hat{H}] = 0$. Applying *Theorem B*, we obtain:

Corollary B: The two-time function $C_{\hat{G}}$ is invariant under discrete translations of this form:

$$C_{\hat{G}}(t - l, t) = C_{\hat{G}}(t, t + l) \ , \tag{23}$$

implying that it is fixed everywhere by giving $C_{\hat{G}}(t, t + l)$ for all t in an interval $[t_0, t_0 + l[$. •

The wave function normalization, $\psi^{*\alpha}\psi^{\alpha} = 1$, then arises here from the coincidence limit of a two-time function with the property $C_{\hat{1}}(t, t + l) \equiv 1$, for all t:

$$1 = \lim_{l \to 0} C_{\hat{1}}(t, t + l) = \psi^{*\alpha}(t)\psi^{\alpha}(t) \ , \tag{24}$$

which is consistent with Eq. (21) and essential for the probability interpretation in QM. An analogous *equal-time* constraint, in general, does not exist on the CA level of description. *E.g.*, $\psi_n^{*\alpha}\psi_n^{\alpha} = x_n^{\alpha}x_n^{\alpha} + p_n^{\alpha}p_n^{\alpha} = 1$, instead of Eq. (13), is compatible only with rather trivial evolution, since all variables are integer-valued.

It is remarkable how properties of Hamiltonian CA produce familiar QM results, even if modified by the finite scale l. The operators or matrices that generate the QM conservation laws do so for the bandwidth limited continuum theory as well, as stated by *Theorem B*. Since the *same* vanishing commutator is responsible for the CA conservation laws, Eqs. (12)–(13), they correspond to each other one-to-one. Yet the QM symmetry transformations, generally, comprise a larger set than the admissible discrete ones for CA, which have to respect complex integer valuedness of the dynamical variables.

These observations leave us with an intriguing question: *What, if any, would be physical reasons for the existence of Hamiltonian CA conservation laws that are not tied to symmetries, which are fully developed only in the continuum limit?*

6. Discussion

It will be important to extend the CA–QM map to relativistic QM and QFT. Since wave equations and functional Schrödinger equation are linear and have a Hamiltonian formulation, it should be possible to employ a generalized *Sampling Theorem* for fields. It has been shown indeed that the d'Alembert operator can be covariantly regularized by imposing a finite bandwidth of its spectrum [8], a useful ingredient. – Conversely, given a *Hamiltonian* CA, one could invoke the path integral for classical systems [22] plus reconstruction formulae, attempting to obtain a relativistic bandwidth limited quantum (field) theory. – Other constructions of CA for *relativistic models* have appeared, which either incorporate QM features from the outset, *e.g.*, for the Dirac equation [23], or derive them, *e.g.*, for bosonic QFT and a superstring model [24]; see also references there. All these models have been *noninteracting*.

Lack of interactions there might be dictated by additional restrictions, such as locality, when placing a CA, say at Planck scale, into physical spacetime as experienced at the scales where QM is tested. It is remarkable that arbitrary QM N-level systems can be described by $2N-1$ nonrelativistic coupled oscillators in one fictitious space dimension [25]. Could it be that fundamental CA exist in an abstract space and QM and spacetime emerge together from there?

The nonrelativistic Hamiltonian CA *do incorporate interactions* through the matrix elements $H_{\alpha\beta}$. Their x^α, p^α- variables can be embedded into two-dimensional phase space, similarly as in Ref. [25]. Yet other interpretations are possible, such as α labelling sites of a d-dimensional lattice or elements of the Hilbert space arrived at in the QM description. This freedom appears in the nonrelativistic situation studied here without reference to gravitation or dynamical spacetime.

Interactions are also incorporated in a statistical theory of certain matrix models, which leads to QM behaviour emerging from a Gibbs distribution [26]. Similarly as in Refs. [24], this assumes a very particular form of dynamics and it remains to be seen whether gauge theories as, for example, in the Standard Model can be recovered. – In distinction, we do not make any assumptions about particular interactions but explore the mapping between structural features of Hamiltonian CA and of QM. It will be challenging to identify principles that distinguish a physically relevant Hamiltonian and related conservation laws within an "ontology" of CA.

In Ref. [1], we have discussed how the essential features of *entanglement* in QM and *apparent nonlocality* [27] come into play in our approach. It will be most interesting to reconsider questions of entanglement and locality in a relativistic generalization of the present theory. – Observables, measurements, and Born rule can be discussed in the bandwidth limited theory with help of Heslot's work [9] and implications for CA *per se* deserve further study.

We remark that Hamiltonian CA might be useful to simulate complex QM systems by the mapping on computer friendly *integer-valued* dynamical variables. A few hints in this direction have been mentioned in Ref. [1].

Let us also draw attention to the differences between the presently introduced Hamiltonian CA and quantum cellular or *quantum lattice-gas automata* (QLGA). The QLGA have recently attracted attention, since they are, by construction, discretizations of the Schrödinger equation [28, 29], cf. also [23]. They are specifically made to reproduce a quadratic kinetic energy term in the Schrödinger equation in the continuum limit in configuration space. This involves judicious choice of transformation matrices, i.e. implicitly of a number of dimensionless parameters [29]. However, the Hilbert space structure of QM state space with complex wave functions and linearity and unitarity of their evolution have to be incorporated *ab initio*.

In these respects, our approach differs remarkably: It is based on integer-valued dynamical variables and an underlying action principle. This *implies* linearity and unitarity together with all conservation laws. Unlike the case of QLGA and in accordance with the discussion in Sect. II. of the role of the discreteness scale l, the Hamiltonian CA here provide a *discrete deformation* of QM that reduces to it for $l \to 0$.

7. Conclusion

A map between Hamiltonian cellular automata (CA) and quantum mechanics (QM) has been constructed by combining elements of *discrete mechanics* [7], *sampling theory* [8], and *Hamiltonian formulation* of QM [9]. Thus, structural features of QM, the Schrödinger picture of evolution and conservation laws in particular, can be seen to originate in integer-valued CA incorporating a fundamental scale.

The postulated action principle refers to arbitrary integer variations of the dynamical CA variables. This enforces the *linearity* of the theory, as we have demonstrated [1].

Consequently, the separability assumption mentioned in Sect. I., which underlies the derivation of the linearity of QM *within* its formal framework, can here be substituted as follows: "*... the dynamics we are considering can be independent of something else in the universe*" [2], if and only if *the* CA *action is stationary under arbitrary integer variations.*

This provides a new view of linearity and the superposition principle in quantum mechanics with manifold consequences and generalizations to be explored.

Acknowledgments

I wish to thank G. 't Hooft for discussions and correspondence and G. Grössing and his collaborators and J. Walleczek for organizing the symposium EmQM13 "Emergent Quantum Mechanics" and for the kind hospitality in Vienna supported by Fetzer Franklin Fund.

References

[1] Elze H-T 2014 *Phys. Rev.* A **89** 012111 (*Preprint* arXiv:1312.1615)
[2] Jordan T F 2006 *Phys. Rev.* A **73** 022101; *do.* 2009 *J. Phys.: Conf. Ser.* **196** 012010
[3] Weinberg S 1989 *Ann. Phys. (N.Y.)* **194** 336; *do.* 1989 *Phys. Rev. Lett.* **62** 485
[4] Gisin N 1990 *Phys. Lett.* A **143** 1
[5] Polchinski J 1991 *Phys. Rev. Lett.* **66** 397
[6] For a relativistic nonlinear generalization, see: Elze H-T 2008 *Int. J. Theor. Phys.* **47** 455
[7] Lee T D 1983 *Phys. Lett.* **122**B 217; *do.* 1987 *J. Statist. Phys.* **46** 843
[8] Kempf A 2010 *New J. Phys.* **12** 115001; *do.* 2009 *Phys. Rev. Lett.* **103** 231301; *do.* 2004 *Phys. Rev. Lett.* **92** 221301; *do.* 2000 *Phys. Rev. Lett.* **85** 2873
[9] Heslot A 1985 *Phys. Rev.* D **31** 1341
[10] The Hamiltonian approach [9] has been applied to quantum-classical hybrids in: Elze H-T 2012 *Phys. Rev.* A **85** 052109; Buric N, Mendas I, Popovic D B, Radonjic M and Prvanovic S 2012 *Phys. Rev.* A **86** 034104
[11] Hossenfelder S 2013 *Living Rev. Relativity* **16** 2
[12] An inspiring discussion about the number of *fundamental constants* in physics is: Duff M J, Okun L B and Veneziano G 2002 *JHEP* **03** 023
[13] D'Innocenzo A, Renna L and Rotelli P 1987 *Eur. J. Phys.* **8** 245
[14] Hong-Bin Z, Li-Qun C and Rong-Wan L 2005 *Chinese Phys.* **14** 1063
[15] Capobianco S and Toffoli T 2011 Can anything from Noether's theorem be salvaged for discrete dynamical systems? *Preprint* arXiv:1103.4785v2
[16] Elze H-T 2013 *EPJ Web of Conferences* **58** 01013
[17] A fundamental scale has been introduced similarly for dynamical *causal sets*, a kind of stochastic CA, see, *e.g.*: Sorkin R D 2005 *Braz. J. Phys.* **35** 280
[18] Shannon C E 1949 *Proc. IRE* **37** 10-21
[19] Jerri A J 1977 *Proc. IEEE* **65** 1565
[20] Strohmer T and Tanner J 2005 *Sampling Theory in Signal and Image Processing* **41**
[21] *E.g.*, if α labels momentum modes for a spatial lattice.
[22] Elze H-T, Gambarotta G and Vallone F 2011 *Int. J. Qu. Inf. (IJQI)* **9** (Suppl. 1) 203
[23] Feynman R P 1982 *Int. J. Theor. Phys.* **21** 467; Bialynicki-Birula I 1994 *Phys. Rev.* D **49** 6920; D'Ariano G M and Perinotti P 2013 Derivation of the Dirac equation from principles of information processing *Preprint* arXiv:1306.1934; Arrighi P, Forets M and Nesme V 2013 The Dirac equation as a quantum walk: higher dimensions, observational convergence 2013 *Preprint* arXiv:1307.3524
[24] 't Hooft G 2013 *Found. Phys.* **43** 597; *do.* 2013 The Fate of the Quantum *Preprint* arXiv:1308.1007
[25] Skinner T E 2013 *Phys. Rev.* A **88** 012110

[26] Adler S L 2004 *Quantum Theory as an Emergent Phenomenon: The Statistical Mechanics of Matrix Models as the Precursor of Quantum Field Theory* (Cambridge: Cambridge Univ. Press)
[27] Englert B-G 2013 *Eur. Phys. J.* D **67** 238
[28] Meyer D A 1997 *Phys. Rev.* E **55** 5261; *do.* 1997 *Int. J. Mod. Phys.* C **8** 717
[29] Bogoshian B M and Taylor IV W 1998 *Phys. Rev.* E **57** 54

Nonlinear Riccati equations as a unifying link between linear quantum mechanics and other fields of physics

Dieter Schuch
Institut für Theoretische Physik, J.W. Goethe-Universität Frankfurt am Main, Max-von-Laue-Str. 1, D-60438 Frankfurt am Main, Germany

E-mail: Schuch@em.uni-frankfurt.de

Abstract. Theoretical physics seems to be in a kind of schizophrenic state. Many phenomena in the observable macroscopic world obey nonlinear evolution equations, whereas the microscopic world is governed by quantum mechanics, a fundamental theory that is supposedly linear. In order to combine these two worlds in a common formalism, at least one of them must sacrifice one of its dogmas. I claim that linearity in quantum mechanics is not as essential as it apparently seems since quantum mechanics can be reformulated in terms of nonlinear Riccati equations. In a first step, it will be shown where complex Riccati equations appear in time-dependent quantum mechanics and how they can be treated and compared with similar space-dependent Riccati equations in supersymmetric quantum mechanics. Furthermore, the time-independent Schrödinger equation can also be rewritten as a complex Riccati equation. Finally, it will be shown that (real and complex) Riccati equations also appear in many other fields of physics, like statistical thermodynamics and cosmology.

1. Introduction

In the beginning of the 20th century, one of the biggest mysteries in physics was the observation of material systems behaving in certain experiments (as expected) like particles, in others, however, (unexpectedly) like waves. This puzzle of wave-particle duality was finally solved by quantum mechanics, developed around the same time by Heisenberg and Schrödinger, whereby Schrödinger's wave mechanical formulation [1] was more popular from the beginning because it used partial differential equations, which were more familiar to most physicists than Heisenberg's matrix mechanics, though both are essentially equivalent. As Schrödinger's original formulation is based on classical Hamilton–Jacobi mechanics it bears similar features to its classical counterpart, like conservation of energy (the operator corresponding to the Hamiltonian function is Hermitian and a constant of motion) and time-reversal symmetry of the dynamics (the time evolution is described by unitary transformations). Since Schrödinger's equation is linear and has features of a wave equation (for a complex quantity!) the superposition principle applies. This is advantageous for computational purposes (besides being suitable for the above-mentioned wave properties of material systems). As almost the last 90 years have shown, quantum mechanics is not only scientifically, but also economically, the most successful and influential theory so far.

Towards the end of the 20th century, another development in physics became quite popular, namely nonlinear (NL) dynamics [2] (also because of the aesthetically-appealing pictures of so-called fractals [3] that were actually merely a side-product). This theory is able to describe many of the phenomena observed in our every day surroundings like the weather, growth processes, etc. It also includes such basic experiences like evolution with a direction of time and dissipation of energy; both aspects are, as mentioned above, absent in quantum (and classical) mechanics in the original canonical form. In addition, the laws of NL dynamics are usually scale invariant; the size of the system does not matter only relative changes are relevant. There are phenomena like self-similarity, fractals, Mandelbrot sets, etc. [4].

Apart from self-similarity, i.e., finding the same structures at different scales, a striking property of fractals is the appearance of spiral forms, i.e., structures whose radius changes while the angle rotates. This combination of radial and angular motion is well known in nature as anyone can verify by looking at the shell of a nautilus or the horn of a ram. In these cases one actually has a "frozen" picture of the evolution and the radius is in fact a measure of time (in the case of the nautilus, the larger the distance from the centre, the older the part of the shell is; with the ram, the opposite applies). The shape of these spirals usually takes the form of a logarithmic spiral, where the radius grows (or shrinks) exponentially.

So, at the beginning of the 21st century, we are faced with a rather schizophrenic situation:

- Quantum mechanics (and other so-called fundamental theories) are

- reversible
- conservative
- linear.

- Observable (macroscopic) nature (as described, e.g., by NL dynamics) is

- irreversible
- dissipative
- nonlinear.

- In addition, there are fields of physics that are not naturally compatible with quantum mechanics, like

- thermodynamics
- cosmology.

Now the question arises: is it possible to find a formulation that is able to unify all these different aspects of physics? At first sight, the answer seems to be negative since linearity of quantum mechanics, by definition, contradicts NL dynamics. But, could it be that quantum mechanics is only a NL theory in disguise (in other words a linearized form of an underlying, more fundamental NL theory)? It might even be only one example of many that can be traced back to the same NL theory. So the question is not so much one of quantum mechanics emerging from a more fundamental "sub-quantum" theory, but rather one of being a (linearized) version of a formal NL theory where this theory applies to many other (in the luckiest case all) fields of physics but appearing in different forms concealing the similarity of the underlying (NL) structure.

In the following, it shall be shown how a (complex) Riccati equation might be a candidate for such a fundamental NL theory. Not only does this equation possess a kind of superposition principle, due to its linearizability, but it also has properties usually common only in NL systems, like sensitivity to initial conditions, bifurcation, and so on.

In Section 2, it will be shown that complex Riccati equations already occur in conventional time-dependent (TD) quantum mechanics and formal similarities (pertaining to the sensitivity to initial conditions) with supersymmetric quantum mechanics will be emphasized. Section 3 presents a short overview of similar formal structures that also exist in time-independent (TI) quantum mechanics. Dissipative versions of both aspects exist, but since they have been discussed in previous articles, details will not be given here but can be found in [5]. Section 4 comprises a comparison of the formal structures found in Section 2 with similar ones in statistical thermodynamics and cosmology.

2. Complex Riccati equations related to the time-dependent Schrödinger equation

In the following, one-dimensional problems with exact analytic solutions of the TD Schrödinger equation (SE) in the form of Gaussian wave packets will be considered, particularly the free motion (potential $V(x) = 0$) and the harmonic oscillator (HO) ($V = \frac{m}{2}\omega^2 x^2$) with constant frequency, $\omega = \omega_0$, or TD frequency, $\omega = \omega(t)$. In these cases, the solution of the TDSE (here for the HO, the case $V = 0$, in the following, is always obtained in the limit $\omega \to 0$)

$$i\hbar\frac{\partial}{\partial t}\Psi(x,t) = \left\{-\frac{\hbar^2}{2m}\frac{\partial^2}{\partial x^2} + \frac{m}{2}\omega^2 x^2\right\}\Psi(x,t) \tag{1}$$

(where $\hbar = \frac{h}{2\pi}$ with $h =$ Planck's constant) can be written as

$$\Psi(x,t) = N(t)\exp\left\{i\left[y(t)\,\tilde{x}^2 + \frac{<p>}{\hbar}\,\tilde{x} + K(t)\right]\right\} \tag{2}$$

with the shifted coordinate $\tilde{x} = x - <x>= x - \eta(t)$, where the mean value $<x>= \int_{-\infty}^{+\infty}\Psi^* x\Psi dx = \eta(t)$ corresponds to the classical trajectory, $<p>= m\dot{\eta}$ represents the classical momentum and the coefficient of the quadratic term in the exponent, $y(t) = y_R(t) + iy_I(t)$, is a complex function of time. The (possibly TD) normalization factor $N(t)$ and the purely TD function $K(t)$ in the exponent are not relevant to the following discussion.

The equations of motion for $\eta(t)$ and $y(t)$, or $\left(\frac{2\hbar}{m}y = \mathcal{C}\right)$ that are obtained by inserting WP (2) into Eq. (1) are important for our purpose and are given by

$$\ddot{\eta} + \omega^2\eta = 0\,, \tag{3}$$

and

$$\dot{\mathcal{C}} + \mathcal{C}^2 + \omega^2 = 0\,, \tag{4}$$

where overdots denote derivatives with respect to time. The Newtonian equation (3) simply means that the maximum of the WP, located at $x =<x>= \eta(t)$, follows the classical trajectory. The equation for the quantity $\frac{2\hbar}{m}y(t) = \mathcal{C}$ has the form a of a *complex* NL Riccati equation and describes the time-dependence of the WP width that is related to the position uncertainty via $y_I = \frac{1}{4<\tilde{x}^2>}$ with $<\tilde{x}^2>=<x^2> - <x>^2$ being the mean square deviation of position.

The dynamics of this quantity can be described more conveniently by introducing a new (real) variable $\alpha(t)$ via $\mathcal{C}_I = \left(\frac{2\hbar}{m}y_I\right) = \frac{1}{\alpha^2}$. Inserting this into the imaginary part of Eq. (4) allows one to determine the real part of the variable as $\mathcal{C}_R = \left(\frac{2\hbar}{m}y_R\right) = \frac{\dot{\alpha}}{\alpha}$, which, when inserted into the real part of (4) together with the above definition of $\mathcal{C}_I$, finally turns the complex Riccati equation into the real NL so-called Ermakov equation [1] for $\alpha(t)$,

$$\ddot{\alpha} + \omega^2\alpha = \frac{1}{\alpha^3}\,. \tag{5}$$

[1] This equation was studied already in 1874 by Adolph Steen [6]. However, Steen's work was ignored by mathematicians and physicists for more than a century because it was published in Danish in a journal usually not containing many articles on mathematics. An English translation of the original paper [7] is available and generalizations can be found in [8].

It had been shown by Ermakov [9] in 1880, i.e., 45 years before quantum mechanics was formulated by Schrödinger and Heisenberg, that from the pair of equations (3) and (5), coupled via ω^2, by eliminating ω^2 from the equations, a dynamical invariant, the Ermakov-invariant

$$I_L = \frac{1}{2}\left[(\dot{\eta}\alpha - \eta\dot{\alpha})^2 + \left(\frac{\eta}{\alpha}\right)^2\right] = \text{const.} \tag{6}$$

can be obtained (this invariant was rediscovered by several authors, also in a quantum mechanical context; see, e.g., [10, 11, 12]).

In the following, the remarkable properties of this invariant shall not be further considered (for details, see [13]), but a different way of treating the (inhomogeneous) Riccati equation shall be discussed. Instead of transforming it into the (real) NL Ermakov equation (5), it can be solved directly by transforming it into a homogeneous NL (complex) Bernoulli equation if a particular solution $\tilde{\mathcal{C}}$ of the Riccati equation is known. The general solution of Eq. (4) is then given by $\mathcal{C} = \tilde{\mathcal{C}} + \mathcal{V}(t)$ where $\mathcal{V}(t)$ fulfils the Bernoulli equation

$$\dot{\mathcal{V}} + 2\,\tilde{\mathcal{C}}\mathcal{V} + \mathcal{V}^2 = 0\,. \tag{7}$$

The coefficient of the linear term depends on the particular solution $\tilde{\mathcal{C}}$. Equation (7) can be linearized via $\mathcal{V} = \frac{1}{\kappa(t)}$ to yield

$$\dot{\kappa} - 2\,\tilde{\mathcal{C}}\,\kappa = 1\,, \tag{8}$$

which can be solved straightforwardly. For constant $\tilde{\mathcal{C}}$, $\kappa(t)$ can be expressed in terms of exponential or hyperbolic functions (for real $\tilde{\mathcal{C}}$) or trigonometric functions (for imaginary $\tilde{\mathcal{C}}$). In this case, $\mathcal{C}$ can be written as

$$\mathcal{C}(t) = \tilde{\mathcal{C}} + \frac{e^{-2\,\tilde{\mathcal{C}}t}}{\frac{1}{2\tilde{\mathcal{C}}}\left(1 - e^{-2\,\tilde{\mathcal{C}}t}\right) + \kappa_0}\,. \tag{9}$$

For $\tilde{\mathcal{C}}$ being TD, $\kappa(t)$ and hence $\mathcal{V}$ can be expressed in terms of $\mathcal{I}(t) = \int^t dt'\, e^{-\int^{t'} dt''\, 2\tilde{\mathcal{C}}(t'')}$. So the general solution of Eq. (4) can be written as

$$\mathcal{C}(t) = \tilde{\mathcal{C}} + \frac{d}{dt}\ln\left[\kappa_0 + \mathcal{I}(t)\right]\,, \tag{10}$$

defining a one-parameter family of solutions depending on the (complex) initial value of $\kappa_0 = \mathcal{V}_0^{-1}$ as parameter. To realize the strong qualitative influence of the initial value κ_0 on the solution of the Riccati equation (which is not surprising since this is a NL differential equation), a comparison with supersymmetric (SUSY) quantum mechanics [14, 15, 16, 17] might be quite useful.

In SUSY quantum mechanics, the Hamiltonian can be represented by a 2×2 diagonal matrix where the potentials V_i of the Hamiltonian operators $H_i = -\frac{\hbar^2}{2m}\frac{d^2}{dx_2} + V_i$ (i= 1,2) on the diagonal are determined by a (real) Riccati equation for the so-called "superpotential" $W(x)$,

$$V_{1/2} = \frac{1}{2}\left[W^2 \mp \frac{\hbar}{\sqrt{m}}\frac{d}{dx}W\right]\,, \tag{11}$$

and differ only by the sign of the term depending on the derivative of $W(x)$. Again, one can try to solve this Riccati equation by reducing it to a Bernoulli equation using a particular solution

$\tilde{W}$. In the case of the HO, this particular solution is given by $\tilde{W} = \sqrt{m}\omega_0\, x$, leading to the two potentials

$$\tilde{V}_{1/2} = \frac{m}{2}\,\omega_0^2\,x^2 \mp \frac{\hbar}{2}\,\omega_0\,, \tag{12}$$

i.e., essentially the parabolic harmonic potential, only shifted by minus/plus the ground state energy $\frac{\hbar}{2}\omega_0$. The general solution can again be written in the form $W(x) = \tilde{W}(x) + \Phi(x)$ where $\Phi(x)$ must now fulfil the Bernoulli equation

$$\frac{\hbar}{\sqrt{m}}\,\frac{d}{dx}\Phi + 2\,\tilde{W}\,\Phi + \Phi^2 = 0 \tag{13}$$

(written here for the plus sign of the derivative). This can be solved in the same way by linearization to finally yield the general solution

$$W(x) = \tilde{W}(x) + \frac{\hbar}{\sqrt{m}}\,\frac{d}{dx}\,\ln\left[\varepsilon + \mathcal{I}(x)\right]\,. \tag{14}$$

The integral $\mathcal{I}(x)$ is formally identical to the one in the TD case, only t must be replaced by x and $\tilde{C}$ by $\tilde{W}$. Also this solution depends on a (this time real) parameter ε (corresponding to $\Phi^{-1}(0)$). This generalized $W(x)$ gives rise to a one-parameter family of isospectral potentials (e.g., for i = 1)

$$V_1(x;\varepsilon) = \tilde{V}_1(x) - \frac{\hbar^2}{m}\,\frac{d^2}{dx^2}\,\ln\left[\varepsilon + \mathcal{I}(x)\right]\,, \tag{15}$$

i.e., potentials with different shapes, but the same energy spectrum, namely that of the HO, only with ground state energy equal to zero (apart from $\varepsilon = 0$, where this state is missing).

The shape of the potentials is now, however, unlike the parabolic harmonic potential, no longer symmetric under the exchange $x \to -x$; there is even a second minimum showing up for negative x whose depth increases with decreasing ε (for details, see [16]). Only for $\varepsilon \to \infty$, the ln-term vanishes and the parabolic potential $\tilde{V}_1(x)$ is re-gained. So, obviously the parameter ε can have drastic qualitative consequences for the solution of the Riccati equation. The same can also happen in the above-mentioned TD case when κ_0 is varied.

Finally, a third way to treat the complex Riccati equation (4) shall be mentioned. Using the ansatz

$$\mathcal{C} = \left(\frac{2\hbar}{m}y\right) = \frac{\dot{\lambda}}{\lambda}\,, \tag{16}$$

with complex $\lambda(t)$ linearizes Eq. (4) to yield the complex Newtonian equation

$$\ddot{\lambda} + \omega^2(t)\lambda = 0\,. \tag{17}$$

Writing λ in terms of real and imaginary parts, i.e., $\lambda = u + i\;z$ would just result in two identical Newtonian equations for u and z where it can be shown [18] that, up to a constant factor, z is identical to the classical trajectory $\eta(t)$. More information can be gained by writing λ in polar form as $\lambda = \alpha e^{i\varphi}$. From definition (16) one then obtains

$$\mathcal{C} = \frac{\dot{\alpha}}{\alpha} + i\,\dot{\varphi}\,. \tag{18}$$

Inserting $\mathcal{C}$ in this form into the complex Riccati equation turns the imaginary part of this equation into

$$\dot{\varphi} = \frac{1}{\alpha^2}\,, \tag{19}$$

which looks like the conservation of angular momentum for the motion of $\lambda(t)$ in the complex plane (and also agrees with the definition of $\mathcal{C}_I$ at the beginning of this section). The real part of the Riccati equation then turns simply into the Ermakov equation (5) for $\alpha(t)$.

3. Complex Riccati equations related to the time-independent Schrödinger equation

We have seen in the TD case that the real and imaginary parts, or phase φ and amplitude α, of the complex variable $\lambda(t) = \alpha\, e^{i\varphi}$ which fulfils the linear equation (17), obtained via Eq. (16) from the Riccati equation (4), are not independent of each other but coupled via the conservation law (19). A similar situation exists when considering the TISE, but now in the space-dependent case.

This can be shown using Madelung's hydrodynamic formulation of quantum mechanics [19] where the wave function is written in polar form as

$$\Psi(\mathbf{r},t) = \varrho^{1/2}(\mathbf{r},t)\ \exp\left(\frac{i}{\hbar}S(\mathbf{r},t)\right) \tag{20}$$

with the square root of the probability density $\varrho = \Psi^*\Psi$ as amplitude and $\frac{1}{\hbar}S$ as phase ($\mathbf{r}$ is the position vector in three dimensions).

Inserting this form into the TDSE (1) (now in three dimensions), and replacing $\frac{\partial}{\partial x}$ by the nabla operator ∇), leads to a modified Hamilton–Jacobi equation for the phase,

$$\frac{\partial}{\partial t}S + \frac{1}{2m}(\nabla S)^2\ + V\ - \frac{\hbar^2}{2m}\frac{\Delta\varrho^{1/2}}{\varrho^{1/2}}\ =\ 0\ , \tag{21}$$

and a continuity equation for the amplitude,

$$\frac{\partial}{\partial t}\varrho\ +\ \frac{1}{m}\nabla(\varrho\ \nabla S) = 0\ . \tag{22}$$

Already here, the coupling of phase and amplitude can be seen clearly since the Hamilton–Jacobi equation for the phase S contains a term (misleadingly called "quantum potential", V_{qu}) depending on ϱ, and the continuity equation for the density ϱ contains ∇S. It can be shown that also in the TI case this coupling is not arbitrary but related to a conservation law.

In 1994, G. Reinisch [20] did this in a NL formulation of TI quantum mechanics. Since in this case $\frac{\partial}{\partial t}\varrho = 0$ and $\frac{\partial}{\partial t}S = -E$ are valid, the continuity equation (22) (we now use the notation $\varrho^{1/2} = |\Psi| = a$) turns into

$$\nabla(a^2\nabla S)\ =\ 0 \tag{23}$$

and the modified Hamilton–Jacobi equation into

$$-\ \frac{\hbar^2}{2m}\Delta a\ +\ (V-E)\ a\ =\ -\ \frac{1}{2m}(\nabla S)^2 a\ . \tag{24}$$

Equation (23) is definitely fulfilled for $\nabla S = 0$, turning (24) into the usual TISE for the real wave function $a = |\Psi|$ with position-independent phase S. (N.B.: the kinetic energy term divided by a is just identical to V_{qu}!)

However, Eq. (23) can also be fulfilled for $\nabla S \neq 0$ if only the conservation law

$$\nabla S\ =\ \frac{C}{a^2} \tag{25}$$

is fulfilled with constant (or, at least, position-independent) C.

This relation now shows explicitly the coupling between phase and amplitude of the wave function and is equivalent to Eq. (19) in the TD case. Inserting (25) into the rhs of Eq. (24) changes this into the Ermakov equation

$$\Delta a\ +\ \frac{2m}{\hbar^2}(E-V)\ a\ =\ (\frac{1}{\hbar}\nabla S)^2\ a\ =\ \left(\frac{C}{\hbar}\right)^2\frac{1}{a^3}\ , \tag{26}$$

equivalent to Eq. (5) in the TD case.

Returning to the method described in [20], so far the energy E occurring in Eq. (26) is still a free parameter that can take any value. However, solving this equation numerically for arbitrary values of E leads, in general, to solutions a that diverge for increasing$x$. Only if the energy E is appropriately tuned to any eigenvalue E_n of the TISE (see Eq. (28), below) does this divergence disappear and normalizable solutions can be found. So, the quantization condition that is usually obtained from the requirement of the truncation of an infinite series in order to avoid divergence of the wave function is, in this case, obtained from the requirement of nondiverging solutions of the NL Ermakov equation (26) by variation of the parameter E. This has been numerically verified in the case of the one-dimensional HO and the Coulomb problem and there is the conjecture that this property is "universal" in the sense that it does not depend on the potential V (see [20, 21]).

The corresponding complex Riccati equation is now given by

$$\nabla\left(\frac{\nabla\Psi}{\Psi}\right) + \left(\frac{\nabla\Psi}{\Psi}\right)^2 + \frac{2m}{\hbar^2}(E-V) = 0 \tag{27}$$

with the complex variable $\mathcal{C} = \left(\frac{\nabla\Psi}{\Psi}\right) = \frac{\nabla a}{a} + i\frac{1}{\hbar}\nabla S$ which corresponds to $\left(\frac{2\hbar}{m}y\right) = \frac{\dot\lambda}{\lambda} = \frac{\dot\alpha}{\alpha} + i\,\dot\varphi$ in the TD problem.

It is possible to show straightforwardly that Eq. (27) can be linearized to yield the usual TISE

$$-\frac{\hbar^2}{2m}\Delta\Psi + V\,\Psi = E\,\Psi\,, \tag{28}$$

but in this form, the information on the coupling of phase and amplitude, expressed by Eq. (25) and originating from the quadratic NL term in Eq. (27), gets lost.

4. Riccati equations in other fields of physics

Let us return to solution (9) of the Riccati equation (4), but now for real $\mathcal{C}$, and rewrite it in the form

$$\mathcal{C}(t) = \tilde{\mathcal{C}} + \frac{2\tilde{\mathcal{C}}\,e^{-2\tilde{\mathcal{C}}t}}{\kappa_0\,2\tilde{\mathcal{C}} + \left(1-e^{-2\tilde{\mathcal{C}}\,t}\right)} = \tilde{\mathcal{C}} + \frac{2\tilde{\mathcal{C}}}{\kappa_0\,2\tilde{\mathcal{C}}\;e^{2\tilde{\mathcal{C}}t} + \left(e^{2\tilde{\mathcal{C}}\,t}-1\right)}\,. \tag{29}$$

For the choice $\kappa_0 = 0$, this turns into

$$\mathcal{C}(t) = \tilde{\mathcal{C}} + \frac{2\tilde{\mathcal{C}}}{\left(e^{2\tilde{\mathcal{C}}\,t}-1\right)} = \tilde{\mathcal{C}}\,\coth\,\tilde{\mathcal{C}}t\;. \tag{30}$$

Replacing time t by $t \to \frac{1}{kT} = \beta$ (with k = Boltzmann's constant) and the constant particular solution by $\tilde{\mathcal{C}} = \frac{\hbar}{2}\,\omega$, one obtains

$$\frac{\hbar}{2}\,\omega + \frac{\hbar\,\omega}{e^{\hbar\omega/kT}-1} = \frac{\hbar}{2}\,\omega\,\coth\left(\frac{\hbar\omega}{2kT}\right) = \,<E>_{th} \tag{31}$$

which is the expression known from statistical thermodynamics for the average energy of a single oscillator in thermal equilibrum [22]. The first term on the lhs is just the ground state energy of the harmonic oscillator, the second is equal to Planck's distribution function for the black body radiation. This type of relation between Eq. (31) and the Riccati equation has also been found by Rosu et al [23]. Equation (31) can also be expressed in terms of the partition function $Z = \sum_n e^{-n\hbar\omega\beta} = \frac{1}{1-e^{\hbar\omega\beta}}$ as

$$< E >_{th} = \frac{\hbar}{2}\,\omega + \frac{\frac{\partial}{\partial\beta} Z^{-1}}{Z^{-1}} = \frac{\hbar}{2}\,\omega - \frac{\partial}{\partial\beta}\ln Z\ . \qquad (32)$$

The Riccati equation corresponding to solution (30) in the form (31) can be written as

$$\mathcal{C}' + \mathcal{C}^2 - \tilde{\mathcal{C}}^2 = 0 \qquad (33)$$

with $\mathcal{C} = \mathcal{C}(\beta)$ depending on the variable $\beta = \frac{1}{kT}$, prime denoting derivative with respect to this variable and $\tilde{\mathcal{C}}$, as mentioned above, being the ground state energy, $\tilde{\mathcal{C}} = \frac{\hbar}{2}\omega$. The inverse quantity of $\mathcal{C}$, multiplied by $-\tilde{\mathcal{C}}^2$, i.e., $\mathcal{K} = -\tilde{\mathcal{C}}^2\,\mathcal{C}^{-1}$ fulfils

$$-\mathcal{K}' + \mathcal{K}^2 - \tilde{\mathcal{C}}^2 = 0\ , \qquad (34)$$

i.e., also a Riccati equation but now with negative sign for the derivative term and the solution

$$\mathcal{K}\left(\frac{1}{kT}\right) = \frac{\hbar}{2}\,\omega - \frac{\hbar\,\omega}{e^{\hbar\omega/kT}+1} = \frac{\hbar}{2}\,\omega\,\tanh\left(\frac{\hbar\omega}{2kT}\right)\ . \qquad (35)$$

¿From the ground state energy in this case a term is subtracted that represents a Fermi–Dirac distribution, whereas in the solution for $\mathcal{C}$, a term was added to the ground state energy that represents a Bose–Einstein distribution. So both quantum statistics can be obtained from the solution of the Bernoulli equations derived from the Riccati equations for $\mathcal{C}\left(\frac{1}{kT}\right)$ and (essentially) its inverse quantity.

Another connection between Bose–Einstein systems and a Riccati equation (in this case a complex one) can be found if a Bose–Einstein condensate (BEC) is described in the mean field approximation by a macroscopic wave packet for the condensate, Ψ, which obeys the Gross–Pitaevskii equation

$$i\hbar\frac{\partial}{\partial t}\Psi = \left\{-\frac{\hbar^2}{2m}\Delta + V(\mathbf{r},t) + g|\Psi|^2\right\}\Psi \qquad (36)$$

where $V(\mathbf{r},t)$ represents the trapping potential and shall be given by $V(\mathbf{r},t) = \frac{m}{2}\omega^2(t)r^2$ with $r = |\mathbf{r}| =$ absolute value of the vector $\mathbf{r}$ and TD frequency $\omega = \omega(t)$. Although this equation cannot be solved analytically, the dynamics of the BEC described by it can be obtained from a set of coupled differential equations for so-called moments that essentially represent the width, radial momentum and energy of the wave packet (for details, see [24]). The important point is that these three moments fulfil a closed set of three TD differential equations that (with the help of a conserved quantity K that corresponds to the conserved angular momentum in the complex plane, discussed in section 2) can be reduced to one equation for the square root of the width (denoted by $X(t)$ and corresponding to $\alpha(t)$ in section 2) with the form of an Ermakov equation,

$$\ddot{X} + \omega^2(t)X = \frac{k}{X^3}\ . \qquad (37)$$

Since Eq. (37) is, as we have seen in section 2, only a different way of writing a complex Riccati equation by making use of a conservation law, also this BEC can be described by a, now complex, Riccati equation.

Finally, another example shall be given where a physical system can be described by a (complex) Riccati equation, or its equivalent, a real Ermakov equation, now not for a microscopic but for a really macroscopic system. For this purpose, we switch to cosmology and apply to Einstein's field equations the cosmological principle, i.e., the assumption of a spatially-homogeneous and isotropic universe. In this case, the Robertson–Walker metric applies [25] that contains the scale factor $a(t)$ (so to say, the radius of the universe) and the curvature k that can attain the values 0 for a flat or $\pm$ 1 for a closed or open universe. This finnaly leads to the Friedman–Lemaitre equations

$$\left(\frac{\dot{a}}{a}\right)^2 = H^2 = \frac{2}{3}\varrho - \frac{k}{a^2} \tag{38}$$

and

$$\dot{\varrho} = -3H(\varrho + p) \tag{39}$$

with $H = \frac{\dot{a}}{a}$ = Hubble parameter; overdots denote derivatives with respect to cosmic proper time, ϱ = energy density and p = pressure (where $4\pi G = c = 1$ and the cosmological constant $\Lambda = 0$ have been set).

Assuming the matter source as a self-interacting scalar field $\Phi = \Phi(t)$, the energy density ϱ and pressure p can be written as

$$\varrho = \frac{1}{2}\left(\frac{d}{dt}\Phi\right)^2 + U(t)\,, \tag{40}$$

$$p = \frac{1}{2}\left(\frac{d}{dt}\Phi\right)^2 - U(t)\,. \tag{41}$$

Taking the time-derivative of (38) and inserting (39) using (40) and (41) leads to

$$\frac{d}{dt}\left(\frac{\dot{a}}{a}\right) = -\left(\frac{d}{dt}\Phi\right)^2 + \frac{k}{a^2}\,. \tag{42}$$

Introducing a new time variable (conformal time τ) via $\frac{d}{dt} = a\,\frac{d}{d\tau}$, Eq. (42) can be rewritten as an Ermakov equation,

$$\frac{d^2}{d\tau^2}a + \left(\frac{d}{d\tau}\Phi\right)^2 a = \frac{k}{a^3}\,, \tag{43}$$

which is equivalent to a Riccati equation for the complex quantity $\mathcal{C} = \left(\frac{\frac{d}{d\tau}a}{a} + i\,\frac{1}{a^2}\right)$. The above derivation of the Ermakov equation in comparison with the afore-mentioned BEC had been given by Lidsey in [26]. Everything said about the complex or real Ermakov equation in section 2 can also be applied to this system, e.g., corresponding creation and annihilation operators and coherent states can be defined (see [13]).

5. Conclusions and perspectives

It has been shown in section 2 that the information about the dynamics of a quantum mechanical wave packet, i.e. the equations of motion for the time-evolution of its maximum and width can not only be obtained from the solution of the TDSE, but equally well from a complex NL Riccati equation. The square root of the inverse of the imaginary part of the quantity fulfilling this equation, $\alpha(t)$, is (up to a constant factor) just the width of the wave packet, fulfilling a so-called Ermakov equation. The classical particle aspect that is reflected by the Newtonian equation determining the motion of the wave packet maximum can also be obtained, using that the complex Riccati equation can be linearized by the ansatz $\mathcal{C} = \frac{\dot{\lambda}}{\lambda}$ to a Newtonian equation for the complex quantity $\lambda(t)$, which can be written as $\lambda = u + z = \alpha\, e^{i\varphi}$. From the knowledge of α (as solution of the Ermakov equation) and with $\dot{\varphi} = \frac{1}{\alpha^2}$, $\varphi(t)$ can be determined and hence

$\lambda(t)$ is obtained. Knowing that $z \propto \eta(t)$, the imaginary part of λ provides (up to a constant factor) immediately the classical trajectory $\eta(t)$.

Formal similarities exist between this TD problem and SUSY quantum mechanics, ehere (real) position dependent Riccati equations play a similar role and the construction of isospectral potentials show the importance of parameters like initial conditions.

In Section 3 it has been shown that the Riccati formalism established for the TDSE can also be applied to the TISE if the TD complex quantity $C = \frac{\dot{\lambda}}{\lambda}$ is replaced by the space-dependent complex quantity $\frac{\nabla\Psi}{\Psi}$ for a complex wave function Ψ. In a certain way, this looks like a complex version of SUSY where not only the logarithmic derivative of the ground state is considered as a variable for a Riccati equation, but also any (even complex) excited state may fulfil a now complex Riccati equation.

The flexibility of the Riccati formalism is illustrated in section 4 where we see that it is not only restricted to quantum mechanics. Replacing time t with (actually imaginary) "time" $\frac{\hbar}{kT}$ leads to well-known expressions from statistical thermodynamics as solutions of Riccati equations that are also able to distinguish between bosonic and fermionic properties. While this example was still dealing with real Riccati equations, in the next example it was shown that also the dynamics of a BEC can be described by a complex Riccati or the corresponding Ermakov equation. Finally, we went from the microscopic scale to the really macroscopic one, to cosmology and demonstrated that the Friedman–Lemaitre equations, under certain assumptions, describe the dynamics of our universe and can be written in terms of a real Ermakov or equivalent complex Riccati equation.

There are also many more examples, from electrodynamics via quantum optics to soliton theory, etc., where the Riccati equation allows for a unifying formulation with the fields mentioned above as well as others. In conclusion therefore, one can say that the NL version of quantum mechanics based on a (complex) Riccati equation is able to cover all phenomena of standard quantum mechanics (including the superposition principle, due to its linearisablity). Due to the sensitivity of NL differential equations to the initial conditions, it can further provide additional information that gets lost in the linearised form. The complex form of the Riccati equation also supplies (via its imaginary part) a new conservation law that resembles the conservation of angular momentum, but now for the motion in the complex plane; something closely related to the quantum property spin.

It should also be mentioned again that effects like dissipation and irreversibility can easily be included into the Riccai formalism. In the TD case, only an additional term linear in C appears in Eq. (4). Since in the solution via the homogeneous Bernoulli equation (7) a linear term already occurs, only the coefficient of this term changes. However, in the TD quantum mechanical case, this leads to bifurcation, i.e., two qualitatively different solutions for the dynamics of the wave packet width, a phenomena well-known in NL dynamics but alien to quantum mechanics (for details see [27]). For the damped HO, $\lambda(t)$ is no longer moving on a circle in the complex plane but on a spiral with the radius decreasing exponentially.

So maybe it is not so much the question of quantum mechanics emerging from an underlying theory that possesses properties like non-equilibrium thermodynamics, hybrid-mechanics or other structures, but that of quantum mechanics, thermodynamics, nonlinear dynamics and other fields of physics somehow being traced back to a formalism based on complex NL Riccati equations or generalizations thereof.

Acknowledgments
The author would like to thank the organizers and sponsors, particularly Gerhard Grössing, for the generous invitation to participate in this very interesting and stimulating conference.

Journal of Physics: Conference Series **504** (2014) 012005 doi:10.1088/1742-6596/504/1/012005

References

[1] Schrödinger E 1926 *Ann. d. Phys.* **79** 361
[2] Schuster H G 1984 *Deterministic Chaos: An Introduction* (Weinheim: Physik Verlag) 112
[3] Peitgen H-O and Richter P H 1986 *The Beauty of Fractals* (Berlin: Springer-Verlag)
[4] Großmann S 1989 *Ordnung und Chaos* ed W Gerok (Stuttgart: Wissenschaftliche Verlagsgesellschaft) p 101
[5] Schuch D 2007 *J. Math. Phys.* **48** 122701; Schuch D 2012 *J, Phys.: Conf. Series* **361** 012020
[6] Steen A 1874 *Overs. over d. K. Danske Vidensk. Selsk. Forh.* 1
[7] Redheffer R 1999 *Aequationes Math.* **58** 60
[8] Redheffer R and Redheffer I 2001 *Aequationes Math.* **61** 131
[9] Ermakov V P 1880 *Univ. Jzv. Kiev* **20** No 9, 1
[10] Milne W E 1930 *Phys. Rev.* **35** 863
[11] Pinney E 1950 *Proc. Amer. Math. Soc.* **1** 681
[12] Lewis H R 1967 *Phys. Rev. Lett.* **18** 510
[13] Schuch D and Moshinsky M 2006 *Phys. Rev.* A **73** 062111; Castaños O, Schuch D and Rosas-Ortiz O 2012 *J. Phys. A: Math. Gen.* **46** 075304
[14] Mielnik B and Rosas-Ortiz O 2004 *J. Phys. A: Math. Gen.* **37** 10007
[15] Andrianov A A and Cannata F 2004 *J. Phys. A: Math. Gen.* **37** 10297
[16] Cooper F, Khare A and Sukhatme U 2001 *Supersymmetry in Quantum Mechanics* (Singapore: World Scientific)
[17] Khare A and Sukhatme U 1989 *J. Phys. A: Math. Gen.* **22** 2847; Keung W-Y, Sukhatme U P, Wang Q and Imbo T D 1989 *J. Phys. A: Math. Gen.* **22** L 987
[18] Schuch D 2008 *SIGMA* **4** 043
[19] Madelung E 1926 *Z. Physik* **40** 322
[20] Reinisch G 1994 *Physica* A **206** 229
[21] Reinisch G 1997 *Phys. Rev.* A **56** 3409
[22] Feynman R P 1972 *Statistical Macanics – A Set of Lectures* (Reading: Benjamin) 2.5
[23] Rosu H C and Aveces de la Cruz F 2002 *Phys. Scripta* **65** 377; Rosu H 2002 `arXiv:math-ph/020813v1`
[24] Garcia-Ripoll J J, Perez-Garcia V M and Torres P 1999 *Phys. Rev. Lett.* **83** 1715
[25] Sexl R U and Urbantke H K 1975 *Gravitation und Kosmologie* (Mannheim: Bibliographisches Institut)
[26] Lidsey J E 2004 *Class. Quantum Grav.* **21** 777
[27] Schuch D 2002 *Phys. Lett. A* **294** 31; Schuch D 2002 *J. Phys. A: Math. Gen.* **35** 1

Journal of Physics: Conference Series **504** (2014) 012006 doi:10.1088/1742-6596/504/1/012006

Relational causality and classical probability: Grounding quantum phenomenology in a superclassical theory

G Grössing[1], **S Fussy**[1], **J Mesa Pascasio**[1,2], **H Schwabl**[1],

[1] Austrian Institute for Nonlinear Studies, Akademiehof, Friedrichstr. 10, 1010 Vienna, Austria
[2] Institute for Atomic and Subatomic Physics, Vienna University of Technology, Operng. 9, 1040 Vienna, Austria

E-mail: ains@chello.at

Abstract. By introducing the concepts of "superclassicality" and "relational causality", it is shown here that the velocity field emerging from an n-slit system can be calculated as an average classical velocity field with suitable weightings per channel. No deviation from classical probability theory is necessary in order to arrive at the resulting probability distributions. In addition, we can directly show that when translating the thus obtained expression for said velocity field into a more familiar quantum language, one immediately derives the basic postulate of the de Broglie-Bohm theory, i.e. the guidance equation, and, as a corollary, the exact expression for the quantum mechanical probability density current. Some other direct consequences of this result will be discussed, such as an explanation of Born's rule and Sorkin's first and higher order sum rules, respectively.

1. Introduction

In 1951, Richard Feynman published a paper [ ] claiming that classical probability theory was not applicable for the description of quantum phenomena, but that instead separate "laws of probabilities of quantum mechanics" were required. In describing the propagations of electrons from a source S to a location X on the screen in a double-slit experiment, Feynman wrote: "We might at first suppose (since the electrons behave as particles) that

I. Each electron which passes from S to X must go either through hole 1 or hole 2. As a consequence of I we expect that:

II. The chance of arrival at X should be the sum of two parts, P_1, the chance of arrival coming through hole 1, plus P_2, the chance of arrival coming through hole 2." However, we apparently know from experiment that this is not so (Fig. 1). Feynman concludes: "Hence experiment tells us definitely that $P \neq P_1 + P_2$ or that II is false. (...) We must conclude that when both holes are open it is not true that the particle goes through one hole or the other."

However, it was shown very clearly in several later papers, notably by B. O. Koopman in 1955 [ ] and by L. E. Ballentine in 1986 [ ], that Feynman's reasoning was not conclusive. In fact, Ballentine concluded that Feynman's "argument draws its radical conclusion from an incorrect application of probability theory", as one has to use conditional probabilities to account for the different experimental contexts: $P_1(X|C_1)$, $P_2(X|C_2)$, $P(X|C_3)$, where the contexts C_1, C_2 and C_3 refer to left, right and both slit(s) open, respectively. Then, it is clear from the

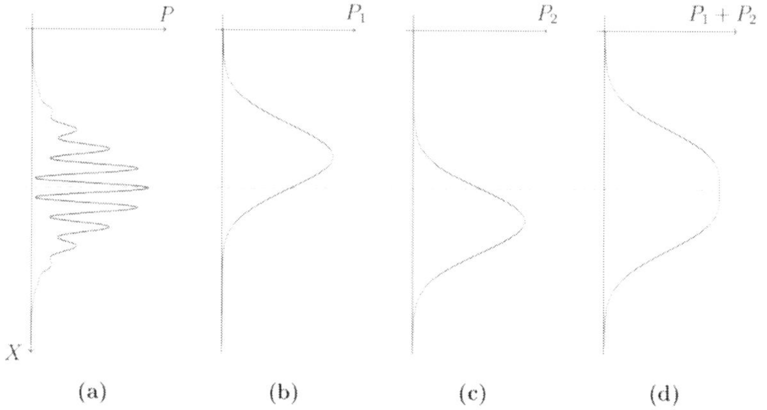

Figure 1.1: Scheme of interference at a double slit: probability distributions for (a) both slits open: P, (b) slit 2 closed: P_1, (c) slit 1 closed: P_2, (d) the sum of (b) and (c): $P_1 + P_2$.

experimental data that $P(X|C_3) \neq P_1(X|C_1) + P_2(X|C_2)$. Accordingly, Ballentine also proved that "the quantum mechanical superposition principle for amplitudes is in no way incompatible with the formalism of probability theory" [ ]. Moreover, as one can easily show with the inclusion of the wave picture for electrons (see Eq. (1.3) below), one correctly obtains that $P(X|C_3) = P_1(X|C_3) + P_2(X|C_3)$.

Note that the focus purely on context-free "particles" in quantum mechanics has been the source of much confusion. We recall the clear words of John Bell in this regard: "While the founding fathers agonized over the question 'particle' or 'wave', de Broglie in 1925 proposed the obvious answer 'particle' and 'wave'. Is it not clear from the smallness of the scintillation on the screen that we have to do with a particle? And is it not clear, from the diffraction and interference patterns, that the motion of the particle is directed by a wave? De Broglie showed in detail how the motion of a particle, passing through just one of two holes in a screen, could be influenced by waves propagating through both holes. And so influenced that the particle does not go where the waves cancel out, but is attracted to where they cooperate. This idea seems to me so natural and simple, to resolve the wave-particle dilemma in such a clear and ordinary way, that it is a great mystery to me that it was so generally ignored." [ ] Particularly since the impressive Paris experiments in the classical domain performed by Yves Couder, Emmanuel Fort and co-workers [–], it is very suggestive that some sort of combination of "particle"- and wave-mechanics is at work also in the quantum domain. Essentially, this amounts to obtaining particle distributions from calculating classical-like wave intensity distributions.

With wave-field intensities P_i for each slit i given by the squared amplitude R_i, i.e. with $P_i = R_i^2$, and with the phase difference φ, one obtains in the classical double slit scenario for the intensity after slit 1

$$P(1) = P_1 + R_1 R_2 \cos\varphi \tag{1.1}$$

and for the intensity after slit 2

$$P(2) = P_2 + R_2 R_1 \cos\varphi \tag{1.2}$$

where the first expression on the r.h.s. refers to the intensity from the slit *per se* and the second expression refers to interference with the other slit, respectively. This provides the total intensity as

$$P = P(1) + P(2) = P_1 + P_2 + 2R_1 R_2 \cos\varphi. \tag{1.3}$$

In the following we try to shed more light on these issues by combining results from the new field of "Emergent Quantum Mechanics" [ ] with concepts of systems theory which we denote as "relational causality" [ ]. Since the physics of different scales is concerned, like, e.g., sub-quantum and classical macro physics, we denote our sub-quantum theory as "superclassical". (Both relational causality and superclassicality are to be defined more specifically below in Section 2.) We consider the quantum itself as an emergent system understood as off-equilibrium steady state oscillation maintained by a constant throughput of energy provided by the ("classical") zero-point energy field. Starting with this concept, our group was able to assess phenomena of standard quantum mechanics like Gaussian dispersion of wave packets, superposition, double slit interference, Planck's energy relation, or the Schrödinger equation, respectively, as the emergent property of an underlying sub-structure of the vacuum combined with diffusion processes reflecting the stochastic parts of the zero-point field [–].

In Section 3 we contrast the well-known physics behind the double slit with an emergent vector field representation of the observed interference field. The essential parts of our superclassical approach are presented and the velocity field corresponding to the guiding equation of the de Broglie-Bohm theory is derived. The explanation and validity of Born's rule is analyzed in Section 4 by means of a three slit configuration. In Section 5 we summarize our results and give an outlook on possible limits of the validity of present-day quantum mechanics from the perspective of our sub-quantum approach.

2. Introducing "superclassicality" and "relational causality"

In quantum mechanics, as well as in our quantum-like modeling via an emergent quantum mechanics approach, one can write down a formula for the total intensity distribution P which is identical to (1.3). For the general case of n slits, it holds with phase differences $\varphi_{ij} = \varphi_i - \varphi_j$ that

$$P = \sum_{i=1}^{n} \left(P_i + \sum_{j=i+1}^{n} 2 R_i R_j \cos \varphi_{ij} \right), \tag{2.1}$$

where the phase differences are defined over the whole domain of the experimental setup. Apart from the role of the relative phase with important implications for the discussions on nonlocality [ ], there is one additional ingredient that distinguishes (2.1) from its classical counterpart (1.3), namely the "dispersion of the wavepacket". As in our model the "particle" is actually a "bouncer" in a fluctuating wave-like environment, i.e. analogously to the bouncers of Couder and Fort's group, one does have some (e.g. Gaussian) distribution, with its center following the Ehrenfest trajectory in the free case, but one also has a diffusion to the right and to the left of the mean path which is just due to that stochastic bouncing. Thus the total velocity field of our bouncer in its fluctuating environment is given by the sum of the forward velocity $\mathbf{v}$ and the respective diffusive velocities $\mathbf{u}_L$ and $\mathbf{u}_R$ to the left and the right. As for any direction i the diffusion velocity $\mathbf{u}_i = D\frac{\nabla_i P}{P}$ does not necessarily fall off with the distance, one has long effective tails of the distributions which contribute to the nonlocal nature of the interference phenomena [ ]. In sum, one has essentially three velocity (or current) channels per slit in an $n-$slit system.

Earlier we showed that this phenomenon can also be understood as a variant of anomalous diffusion termed "ballistic diffusion": a Brownian-type displacement with a time-dependent diffusivity (e.g., $D_{\mathrm{t}} = \frac{D^2}{\sigma_0^2} t$ in the case of a Gaussian wave packet with standard deviation σ), leading to a "classically" obtained total velocity field

$$\mathrm{v_{tot}}\,(t) = \mathrm{v}\,(t) + \left[\mathrm{x_{tot}}\,(t) - \mathrm{v}(t)\,t\right] \frac{D_{\mathrm{t}}}{\sigma^2}, \tag{2.2}$$

which is very practical to use in a computer simulation tool [,]. Moreover, ballistic diffusion is a signature of what has in recent years become known as "superstatistics" [ ]. Actually, the prototype of a phenomenon amenable to superstatistics is just a Brownian particle moving through a thermally changing environment: Combining in one formalism a relatively fast dynamics (e.g. particle velocity) on a small scale and slow changes (e.g. due to temperature fluctuations) on a large scale leads to a superposition of two statistics, i.e. superstatistics. One of the main features of superstatistics consists in *emergent properties arising on intermediate scales*, which may be *completely unexpected* if one looks only at either the small or the large scale.

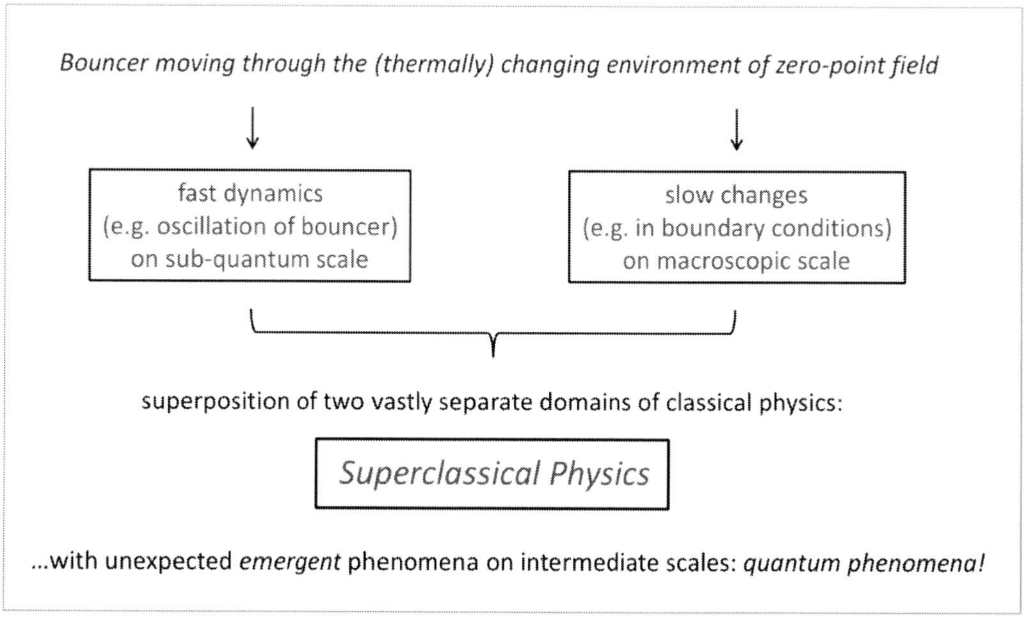

Figure 2.1: Scheme of *Superclassical Physics* as applied to our bouncer model

The phenomenon just described, however, also provides us with the opportunity of solving a terminological problem with regard to sub-quantum theories such as ours. For, although we consider our approach as based on modern, "21st century classical physics", the latter must not be confused with the term "classical physics" as referring to the state-of-art of, say, the early 20th century, including e.g. the time of the first experimental evidence of a zero-point energy (viz. Mulliken's work from 1924 [ ], which itself was before the advent of quantum theory proper). Acknowledging both that a) a zero-point field can in principle be considered a "classical" one and that b) quantum theory is characterized by some decisively unexpected features when considered from a classical viewpoint, the phenomenon of superstatistics thus suggests the following analogy for the modeling of quantum systems: quantum behavior can be seen to emerge from the interplay of classical processes at very small ("sub-quantum") and large (macroscopic) scales. A *combination of the physics on these vastly different scales* we call "superclassical" physics. The whole system under study then is characterized by *processes of emergence through the co-evolution of microscopic, local processes* (like the oscillations of a bouncer) *and of macroscopic processes* (like the time evolution of the experimental boundary conditions). In other words, starting from physical processes at such two vastly different classical scales, their superposition (i.e. within the framework of superclassical physics) makes possible new types of phenomena due to *emergent features unexpected from either the very small or the large scale physics* (Fig. 2.1; compare also the related concept of "emergent relativity" as discussed, e.g., by Jizba and Scardigli, this volume, [ ], and [ ]). In our development of an emergent quantum mechanics, we shall thus further on describe it as a superclassical approach, in order to avoid confusions when using "classical" explanations.

As mentioned, our superclassical approach is particularly suited to account for the emergent processes involved, i.e. for processes of a type that is well-known, for example, from the physics of Rayleigh-Bénard cells. The latter appear in a fluid subjected to a temperature gradient (like its container being heated from below) producing convection rolls, with the emergent particle trajectories strongly depending on the boundary conditions: The form of the convection rolls can be radically changed by changing the boundaries of the container. Generally, emergence of this type is characterized by *relational causality*, i.e. the co-evolution of bottom-up and top-down processes (Fig. 2.2).

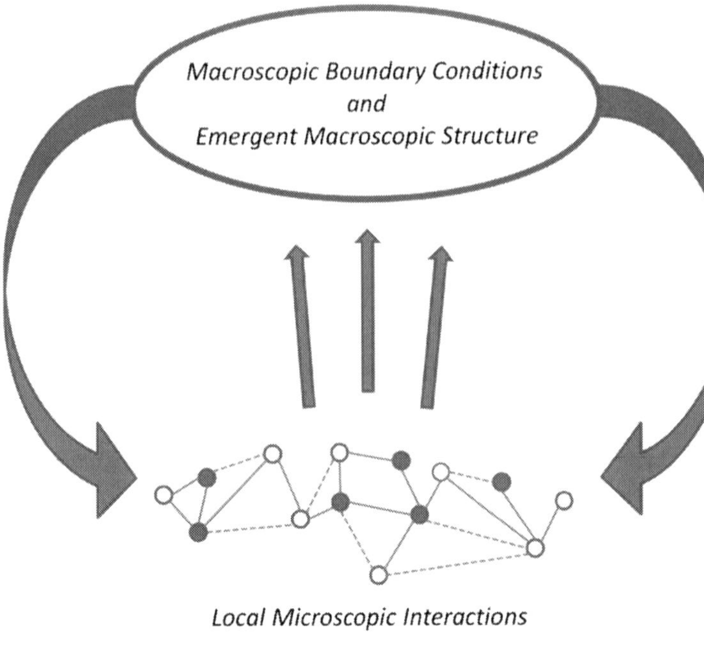

Figure 2.2: Scheme of *Relational Causality*: Co-evolution of bottom-up and top-down processes (after [ ])

In our emergent quantum mechanics approach, we have given a superclassical explanation of interference effects at a double slit [ ], thereby arriving at expressions for the probability density current and the corresponding velocity field completely equivalent to the *guidance equation*, which is the central postulate of the de Broglie-Bohm interpretation [,]. Our claims, to be substantiated in this paper, consist in the assertion that the guidance equation is of some "invisible hand" type, i.e. somehow mysteriously reaching out from configuration space in order to guide particles in real three-space. Instead, we shall argue, the guidance equation is completely understandable in real coordinate space, once the concept of emergence is taken seriously and introduced within a superclassical approach.

Note that a basic characteristic of emergent systems can be described via the above-mentioned relational causality (for a similar view, see [ ]). Consider for example the following computer simulations (Figs. 2.3a and 2.3b). Similar to the experiments with droplets by Couder and

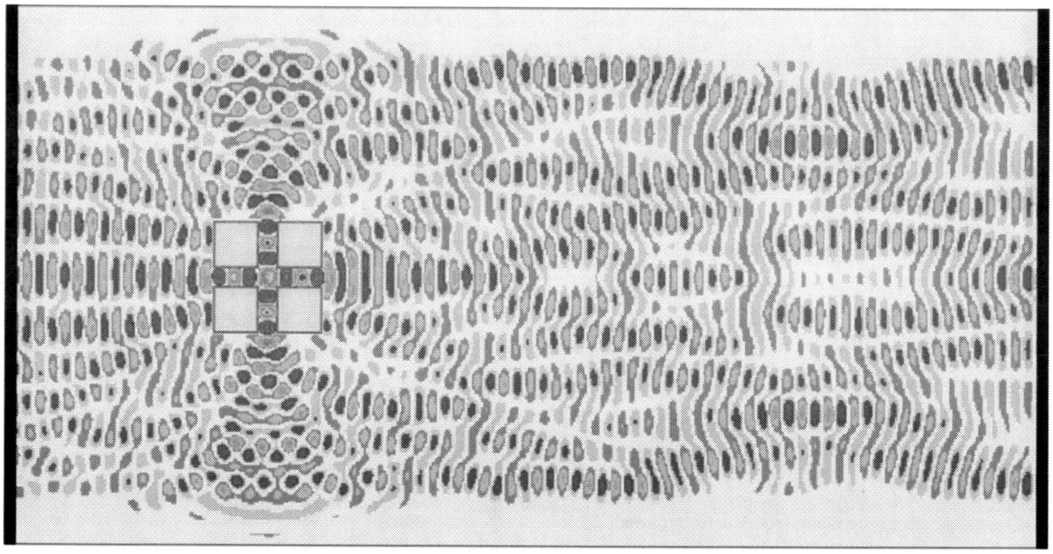

(a) Illustration of *relational causality* under varying boundary conditions as described in the text, with standing waves between plane mirrors (black).

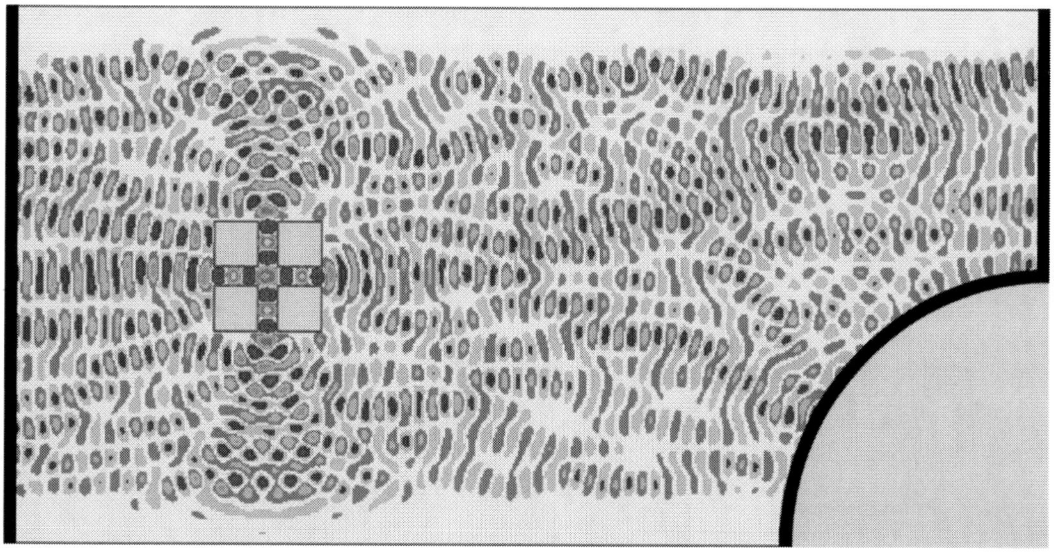

(b) Same as (a), with modified standing waves due to modified geometry of mirrors. Note the changes in the wave patterns around the squares despite the fact that the local physics there is kept unchanged. These changes are solely due to the altered boundary conditions.

Figure 2.3

Fort's group, we insert onto a fluid surface a droplet in the center of the square constituted by four smaller squares of solid material, such that due to microscopic fluctuations the developing bouncer/walker will propagate along one of the four narrow paths until it finally escapes from the region of the square through one of the four slits into the open area. With the bouncer creating waves that will be reflected from the walls (drawn in black), after some time the whole system will develop standing waves between said walls, which act as physically effective boundary conditions. In Fig. 2.3a the walls act as simple plane mirrors, whereas in Fig. 2.3b the reflecting part on the right contains a circular structure, with the effect that the pattern of standing waves is altered. The figures display intensity distributions of the wave field which, accordingly, coincide with probability density distributions of finding the droplet at a specific location. Note that although the local physics in the vicinity of the square would be identical for all four particle trajectories emerging from one of the four slits *if* one disregarded the context, the pattern of the probability density field is different in both cases. This is of course due to the fact that one cannot, in principle, neglect the context, i.e. the boundary conditions create different standing wave patterns which in turn interfere with the otherwise identical local processes. *Relational causality* means that one must consider the *co-evolution of processes stemming from the local slits and those from the global environment*, the latter including the macroscopic boundary conditions. One can thus also speak of a "confluence" of different currents, i.e. those pertaining to the four slits and those coming from the larger environment. Therefore, if one records the bouncers' trajectories and collects them in a synoptic manner, the whole velocity field will be a decisively *emergent* one. It is the totality of the whole experimental arrangement, and not just the classically local influences, that results in the behavior of the particle trajectories.

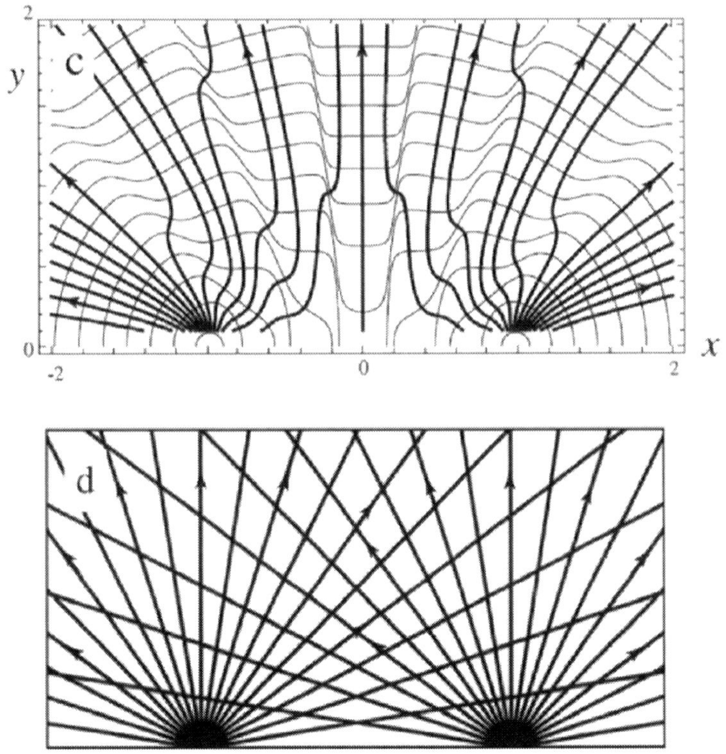

Figure 2.4: Quantum case streamlines (top) vs. geometrical rays (bottom). From [ ]. Here it is demonstrated that by bringing together two sources, the temporal development of the velocity field (and the corresponding "trajectories") in the quantum case can be understood only if considered as the result of a complex interplay of influences, updated at each instant of the time evolution, i.e. as the result of emergent processes.

From this perspective, it is not very surprising that any relic of context-free modeling should run into difficulties. Ironically, it was an orthodox position that has argued against the consequences of the more holistic approach acknowledging the context, i.e. when papers on apparently "surreal trajectories" were published against the assertion of the de Broglie-Bohm model that particles would not just follow trajectories as expected from context-free propagation obeying simple classically-local momentum conservation (see, e.g., the assertions of Scully [ ]). The latter would apply only to geometrical rays, whereas the quantum case streamlines are more complicated, a situation well-known from optics (see Fig. 2.4 for illustration), and recently also confirmed via weak measurements in a double-slit experiment involving photons [ ].

Having thus introduced the notions of superclassical physics and relational causality, we are now ready to reconsider interference at the double slit. Recall that to arrive at the *classical* double slit formula for the intensities on the screen one had to just consider the mean velocity $\mathbf{v}$, i.e. there was just one channel (of the velocity, or the probability density current, respectively) per slit. Because the currents represent wave propagations, this has led to the total intensity of (1.3), $P = P(1) + P(2) = R_1^2 + R_2^2 + 2R_1R_2\cos\varphi$. Now, however, we are going to look for the *superclassical* interference formula. As mentioned, we now have to deal with three channels per slit, due to the additional sub-quantum diffusion velocities $\mathbf{u}_L$ and $\mathbf{u}_R$, next to $\mathbf{v}$, and *all three co-evolving*. Surprisingly, this is all that is needed to arrive at the well-known quantum mechanical results which usually are obtained only via complex-valued probability amplitudes.

3. A superclassical derivation of the guidance equation

Considering particles as oscillators ("bouncers") coupling to regular oscillations of the vacuum's zero-point field, which they also generate, we have shown how a quantum can be understood as an emergent system. In particular, with the dynamics between the oscillator on the one hand, and the "bath" of its thermal environment as constrained by the experiment's boundary conditions on the other hand, one can explain not only Gaussian diffraction at a single slit [5], but also the well-known interference effects at a double slit [12, 27]. As already mentioned, we have also shown that the spreading of a wave packet can be exactly described by combining the convective with the orthogonal diffusive velocity fields. The latter fulfill the condition of being unbiased w.r.t. the convective velocities, i.e. the orthogonality relation for the *averaged* velocities derived in [15] is $\overline{\mathbf{v}\mathbf{u}} = 0$, since any fluctuations $\mathbf{u} = \delta(\nabla S/m)$ are shifts along the surfaces of action $S = \text{const}$.

To account for the different velocity channels $i = 1, \ldots, 3n$, n being the number of slits, we now introduce for general cases the generalized velocity vectors $\mathbf{w}_i$, which in the case of $n = 2$ are

$$\mathbf{w}_1 := \mathbf{v}_1, \quad \mathbf{w}_2 := \mathbf{u}_{1\mathrm{R}}, \quad \mathbf{w}_3 := \mathbf{u}_{1\mathrm{L}} \tag{3.1}$$

for the first channel, and

$$\mathbf{w}_4 := \mathbf{v}_2, \quad \mathbf{w}_5 := \mathbf{u}_{2\mathrm{R}}, \quad \mathbf{w}_6 := \mathbf{u}_{2\mathrm{L}} \tag{3.2}$$

for the second channel. The associated amplitudes $R(\mathbf{w}_i)$ for each channel are taken to be the same, i.e. $R(\mathbf{w}_1) = R(\mathbf{w}_2) = R(\mathbf{w}_3) = R_1$, and $R(\mathbf{w}_4) = R(\mathbf{w}_5) = R(\mathbf{w}_6) = R_2$.

Now, relational causality manifests itself in that the total wave intensity field consists of the sum of all local intensities in each channel, and the local intensity in each channel is the result of the interference with the total intensity field. Thus, any change in the local field affects the total field, and *vice versa*: Any change in the total field affects the local one. In order to completely accommodate the totality of the system, we therefore need to define a "wholeness"-related local wave intensity $P(\mathbf{w}_i)$ in one channel (i.e. $\mathbf{w}_i$) upon the condition that the totality of the superposing waves is given by the "rest" of the $3n - 1$ channels. Concretely, we account for the phase-dependent amplitude contributions of the total system's wave field projected on one channel's amplitude $R(\mathbf{w}_i)$ at the point $(\mathbf{x}, t)$ with a *conditional probability density* $P(\mathbf{w}_i)$. The expression for $P(\mathbf{w}_i)$ thus constitutes the representation of relational causality within our ansatz. Moreover, as usual one can define a local current $\mathbf{J}(\mathbf{w}_i)$ per channel as the corresponding "local" intensity-weighted velocity $\mathbf{w}_i$. Since the two-path set-up has $3n = 6$ velocity vectors at each point (cf. Eqs. (3.1) and (3.2)), we thus obtain for the partial intensities and currents, respectively, i.e. for each channel component i,

$$P(\mathbf{w}_i) = R(\mathbf{w}_i)\hat{\mathbf{w}}_i \cdot \sum_{j=1}^{6} \hat{\mathbf{w}}_j R(\mathbf{w}_j) \tag{3.3}$$

$$\mathbf{J}(\mathbf{w}_i) = \mathbf{w}_i P(\mathbf{w}_i), \quad i = 1, \ldots, 6, \tag{3.4}$$

with

$$\cos\varphi_{i,j} := \hat{\mathbf{w}}_i \cdot \hat{\mathbf{w}}_j. \tag{3.5}$$

Consequently, the total intensity and current of our field read as

$$P_{\text{tot}} = \sum_{i=1}^{6} P(\mathbf{w}_i) = \left(\sum_{i=1}^{6} \hat{\mathbf{w}}_i R(\mathbf{w}_i) \right)^2 \tag{3.6}$$

$$\mathbf{J}_{\text{tot}} = \sum_{i=1}^{6} \mathbf{J}(\mathbf{w}_i) = \sum_{i=1}^{6} \mathbf{w}_i P(\mathbf{w}_i), \tag{3.7}$$

leading to the *emergent total velocity*

$$\mathbf{v}_{\text{tot}} = \frac{\mathbf{J}_{\text{tot}}}{P_{\text{tot}}} = \frac{\sum_{i=1}^{6} \mathbf{w}_i P(\mathbf{w}_i)}{\sum_{i=1}^{6} P(\mathbf{w}_i)}. \tag{3.8}$$

Returning now to our previous notation for the six velocity components $\mathbf{v}_i$, $\mathbf{u}_{i\text{R}}$, $\mathbf{u}_{i\text{L}}$, $i = 1, 2$, the partial current associated with $\mathbf{v}_1$ originates from building the scalar product of $\hat{\mathbf{v}}_1$ with all other unit vector components and reads as

$$\mathbf{J}(\mathbf{v}_1) = \mathbf{v}_1 P(\mathbf{v}_1) = \mathbf{v}_1 R_1 \hat{\mathbf{v}}_1 \cdot (\hat{\mathbf{v}}_1 R_1 + \hat{\mathbf{u}}_{1\text{R}} R_1 + \hat{\mathbf{u}}_{1\text{L}} R_1 + \hat{\mathbf{v}}_2 R_2 + \hat{\mathbf{u}}_{2\text{R}} R_2 + \hat{\mathbf{u}}_{2\text{L}} R_2). \tag{3.9}$$

Since trivially

$$\hat{\mathbf{u}}_{i\text{R}} R_i + \hat{\mathbf{u}}_{i\text{L}} R_i = 0, \quad i = 1, 2, \tag{3.10}$$

Eq. (3.9) leads to

$$\mathbf{J}(\mathbf{v}_1) = \mathbf{v}_1 \left(R_1^2 + R_1 R_2 \cos\varphi \right), \tag{3.11}$$

which results from the representation of the emerging velocity fields, since we get the cosine of the phase difference φ as a natural result of the scalar product of the velocity vectors $\mathbf{v}_i$. The non-zero residua of the other vector fields yield

$$\mathbf{J}(\mathbf{u}_{1\text{R}}) = u_{1\text{R}} P(\mathbf{u}_{1\text{R}}) = \mathbf{u}_{1\text{R}} \left(R_1 \hat{\mathbf{u}}_{1\text{R}} \cdot \hat{\mathbf{v}}_2 R_2 \right) = \mathbf{u}_{1\text{R}} R_1 R_2 \cos\left(\frac{\pi}{2} - \varphi \right) = \mathbf{u}_{1\text{R}} R_1 R_2 \sin\varphi \tag{3.12}$$

and

$$\mathbf{J}(\mathbf{u}_{1\text{L}}) = \mathbf{u}_{1\text{L}} P(\mathbf{u}_{1\text{L}}) = \mathbf{u}_{1\text{L}} \left(R_1 \hat{\mathbf{u}}_{1\text{L}} \cdot \hat{\mathbf{v}}_2 R_2 \right) = \mathbf{u}_{1\text{L}} R_1 R_2 \cos\left(\frac{\pi}{2} + \varphi \right) = -\mathbf{u}_{1\text{L}} R_1 R_2 \sin\varphi. \tag{3.13}$$

Analogously, we obtain for the convective velocity vector field of the second channel

$$\mathbf{J}(\mathbf{v}_2) = \mathbf{v}_2 P(\mathbf{v}_2) = \mathbf{v}_2 \left(R_2^2 + R_1 R_2 \cos\varphi \right). \tag{3.14}$$

The corresponding diffusive velocity vector fields read as

$$\mathbf{J}(\mathbf{u}_{2\mathrm{R}}) = \mathbf{u}_{2R} P(\mathbf{u}_{2\mathrm{R}}) = \mathbf{u}_{2R} \left(R_2 \hat{\mathbf{u}}_{2\mathrm{R}} \cdot \hat{\mathbf{v}}_1 R_1 \right) = \mathbf{u}_{2R} R_1 R_2 \cos\left(\frac{\pi}{2} + \varphi \right) = -\mathbf{u}_{2R} R_1 R_2 \sin\varphi, \tag{3.15}$$

$$\mathbf{J}(\mathbf{u}_{2\mathrm{L}}) = \mathbf{u}_{2\mathrm{L}} P(\mathbf{u}_{2\mathrm{L}}) = \mathbf{u}_{2\mathrm{L}} \left(R_2 \hat{\mathbf{u}}_{2\mathrm{L}} \cdot \hat{\mathbf{v}}_1 R_1 \right) = \mathbf{u}_{2\mathrm{L}} R_1 R_2 \cos\left(\frac{\pi}{2} - \varphi \right) = \mathbf{u}_{2\mathrm{L}} R_1 R_2 \sin\varphi. \tag{3.16}$$

Note that the nontrivial sine contributions to the total current stem from the projections between the diffusive velocities $\mathbf{u}_{1\mathrm{R(L)}}$ of the first channel on the unit vector $\hat{\mathbf{v}}_2$ of the convective velocity of the second channel, and *vice versa*. Combining all terms, we obtain with Eq. (3.7) the result for the total current

$$\begin{aligned} \mathbf{J}_{\mathrm{tot}} &= \mathbf{v}_1 P(\mathbf{v}_1) + \mathbf{u}_{1R} P(\mathbf{u}_{1\mathrm{R}}) + \mathbf{u}_{1\mathrm{L}} P(\mathbf{u}_{1\mathrm{L}}) + \mathbf{v}_2 P(\mathbf{v}_2) + \mathbf{u}_{2R} P(\mathbf{u}_{2\mathrm{R}}) + \mathbf{u}_{2\mathrm{L}} P(\mathbf{u}_{2\mathrm{L}}) \\ &= R_1^2 \mathbf{v}_1 + R_2^2 \mathbf{v}_2 + R_1 R_2 (\mathbf{v}_1 + \mathbf{v}_2) \cos\varphi + R_1 R_2 \left([\mathbf{u}_{1\mathrm{R}} - \mathbf{u}_{1\mathrm{L}}] - [\mathbf{u}_{2\mathrm{R}} - \mathbf{u}_{2\mathrm{L}}] \right) \sin\varphi. \end{aligned} \tag{3.17}$$

The resulting diffusive velocities $\mathbf{u}_{i\mathrm{R}} - \mathbf{u}_{i\mathrm{L}}$ are identified with the effective diffusive velocities $\mathbf{u}_i$ for each channel. Note that *one* of those velocities, $\mathbf{u}_{i\mathrm{R}}$ or $\mathbf{u}_{i\mathrm{L}}$, respectively, is always zero, so that the product of said difference with $\sin\varphi$ guarantees the correct sign of the last term in Eq. (3.17). Thus we obtain the final expression for the total density current built from the remaining $2n = 4$ velocity components

$$\mathbf{J}_{\mathrm{tot}} = R_1^2 \mathbf{v}_1 + R_2^2 \mathbf{v}_2 + R_1 R_2 (\mathbf{v}_1 + \mathbf{v}_2) \cos\varphi + R_1 R_2 (\mathbf{u}_1 - \mathbf{u}_2) \sin\varphi. \tag{3.18}$$

The obtained total density current field $\mathbf{J}_{\mathrm{tot}}(\mathbf{x}, t)$ spanned by the various velocity components $\mathbf{v}_i(\mathbf{x}, t)$ and $\mathbf{u}_{i\mathrm{R(L)}}(\mathbf{x}, t)$ we have denoted as the "path excitation field" [12]. It is built by the sum of all its partial currents, which themselves are built by an amplitude weighted projection of the total current.

Summing up the probabilities associated with each of the partial currents we obtain according to the ansatz (3.3) and the relations (3.6) and (3.10)

$$\begin{aligned} P_{\mathrm{tot}} &= \left(R_1 \hat{\mathbf{v}}_1 + R_1 \hat{\mathbf{u}}_{1\mathrm{R}} + R_1 \hat{\mathbf{u}}_{1\mathrm{L}} + R_2 \hat{\mathbf{v}}_2 + R_2 \hat{\mathbf{u}}_{2\mathrm{R}} + R_2 \hat{\mathbf{u}}_{2\mathrm{L}} \right)^2 \\ &= \left(R_1 \hat{\mathbf{v}}_1 + R_2 \hat{\mathbf{v}}_2 \right)^2 = R_1^2 + R_2^2 + 2 R_1 R_2 \cos\varphi = P(\mathbf{v}_1) + P(\mathbf{v}_2). \end{aligned} \tag{3.19}$$

The total velocity $\mathbf{v}_{\mathrm{tot}}$ according to Eq. (3.8) now reads as

$$\mathbf{v}_{\mathrm{tot}} = \frac{R_1^2 \mathbf{v}_1 + R_2^2 \mathbf{v}_2 + R_1 R_2 (\mathbf{v}_1 + \mathbf{v}_2) \cos\varphi + R_1 R_2 (\mathbf{u}_1 - \mathbf{u}_2) \sin\varphi}{R_1^2 + R_2^2 + 2 R_1 R_2 \cos\varphi}. \tag{3.20}$$

The trajectories or streamlines, respectively, are obtained according to $\dot{\mathbf{x}} = \mathbf{v}_{\mathrm{tot}}$ in the usual way by integration. As first shown in [12], by re-inserting the expressions for convective and diffusive velocities, respectively, i.e. $\mathbf{v}_{i,\mathrm{conv}} = \frac{\nabla S_i}{m}$, $\mathbf{u}_i = -\frac{\hbar}{m} \frac{\nabla R_i}{R_i}$, one immediately identifies Eq. (3.20) with the Bohmian guidance equation and Eq. (3.18) with the quantum mechanical pendant for the probability density current [28]. The latter can be seen as follows. Upon employment of the Madelung transformation for each path j ($j = 1$ or 2),

$$\psi_j = R_j \mathrm{e}^{\mathrm{i} S_j / \hbar}, \tag{3.21}$$

and thus $P_j = R_j^2 = |\psi_j|^2 = \psi_j^* \psi_j$, with $\varphi = (S_1 - S_2)/\hbar$, and recalling the usual trigonometric identities such as $\cos\varphi = \frac{1}{2}\left(\mathrm{e}^{\mathrm{i}\varphi} + \mathrm{e}^{-\mathrm{i}\varphi} \right)$, one can rewrite the total average current (3.18) immediately as

$$\begin{aligned}\mathbf{J}_{\text{tot}} &= P_{\text{tot}}\mathbf{v}_{\text{tot}} \\ &= (\psi_1+\psi_2)^*(\psi_1+\psi_2)\frac{1}{2}\left[\frac{1}{m}\left(-\mathrm{i}\hbar\frac{\nabla(\psi_1+\psi_2)}{(\psi_1+\psi_2)}\right)+\frac{1}{m}\left(\mathrm{i}\hbar\frac{\nabla(\psi_1+\psi_2)^*}{(\psi_1+\psi_2)^*}\right)\right] \\ &= -\frac{\mathrm{i}\hbar}{2m}\left[\Psi^*\nabla\Psi-\Psi\nabla\Psi^*\right]=\frac{1}{m}\mathrm{Re}\left\{\Psi^*(-\mathrm{i}\hbar\nabla)\Psi\right\},\end{aligned} \tag{3.22}$$

where $P_{\text{tot}} = |\psi_1+\psi_2|^2 =: |\Psi|^2$. The last two expressions of (3.22) are the exact formulations of the quantum mechanical probability current, here obtained just by a re-formulation of (3.18). In fact, it is a simple exercise to insert the wave functions (3.21) into (3.22) to re-obtain (3.18).

Note that it is straightforward to extend this derivation to the many-particle case. As the individual terms in the expressions for the total current and total probability density, respectively, are purely additive also for N particles, a fact that is well-known also from Bohmian theory, the above-mentioned "translation" into orthodox quantum language is straightforward, with the effect that the currents' nabla operators just have to be applied at all of the locations $\mathbf{x}$ of the respective N particles, thus providing the quantum mechanical formula $\mathbf{J}_{\text{tot}}(N) = \frac{1}{m}\mathrm{Re}\left\{\Psi^*(-\mathrm{i}\hbar\nabla_N)\Psi\right\}$, where Ψ now is the total N-particle wave function.

Again we emphasize that our result was obtained solely out of kinematic relations by applying the superclassical rules introduced above on the basis of a relational causality, i.e. without invoking complex ψ functions or the like. Moreover, as opposed to the Bohmian theory, we obtained our results not in configuration space but in ordinary coordinate space. *What looks like the necessity to superpose wave functions in configuration space, which then are imagined to guide the particles by some invisible hand, can equally be obtained by superpositions of all relational amplitude configurations of waves in real space, i.e. by understanding the resulting system's evolutions as processes of emergence.*

Thus, with $\mathbf{w}_i = \frac{\mathbf{J}(\mathbf{w}_i)}{P(\mathbf{w}_i)}$ and the classical composition principles of Eqs. (3.6) and (3.7) we have shown that the total velocity field is given in the simple form of a *(super)classical average velocity field*:

$$\mathbf{v}_{\text{tot}} = \frac{\mathbf{J}_{\text{tot}}}{P_{\text{tot}}} = \frac{\sum_i \mathbf{J}(\mathbf{w}_i)}{\sum_i P(\mathbf{w}_i)} = \frac{\sum_i \mathbf{w}_i P(\mathbf{w}_i)}{\sum_i P(\mathbf{w}_i)}\,. \tag{3.23}$$

In other words, the guidance equation postulated by de Broglie-Bohm is here derived and explained via relational causality, with $\mathbf{v}_{\text{tot}}$ being an *emergent velocity field.*

4. Three-slit interference, Born's rule, and Sorkin's sum rules

The extension to three slits, beams, or probability current channels, respectively, is straightforward. We just introduce a third emergent propagation velocity $\mathbf{v}_3$ and its corresponding diffusive velocities $\mathbf{u}_{3\mathrm{L(R)}}$. The phase shift of the third beam is denoted as χ and represents the angle between the second and the third beam in our geometric representation of the path excitation field. According to Born's rule the probability of even a single particle passing any of the three slits splits into a sum of probabilities passing the slits pairwise, i.e. going along both A and B, B and C, or A and C, but never passing A, B and C simultaneously.

Interference phenomena have recently been analyzed thoroughly for the cases of only one open slit up to n open slits by Sorkin [ ]. For a double slit setup the interference term is non-zero, i.e. $I_{AB} := P_{AB} - P_A - P_B \neq 0$, with $P_{A(B)}$ being the detection probability with only one slit/path A or B, respectively, of a total of n slits/paths open, and P_{AB} for both slits A and B open. This "first order sum rule" is to be contrasted with Sorkin's results for the following, so-called "second order sum rule" [29]:

$$
\begin{aligned}
I_{ABC} := & P_{ABC} - P_{AB} - P_{AC} - P_{BC} + P_A + P_B + P_C \\
= & P_{ABC} - (P_A + P_B + P_C + I_{AB} + I_{AC} + I_{BC}) = 0.
\end{aligned} \tag{4.1}
$$

This result is remarkable insofar as it can be inferred that interference terms theoretically always originate from pairings of paths, but never from triples etc. Any violation of this second order sum rule, i.e. $I_{ABC} \neq 0$, and thus of Born's rule would have dramatic consequences for quantum theory like a modification of the Schrödinger equation, for example.

Returning to our model, the total probability density current for three paths is calculated according to the rules set up in Section 3. We adopt the notations of the two slit system also for three slits, i.e. now employing nine velocity contributions: $\mathbf{v}_i$, $\mathbf{u}_{i\mathrm{R(L)}}$, $i = 1, 2, 3$. Analogously, the three generally different amplitudes are denoted as $\mathrm{R}(\mathbf{v}_i) = \mathrm{R}(\mathbf{u}_{i\mathrm{R}}) = \mathrm{R}(\mathbf{u}_{i\mathrm{L}}) = R_i$, $i = 1, 2, 3$. We keep the definition of φ as $\varphi := \arccos(\hat{\mathbf{v}}_1 \cdot \hat{\mathbf{v}}_2)$, and we define the second angle as $\chi := \arccos(\hat{\mathbf{v}}_2 \cdot \hat{\mathbf{v}}_3)$. Similarly to Eq. (3.10), the diffusive velocities $\mathbf{u}_{i\mathrm{R}} - \mathbf{u}_{i\mathrm{L}}$ combine to $\mathbf{u}_i$, $i = 1, 2, 3$, thus ending up with $2n = 6$ effective velocities. Therefore we obtain, analogously to the calculation in the previous section,

$$
\begin{aligned}
\mathbf{J}_{\mathrm{tot}} = & R_1^2 \mathbf{v}_1 + R_2^2 \mathbf{v}_2 + R_3^2 \mathbf{v}_3 + R_1 R_2 (\mathbf{v}_1 + \mathbf{v}_2) \cos\varphi + R_1 R_2 (\mathbf{u}_1 - \mathbf{u}_2) \sin\varphi \\
& + R_1 R_3 (\mathbf{v}_1 + \mathbf{v}_3) \cos(\varphi + \chi) + R_1 R_3 (\mathbf{u}_1 - \mathbf{u}_3) \sin(\varphi + \chi) \\
& + R_2 R_3 (\mathbf{v}_2 + \mathbf{v}_3) \cos\chi + R_2 R_3 (\mathbf{u}_2 - \mathbf{u}_3) \sin\chi
\end{aligned} \tag{4.2}
$$

and

$$
\begin{aligned}
P_{\mathrm{tot}} & = R_1^2 + R_2^2 + R_3^2 + 2R_1 R_2 \cos\varphi + 2R_1 R_3 \cos(\varphi + \chi) + 2R_2 R_3 \cos\chi \\
& = P(\mathbf{v}_1) + P(\mathbf{v}_2) + P(\mathbf{v}_3).
\end{aligned} \tag{4.3}
$$

In analogy to the double slit case (cf. Eq. (3.19)) we obtain a *classical* Kolmogorov sum rule for the probabilities on the one hand, but also the complete interference effects for the double, three- and, as we have shown in [27], for the n-slit cases, on the other. However, the particular probabilities $P(\mathbf{v}_i)$ in Eqs. (3.19) and (4.3), do not correspond to the probabilities of the assigned slits if solely opened, i.e. $P_{AB}(\mathbf{v}_1) = \left(R_1^2 + R_1 R_2 \cos\varphi\right) \neq P_A(\mathbf{v}_1) = R_1^2$. Consequently, each of the probability summands in said equations does *not* correspond to an independent probability of the respective slit if solely opened, a fact that was already clarified in our discussion of the issue of contexts in Section 1 .

Finally, we obtain for the cases of one (i.e. $n = A$), two and three open slits, respectively,

$$
I_A = P_A(\mathbf{v}_1) = R_1^2, \tag{4.4}
$$

$$
I_{AB} = P_{AB} - P_A(\mathbf{v}_1) - P_B(\mathbf{v}_2) = 2R_1 R_2 \cos\varphi, \tag{4.5}
$$

$$
I_{ABC} = P_{ABC} - P_{AB} - P_{AC} - P_{BC} + P_A(\mathbf{v}_1) + P_B(\mathbf{v}_2) + P_C(\mathbf{v}_3) = 0\,, \tag{4.6}
$$

where P_{AB} is assigned to P_{tot} of Eq. (3.19) and P_{ABC} to P_{tot} of Eq. (4.3). In the double slit case, e.g., with slits A and B open, we obtain the results of (3.19). If B were closed and C were open instead, we would get the analogous result, i.e. $\mathbf{v}_2$ and φ replaced by $\mathbf{v}_3$ and $\varphi_{1,3}$. If all three slits A, B, C are open, we can use the pairwise permutations of the double slit case, i.e. $A \wedge B$, $A \wedge C$, or $B \wedge C$, respectively, with $\varphi_{1,3}$ identified with $(\varphi + \chi)$, etc. Thus we conclude that in our model the addition of "sub-probabilities" indeed works and provides the correct results.

Summarizing, with our superclassical model emerging out of a sub-quantum scenario we arrive at the same results as standard quantum mechanics fulfilling Sorkin's sum rules [29]. However, whereas in standard quantum mechanics Born's rule originates from building the squared absolute values of additive ψ functions representing the probability amplitudes for different paths, in our case we obtain the pairing of paths as a natural consequence of the *pairwise* selection of unit vectors of all existing velocity components constituting the probability currents. Thus we obtain *all* possible pathways within an n-slit setup by our projection method. The sum rules, Eqs. (3.3) through (3.8), guarantee that each partial contribution, be it from the velocity contributions within a particular channel or from different channels, accounts for the final total current density for each point between source and detector. Since for only one slit open the projection rule (3.3) trivially leads to a linear relation between P and R^2, the asymmetry between the latter quantities, due to the nonlinear projection rule, becomes effective for $n \geq 2$ slits open. Consequently, the violation of the first order sum rule (4.5), i.e. $I_{AB} \neq 0$, represents a *natural* result of our principle of relational causality. Moreover, as we have argued above, the opening of an additional slit solely adds pairwise path combinations. As all higher interference terms have already incorporated said asymmetry, the result can finally be reduced to the double slit case, thus yielding a zero result as in Eq. (4.6) according to Sorkin's analysis.

This is a further hint that our model can reproduce all phenomena of standard quantum theory with the option of giving a deeper reasoning to principles like Born's rule or the hierarchical sum-rules, respectively.

5. Conclusions and outlook

We have previously shown in a series of papers [[illegible]–[illegible], [illegible], [illegible]] that phenomena of standard quantum mechanics like Gaussian dispersion of wave packets, superposition, double slit interference, Planck's energy relation, or the Schrödinger equation can be assessed as the emergent property of an underlying sub-structure of the vacuum combined with diffusion processes reflecting also the stochastic parts of the zero-point field, i.e. the zero point fluctuations. (For similar approaches see the works of Cetto and de la Peña [[illegible], and this volume], Nieuwenhuizen [this volume], or Khrennikov *et al.* [[illegible]].) Thus we obtain the quantum mechanical dynamics as an averaged behavior of sub-quantum processes. The inclusion of relativistic physics has not been considered yet, but should be possible in principle.

By introducing the concepts of superclassicality and relational causality, we have in this paper shown that quantum phenomenology can be meaningfully grounded in a superclassical approach relying solely on classical probability theory. Apart from an application for a deeper understanding of Born's rule, the central result of this work is a demonstration that the guidance equation can be derived and explained within ordinary coordinate space. We have proven the identity of our emergent velocity field $\mathbf{v}_{\text{tot}}$ with the corresponding Bohmian one, $\mathbf{v}_{\text{tot(Bohm)}}$, and the orthodox quantum mechanical one, $\mathbf{v}_{\text{tot(QM)}}$, respectively:

$$
\begin{aligned}
\mathbf{v}_{\text{tot(emergent)}} &= \frac{\sum\limits_i \mathbf{w}_i P(\mathbf{w}_i)}{\sum\limits_i P(\mathbf{w}_i)} \\
= \mathbf{v}_{\text{tot(Bohm)}} &= \frac{R_1^2 \mathbf{v}_1 + R_2^2 \mathbf{v}_2 + R_1 R_2 \left(\mathbf{v}_1 + \mathbf{v}_2\right) \cos\varphi + R_1 R_2 \left(\mathbf{u}_1 - \mathbf{u}_2\right) \sin\varphi}{R_1^2 + R_2^2 + 2 R_1 R_2 \cos\varphi} \qquad (5.1) \\
= \mathbf{v}_{\text{tot(QM)}} &= \frac{1}{\mathrm{m}|\Psi|^2} \mathrm{Re}\left\{\Psi^* (-i\hbar\nabla)\Psi\right\}, \quad \text{with } \Psi = \textstyle\sum_j \psi_j .
\end{aligned}
$$

Finally, with our superclassical theory we can also enquire into the possible limits of present-day quantum theory. For example, the latter is expected to break down at the time scales of our bouncer's oscillation frequency, e.g., for the electron $\omega \approx \mathcal{O}\left(10^{21}\,\mathrm{Hz}\right)$. As we have seen, at the emerging quantum level, i.e. at times $t \ggg 1/\omega$, we obtain exact results strongly suggesting the validity of Born's rule, for example. However, approaching said sub-quantum regions by increasing the time resolution to the order of $t \approx 1/\omega$ suggests a possibly gradual breakdown of said rule, since the averaging of the diffusive and convective velocities and their mutual orthogonality of the averaged velocities is not reliable any more. In principle, this should eventually be testable in experiment. Moreover, upon introduction of a new bias with respect to the velocities $\mathbf{v}$ and $\mathbf{u}_{L(R)}$, either in the average orthogonality condition, or between the different velocity channels, the question may be of relevance whether these would lead to the collapse of the superposition principle, as the assumed sub-quantum nonlinearities would then become manifest. We have not discussed the important issue of nonlocality and possible consequences with regard to the non-signaling principle here, and refer the reader to the paper by Jan Walleczek and Gerhard Grössing (this volume) for consideration of some of the topics in question.

Acknowledgments

We thank Hans De Raedt for pointing out ref. [ ] to us, Thomas Elze for the critical reading of an earlier version of this paper, Jan Walleczek for many enlightening discussions, and the Fetzer Franklin Fund of the John E. Fetzer Memorial Trust for partial support of the current work.

References

[1] R. P. Feynman, "The concept of probability in quantum mechanics," in *Proceedings of the Second Berkeley Symposium on Mathematical Statistics and Probability*, pp. 533–541. University of California, Berkeley, CA, 1951.

[2] B. O. Koopman, "Quantum theory and the foundations of probability," in *Applied Probability*, L. A. MacColl, ed., pp. 97–102. McGraw-Hill, New York, 1955.

[3] L. E. Ballentine, "Probability theory in quantum mechanics," *Am. J. Phys.* **54** (1986) 883.

[4] J. S. Bell, *Speakable and Unspeakable in Quantum Mechanics*. Cambridge University Press, Cambridge, 2 ed., 2004.

[5] Y. Couder and E. Fort, "Single-particle diffraction and interference at a macroscopic scale," *Phys. Rev. Lett.* **97** (2006) 154101.

[6] Y. Couder and E. Fort, "Probabilities and trajectories in a classical wave-particle duality," *J. Phys.: Conf. Ser.* **361** (2012) 012001.

[7] E. Fort, A. Eddi, A. Boudaoud, J. Moukhtar and Y. Couder, "Path-memory induced quantization of classical orbits," *PNAS* **107** (2010) 17515–17520.

[8] G. Grössing, ed., *Emergent Quantum Mechanics 2011*. No. 361/1. IOP Publishing, Bristol, 2012. http://iopscience.iop.org/1742-6596/361/1.

[9] J. Walleczek, "Mission statement." Http://www.fetzer-franklin-fund.org/mission/, 2012.

[10] G. Grössing, "The vacuum fluctuation theorem: Exact Schrödinger equation via nonequilibrium thermodynamics," *Phys. Lett. A* **372** (2008) 4556–4563, quant-ph/0711.4954.

[11] G. Grössing, "On the thermodynamic origin of the quantum potential," *Physica A* **388** (2009) 811–823, quant-ph/0808.3539.

[12] G. Grössing, S. Fussy, J. Mesa Pascasio and H. Schwabl, "An explanation of interference effects in the double slit experiment: Classical trajectories plus ballistic diffusion caused by zero-point fluctuations," *Ann. Phys.* **327** (2012) 421–437, quant-ph/1106.5994.

[13] G. Grössing, S. Fussy, J. Mesa Pascasio and H. Schwabl, "'Systemic nonlocality' from changing constraints on sub-quantum kinematics," *J. Phys.: Conf. Ser.* **442** (2013) 012012, quant-ph/1303.2867.

[14] G. Grössing, S. Fussy, J. Mesa Pascasio and H. Schwabl, "Elements of sub-quantum thermodynamics: Quantum motion as ballistic diffusion," *J. Phys.: Conf. Ser.* **306** (2011) 012046, physics.gen-ph/1005.1058.

[15] G. Grössing, S. Fussy, J. Mesa Pascasio and H. Schwabl, "Emergence and collapse of quantum mechanical superposition: Orthogonality of reversible dynamics and irreversible diffusion," *Physica A* **389** (2010) 4473–4484, quant-ph/1004.4596.

[16] C. Beck, "Superstatistics: Theoretical concepts and physical applications," in *Anomalous Transport: Foundations and Applications*, R. Klages, G. Radons and I. M. Sokolov, eds., pp. 433–457. Wiley-VCH, 2008.

Journal of Physics: Conference Series **504** (2014) 012006 doi:10.1088/1742-6596/504/1/012006

[17] R. S. Mulliken, "The band spectrum of boron monoxide," *Nature* **114** (1924) 349–350.
[18] P. Jizba and F. Scardigli, "Quantum mechanics and local Lorentz symmetry violation," *J. Phys.: Conf. Ser.* **361** (2012) 012026.
[19] P. Jizba and F. Scardigli, "Emergence of special and doubly special relativity," *Phys. Rev. D* **86** (2012) 025029, hep-th/1105.3930.
[20] J. Walleczek, ed., *Self-Organized Biological Dynamics and Nonlinear Control*. Cambridge University Press, New York, 2000.
[21] D. Bohm and B. J. Hiley, *The Undivided Universe: An Ontological Interpretation of Quantum Theory*. Routledge, London, 1993.
[22] P. R. Holland, *The Quantum Theory of Motion*. Cambridge University Press, Cambridge, 1993.
[23] G. F. Ellis, "On the limits of quantum theory: Contextuality and the quantum-classical cut," *Ann. Phys.* **327** (2012) 1890–1932, quant-ph/1108.5261.
[24] M. V. Berry, "Optical currents," *J. Opt. A: Pure Appl. Opt.* **11** (2009) 094001.
[25] M. O. Scully, "Do Bohm trajectories always provide a trustworthy physical picture of particle motion?," *Phys. Scr.* **T76** (1998) 41–46.
[26] S. Kocsis, B. Braverman, S. Ravets, M. J. Stevens, R. P. Mirin, L. K. Shalm, and A. M. Steinberg, "Observing the average trajectories of single photons in a two-slit interferometer," *Science* **332** (2011) 1170–1173.
[27] S. Fussy, J. Mesa Pascasio, H. Schwabl and G. Grössing, "Born's rule as signature of a superclassical current algebra," *Ann. Phys.* (2014) , quant-ph/1308.5924. In press.
[28] A. S. Sanz and S. Miret-Artés, "A trajectory-based understanding of quantum interference," *J. Phys. A: Math. Gen.* **41** (2008) 435303, quant-ph/0806.2105.
[29] R. D. Sorkin, "Quantum mechanics as quantum measure theory," *Mod. Phys. Lett. A* **09** (1994) 3119–3127, gr-qc/9401003.
[30] G. Grössing, J. Mesa Pascasio and H. Schwabl, "A classical explanation of quantization," *Found. Phys.* **41** (2011) 1437–1453, quant-ph/0812.3561.
[31] A. M. Cetto and L. de la Peña, "Quantization as an emergent phenomenon due to matter-zeropoint field interaction," *J. Phys.: Conf. Ser.* **361** (2012) 012013
[32] A. Khrennikov, B. Nilsson and S. Nordebo, "Classical signal model reproducing quantum probabilities for single and coincidence detections" *J. Phys.: Conf. Ser.* **361** (2012) 012030

Journal of Physics: Conference Series **504** (2014) 012007 doi:10.1088/1742-6596/504/1/012007

Emergence of quantization: the spin of the electron

A M Cetto, L de la Peña and A Valdés-Hernández

Instituto de Física, Universidad Nacional Autónoma de México,
Ap. postal 20-364, 01000 México DF, Mexico

E-mail: ana, luis, andreavh@fisica.unam.mx

Abstract. In previous papers, the quantum behavior of matter has been shown to emerge as a result of its permanent interaction with the random zero-point radiation field. Fundamental results, such as the Schrödinger and the Heisenberg formalism, have been derived within this framework. Further, the theory has been shown to provide the basic QED formulas for the radiative corrections, as well as an explanation for entanglement in bipartite systems.

This paper addresses the problem of spin from the same perspective. The zero-point field is shown to produce a helicoidal motion of the electron, through the torque exerted by the electric field modes of a given circular polarization, which results in an intrinsic angular momentum, of value $\hbar/2$. Associated with it, a magnetic moment with a g-factor of 2 is obtained. This allows us to identify the spin of the electron as a further emergent property, generated by the action of the random zero-point field.

1. Introduction

In previous work, the quantum behavior of matter has been shown to emerge as a result of its continuous interaction with the random zero-point radiation field (ZPF). This has been, in essence, the program of stochastic electrodynamics (SED). Fundamental quantum results, such as the Schrödinger and the Heisenberg formalism, have been recently obtained within this framework [1]. Contact has been made with (nonrelativistic) quantum electrodynamics, by deriving the main formulas for the radiative corrections, notably the Lamb shift and the atomic lifetimes [1], [2]. Further, the study of bipartite systems has shown the ZPF to lie at the origin of entanglement and the symmetrization postulate [3].

Another most fundamental problem in quantum theory relates to the nature of the spin of the electron: is it something inherent, as is usually considered, or is it a result of the dynamics, as has been postulated time and again? A theory intended to explain the genesis of quantization should be expected to provide an answer to this query, instead of *a priori* taking the spin as one more innate property of the particle, like the mass or the electric charge.

During the initial period of SED [4] the electron was in fact taken as a spinless particle, the only exception being the analysis by Braffort and Taroni [5] of some effects due to spin. As of 1981, a number of models for spin as an acquired property have been proposed [6]-[12] (there were of course earlier models, such as that of Huang [13]). Though based on classical models, the various SED calculations exhibit the ZPF as the source of a kind of (nonrelativistic) zitterbewegung that gives rise to an intrinsic angular momentum of the electron, with a mean square value of order $\hbar^2$ and projections of order $\hbar$.

This paper re-addresses the problem of spin from the perspective offered by present SED. Before entering into the subject matter, a brief introduction to recent work on SED is provided, focusing on those results that will be used for the present derivations. It then will be shown that in addition to giving rise to position, momentum and energy fluctuations, the ZPF induces a helicoidal motion, as a result of the torque exerted by the random Lorentz (electric) force on the particle. The field modes of a given circular polarization are shown to give rise to an (intrinsic) spin angular momentum of the electron, of value $\hbar/2$. Additionally, the corresponding magnetic moment with a g-factor of 2 is derived. The paper ends with some brief additional remarks on the physical meaning of the results obtained.

2. Emergence of quantum mechanics

2.1. *The original stochastic problem*

This section contains the basic steps that take us from the original dynamical equation for the particle embedded in the random ZPF, to the Schrödinger formulation of quantum mechanics [1]. The system under study is a particle of mass m and charge e —typically an electron—, subject to an external (binding) potential V in addition to the stationary ZPF. The motion of the particle is described by

$$m\ddot{\boldsymbol{x}} = \boldsymbol{f}(\boldsymbol{x}) + m\tau\dddot{\boldsymbol{x}} + e\boldsymbol{E}(t), \tag{1}$$

where $m\tau\dddot{\boldsymbol{x}}$, with $\tau = 2e^2/3mc^3$, represents the reaction force due to Larmor radiation, and $\boldsymbol{f}(\boldsymbol{x})$ is the force due to the potential V. A nonrelativistic description is assumed to be sufficient, so that the magnetic term of the Lorentz force is neglected. The random field $\boldsymbol{E}(t)$ is taken in the long-wavelength approximation, anticipating that those field modes that become eventually dominant have wavelengths much larger than the characteristic dimensions of the motion. The random field has zero mean value, i.e., $\overline{\boldsymbol{E}(t)}^{(i)} = 0$, where $\overline{(\cdot)}^{(i)}$ denotes the average over all realizations (i) of the field, and a spectral energy density corresponding to a mean energy $\hbar\omega/2$ per frequency mode, i. e.,

$$\rho_0(\omega) = \frac{\omega^2}{\pi^2 c^3}\frac{1}{2}\hbar\omega = \frac{\hbar\omega^3}{2\pi^2 c^3}. \tag{2}$$

This expression corresponds to the autocorrelation function

$$\overline{E_i(t)E_j(t')}^{(i)} = \delta_{ij}\varphi(t-t'), \tag{3}$$

where

$$\varphi(t-t') = \frac{4\pi}{3}\int_0^\infty \rho_0(\omega)\cos\omega(t-t')d\omega. \tag{4}$$

If there is an extra component due to an external field, as could be a thermal equilibrium radiation or any other excitation of the field modes, the corresponding contribution must be added to $\rho_0(\omega)$ in equation (2).

2.2. *Statistical description from phase space to configuration space*

The detailed motion of a *single* particle depends on the specific realization (i) of the field, which is unknown. Therefore a statistical description is made, by constructing the equation of evolution for the particle phase-space probability density $Q(\boldsymbol{x}, \boldsymbol{p}, t)$, taking the ZPF as given and starting from the dynamical equations obtained from (1),

$$m\dot{\boldsymbol{x}} = \boldsymbol{p}, \quad \dot{\boldsymbol{p}} = \boldsymbol{f}(\boldsymbol{x}) + m\tau\dddot{\boldsymbol{x}} + e\boldsymbol{E}(t). \tag{5}$$

Through a standard projection procedure [14], the generalized Fokker-Planck equation (GFPE)

$$\frac{\partial Q}{\partial t} + \frac{1}{m}\frac{\partial}{\partial x_i}p_i Q + \frac{\partial}{\partial p_i}f_i Q + m\tau\frac{\partial}{\partial p_i}\dddot{x}_i Q = e^2\frac{\partial}{\partial p_i}\hat{\mathcal{D}}_i Q \tag{6}$$

Journal of Physics: Conference Series **504** (2014) 012007 doi:10.1088/1742-6596/504/1/012007

is obtained, with the diffusion operator $\hat{\boldsymbol{\mathcal{D}}}(t)$ defined by means of the expression

$$\hat{\mathcal{D}}_i(t)Q = \hat{P}E_i\hat{G}\frac{\partial}{\partial p_j}E_j\sum_{k=0}^{\infty}\left[e\hat{G}\frac{\partial}{\partial p_l}(1-\hat{P})E_l\right]^{2k}Q, \tag{7}$$

and the projection operator $\hat{P}$ and the evolution operator $\hat{G}$ given by

$$\hat{P}A = \overline{A}^{(i)}, \quad \hat{G}A(\boldsymbol{x},\boldsymbol{p},t) = \int_{-\infty}^{t} e^{-\hat{L}(t-t')}A(\boldsymbol{x},\boldsymbol{p},t')dt', \tag{8}$$

where $\hat{L}$ is

$$\hat{L} = \frac{1}{m}\frac{\partial}{\partial x_i}p_i + \frac{\partial}{\partial p_i}\left(f_i + m\tau\dddot{x}_i\right). \tag{9}$$

The operator $e^{-\hat{L}(t-t')}$ in equation (8) makes $\boldsymbol{x}(t')$, $\boldsymbol{p}(t')$ ($t' < t$) evolve towards $\boldsymbol{x}(t)$, $\boldsymbol{p}(t)$ as final conditions, following a deterministic path.

Right after particle and field start to interact, the system is far from equilibrium. In this initial regime the main effect of the ZPF on the particle is due to the high-frequency modes, which produce violent accelerations and randomize the motion. Eventually, however (after a transient period which is estimated to be of the order of $\hbar/mc^2 \simeq 10^{-20}$ s for an electron), the interplay between the electric field force and radiation reaction is expected to drive the system close to equilibrium. In this new (time-reversible) regime the Markovian approximation applies, and the GFPE (6) reduces to a true Fokker-Planck equation,

$$\frac{\partial Q}{\partial t} + \frac{1}{m}\frac{\partial}{\partial x_i}p_iQ + \frac{\partial}{\partial p_i}\left(f_i + m\tau\dddot{x}_i\right)Q = \frac{\partial}{\partial p_i}D_{ij}^{pp}\frac{\partial Q}{\partial p_j} + \frac{\partial}{\partial p_i}D_{ij}^{px}\frac{\partial Q}{\partial x_j}, \tag{10}$$

with the diffusion coefficients given to lowest order in e^2 by

$$D_{ij}^{pp} = e^2\int_{-\infty}^{t}dt'\varphi(t-t')\frac{\partial p_j}{\partial p_i'}, \quad D_{ij}^{px} = e^2\int_{-\infty}^{t}dt'\varphi(t-t')\frac{\partial x_j}{\partial p_i'}. \tag{11}$$

A statistical description *in configuration* space is made in terms of a hierarchy of equations, obtained by multiplying (10) successively by p_i^k ($k = 1, 2, \ldots$, where p_i stands for the different components of the momentum $\boldsymbol{p}$) and integrating over momentum space (assuming the integrand to vanish in the limits). In the time-reversible regime, the radiative terms become small and tend to balance each other in the mean, so any additional effect they may have on the dynamics becomes negligible. When the radiationless approximation is made (i. e., to zero order in e^2), the first two equations of the hierarchy can be recast in the form of a Schrödinger-like equation for a complex function $\psi(\boldsymbol{x}, t)$,

$$-\frac{2\eta^2}{m}\boldsymbol{\nabla}^2\psi + V\psi = 2i\eta\frac{\partial\psi}{\partial t}, \tag{12}$$

and the corresponding complex conjugates, with $\psi^*\psi = \int_{-\infty}^{\infty}d^3pQ(\boldsymbol{x},\boldsymbol{p},t) = \rho(\boldsymbol{x},t)$, and η a free parameter. The value of this parameter is determined by the energy-balance condition that must hold in the equilibrium regime (see below).

2.3. Some important relations for average values

Equation (6) (or its Markovian version, (10)) contains a wealth of statistical information about the dynamics of the system, part of which is lost in the transition to the reduced, radiationless description in configuration space. To recover some of this information we take the (phase-space) average of a generic function $\mathcal{G}(\boldsymbol{x},\boldsymbol{p})$ that has no explicit time dependence. Equation (6) multiplied from the left by $\mathcal{G}$ and integrated over the entire phase space gives thus

$$\frac{d}{dt}\langle\mathcal{G}\rangle=\left\langle \dot{x}_i\frac{\partial\mathcal{G}}{\partial x_i}\right\rangle+\left\langle f_i\frac{\partial\mathcal{G}}{\partial p_i}\right\rangle+m\tau\left\langle \dddot{x}_i\frac{\partial\mathcal{G}}{\partial p_i}\right\rangle-e^2\left\langle\frac{\partial\mathcal{G}}{\partial p_i}\hat{\mathcal{D}}_i\right\rangle. \tag{13}$$

In particular, for any $\mathcal{G}=\xi(\boldsymbol{x},\boldsymbol{p})$ representing a classical ('radiationless') integral of the motion, we have

$$\frac{d}{dt}\langle\xi\rangle=m\tau\left\langle \dddot{x}_i\frac{\partial\xi}{\partial p_i}\right\rangle-e^2\left\langle\frac{\partial\xi}{\partial p_i}\hat{\mathcal{D}}_i\right\rangle. \tag{14}$$

In the Markovian limit mentioned above, this equation takes the form

$$\frac{d}{dt}\langle\xi\rangle=m\tau\left\langle \dddot{x}_i\frac{\partial\xi}{\partial p_i}\right\rangle+\left\langle D^{pp}_{ij}\frac{\partial^2\xi}{\partial p_i\partial p_j}\right\rangle+\left\langle D^{px}_{ij}\frac{\partial^2\xi}{\partial p_i\partial x_j}\right\rangle, \tag{15}$$

with the diffusion coefficients given by (11). For instance, for the particle Hamiltonian (defined without the radiative terms, i. e., to zero order in e^2),

$$H=\frac{1}{2m}\boldsymbol{p}^2+V(\boldsymbol{x}), \tag{16}$$

equation (15) gives

$$\frac{d}{dt}\langle H\rangle=\tau\langle\dddot{\boldsymbol{x}}\cdot\boldsymbol{p}\rangle+\frac{1}{m}\langle\mathrm{tr}\boldsymbol{D}^{pp}\rangle. \tag{17}$$

This equation gives the contributions of the dissipative and diffusive terms to the energy balance. Stationarity is reached when the terms cancel each other, which should occur when both field and particle are in their ground state, i.e.,

$$\tau\langle\dddot{\boldsymbol{x}}\cdot\boldsymbol{p}\rangle_0=-\frac{1}{m}\langle\mathrm{tr}\boldsymbol{D}^{pp}\rangle_0. \tag{18}$$

Notice that for the calculation of these terms (to the lowest order of approximation) one must use the solutions of the Schrödingerlike equation (12), since the system is already in the time-reversible regime. This means, in particular, that in equations (11) the following replacements must be made:

$$\frac{\partial p_j}{\partial p_i'}\to\frac{1}{2i\eta}[\hat{x}_i',\hat{p}_j],\quad\frac{\partial x_j}{\partial p_i'}\to\frac{1}{2i\eta}[\hat{x}_i',\hat{x}_j]. \tag{19}$$

Explicit calculation gives that for the energy-balance condition (18) to be satisfied, the parameter η *must* have precisely the value $\eta=\hbar/2$ [1]. This is, then, the point of entry of Planck's constant into the Schrödinger equation, obtained from (12),

$$-\frac{\hbar^2}{2m}\boldsymbol{\nabla}^2\psi+V\psi=i\hbar\frac{\partial\psi}{\partial t}. \tag{20}$$

The time-reversible and radiationless regime, in which the mechanical system is correctly described by this equation, is therefore called the *quantum regime*. Notice that, from this perspective, $\hbar$ provides a direct measure of the intensity of the fluctuations impressed by the ZPF upon the (quantum) particle.

Journal of Physics: Conference Series **504** (2014) 012007 doi:10.1088/1742-6596/504/1/012007

Let us now go back to equation (15), and apply it to the angular momentum $\boldsymbol{L}$, which is a 'classical' constant of the motion for any central-force problem. By taking $\xi = L_{ij} = x_i p_j - x_j p_i$ $(i \neq j)$, we get

$$\frac{d}{dt}\langle \boldsymbol{L}\rangle = m\tau \langle \boldsymbol{x} \times \dddot{\boldsymbol{x}}\rangle - \langle \boldsymbol{D}^A\rangle, \tag{21}$$

where $\boldsymbol{D}^A$ is the antisymmetric tensor with components $D^{px}_{ij} - D^{px}_{ji}$. This equation tells us that for the angular momentum to be conserved, the radiative terms must cancel each other,

$$m\tau \langle \boldsymbol{x} \times \dddot{\boldsymbol{x}}\rangle = \langle \boldsymbol{D}^A\rangle. \tag{22}$$

Finally, for the square of the angular momentum $\boldsymbol{L}^2$, equation (14) gives

$$\frac{d}{dt}\langle \boldsymbol{L}^2\rangle = m\tau \left\langle \dddot{x}_i \frac{\partial \boldsymbol{L}^2}{\partial p_i}\right\rangle + \left\langle D^{pp}_{ij} \frac{\partial^2 \boldsymbol{L}^2}{\partial p_i \partial p_j}\right\rangle + \left\langle D^{px}_{ij} \frac{\partial^2 \boldsymbol{L}^2}{\partial p_i \partial x_j}\right\rangle, \tag{23}$$

so that for $\boldsymbol{L}^2$ to be conserved, we must have

$$m\tau \left\langle \dddot{x}_i \frac{\partial \boldsymbol{L}^2}{\partial p_i}\right\rangle = -\left\langle D^{pp}_{ij} \frac{\partial^2 \boldsymbol{L}^2}{\partial p_i \partial p_j}\right\rangle - \left\langle D^{px}_{ij} \frac{\partial^2 \boldsymbol{L}^2}{\partial p_i \partial x_j}\right\rangle. \tag{24}$$

These latter results, first proposed in [9], will be used in section 3.1.

2.4. Linear response of the particle in the stationary regime

A close analysis of the stationary states attained by the particle in the time-reversible regime described above, has proved particularly revealing of some general properties of quantum systems. For such analysis a more straightforward approach, complementary to the previous one, has been developed [1], starting again from the same equation of motion (1), and assuming that any stationary solution of it (characterized by the index α) can be written as an expansion of the form

$$x_\alpha(t) = \sum_\beta \tilde{x}_{\alpha\beta} a_{\alpha\beta} e^{i\omega_{\alpha\beta} t}, \tag{25}$$

and similarly for any dynamical variable,

$$A_\alpha(t) = \sum_\beta \tilde{A}_{\alpha\beta} a_{\alpha\beta} e^{i\omega_{\alpha\beta} t}, \tag{26}$$

where $a_{\alpha\beta}$ stands for the random amplitude of the ZPF mode of frequency $\omega_{\alpha\beta}$,

$$E_\alpha(t) = \sum_\beta \tilde{E}_{\alpha\beta} a_{\alpha\beta} e^{i\omega_{\alpha\beta} t}. \tag{27}$$

In these equations the coefficients $\tilde{x}_{\alpha\beta}$, $\tilde{A}_{\alpha\beta}$ are in principle functions of the amplitudes $a_{\alpha\beta}$, and hence stochastic variables. By introducing these expansions into equation (1), the system is found to respond *resonantly* to certain field modes $(\alpha\beta)$; the rest of the ZPF (to be neglected in the expansions) represents just a background noise. The values of the corresponding $\omega_{\alpha\beta}$ (called relevant frequencies), as well as of the coefficients $\tilde{x}_{\alpha\beta}$, depend on the specific problem. By imposing the condition of ergodicity on the stationary solutions, the $\{\tilde{x}_{\alpha\beta}\}$ turn out to be independent of the $\{a_{\alpha\beta}\}$, which means, according to (25), that the response of the system to the respective field modes is *linear* in the field variables. Further, those modes controlling the stationary states are found to satisfy certain properties, which can be summarized in terms of the *chain rule* for the random field coefficients, $a_{\alpha\beta'} a_{\beta'\beta''} a_{\beta''\beta'''} \cdots a_{\beta(n-1)\beta} = a_{\alpha\beta}$, whence

$a_{\alpha\beta} = e^{i\varphi_{\alpha\beta}}$ with $\varphi_{\alpha\beta} = \phi_\alpha - \phi_\beta$ a random phase. Similarly for the corresponding relevant frequencies, the relation $\omega_{\alpha\beta'} + \omega_{\beta'\beta''} + ... + \omega_{\beta(n-1)\beta} = \omega_{\alpha\beta}$ applies, which means that $\omega_{\alpha\beta}$ is of the form

$$\omega_{\alpha\beta} = \Omega_\alpha - \Omega_\beta. \tag{28}$$

As a result of these properties, the coefficients in the above expansions satisfy a matrix algebra: thus, for instance, if $\hat{x}$ is the matrix with elements $\tilde{x}_{\alpha\beta}$, we have

$$(x^n)_\alpha = \sum_\beta (\widetilde{x^n})_{\alpha\beta} a_{\alpha\beta} e^{i\omega_{\alpha\beta}t}, \tag{29}$$

with $(\widetilde{x^n})_{\alpha\beta}$ given by the element $\alpha\beta$ of the corresponding matrix product, $(\widetilde{x^n})_{\alpha\beta} = (\hat{x}^n)_{\alpha\beta}$.

It is clear that the time-dependence of $x_\alpha(t)$ can be transferred to every single factor $\tilde{x}_{\alpha\beta}$, so that the evolving matrix $\hat{x}(t)$ has as elements the coefficients $\tilde{x}_{\alpha\beta} e^{i\omega_{\alpha\beta}t}$. This allows to write the equation satisfied by the stationary solutions in the form

$$m\frac{d^2\hat{x}(t)}{dt^2} = \hat{f}(t) + m\tau\frac{d^3\hat{x}(t)}{dt^3} + e\hat{E}(t). \tag{30}$$

In the radiationless approximation, the last two terms are neglected and one is left with

$$m\frac{d^2\hat{x}(t)}{dt^2} = \hat{f}(t). \tag{31}$$

With these results one may construct the law of evolution for the matrix $\hat{A}(t)$ associated with the (generic) dynamical variable A; the outcome is

$$i\frac{d\hat{A}(t)}{dt} = \left[\hat{A}(t), \hat{\Omega}\right], \tag{32}$$

with the matrix elements of $\hat{\Omega}$ given by $\Omega_{\alpha\beta} = \Omega_\alpha \delta_{\alpha\beta}$. Notice that, as with the previous derivation presented in section 2.2, the radiationless approximation has deleted any explicit trace of the ZPF. No stochastic variable is contained in equations (30)-(32); instead, they are expressed in terms of operators. In section 2.3, the missing value of the paramerter η in equation (12) was obtained by imposing the energy-balance condition; this led unequivocally to the Schrödinger equation (20). In the present case, it is again the scale of the solution what is missing in equation (32). The loss is repaired by, first, finding that the canonical commutator $[\hat{x}, \hat{p}]$, with

$$\hat{p} = m\frac{d\hat{x}(t)}{dt} = -im\left[\hat{x}(t), \hat{\Omega}\right], \tag{33}$$

has the *universal* form $[\hat{x}, \hat{p}] = C\mathbb{I}$, with C constant, and, second, finding the value of C. This is readily achieved by applying the above equations to the harmonic oscillator of natural frequency ω_0 in its ground state, when it is in equilibrium with the ZPF mode of the same frequency. One thus obtains $C = i\hbar$, whence

$$[\hat{x}, \hat{p}] = i\hbar\mathbb{I}. \tag{34}$$

The value of the commutator represents therefore a direct measure of the intensity of the fluctuations impressed upon the particle by the ZPF, just as was concluded in section 2.3. Finally, from the above results it can be readily seen that the relevant frequencies are given by

$$\omega_{\alpha\beta} = \Omega_\alpha - \Omega_\beta = \hbar^{-1}\left(\mathcal{E}_\alpha - \mathcal{E}_\beta\right), \tag{35}$$

where $\mathcal{E}_\alpha$ is the eigenvalue of the Hamiltonian in state α, and the final form of equation (32) is therefore

$$i\hbar\frac{d\hat{A}(t)}{dt} = \left[\hat{A}(t), \hat{H}\right], \tag{36}$$

i. e., the Heisenberg equation for the operator $\hat{A}(t)$.

3. Revealing the spin of the electron

To disclose the rotational effect of the ZPF on the particle, let us briefly go back to the original (stochastic) equation of motion (1) and rewrite it as

$$\dot{\boldsymbol{p}} = \boldsymbol{f} + m\tau\dddot{\boldsymbol{x}} + e\boldsymbol{E}(t), \tag{37}$$

where $\boldsymbol{p} = m\dot{\boldsymbol{x}}$. We shall consider that there is no external torque: the (central) force can then be written as $\boldsymbol{f} = g(r)\boldsymbol{x}$, $r = |\boldsymbol{x}|$. By taking the vector product of this equation with $\boldsymbol{x}$ we get

$$\frac{d}{dt}\boldsymbol{L} = \boldsymbol{x} \times \dot{\boldsymbol{p}} = m\tau\boldsymbol{x} \times \dddot{\boldsymbol{x}} + e\boldsymbol{x} \times \boldsymbol{E}, \tag{38}$$

which gives the instantaneous change of the angular-momentum due to the torque exerted by both radiation reaction and the random ZPF. The average of this expression over the field realizations is

$$\frac{d}{dt}\overline{\boldsymbol{L}}^{(i)} = m\tau\overline{(\boldsymbol{x} \times \dddot{\boldsymbol{x}})}^{(i)} + e\overline{(\boldsymbol{x} \times \boldsymbol{E})}^{(i)}. \tag{39}$$

Since in the quantum regime the dynamical variables satisfy the condition of ergodicity, according to our discussion in section 2.4, this equation is equivalent to (21). Equation (39), however, has the advantage of allowing us to directly identify the last term as the effective torque exerted by the Lorentz force of the ZPF on the particle. For the angular momentum to be conserved, we must have (compare with equation (22))

$$m\tau\overline{(\boldsymbol{x} \times \dddot{\boldsymbol{x}})}^{(i)} = -e\overline{(\boldsymbol{x} \times \boldsymbol{E})}^{(i)}. \tag{40}$$

Let us assume that the system is in its ground state, so that there is no orbital angular momentum. Making the usual substitution $\dot{\boldsymbol{p}} \rightarrow \boldsymbol{f}$ (which is valid to zero order in $\tau \sim e^2$) and writing $\boldsymbol{f}$ as $g(r)\boldsymbol{x}$, we get explicitly

$$\tau\overline{g(r)\boldsymbol{L}}^{(i)}\Big|_0 = -me\,\overline{(\boldsymbol{x} \times \boldsymbol{E})}^{(i)}\Big|_0. \tag{41}$$

Since only the fluctuating component of $\boldsymbol{x}$ can contribute to the average $\overline{(\boldsymbol{x} \times \boldsymbol{E})}^{(i)}$ (because $\boldsymbol{E}(t)$ is purely random), it is clear that all the angular momentum thus generated is due to the random motion around the mean trajectory followed by the particle; thus, it is independent of the system of coordinates, and has an 'internal' nature, usualy taken as intrinsic.

3.1. The electron's intrinsic angular momentum

The above discussion suggests looking for a procedure that can bring to the surface the 'intrinsic' angular momentum acquired by the electron through its interaction with the ZPF. For this purpose we recall the experimental observation that the interaction of the electron with the radiation field takes place via the circular polarized modes of the field (or modes of a certain helicity). This is known to be the case for the photonic field, which, from the perspective of SED, is the excited state of the radiation field, additional to the zero-point component. Now, it is natural to assume that the modes of the field in its ground state, i. e., of the ZPF, interact in a similar way with the electron. To analyze the effect of such interaction, we should therefore consider the ZPF as composed of modes of both right- and left-handed circular polarization. When the ensemble of field modes is considered in its entirety, as is usually done, any such effect is concealed; yet by focusing on one of the two subensembles of a given polarization, the effective rotation induced on the corresponding particles should be disclosed.

Let us therefore consider a situation in which the particle is in a stationary state, under the action of the background field of a given circular polarization with respect to an axis $\boldsymbol{k}$. To start with, we shall consider the particle in its ground state, so as to ensure that there is no orbital angular momentum. We therefore take

$$\langle L_{ij}\rangle_0 = \langle 0| (\hat{x}_i \hat{p}_j - \hat{x}_j \hat{p}_i) |0\rangle \tag{42}$$

and analyze separately the contributions arising from each of the two circular polarizations, characterized by the (circularly polarized) vectors

$$\boldsymbol{\epsilon}_{\boldsymbol{k}\pm} = \frac{1}{\sqrt{2}} (\boldsymbol{\epsilon}_{\boldsymbol{k}i} \pm i \boldsymbol{\epsilon}_{\boldsymbol{k}j}), \tag{43}$$

with $\boldsymbol{\epsilon}_{\boldsymbol{k}i}$, $\boldsymbol{\epsilon}_{\boldsymbol{k}j}$ unit Cartesian vectors orthogonal to the axis $\boldsymbol{k}$.

Since according to the results reported in section 2.4, the response of the particle to the field is linear, to describe its motion under the action of a circular field mode one should write the variable $\boldsymbol{x}$ in cylindrical coordinates, i. e. $\boldsymbol{x} = x^+ \boldsymbol{\epsilon}_{\boldsymbol{k}+} + x^- \boldsymbol{\epsilon}_{\boldsymbol{k}-} + x_k \hat{\boldsymbol{k}}$, with

$$x^{\pm} = \frac{1}{\sqrt{2}} (x_i \mp i x_j), \tag{44a}$$

$$x_i = \frac{1}{\sqrt{2}} \left(x^+ + x^-\right), \quad x_j = i\frac{1}{\sqrt{2}} \left(x^+ - x^-\right). \tag{44b}$$

Taking into account that $x^{\pm}_{n0} = \left(x^{\mp}_{0n}\right)^*$, equation (42) becomes explicitly

$$\langle L_{ij}\rangle_0 = m \sum_n \omega_{n0} \left(x^+_{0n} x^-_{n0} - x^-_{0n} x^+_{n0}\right) = m \sum_n \omega_{n0} \left(\left|x^+_{0n}\right|^2 - \left|x^-_{0n}\right|^2\right). \tag{45}$$

In the ground state, $\langle L_{ij}\rangle_0 = 0$; hence the two sums on the right-hand side contribute with equal magnitude and opposite sign to the k-component of the total angular momentum, as should be the case for a nonpolarized vacuum. Taken separately, these contributions are

$$\langle L_{ij}\rangle_0^+ = m \sum_n \omega_{n0} \left|x^+_{0n}\right|^2, \quad \langle L_{ij}\rangle_0^- = -m \sum_n \omega_{n0} \left|x^-_{0n}\right|^2. \tag{46}$$

Since, on the other hand, the mean value of the commutator (34) gives the sum rule

$$m \sum_n \omega_{n0} \left|x^{\pm}_{0n}\right|^2 = \frac{\hbar}{2}, \tag{47}$$

the size of each separate contribution to (45) is just $\hbar/2$. In order to distinguish these contributions from the (orbital) component of the angular momentum we write $\langle S_{ij}\rangle^{\pm}$ instead of $\langle L_{ij}\rangle_0^{\pm}$; thus $\langle L_{ij}\rangle_0 = \langle S_{ij}\rangle^+ + \langle S_{ij}\rangle^-$, with

$$\langle S_{ij}\rangle^{\pm} = \pm\frac{\hbar}{2}. \tag{48}$$

Direct calculation of the square of the angular momentum, using (24), is more cumbersome. As a simple expedient, let us carry out this calculation for the isotropic harmonic oscillator in its ground state. The (radiationless) approximation $\dot{p}_i = f_i$ allows us to write in this case $m\dddot{x}_i = -\omega_0^2 p_i$. Further, since $\boldsymbol{L}^2 = \boldsymbol{x}^2 \boldsymbol{p}^2 - (\boldsymbol{x}\cdot\boldsymbol{p})^2$, we have

$$p_i \frac{\partial \boldsymbol{L}^2}{\partial p_i} = 2\boldsymbol{L}^2, \quad \frac{\partial^2 \boldsymbol{L}^2}{\partial p_i \partial p_i} = 4\boldsymbol{x}^2, \quad \frac{\partial^2 \boldsymbol{L}^2}{\partial x_i \partial p_i} = 0.$$

With these results equation (24) transforms into

$$\left\langle \boldsymbol{L}^2\right\rangle_0 = \frac{mc^3}{e^2\omega_0^2} \left\langle D^{pp} \boldsymbol{x}^2\right\rangle_0, \tag{49}$$

with D^{pp} given by equation (11). Now in the case of the harmonic oscillator, using equations (2), (4) and (19) we obtain a constant value for the diffusion coefficient,

$$D^{pp} = \frac{\hbar e^2 \omega_0^3}{c^3}, \tag{50}$$

whence using $\langle \boldsymbol{x}^2 \rangle_0 = 3\langle \hat{x}^2 \rangle_0 = 3\hbar/(2m\omega_0)$, equation (49) becomes finally

$$\left\langle \boldsymbol{L}^2 \right\rangle_0 = m\hbar\omega_0 \left\langle \boldsymbol{x}^2 \right\rangle_0 = \frac{3}{2}\hbar^2. \tag{51}$$

This result was obtained for the first time by Marshall in 1965 [15], and taken as an additional contribution to the orbital angular momentum due to the ZPF. In [9] the same result was obtained, but interpreted as an intrinsic (spin) angular momentum of double the correct value. However, in line with the present approach, we separate again the full ensemble into two subensembles corresponding to the different circular polarizations, thus obtaining

$$\left\langle \boldsymbol{L}^2 \right\rangle_0 = \left\langle \boldsymbol{L}^2 \right\rangle_0^+ + \left\langle \boldsymbol{L}^2 \right\rangle_0^-, \tag{52}$$

with each partial contribution to the mean square angular momentum given by

$$\left\langle \boldsymbol{L}^2 \right\rangle_0^+ = \left\langle \boldsymbol{L}^2 \right\rangle_0^- = \tfrac{3}{4}\hbar^2. \tag{53}$$

Using the notation introduced above (see equation (48)), which identifies this as an intrinsic angular momentum, we write

$$\left\langle \boldsymbol{S}^2 \right\rangle^+ = \left\langle \boldsymbol{S}^2 \right\rangle^- = \tfrac{3}{4}\hbar^2. \tag{54}$$

The fact that this result does not depend on the oscillator's frequency ω_0, suggests that it holds in the general case, and for the free particle in particular. Therefore, we conclude that when the transformation (44) possesses physical meaning, so that the decompositions $\langle L_{ij} \rangle_0 = \langle L_{ij} \rangle_0^+ + \langle L_{ij} \rangle_0^-$ and $\left\langle \boldsymbol{L}^2 \right\rangle_0^+ = \left\langle \boldsymbol{L}^2 \right\rangle_0^+ + \left\langle \boldsymbol{L}^2 \right\rangle_0^-$ make sense, equations (48) and (54) tell us that there exists an angular momentum that does *not* correspond to an orbital motion of the particle and can therefore be considered as *intrinsic*. For an electron, which (as stated above) interacts with the radiation field via its circular polarized modes, the transformation (44) is indeed physically meaningful, and the angular momentum thus induced can therefore be identified with its spin.

3.2. General derivation of the spin angular momentum

In the preceding section we have disclosed the existence of the spin angular momentum for an electron in its ground state. Let us now extend our analysis to the general case, including excited states with orbital angular momentum.

According to the above discussion, we should separate the contributions to the angular momentum arising from the two circular polarizations of the field. Denoting with the index n (or k) the set of quantum numbers that characterize a state of the particle, including the orbital angular momentum and its projection along the z-axis, we have (for simplicity in the writing we use $x_i = x$, $x_j = y$, and $x_k = z$)

$$\begin{aligned} \left\langle \hat{L}_z \right\rangle_n &= \langle n | \hat{L}_z | n \rangle = \sum\nolimits_k (x_{nk} p_{ykn} - y_{nk} p_{xkn}) = \\ &= im \sum\nolimits_k \omega_{kn} (x_{nk} y_{kn} - y_{nk} x_{kn}). \end{aligned} \tag{55}$$

Under the same procedure that led to (45), equation (55) transforms into

$$\left\langle \hat{L}_z \right\rangle_n = m \sum_k \omega_{kn} \left(\left|x^+_{nk}\right|^2 - \left|x^-_{nk}\right|^2 \right). \tag{56}$$

This expression can be rewritten as

$$\left\langle \hat{L}_z \right\rangle_n = \langle O_z \rangle_n^+ + \langle O_z \rangle_n^- , \tag{57}$$

with $\langle O_z \rangle_n^\sigma$ given by ($\sigma = \pm$)

$$\langle O_z \rangle_n^\sigma = \sigma m \sum_k \omega_{kn} \left|x^\sigma_{nk}\right|^2 . \tag{58}$$

Using again the sum rule $m \sum_k \omega_{kn} |x_{nk}|^2 = m \sum_k \omega_{kn} |y_{nk}|^2 = \frac{1}{2}\hbar$, one obtains from (58)

$$\hbar = m \sum_k \omega_{kn} \left(\left|x^+_{nk}\right|^2 + \left|x^-_{nk}\right|^2 \right) = \langle O_z \rangle_n^+ - \langle O_z \rangle_n^- , \tag{59}$$

which combined with (57) gives

$$\langle O_z \rangle_n^\sigma = \tfrac{1}{2} \left\langle \hat{L}_z \right\rangle_n + \sigma \tfrac{1}{2}\hbar. \tag{60}$$

This quantity $\langle O_z \rangle_n^\sigma$ contains, for every polarization state σ, both the corresponding part of the orbital angular momentum, and the spin associated with that state.

To construct the operator associated with the vector $\boldsymbol{S}$ introduced in section 3.1, we observe that the mean value $\langle \hat{L}_z \rangle_n$ does not depend on σ, whereas the term $\sigma\hbar/2$ does not depend on n. This indicates that the operator $\hat{L}_z$ and the operator to be associated with $\sigma\hbar/2$ (which we shall call $\hat{\Sigma}_z$) belong to different Hilbert spaces. In order to express $\langle O_z \rangle_n^\sigma$ as the average of an operator, we must therefore extend the Hilbert space to include the dichotomous variable σ in addition to the quantum index n. The result is the product space $\mathcal{H} = \mathcal{H}_n \otimes \mathcal{H}_2$, with $\mathcal{H}_2$ a bidimensional vector space spanned by an orthonormal basis having as elements the vectors $\{|\sigma\rangle\} = \{|+\rangle, |-\rangle\}$. In terms of $|n\sigma\rangle = |n\rangle \otimes |\sigma\rangle$, equation (60) rewrites as

$$\langle O_z \rangle_n^\sigma = \tfrac{1}{2} \langle n\sigma | \hat{L}_z | n\sigma \rangle + \tfrac{1}{2}\hbar \langle n\sigma | \hat{\Sigma}_z | n\sigma \rangle , \tag{61}$$

with $\hat{\Sigma}_z$ an operator that has $|\sigma\rangle$ as eigenvector,

$$\langle n\sigma | \hat{\Sigma}_z | n\sigma \rangle = \langle \sigma | \hat{\Sigma}_z | \sigma \rangle = \sigma. \tag{62}$$

Expressing $\hat{\Sigma}_z$ in terms of the Pauli matrices gives

$$\hat{\boldsymbol{\Sigma}}_z = a_0 \mathbb{I} + a_z \hat{\sigma}_z + a^+ \hat{\sigma}_+ + a^- \hat{\sigma}_- , \tag{63}$$

where $a^\pm = (a_x \mp i a_y)/\sqrt{2}$, and $\hat{\sigma}_+ = \sqrt{2}\,|+\rangle\langle -|$, $\hat{\sigma}_- = \sqrt{2}\,|-\rangle\langle +|$ are ladder operators. Condition (62) imposed on $\hat{\boldsymbol{\Sigma}}_z$ gives $a_0 = 0$, $a_z = 1$. Further, since we are here considering the variables (x^+, x^-, z), the polarization vectors (43) fix $\hat{\boldsymbol{z}}$ as the preferred axis, whence $a^\pm = 0$, $\hat{\boldsymbol{\Sigma}}_z = \hat{\sigma}_z$, and equation (61) becomes

$$\langle O_z \rangle_n^\sigma = \langle n\sigma | \left(\tfrac{1}{2}\hat{\boldsymbol{L}} + \hat{\boldsymbol{S}} \right) \cdot \hat{\boldsymbol{z}} \, | n\sigma \rangle , \tag{64}$$

with $\hat{\boldsymbol{S}}$ the vector operator defined as $\hat{\boldsymbol{S}}_z = \hbar \frac{1}{2} \hat{\Sigma}_z$, i.e.,

$$\hat{\boldsymbol{S}} = \tfrac{1}{2}\hbar\hat{\boldsymbol{\sigma}}. \tag{65}$$

The identification of the operator $\hat{\boldsymbol{S}}$ with the spin of the electron is thus justified. The independence of $\langle \hat{L}_z \rangle_n$ from σ and of $\langle \hat{\Sigma}_z \rangle$ from n, indicates that under the present conditions, the fluctuations associated with the spin are not correlated with those that characterize the kinematics of the particle in the configuration space: $\boldsymbol{L}$ and $\boldsymbol{S}$ constitute independent dynamical variables. Of course the spaces of the spin and the orbital angular momentum may become connected by the presence of magnetic fields.

It should be stressed that even if $\hat{\boldsymbol{L}}$ and $\hat{\boldsymbol{S}}$ are combined under one expression for the angular motions, e. g. equation (64), the spin is not an orbital angular momentum. Indeed, there are fundamental differences between $\hat{\boldsymbol{L}}$ and $\hat{\boldsymbol{S}}$. In particular, the mean value of $\hat{L}_z$, say, can be freely determined by adjusting external parameters, and may acquire a whole spectrum of values. However, only the sign of the projection $\hat{S}_z$ can be subject to external adjustment; its absolute value is determined by the fundamental commutator, which in its turn is fixed by the ZPF. It is because of the *universal* value of the commutator that the spin of the electron is the same for all electrons under all circumstances, which reinforces its 'intrinsic' nature.

The connection of the commutator $[\hat{x}, \hat{p}_x] = i\hbar$ with the spin of the electron deserves a further comment. As pointed out in previous sections, Planck's constant $\hbar$ is a direct measure of the size of the fluctuations, both those of the ZPF and those impressed by it on the particle. Specifically, since the commutator implies that the fluctuations of x and p_x have a minimum value adjusted to the rule $\sigma_x^2 \sigma_{p_x}^2 \big|_{\min} = \hbar^2/4$ (σ_x^2, $\sigma_{p_x}^2$ are variances), one may write the *numerical* relation $\left| \langle \pm | \hat{S}_z | \pm \rangle \right| = \hbar/2 = \sigma_x \sigma_{p_x} \big|_{\min}$, which emphasizes the fact that the *value* of the electron spin is determined by the irreducible fluctuations of the phase-space variables x, p_x around the instantaneous position of the particle.

3.3. The spin gyromagnetic factor

It was discovered experimentally that the g-factor associated with the spin magnetic moment of the electron has an approximate value $g_S = 2$, whereas for the orbital magnetic moment the g-factor is $g_L = 1$. This characteristic value of g_S is incorporated into nonrelativistic quantum theory by hand, usually without further elaboration. The issue is normally resolved by resorting to the Dirac equation, which predicts $g_S = 2$. Given that the present theory produces the electron spin, it seems appropriate to investigate the value predicted by it for g_S. For this purpose consider the electron acted on, in addition to the external force $\boldsymbol{f}(\boldsymbol{x})$, by a static uniform magnetic field $\boldsymbol{B} = B\hat{\boldsymbol{z}}$. The contribution of the orbital angular momentum $\boldsymbol{L}$ to the Hamiltonian is given by

$$\hat{H} = -\hat{\boldsymbol{\mu}} \cdot \boldsymbol{B} = -\mu_z B, \tag{66}$$

where $\hat{\boldsymbol{\mu}} = -(g_L \mu_0 \hat{\boldsymbol{L}})/\hbar$ is the magnetic moment due to $\hat{\boldsymbol{L}}$, $\mu_0 = |e|\,\hbar/(2mc)$ is the Bohr magneton (with $-e = |e|$), and $g_L = 1$. Therefore the mean energy is

$$\mathcal{E} = \frac{\mu_0}{\hbar} B \langle \hat{L}_z \rangle. \tag{67}$$

Consider now a situation in which the spin projection along $\hat{\boldsymbol{z}}$ has a well-defined value, say $\langle \hat{S}_z \rangle = +\hbar/2$. This means that one should consider only the subensemble that corresponds to $\sigma = +$. Resorting to equation (60) to write the corresponding contribution to $\langle L_z \rangle$ as $(\langle \hat{L}_z \rangle + \hbar)/2$, the component of $\mathcal{E}$ of interest is

$$\mathcal{E}^+ = \frac{\mu_0 B}{\hbar} \left(\frac{1}{2} \langle \hat{L}_z \rangle + \frac{\hbar}{2} \right) = \frac{\mu_0 B}{\hbar} \left(\frac{1}{2} \langle \hat{L}_z \rangle + \langle \hat{S}_z \rangle \right)^+ . \tag{68}$$

An analogous result holds for the subensemble with $\sigma = -$, for which $\langle \hat{S}_z \rangle = -\hbar/2$,

$$\mathcal{E}^- = \frac{\mu_0 B}{\hbar} \left(\frac{1}{2} \langle \hat{L}_z \rangle - \frac{\hbar}{2} \right) = \frac{\mu_0 B}{\hbar} \left(\frac{1}{2} \langle \hat{L}_z \rangle + \langle \hat{S}_z \rangle \right)^- . \tag{69}$$

The corresponding Hamiltonian describing the total magnetic interaction of the electron follows from the sum of these contributions; it is therefore

$$\hat{H}_{LS} = \frac{\mu_0 B}{\hbar}\left(\hat{L}_z + 2\hat{S}_z\right). \tag{70}$$

This contains the correct g-factor of 2 for the spin of the electron, in the radiationless approximation. It is clear from the derivation that this value is determined by the two degrees of freedom associated with the polarization of the ZPF.

Notice that the result (70) gives a precise meaning to the operator appearing in equation (64). Indeed, from this latter equation we can write $\hat{\boldsymbol{O}} = (\hat{\boldsymbol{L}} + 2\hat{\boldsymbol{S}})/2$, whence

$$\hat{H}_{LS} = \frac{\mu_0}{\hbar}\boldsymbol{B}\cdot\left(\hat{\boldsymbol{L}} + 2\hat{\boldsymbol{S}}\right) = -\hat{\boldsymbol{\mu}}\cdot\boldsymbol{B}, \tag{71}$$

with $\hat{\boldsymbol{\mu}} = -\frac{2\mu_0}{\hbar}\hat{\boldsymbol{O}}$, which directly relates $\hat{\boldsymbol{O}}$ with the *total* magnetic-moment operator of the atomic electron.

4. Final remarks

The present results give strong support to the representation of the electron spin as an acquired angular momentum. They reaffirm the image, suggested in previous SED work, of spin as a helicoidal motion around the local mean path followed by the particle, produced by the high-frequency (circular polarized) modes of the fluctuating vacuum.

If, according to the results presented, a charged particle acquires spin 1/2, the question arises as to whether a similar effect shouldn't be expected to appear in the case of scalar bosons. A possible answer to this question is that the spin 1/2 is acquired by elementary particles, such as the electron, whereas composite particles may acquire it or not, depending on their internal structure. This suggests that bosons are in general composite structures, with an even number of elements (if of fermion type).

Acknowledgments

This work was supported by DGAPA-UNAM through project PAPIIT IN106412.

References

[1] A. M. Cetto, L. de la Peña and A. Valdés-Hernández, *J. Phys. JPCS* **361** (2012) 012013 and references therein; L. de la Peña, A. M. Cetto and A. Valdés-Hernández, *Int. J. Modern Phys.* A (to be published).
[2] A. M. Cetto and L. de la Peña, *Phys. Scripta* T**151** (2002) 014009; A. M. Cetto, L. de la Peña and A. Valdés-Hernández, *Rev. Mex. Fis.* **59** (2013) 433; arXiv:1301.6200 (quant-ph), 26 Jan 2013.
[3] A. Valdés-Hernández, *Investigación del origen del enredamiento cuántico desde la perspectiva de la electrodinámica estocástica lineal*, Ph. D. thesis, UNAM (Mexico, 2010); A. Valdés-Hernández, L. de la Peña and A. M. Cetto, *Found. Physics* **41** (2011) 843.
[4] L. de la Peña and A. M. Cetto, *The Quantum Dice* (Kluwer, Dordrecht, 1996) and references therein.
[5] P. Braffort and A. Taroni, *C. R. Acad. Sc. Paris* **264** (1967) 1437.
[6] S. M. Moore and J. A. Ramírez, *Lett. Nuovo Cim.* **33** (1982) 87; S. M. Moore, *Lett. Nuovo Cim.* **40** (1984) 385.
[7] G. Cavalleri, *Lett. Nuovo Cim.* **43** (1985) 285.
[8] A. Rueda, *Found. Phys. Lett.* **6** (1993) 139.
[9] A. Jáuregui and L. de la Peña, *Phys. Lett.* A, **86** (1981) 280; L. de la Peña and A. Jáuregui, *Found. Phys.* **12** (1982) 441.
[10] A. Barranco, S. A. Brunini and H. M. Franca, *Phys. Rev.* A **39** (1989), 5492.
[11] D. Hestenes, *Found. Phys.* **15** (1985) 63.
[12] K. Muralidhar, *Apeyron* **18** (2011) 146.
[13] K. Huang, *Am. J. Phys.* **20** (1952) 479.
[14] U. Frisch, in *Probabilistic Methods in Applied Mathematics*, vol. I, A. T. Barucha-Reid, ed. (Academic Press, NY, 1968); see also L. de la Peña and A. M. Cetto, *J. Math. Phys.* **18** (1977) 1612.
[15] T. W. Marshall, *Nuovo Cim.* 38 (1965) 206.

A subquantum arrow of time

Theo M Nieuwenhuizen

Institute for Theoretical Physics, University of Amsterdam, Science Park 904, P.O. Box 94485, 1090 GL Amsterdam, The Netherlands

E-mail: t.m.nieuwenhuizen@uva.nl

Abstract. The outcome of a single quantum experiment is unpredictable, except in a pure-state limit. The definite process that takes place in the apparatus may either be intrinsically random or be explainable from a deeper theory. While the first scenario is the standard lore, the latter implies that quantum mechanics is emergent. In that case, it is likely that one has to reconsider radiation by accelerated charges as a physical effect, which thus must be compensated by an energy input. Stochastic electrodynamics, for example, asserts that the vacuum energy arises from classical fluctuations with energy $\frac{1}{2}\hbar\omega$ per mode. In such theories the stability of the hydrogen ground state will arise from energy input from fluctuations and output by radiation, hence due to an energy throughput. That flux of energy constitutes an arrow of time, which we call the "subquantum arrow of time". It is related to the stability of matter and it is more fundamental than, e.g., the thermodynamic and cosmological arrows.

1. Introduction

Time and tide wait for no man
Proverb

The arrow of time: why do we get older, never younger? has bothered people and philosophers since long. The fact that Nature has an arrow of time while being described by reversible dynamical equations such as classical mechanics, electrodynamics, quantum mechanics and general relativity, poses a deep problem: is there an ultimate cause for the arrow of time? Of the many answers given, we mention the thermodynamic arrow, relying on the second law (entropy increase), and the cosmological arrow, which acknowledges that the Universe is expanding while aging. But is there a still more fundamental one? The aim of the present paper is to investigate whether the *emergence of quantum mechanics* [1] can provide one. To get there, we need first to dwell on recent progress on the foundations of quantum mechanics itself.

2. On quantum measurement and the interpretation of quantum mechanics

To understand Nature
we have become accustomed
to inconceivable concepts

In textbooks quantum measurement is treated by means of postulates for ideal measurements: the collapse postulate and the Born rule. To understand this from a more conceptual level, many attempts have been made. While various approaches stay within standard quantum mechanics, others go beyond it. But in the practice of a laboratory a measurement is an interaction between two systems: the tested system and the apparatus. This being remote from postulates, motivates the study of models aiming to describe realistic ideal measurements. The literature on this subject has been reviewed in [2].

Not having reached consensus on the meaning of quantum mechanics (QM) after nearly a century (as starting point we take the *Dreimännerarbeit* of Born, Jordan and Heisenberg in 1926) has had an enormous impact on the philosophical view of the community. Instead of admitting that convincing arguments have yet to be filtered out, the case has turned into "We did our utmost best to find a solution, but there really is none." This psychologically understandable but unfortunate position has led to remarkable statements such as: "the development from animals to humans has prepared brains incapable to understand the quantum nature of matter"[1, 2]. We do not have to wonder what Einstein would say about that, he has spent half of his life trying to avoid this trap; so did 't Hooft [3]. Indeed, it is never too late to face the fundamental questions rather than to answer them by "talking dynamics" (intricate wording instead of convincing, functioning principles).

Since the only point of contact between quantum theory and Nature lies in measurements, it seems rather obvious [2] that in order to determine the meaning of the theory, one must understand what goes on during a measurement.[3] Only focussing on the measurement postulates, like most approaches do [4], is a *black box* approach, that may be of help for certain questions, but is doomed to fail at the most fundamental level. We expect the same to hold for quantum information, in the sense that it can provide us with constraints on the working of quantum theory, without being able to answer all the fundamental questions that can be answered within QM.

Our basic notion is that an apparatus used to perform a measurement interacts with the tested system, building a special type of interacting quantum systems. [5] The apparatus consists of atoms and the measurement causes a change of its state. On a theoretical level we then have to ask: how can an ideal measurement be modelled? Obviously, the answer lies in the quantum solid state theory for two coupled quantum systems, the tested system and the apparatus. Fortunately, in applications to the solid state the interpretation of quantum mechanics plays a secondary role [6]; in reverse, the involved one can be expected to have an impact on understanding of measurements.

2.1. The Curie-Weiss model for a quantum measurement and its dynamics

Considering a model for a quantum measurement, there are three important issues to be explained: *truncation* of the density matrix (vanishing of Schrödinger cat terms); *registration*, the stable indication of the measurement outcome by the pointer variable, and the possibility of *reduction*, to update our knowledge about the ensemble of tested systems on the basis of measurement outcomes [2]. For the latter, one needs to explain that single measurements have definite outcomes even though quantum mechanics is a wave theory, that is, to be able to speak about individual events, even though quantum mechanics is an ensemble theory. Either way, one needs to solve the *measurement problem*. [4]

[1] The situation being: after we adopt this or that interpretation, we run into oddities. A logical conclusion is, however, not to blame the theory but the making of such assumptions.

[2] A similar statement by Edward Witten about string theory: "String theory is a part of 21st-century physics that fell by chance into the 20th century" is to be regarded as equally remarkable. This theory has not reproduced the standard model of particle physics, and even if it would, for what reason would the necessary set of parameters be singled out? In fact, string theory has more the looks of a *framework*, applicable to different situations, including Fermi liquid theory, than of a fundamental *theory of Nature*.

[3] That this is not obvious to everybody, may be seen from several approaches to the interpretation of QM that try to do away with measurements, sometimes even banning the term. See [4] for a review.

[4] The opinion: "the alleged 'measurement problem' does not exist as a problem of quantum theory" [7] misses the essence for the many interpretations of QM and fails to acknowledge the answer presented in ref. [2] and the

A rather rich but flexible and solvable model is the Curie-Weiss model, introduced in [9] and worked out in detail in [2]. The system S is a spin $\frac{1}{2}$. The aim is to measure its z-component by means of an apparatus A, which consists of an Ising magnet M coupled to a bath B. The magnet starts in a metastable paramagnetic initial state and will be triggered by the measurement to go to one of its two stable ferromagnetic phases with magnetization $\pm m_{\mathrm{F}}$. Hence the measurement relies on a first order phase transition which amplifies the small quantum signal, a situation known from our retina, where a handful of photons can achieve an avalanche of electrons, that sends an electric current to the cortex. Not surprisingly, the measurement basis is determined by the interaction Hamiltonian, that is, by the forces exerted by the apparatus on the tested system.[5]

After coupling S to A, there first occurs a *dephasing* in the off-diagonal terms of the density matrix, as happens in NMR physics. Under suitable conditions the now hidden information is lost before recurrences may occur by a *decoherence* due to the bath, from which it cannot be recovered. Hence a *truncation* of off-diagonal elements of the density matrix occurs, commonly called decay of Schrödinger cat terms, death of Schrödinger cats, and so. It is a *physical process*, due to coupling of S to A, more precisely: first due to the interaction Hamiltonian, then to the bath.

In a second stage of the process, the *registration* of the result occurs. The magnet M goes from its initial paramagnetic state to the stable up–or–down ferromagnetic state, provided that the initial metastable state has been made unstable by a strong enough coupling between S and A. This phase transition is exploited to amplify the small quantum signal.

The third stage of the measurement is a new one, not discussed in the literature until recently, appearing first in version 3 of the arXiv preprint of our review [2].

2.2. The measurement problem: a conundrum for theorists and philosophers

Win the battle only to lose the war
Proverb

Experience shows that experiments yield definite outcomes. In classical physics this is obvious: when one throws a handful of coins on the ground, each one will expose either head or tail. In quantum physics things are not clear, however. Indeed, an unpolarized beam of neutrons can be decomposed as half having spin up and half spin down. However, this can be done *along any axis*, bringing an infinity of possibilities, so the decomposition is a mathematical one, bearing no physics, and, in particular, not allowing to suppose that half of the neutrons can be thought to have spin up and the other half spin down along the z-axis.[6] So no physical interpretation of the unpolarised beam can be given. This of course changes when a measurement is performed.

For the description of measurements, however, one obviously wishes to choose the measurement basis as decomposition basis, that is, one wishes to interpret the final density matrix $\mathcal{D}$ of the combined system at the final time of the measurement, t_{f}, in a straightforward manner. Let us consider the example of the Curie-Weiss model. In the theoretical description of an ideal measurement of a spin $\frac{1}{2}$, the outcome for the spin can be $|\uparrow\rangle$, indicated by the apparatus

general theory of [8].

[5] This answers the "which basis" question frequently occurring in the measurement literature based on postulates rather than models, in particular in those approaches that "do away with measurements".

[6] The combination of a neutron beam polarised in the $+z$ direction with an equal one polarised in the $-z$ direction leads to an unpolarised beam. Now consider the case that the beams have not been mixed *physically*, but only *mathematically*, and that experiments are performed randomly on either beam. An observer aware of the non-mixing would still describe them as two separate beams, with measurements giving $s_z = +1$ and -1 in the way he expects; an observer unaware of the non-mixing would describe the situation by an unpolarised beam. Both observers would give a consistent and testable description of the physical situation.

being in a thermodynamic state with positive magnetisation, described by the mixed[7] density matrix $\mathcal{R}_{\Uparrow}$, or the spin can be or $|\downarrow\rangle$, with the magnetisation in the state $\mathcal{R}_{\Downarrow}$, with the Born probabilities $p_{\uparrow}$ or $p_{\downarrow}$, respectively,

$$\mathcal{D}(t_f) = p_{\uparrow}|\uparrow\rangle\langle\uparrow|\mathcal{R}_{\Uparrow} + p_{\downarrow}|\downarrow\rangle\langle\downarrow|\mathcal{R}_{\Downarrow}. \tag{1}$$

This is indeed the result in the Curie-Weiss model after the truncation and registration phases [9, 2]. (The off-diagonal terms $\uparrow\rangle\langle\downarrow|$ and $\downarrow\rangle\langle\uparrow|$ are formally also present, but their apparatus density operators do not contribute to physical observables, since they are sums over many phase factors.) It is tempting, common practice and even physically expected, to consider each of the two terms of (1) as describing a subensemble, one consisting of events with spin up and pointer variable up, the other one with spin and pointer down. However, due to the above quantum oddity, a mixed density matrix such as (1) allows a mathematical decomposition in any basis, typically having components both on $|\uparrow\rangle$ and $|\downarrow\rangle$. May one thus neglect the multitude of other decompositions? Till recently, there has been no compelling argument to do this, and hence to explain that (1) has the desired physical interpretation. This conundrum is called the *measurement problem*, as formulated by Laloë [10] [8].

Considered by some as the deepest problem of quantum mechanics, the impact of the "unsolvable" measurement problem has been profound. It has led to many attempts to find peace with the situation, of which we mention: the Copenhagen interpretation, the relative state interpretation [11], the many worlds interpretation [12], the mind-body issue [13], the modal interpretation [14, 15], the Bohmian interpretation [16], the consistent histories interpretation [17], the real ensemble[9] interpretation [18] presented at the 2011 EmQM-1 meeting in Vienna, or the road to extensions of QM, notably spontaneous collapse models [19, 20]. Einstein supported the statistical interpretation, although he never specified its precise meaning. [21]

2.3. Answer to the measurement problem: a physical way out of a mathematical embarrassment

A resolution based on a new dynamical effect was offered recently within standard quantum statistical mechanics [2, 8]: Near the end of the measurement, realistic, weak dynamical effects inside the apparatus, that do not change the already attained final value of the pointer, make most decompositions of the density matrix *unstable*. Only the decomposition on the measurement basis is dynamically stable, so this is identified as the physical basis towards which the many possible decompositions all relax. This *relaxation* thus indeed allows to interpret each of the terms of (1) as describing a physical subensemble. Each of the two thus obtained subensembles being *pure* as regarding the tested system, implies that all its members will yield the same measurement outcome. It is via this pure-state connection that we are allowed to make a transition from an *ensemble description* to *individual systems* and *individual measurements*.

Arbitrary subensembles can then be decomposed on this basis, having the form (1) with unknown coefficients $q_{\uparrow}$, $q_{\downarrow}$ for each subensemble, constrained only by their sum. [8] The obtained structure connects to ordinary probabilities in the frequency interpretation. [22]

2.4. Lessons worth taking

Looking back we may see which ingredients were needed to reach the answer. First, it should not be hard to admit that a quantum measurement involves the interaction between two

[7] It is unrealistic to assume that a macroscopic apparatus can be in a pure state, because then a macroscopic number of data should be known; still, this is often assumed in measurement theory.

[8] Of course, the situation is not so severe that experimentalists are hampered by it. They have long considered the problem as irrelevant in practice.

[9] In plain terms, one may say that this idea puts forward that when a certain measurement is done, an ensemble of "little green men" at remote locations in the Universe is doing the same measurement, and employing faster-than-light exchange, they together establish the Born probabilities.

quantum systems, the tested system and the apparatus. Measurements are thus a special type of interacting quantum systems.[10] The apparatus must have some classical features to produce a macroscopic pointer variable, but it is intrinsically quantum as well.[11] Realistic modelling of the measurements should acknowledge that: approaches that try to do away with measurements or apparatuses look devoid of a proper physical setting. We hold the opinion that the elimination of the apparatus in the theory of measurements is no less serious than its elimination in the experiment! [2]

Next, as is fitting with quantum solid state theory, one adopts the *statistical* or *ensemble interpretation* of QM [6]. A quantum state is characterised by a density matrix, which in the limiting case of a pure state is described by a wave function. Either case, it describes *our knowledge* about an ensemble of identically prepared physical systems, [2, 23, 24] so pure states have the same meaning as mixed states, they describe also ensembles, be it purified ones. For the Large Hadron Collider the ensemble could be taken as all collisions within one year of data taking. A quantum measurement describes an ensemble of physical measurements on the elements of the ensemble of systems.

The statistical interpretation must have been very appealing to the founding fathers, it was mentioned in a book by Kemble in 1937 [25]. Indeed, it is in spirit close to statistical mechanics, which was well understood when QM was developed. Still, subtleties related to the understanding of measurements led the founding fathers to more elaborate constructs, unnecessary as we now have pointed out. My late teacher Nico van Kampen also strongly supported the statistical interpretation. [23][12] Arguments against the statistical interpretation based on Bell's theorem [21] will be refuted in section 3.2, where we discuss the contextually loophole.

An essential condition is that the apparatus starts in a *metastable state*, a realisation of the abstract "ready state" often encountered in measurement literature, from which the measurement drives the apparatus into one of the stable states, a transition indicated by a change in the pointer. Hence the small quantum signal is amplified by the phase transition, as happens visibly, e.g., in bubble and droplet chambers. After the measurement the apparatus has to be reset in the metastable state. Mixed initial states avoid the *unitarity paradox*.[13,14]

3. Elements of a subquantum theory

We have thus reached the conclusion that the minimal interpretation of QM is the statistical or ensemble interpretation. The question that immediately comes to the mind is: the ensemble of what? The natural answer is of course: of physical systems. Hence atoms exist, protons and electrons exist, matter exists, and QM makes statistical statements about measurements on ensembles thereof.

Next question is then: what does an electron look like? One line of research connects it to Kerr-Newman black holes in general relativity. An intriguing number of phenomena occurring in this classical theory corresponds to notions known from Dirac theory. [26]

But first comes the question: what causes quantum behaviour of these quasi-classical objects?

[10] A broken apparatus may still undergo a similar interaction with the tested system, but it would no longer act as a useful measurement device.

[11] The old idea that the apparatus is classical focusses too much on its pointer variable; apparatuses are made of atoms and hence intrinsically quantum.

[12] Van Kampen's Theorem IV reads: *Whoever endows ψ with more meaning than is needed for computing observable phenomena is responsible for the consequences.*

[13] The unitarity paradox shows up when the tested system is in a pure state and one (unphysically) assumes the apparatus to be in a pure state as well. Then the total state is pure, which under the unitary quantum dynamics is incompatible with the final state being mixed.

[14] The unitary paradox for a macroscopic black hole is a related misconception, avoidable by realising that we can never have enough information to describe the black hole by a pure state.

3.1. On the quantum vacuum

You were within
and I sought you outside
St. Augustine

Is is known from experiments that the Casimir effect is a real physical effect. There exists a simple analog in ordinary life: Boats in harbours attract each other, they end up lying in pairs, triples and so on. This happens because fewer water surface waves fit between them than on the outside, causing an effective attraction. Likewise, if we suppose that the quantum vacuum is a real physical vacuum, zero point fluctuations are due to real fields, which induce forces on particles, and may, in principle, explain quantum behaviour from classical mechanics. As a task, this is pretty risky: once the quantum vacuum has been defined, there is a theory with no adjustable parameters, which has to explain the very many known quantum effects established already. But if Nature has this structure, things will just work out, one by one.

While the quantum vacuum would not exist inside its excitations, "point" particles, the solitonic nature we presume particles have, would imply that they have a finite size, inside which the vacuum fields would likely also pervade.

Interestingly, quantum-like behaviour is found in the motion of oil drops, which, when tapped, collide with surface waves created earlier. [28, 29] This *proves the possibility* that true quantum behaviour originates from classical stochastic forces. On another track, De Raedt and Michielsen have designed learning algorithms for event-by-event simulation, which reproduce the quantum predictions for many photon and neutron experiments without any quantum input. [30] Both these subjects motivate to continue the search for a subquantum mechanics.

Concerning the vacuum, a further question is: Up to which energy is it filled? A physical picture would be that the vacuum fields get created soon *after*, but not at, the beginning of the Universe. The maximal filling energy (cutoff energy) would typically be below Planck energy, hence there would be quantum behavior up to that energy scale, but not at the Planck scale. Vacuum energy (and pressure) are not expected to come "out of nothing", but be borrowed from (compensated by) gravitation. The cosmological constant could then be protected by energy conservation; still fine tuning seems needed get it to today's small value. [31, 1]

If the quantum vacuum is not filled up to the Planck energy, quantum gravity seems useless, and string theory can only be an effective theory.[15]

3.2. The need for a subquantum theory

A friend in need is a friend indeed
Proverb

When in a laboratory a quantum measurement is performed, we can trust, after nearly a century of tests, that the statistics of the outcomes is given by Born's rule. But at the end of one specific measurement we can ask ourselves another question: what went on in the apparatus to produce this outcome? And a theorist may ask: with which theory should it be described? The answer is embarrassing: *we have no theory to describe individual events.*[16] But Nature

[15] String Theory is a framework that may apply to a diversity of problems. Having assumed a large set of (super)symmetries, it possesses properties that bewilder its practitioners. However, it does not reproduce the Standard Model of particle physics. Even if it achieves to do so, there will remain the question: how to explain this set of fine tuned parameters? Multiverse arguments can be regarded as equivalent to evading an answer. In this perspective, string theory cannot be a fundamental theory of Nature.

[16] The approach of [2, 8] answers the measurement problem and thus allows to make the statement that an individual measurement yields one of the quantum outcomes. One may then hold the opinion that Nature is fundamentally random, and that no more than QM can be known about it. We do not support this position, but expect that the quantum randomness can be traced back to a stochastic vacuum field.

employs it in every measurement, so such a theory must exist. It should be less statistical than QM, but reproduce QM at its statistical level. Very likely it is a hidden variables theory.

Contrary to the current general opinion, hidden variables theories are not ruled out by Bell's theorem, since it suffers from the *contextuality loophole*, which can not be closed [27]. Indeed, detectors also have hidden variables, so the results from different measurement setups can not be combined in a hidden-variables-theory-for-all-detector-positions, so that one can *not derive* the Bell inequality. Hence its violations have no physical implications: absence-of-local-realism arguments based on Bell's theorem should better retire. The same holds for Bell-type arguments against the statistical interpretation.

The sought subquantum theory should in any case work for measurement apparatuses, that is, in the GeV scale for the nuclei, the MeV scale for the electrons, the eV energy scale for the atoms and in the meV scale for their functioning. For the GeV and MeV scales fluctuating strong and weak vacuum fields are expected too. The Planck scale needs not be involved and space-time can just be the good old Minkowski space-time: no extra dimensions, no curvature would be involved at its first level of description. On cosmological scales curvature would enter, though, by the fact that matter attracts other matter and light.

3.3. The classical electron's near field

It has been proposed that particles like an electron or a positron are solitons (solitonic field configurations), affected by vacuum fluctuations that act on them as stochastic forces. [32, 33, 1] Hence a *Stochastic Soliton Mechanics* might underly quantum mechanics [32], being a less statistical theory, in the way that a given realisation of the Langevin equation is less statistical than the Fokker-Planck equation.

The electromagnetic field of a classical point particle (modelling the exterior of a proton, neutron, electron, positron or an ion) with charge $q = Ze$, $Z = 0, \pm 1, \pm 2, \cdots$, and magnetic dipole moment $\mathbf{m}$, moving with non-relativistic speed $\mathbf{v}$, reads

$$\mathbf{E} = \frac{q}{4\pi\epsilon_0}\frac{\hat{\mathbf{r}}}{r^2} + \frac{q}{4\pi\epsilon_0 c^2}\frac{\hat{\mathbf{r}}\times(\hat{\mathbf{r}}\times\dot{\mathbf{v}})}{r}, \quad \mathbf{B} = \frac{\mu_0}{4\pi}\frac{3\hat{\mathbf{r}}(\hat{\mathbf{r}}\cdot\mathbf{m})-\mathbf{m}}{r^3} + \frac{\mu_0 q}{4\pi}\frac{\mathbf{v}\times\hat{\mathbf{r}}}{r^2} - \frac{\mu_0 q}{4\pi c}\frac{\hat{\mathbf{r}}\times\dot{\mathbf{v}}}{r}, \tag{2}$$

where the first terms are the charge, magnetic dipole and Lorentz terms, and the last terms the radiation fields. Clearly, the charge and magnetic dipole terms act as a polarisation of the vacuum, which, as for polarons, travels with the electron.[17] Since $\mathbf{m} = g(q/2m)\mathbf{S}$, the magnetic dipole field carries, in cases with $|g| = \mathcal{O}(1)$, information about the particle spin $\mathbf{S}$. From their ubiquitous role in quantum theory, one expects a major role for this magnetic dipole term in the cause of spin-spin interactions. It is thus conceivable that the magnetic dipole moment plays a role in the EPR paradox, when a pair of electrons or an electron-positron pair separates, with each partner carrying the information about its spin with it in its surrounding $\mathbf{B}$ field-cloud. Hence no "action-at-a-distance" is needed to explain a correlation between spins of the partners, it will be created at the source and just be conserved in the motion of the members of the pair.

The question why neutrons do interact with normal matter, while neutrinos hardly do, gets a simple answer as well. First, from the Compton radius $\hbar/mc = 0.21$ fm for neutrons while equal to 0.13 μ for the case that the neutrino has a mass of 1.5 eV [34], one might be tempted to expect that the neutrinos interact much stronger than neutrons. This is obviously not the case, so the Compton size is not the important aspect. In contrast, even though both have spin $S = \frac{1}{2}$, the absence of a neutrino magnetic moment (it is often estimated to be $10^{-19}\mu_B$), directly explains their inertness in matter compared to neutrons with their $g_n = -3.8$. [18]

[17] Since the classical electron radius ~ 1 fm is very small, it likely plays no role in atoms and molecules.
[18] Neutrons have $g_n = -3.82608545 \pm 0.00000090$, while for protons $g_p = 5.585694713 \pm 0.000000046$.

4. Stochastic electrodynamics and the hydrogen problem

Stochastic Electrodynamics (SED) is a theory that views the world as classical, with a real vacuum field composed of fluctuating electrodynamics fields. Each mode of the vacuum with frequency $\omega = ck$ is supposed to have the zero-point Planck energy $\frac{1}{2}\hbar\omega$, implying a spectrum

$$\rho_{ZP}(\omega)\mathrm{d}\omega = 2\frac{4\pi k^2\mathrm{d}k}{(2\pi)^3}\frac{1}{2}\hbar\omega = \frac{\hbar\omega^3}{2\pi^2c^3}\mathrm{d}\omega, \tag{3}$$

where the factor 2 arises from the polarisation modes. This ω^3 spectrum is expected to hold up to a very large cutoff $\omega_c = E_c/\hbar$.

The theory is known to work well for harmonic oscillator problems, while generally suspected to fail beyond. However, a general structure for 'nonlinear' problems has been put forward in the books by de la Peña, Cetto and Valdés. [35, 36]

Let us thus consider the most known, nonlinear problem: the hydrogen atom in its ground state. Within SED this is the motion of a classical point charge $-e$ in the field of a central charge e. The Newton equation with damping and noise, also called Abraham-Lorentz equation or Brafford-Marshall equation, reads

$$m\ddot{\mathbf{r}} = -\frac{e^2\mathbf{r}}{4\pi\epsilon_0 r^3} + \frac{e^2}{6\pi\epsilon_0 c^3}\dddot{\mathbf{r}} - e\mathbf{E}(\mathbf{r}, t). \tag{4}$$

The stochastic fields are described by a vector potential

$$\mathbf{A} = \sum_{n_1n_2n_3\lambda}\sqrt{\frac{\mu_0\mathcal{E}_\mathbf{n}}{V}}\hat{\varepsilon}_{\mathbf{n}\lambda}[A_{\mathbf{n}\lambda}\sin(\mathbf{k_n}\cdot\mathbf{r} - \omega_\mathbf{n}t) + B_{\mathbf{n}\lambda}\cos(\mathbf{k_n}\cdot\mathbf{r} - \omega_\mathbf{n}t)], \tag{5}$$

where the $\hat{\varepsilon}_{\mathbf{n}\lambda}$ with $\lambda = 1, 2$ are polarisation vectors, the $A_{\mathbf{n}\lambda}$ and $B_{\mathbf{n}\lambda}$ are independent Gaussian random amplitudes with average 0 and variance 1 of the plane waves labeled by $\mathbf{k_n} = 2\pi\mathbf{n}/V^{1/3}$, with $\mathbf{n} = (n_1, n_2, n_3)$ having integer components. This determines the electric field $\mathbf{E} = -\dot{\mathbf{A}}$ and magnetic field $\mathbf{B} = \nabla \times \mathbf{A}$, which together lead to an energy $\frac{1}{2}\hbar\omega$ per mode in the volume V. With Bohr frequency ω_0 defined by twice the Rydberg energy, $\hbar\omega_0 \equiv 2R_y = \alpha^2mc^2$, where $\alpha = e^2/4\pi\epsilon_0\hbar c \approx 1/137$ is the fine structure constant, and the Bohr radius $a_0 = \hbar/\alpha mc$, it is seen that for frequencies $\omega \sim \omega_0$, the wavelength is of order a_0/α, implying that the field is uniform over the extension of the atom. Hence one can adopt the dipole approximation $\mathbf{k}\cdot\mathbf{r} \to 0$ in the $\mathbf{E}$ and $\mathbf{B}$ fields, producing

$$\mathbf{E} = \sum_{n_1n_2n_3\lambda}\sqrt{\frac{\mathcal{E}_\mathbf{n}}{\epsilon_0 V}}\hat{\varepsilon}_{\mathbf{n}\lambda}[A_{\mathbf{n}\lambda}\cos\omega_\mathbf{n}t + B_{\mathbf{n}\lambda}\sin\omega_\mathbf{n}t], \tag{6}$$

and a similar expression for $\mathbf{B}$. To leading order, the latter is not needed, since in (4) we could omit the Lorentz force $-e\,\mathbf{v}\times\mathbf{B}$, since it is smaller than the electric force $-e\mathbf{E}$ by a factor $v/c \sim \omega_0 a_0/c = \alpha$, and we shall only be interested in the leading order in α.

With the equation of motion (4) for the electron and its stochastic field fixed statistcially, the first question is whether it will produce a stable ground state with the statistical properties known from quantum mechanics. In absence of direct analytical methods, this is a problem for simulations. An encouraging answer was given in 2003 by Cole and Zou [37]. Further investigations are being carried out at the University of Amsterdam [38].

5. A subquantum arrow of time

Time will consume all things including itself
Indian proverb

We now have finished the preparations to arrive at the main theme of this work. Though motivated by SED we now focus on a broader class of similar subquantum theories, like [39, 40].

If the equation of motion (4) has a stationary distribution, it will be a state with energy output by the radiation term and energy input from the average behaviour of the stochastic electric field. Hence there is a current of energy involved, an *energy throughput*, which means that there is an arrow of time. We call this the *subquantum arrow of time*. It goes hand-in-hand with the stability of atoms and matter.

This subquantum arrow of time is more fundamental than the entropic arrow of time, since it holds already for each of the involved atoms. It is likely more fundamental than the cosmological arrow of time, since one expects that the expansion of the Universe has only minuscule impact on laboratory physics. The quantum arrow of time related to wave function collapse in a measurement is related to irreversibility in the measurement apparatus [2], so it is akin to the thermodynamic arrow of time and also less fundamental than the subquantum arrow of time.

6. Summary

From the analysis of ideal measurements, we conclude that QM itself describes the statistics of measurement outcomes. The minimal interpretation is the statistical one; other interpretations involve additional assumptions of one type or another, which can be omitted by Occam's razor. To understand ideal measurements *no measurement postulates are needed* and *no extension of QM is needed*, which is a very satisfactory solution to this century-long problem. There is *no role for the observer* (which can be an automated routine such as used at the LHC to pre-select desired events), except that observation allows to update the knowledge about the ensemble ("wave function collapse"). The occurrence of *individual events* and their classical probability structure is *described by quantum theory* as well, due to a new relaxation mechanism which takes place inside the apparatus near the end of the measurement.

We advocate the statistical formulation for the teaching of quantum theory, since it fully explains ideal measurements. The concept of a quantum state is simple to grasp by being in spirit close to states in classical statistical physics. States described by wave functions should be regarded only as special cases, since pure states as well as mixed states describe (sub)ensembles.

Non-intuitive features of quantum mechanics remain concentrated in the non-commutation of the observables representing the physical quantities. The situation may be further specified in a subquantum mechanics, and in particular in Stochastic Electrodynamics. In such theories the atomic stability will be induced by a classical vacuum, which pervades all space, inside and outside atoms. It has probably been created very early in the Universe, but possibly well after the Planck time. This vacuum enables the stability of matter by a throughput of energy, hence it carries a "subquantum" arrow of time, combining the stability of matter with the forward direction of time. The questions "why does matter exist?" and "why is time increasing?" may thus appear to be intimately connected.

References

[1] Th M Nieuwenhuizen, 2012 J. Phys.: Conf. Ser. 361 012036
[2] A E Allahverdyan, R Balian and Th M Nieuwenhuizen, 2013 Phys. Rep. 525, 1
[3] G 't Hooft, 2012 J. Phys.: Conf. Ser. 361 012024
[4] M Schlosshauer, 2004 Rev. Mod. Phys. 76, 1267
[5] W M de Muynck, 2002 *Foundations of quantum mechanics, an empiricist approach*, (Kluwer, Dordrecht)
[6] R Balian, 2007 *From microphysics to macrophysics: Methods and applications of statistical physics, I, II*, (Springer, Berlin)

[7] B G Englert, 2013 Eur. Phys. J. D 67 238
[8] A E Allahverdyan, R Balian and Th M Nieuwenhuizen, 2013 arXiv:1303.7257
[9] A E Allahverdyan, R Balian and Th M Nieuwenhuizen, 2003 Europhys. Lett. **61**, 453
[10] F Laloë, 2012 *Do we really understand quantum mechanics?*, (Cambridge University Press, Cambdrige UK)
[11] H Everett, 1957 Rev. Mod. Phys. 29, 454
[12] B S M De Witt, 1970 Physics Today 23, 30-35
[13] E P Wigner, 1963 American Journal of Physics 31, 6
[14] B C van Fraassen, 1974 Synthese 29, 291
[15] D Dieks and P Vermaas (eds), 1988 *The Modal Interpretation of Quantum Mechanics*, (Kluwer, Dordrecht)
[16] D Bohm, 1952 Physical Review 85: 166-193
[17] R B Griffiths, 2002 *Consistent Quantum Theory* (Cambridge University Press)
[18] L Smolin, 2011 arXiv:1104.2822
[19] G C Ghirardi, A Rimini and T Weber, 1986 Phys. Rev. D 34, 470
[20] G C Ghirardi P Pearle and A Rimini, 1990 Phys. Rev. A 42, 78
[21] A Fine, 1996 *The shaky game*, (Univ. of Chicago Press, Chicago, 2nd edition)
[22] R von Mises, 1957 *Probability, Statistics and Truth* (Macmillan, London)
[23] N G van Kampen, 1988 Physica A, 153, 97
[24] C A Fuchs and A Peres, 2000 Physics Today, 53, 70
[25] E C Kemble, 1958 and 2005 *The Fundamental Principles of Quantum Mechanics with Elementary Applications* (McGraw Hill, 1937) (Dover)
[26] A Burinskiy, 1999 Class. Quant. Grav. 16, 3497; 2004 Grav. Cosmol. 10, 50
[27] Th M Nieuwenhuizen, 2010 Found. Phys. 41, 580
[28] Y Couder and E Fort, 2006 Phys. Rev. Lett. 97 154101; Y Couder and E Fort, 2012 J. Phys.: Conf. Ser. 361 012001
[29] R Brady and R Anderson, 2014 arXiv:1401.4356v1
[30] H De Raedt, S Zhao, S Yuan, F Jin, K Michielsen and S Miyashita, 2010 Physica E 42, 298
[31] Th M Nieuwenhuizen, 2011 Phys. Rev. D 84, 024038
[32] Th M Nieuwenhuizen, 2006 AIP Conf Proc 810, *Quantum Theory: Reconsideration of Foundations – 3*, G Adenier, A Yu Khrennikov and Th M Nieuwenhuizen, eds (AIP, Melville, NY), pp 198
[33] Th M Nieuwenhuizen, 2007 in *Beyond the Quantum*, T M Nieuwenhuizen, V Spicka, B Mehmani, M J Aghdami and A Yu Khrennikov, eds (World Scientic), pp 332-342
[34] Th M Nieuwenhuizen, Europhys. Lett. 2009 86, 59001
[35] L de la Peña and A M Cetto, 1996 *The Quantum Dice: An Introduction to Stochastic Electrodynamics*, (Kluwer, Dordrecht)
[36] L de la Peña, A M Cetto and A Valdés Hernández, *The Emerging Quantum: The physics behind quantum mechanics*, (Springer, Berlin, in press)
[37] D C Cole and Y Zou, 2003 Physics Letters A 317, 14
[38] Th M Nieuwenhuizen, M T P Liska and E F G van Heusden, 2014 in preparation
[39] A Khrennikov, B Nilsson and S Nordebo, 2012 J. Phys.: Conf. Ser. 361 012030
[40] G Grössing, S Fussy, J Mesa Pascasio and H Schwabl, 2012 J. Phys.: Conf. Ser. 361 012008

Entropic Dynamics: an inference approach to quantum theory, time and measurement

Ariel Caticha

Department of Physics, University at Albany-SUNY, Albany, NY 12222, USA.

E-mail: ariel@albany.edu

Abstract. We review the derivation of quantum theory as an application of entropic methods of inference. The new contribution in this paper is a streamlined derivation of the Schrödinger equation based on a different choice of microstates and constraints.

1. An overview

Quantum mechanics involves probabilities in a fundamental way and, therefore, it is a theory of inference. But this has not always been clear. The controversy revolves around the interpretation of the quantum state — the wave function. Does it represent the actual real state of the system — its *ontic* state — or does it represent a state of knowledge about the system — an *epistemic* state?

Examples of ontic interpretations include, to name a few, Bohm's causal interpretation, Everett's many-worlds interpretation, and Nelson's stochastic mechanics [1]. The epistemic interpretations have also had a number of advocates (for example, [2]-[5]) starting, most prominently, with Einstein. The "orthodox" or Copenhagen interpretation lies somewhere in between. On one hand, as described in standard textbooks such as the early classics by Dirac and von Neumann, it regards the quantum state as a complete and objective specification of the properties of the system — an ontic concept that is divorced from the state of knowledge of any rational agent. On the other hand, the founders of quantum theory — Bohr, Heisenberg, Born, etc. — were keenly aware of the epistemological and pragmatic elements in quantum mechanics (see e.g., [6]) but, unfortunately, they wrote at a time when the language, the tools and the rules of quantitative epistemology — the Bayesian and entropic methods of inference — had not yet been sufficiently developed. As a result they did not succeed in drawing a sharp line between the ontic and the epistemic and thereby started a controversy that lingers to this day.

But interpreting quantum theory is not merely a matter of postulating the mathematical formalism and then appending an interpretation to it. For the epistemic view of quantum states to be satisfactory it is not sufficient to state that the probability $|\psi|^2$ represents a state of knowledge; we must also show that changes or updates of the epistemic ψ — which include both the unitary time evolution according to the Schrödinger equation and the projection postulate during measurement — obey the rules of inference. In a truly epistemic interpretation there is no logical room for "quantum" probabilities obeying alternative rules of inference.

Our subject is Entropic Dynamics (ED), a framework in which quantum theory is formulated as an example of entropic inference [7]. ED differs from other approaches in several important respects. For example, in the standard view quantum theory is considered as an extension of classical mechanics and therefore deviations from causality demand an explanation. In the entropic view, on the other hand, quantum mechanics is an example of entropic inference, a scheme designed to handle insufficient information [8]. From the entropic perspective indeterminism requires no explanation: uncertainty and probabilities are the norm. It is certainty and determinism that demand explanations.

ED also differs from other approaches based on information theory. (See *e.g.*, [9]-[15].) In ED the laws of physics are rules for processing information. The information in question possibly originates and might even find its ultimate justification in some sub-quantum dynamics that remains to be discovered. However, as we shall see, once the relevant information has been identified, the remaining details of any such underlying dynamics turn out to be irrelevant for behavior at the quantum level. In ED those irrelevant details are ignored from the start — which is a significant simplification. The situation is somewhat analogous to the laws of thermodynamics which also turns out to be largely independent of microscopic details at the atomic level.

The analogy with thermodynamics has inspired several attempts to explain the emergence of quantum behavior from specific proposals of a sub-quantum dynamics with some additional stochastic element. (See *e.g.*, [1][16]-[21].) In contrast, ED does not assume any underlying mechanics whether classical, deterministic, or stochastic. Both quantum dynamics and its classical limit are derived as examples of entropic inference.

Another difference is that ED naturally leads to an "entropic" notion of time. Time is introduced as a convenient book-keeping device to keep track of the accumulation of change. The task is to develop a model that includes (a) something one might identify as an "instant", (b) a sense in which these instants can be "ordered", (c) a convenient concept of "duration" measuring the separation between instants. The welcome new feature is that entropic time is intrinsically directional. Thus, an arrow of time is generated automatically.

ED offers a new perspective on the notorious problem of measurement (see [3][22][23]). Questions such as "How can a measurement ever yield a definite outcome?" or "Are the values of observables created during the act of measurement?" led von Neumann to postulate a dual mode of wave function evolution, either continuous and deterministic according to the Schrödinger equation, or discontinuous and stochastic during the measurement process. Once one accepts quantum theory as a theory of inference the dichotomy between the two modes disappears. Unitary evolution and discontinuous collapse correspond to two modes of updating probabilities which, as shown in [24], are not intrinsically different; they are special cases within a broader scheme of entropic inference [8].

Yet another distinguishing feature is that in ED the positions of particles have definite values just as they would in classical physics. This implies that the process of observation is essentially classical and measurements of position automatically yield definite outcomes. This solves the problem of measurement because position is the only observable. Indeed, in ED all other observables such as momentum, energy, and so on, are statistical concepts — just like temperature in statistical mechanics. They are not properties of the particles but of their probability distributions. As shown in [25][26] their values are indeed created by the act of measurement [27]. A more detailed treatment is given in [28] and [8].

In order to formulate quantum theory as an entropic dynamics — just as with any other inference problem — we must decide which microstates are the subject of our inference, we must identify the prior probabilities, and we must identify those constraints that represent the information that is relevant to our problem. The new contribution in this paper is an entropic derivation of the Schrödinger equation based on a choice of microstates and constraints that differs and is in some respects more advantageous than the choice adopted in [7].

Journal of Physics: Conference Series **504** (2014) 012009 doi:10.1088/1742-6596/504/1/012009

2. Entropic Dynamics

In this model we consider particles living in flat three-dimensional space. The particles have *definite* positions and it is their unknown values that we wish to infer.[1]

The basic dynamical assumptions are that motion happens and that it is continuous. Thus, short displacements happen and it is their accumulation that leads to motion. We do not explain why motion happens but, given the information that it does, our task is to venture a guess about what to expect.

For simplicity here we will focus on a single particle; the generalization to N particles is straightforward. For a single particle the configuration space is $\mathbb{R}^3$ with metric δ_{ab}. The particle moves from an initial x to an unknown x'. The goal is to to find the probability distribution $P(x'|x)$. To find it maximize the appropriate (relative) entropy,

$$\mathcal{S}[P, Q] = -\int d^3x' \, P(x'|x) \log \frac{P(x'|x)}{Q(x'|x)} \ . \tag{1}$$

The relevant information is introduced through the prior probability $Q(x'|x)$, which reflects our knowledge about which x' to expect *before* we have any information about the motion, and the constraints that specify the family of acceptable posteriors $P(x'|x)$.

The prior We adopt a prior that represents a state of extreme ignorance. Knowledge of x tells us nothing about x'. Such ignorance is represented by a uniform distribution: $Q(x'|x)d^3x'$ is proportional to the volume element d^3x'. (The proportionality constant has no effect on the entropy maximization and can be safely ignored.)

The constraints on the motion The information that motion is continuous is imposed through a constraint. For a short step let $\vec{x}' = \vec{x} + \Delta\vec{x}$. We require that the expected squared displacement,

$$\langle \Delta\vec{x} \cdot \Delta\vec{x} \rangle = \kappa \ , \tag{2}$$

be some small but for now unspecified value κ, which we take to be independent of x in order to reflect the translational symmetry of the configuration space $\mathbb{R}^3$.

If this were the only constraint the resulting motion would be a completely isotropic diffusion. Clearly some information is still missing. The additional piece of relevant information, that once particles are set in motion they tend to persist in it, is expressed by assuming the existence of a "potential" $\phi(x)$ and imposing that the expected displacement $\langle \Delta x^a \rangle$ in the direction of the gradient of ϕ is constrained to be

$$\langle \Delta\vec{x} \rangle \cdot \vec{\nabla}\phi = \kappa' \tag{3}$$

where κ' is another small but for now unspecified position-independent constant.

The seemingly ad hoc introduction of a potential ϕ will not be justified — at least not here. The important point is that we have identified the information needed for inference. Where this information originates and why it turns out to be relevant are, of course, interesting questions but their answers lie elsewhere — at some deeper level of physics. For the purpose of inference no further hypotheses need be made.[2]

[1] In this work ED is developed as a model for the quantum mechanics of particles. The same framework can be deployed to construct models for the quantum mechanics of fields, in which case it is the fields that are objectively "real" and have well-defined albeit unknown values.

[2] In our earlier development of ED [7] the set of microstates involved the positions x and some additional mysterious variables that we called y. The present treatment is simpler in that no y variables need be postulated. This simplification comes at a price. In the y-variable model both the potential ϕ and the appearance of its gradient arise naturally without further assumptions. Thus we have a trade-off. We can simplify the microstates at the expense of the constraints. A deeper justification for ϕ, its geometric significance, and its natural relation to gauge symmetries can be given and will be discussed elsewhere in the context of particles with spin.

Having specified the prior and the constraints the ME method takes over. Varying $P(x'|x)$ to maximize $\mathcal{S}[P,Q]$ in (1) subject to the two constraints plus normalization gives

$$P(x'|x) = \frac{1}{\zeta}\exp[-\frac{1}{2}\alpha\,\Delta\vec{x}\cdot\Delta\vec{x} + \alpha'\Delta\vec{x}\cdot\vec{\nabla}\phi] \ , \tag{4}$$

where ζ is a normalization constant,

$$\zeta(x,\alpha,\alpha') = \int d^3x'\, e^{-\frac{1}{2}\alpha\,\Delta\vec{x}\cdot\Delta\vec{x}+\alpha'\Delta\vec{x}\cdot\vec{\nabla}\phi} \ . \tag{5}$$

The Lagrange multipliers α and α' are determined in the standard way,

$$\partial\log\zeta/\partial\alpha = -\kappa/2 \quad \text{and} \quad \partial\log\zeta/\partial\alpha' = -\kappa' \ . \tag{6}$$

Since both the function ϕ and the constant κ' are so far unspecified, so is the multiplier α'. Without loss of generality, we can absorb α' into ϕ, $\alpha'\phi \to \phi$, which amounts to setting $\alpha' = 1$.

Eq.(4) for $P(x'|x)$ shows that short steps are obtained for large α and that they happen in essentially random directions with a small anisotropic bias along the gradient of ϕ. The distribution $P(x'|x)$ is Gaussian and is conveniently written as

$$P(x'|x) \propto \exp\left[-\frac{\alpha}{2}\left(\Delta\vec{x} - \langle\Delta\vec{x}\rangle\right)^2\right] \ . \tag{7}$$

The displacement $\Delta\vec{x} = \Delta\bar{\vec{x}} + \Delta\vec{w}$ can be expressed as the expected drift plus a fluctuation

$$\langle\Delta\vec{x}\rangle = \Delta\bar{\vec{x}} = \frac{1}{\alpha}\vec{\nabla}\phi \ , \tag{8}$$

$$\langle\Delta w^a\rangle = 0 \quad \text{and} \quad \left\langle\Delta w^a \Delta w^b\right\rangle = \frac{1}{\alpha}\delta^{ab} \ . \tag{9}$$

As $\alpha \to \infty$ the fluctuations become dominant: the drift $\Delta\bar{\vec{x}} \sim \alpha^{-1}$ while $\Delta\vec{w} \sim \alpha^{-1/2}$. This implies that, as in Brownian motion, the trajectory is continuous but not differentiable. Here we see the roots of the uncertainty principle: a particle has a definite position but its velocity, the tangent to the trajectory, is completely undefined.

3. Entropic time

The foundation of all notions of time is dynamics. In ED time is introduced as a book-keeping device to keep track to the accumulation of small changes.

3.1. An ordered sequence of instants

In ED, at least for infinitesimally short steps, change is given by the transition probability $P(x'|x)$ in eq.(7). The nth step takes us from $x = x_{n-1}$ to $x' = x_n$. Using the product rule for the joint probability, $P(x_n, x_{n-1}) = P(x_n|x_{n-1})P(x_{n-1})$, and integrating over x_{n-1}, we get

$$P(x_n) = \int d^3x_{n-1}\, P(x_n|x_{n-1})P(x_{n-1}) \ . \tag{10}$$

This equation is a direct consequence of the laws of probability. However, if $P(x_{n-1})$ happens to be the probability of different values of x_{n-1} *at a given instant labelled* t, then we will interpret $P(x_n)$ as the probability of values of x_n *at the "later" instant* $t' = t + \Delta t$. Accordingly, we write $P(x_{n-1}) = \rho(x,t)$ and $P(x_n) = \rho(x',t')$ so that

$$\rho(x',t') = \int d^3x\, P(x'|x)\rho(x,t) \tag{11}$$

Nothing in the laws of probability that led to eq.(10) forces this interpretation on us — this is an independent assumption about what constitutes time in our model. We use eq.(11) to define what we mean by an instant: *if the distribution* $\rho(x,t)$ *refers to one instant* t, *then the distribution* $\rho(x',t')$ *defines what we mean by the "next" instant* $t' = t + \Delta t$. Thus, eq.(11) allows *entropic time* to be constructed one instant after another.

We can phrase this idea somewhat differently. Once we have decided on the relevant information necessary for predicting future behavior we can imagine all that information codified into an "instant". Thus, we define instants so that *given the present the future is independent of the past.*[3]

3.2. *The arrow of entropic time*

The notion of time as constructed according to eq.(11) is remarkable in that it incorporates an intrinsic directionality: there is an absolute sense in which $\rho(x,t)$ is prior and $\rho(x',t')$ is posterior.

Suppose we wanted to find a time-reversed evolution. We would write

$$\rho(x,t) = \int d^3x'\, P(x|x')\rho(x',t') \,. \tag{12}$$

This is perfectly legitimate but in order to be correct $P(x|x')$ cannot be obtained from eq.(7) by merely exchanging x and x'. According to the rules of probability theory $P(x|x')$ is related to eq.(7) by Bayes' theorem,

$$P(x|x') = \frac{P(x)}{P(x')} P(x'|x) \,. \tag{13}$$

In other words, one of the two transition probabilities, either $P(x'|x)$ or $P(x|x')$, *but not both*, can be given by the maximum entropy distribution eq.(7). The other is related to it by Bayes' theorem. There is no symmetry between the inferential past and the inferential future because there is no symmetry between priors and posteriors.

The puzzle of the arrow of time has a long history (see *e.g.* [29][30]). The standard question has been how can an arrow of time be derived from underlying laws of nature that are symmetric? ED offers a new perspective. The asymmetry is the inevitable consequence of entropic inference. From the point of view of ED the challenge does not consist in explaining the arrow of time, but rather in explaining how it comes about that despite the arrow of time some laws of physics turn out to be reversible. Indeed, even when the derived laws of physics – in our case, the Schrödinger equation – turns out to be fully time-reversible, *entropic time itself only flows forward.*

3.3. *Duration: a convenient time scale*

Having introduced the notion of successive instants we now have to specify the interval Δt between them. This amounts to specifying the multiplier $\alpha(x,t)$ in terms of Δt.

Time is defined so that motion looks simple. For large α the dynamics is dominated by the fluctuations Δw. In order that the fluctuations $\langle \Delta w^a \Delta w^b \rangle$ reflect the symmetry of translations in space and time — a Newtonian time that flows "equably" everywhere and everywhen — we choose α to be independent of x and t, $\alpha(x,t) = C/\Delta t$, where C is some constant.

The extension of ED to several non-identical particles is not our subject here but a quick remark is useful. The extension is achieved by introducing separate constraints, eq.(2), for each particle, each with its own κ_i, and each with its own multiplier $\alpha_i = C_i/\Delta t$. It is convenient

[3] An equation such as (11) is commonly employed to define Markovian behavior in which case it is known as the Chapman-Kolmogorov equation. Markovian processes are such that specifying the state of the system at time t is sufficient to fully determine its state after time t — no additional information about the past is needed. We make no Markovian assumptions. We are concerned with a different problem. We do not use (11) to define Markovian processes; we use it to define time.

to write each of these multipliers α_i as $\alpha_i = m_i/\hbar\Delta t$ in terms of a particle-specific constant m_i and an overall constant $\hbar$ which fixes the units of the m_is relative to the units of time. Thus

$$\alpha = \frac{m}{\hbar\Delta t} \ . \tag{14}$$

With this choice of the multiplier α the dynamics is indeed simple: $P(x'|x)$ in (7) becomes a standard Wiener process. The displacement is

$$\Delta\vec{x} = \vec{b}\Delta t + \Delta\vec{w} \ . \tag{15}$$

where $b^a(x)$ is the drift velocity,

$$\langle\Delta\vec{x}\rangle = \vec{b}\Delta t \quad \text{with} \quad \vec{b} = \frac{\hbar}{m}\vec{\nabla}\phi \ , \tag{16}$$

and Δw^a is a fluctuation,

$$\langle\Delta w^a\rangle = 0 \quad \text{and} \quad \langle\Delta w^a \Delta w^b\rangle = \frac{\hbar}{m}\Delta t\,\delta^{ab} \ . \tag{17}$$

The formal similarity to Nelson's stochastic mechanics [1] is evident but the interpretations are completely different.

Two remarks are in order: one on the nature of clocks and another on the nature of mass.

On clocks: Time is defined so that motion looks simple. In Newtonian mechanics the prototype of a clock is the free particle. Time is defined so that the free particle moves equal distances in equal times. In ED the prototype of a clock is a free particle too. (For sufficiently short times all particles are free.) And time is defined so that the particle undergoes equal fluctuations in equal times.

On mass: The particle-specific constant m will, of course, be called 'mass' and eq.(17) provides its interpretation: mass is an inverse measure of fluctuations.

4. Accumulating changes: the Fokker-Planck equation

Equation is an integral equation for the evolution of the distribution $\rho(x,t)$. As is well-known from diffusion theory [8] it can be written in differential form as a Fokker-Planck equation (FP),

$$\partial_t\rho = -\vec{\nabla}\cdot(\rho\vec{b}) + \frac{\hbar}{2m}\nabla^2\rho \ , \tag{18}$$

which can itself be rewritten as a continuity equation,

$$\partial_t\rho = -\vec{\nabla}\cdot(\rho\vec{v}) \ . \tag{19}$$

The velocity $\vec{v}$ of the probability flow or *current velocity* is

$$\vec{v} = \vec{b} + \vec{u} \quad \text{where} \quad \vec{u} = -\frac{\hbar}{m}\vec{\nabla}\log\rho^{1/2} \ , \tag{20}$$

the *osmotic velocity*, represents the tendency for probability to flow down the density gradient.

Since both $\vec{b}$ and $\vec{u}$ are gradients, it follows that the current velocity is a gradient too,

$$\vec{v} = \frac{\hbar}{m}\vec{\nabla}\Phi \quad \text{where} \quad \Phi = \phi - \log\rho^{1/2} \ . \tag{21}$$

With these results ED reaches a certain level of completion: We figured out what small changes to expect — they are given by $P(x'|x)$ — and time was introduced to keep track of how these small changes accumulate; the net result is diffusion according to the FP equation.

But quantum mechanics is not a *standard* diffusion. The discussion so far has led us to the density $\rho(x,t)$ as the important dynamical object but to construct a wave function, $\Psi = \rho^{1/2}e^{i\Phi}$, we need a second degree of freedom, the phase Φ. The problem is that as long as the potential ϕ is externally prescribed the function Φ in eq.(21) does not represent an independent degree of freedom. The natural solution is to relax this constraint and allow ϕ (or equivalently Φ) to participate in the dynamics. Thus the dynamics will consist of the coupled evolution of $\rho(x,t)$ and $\Phi(x,t)$.

5. Non-dissipative diffusion

To specify the dynamics we follow [31] and impose that the dynamics be non-dissipative, that is, we require the conservation of a certain functional $E[\rho,\phi]$ which will be called "energy".

At first sight it might appear that imposing that some energy $E[\rho,\phi]$ be conserved is natural because it agrees with our classical preconceptions of what physics ought to be like. But classical intuitions are not a good guide here. In the more sophisticated approaches to physics energy is taken to be whatever happens to be conserved as a result of invariance under translations in time. But our dynamics has hardly been defined yet; what, then, is E and why should it be conserved in the first place? Furthermore, if we go back to eq.(15) we see that it is the kind of equation (a Langevin equation) that characterizes a Brownian motion in the limit of *infinite* friction. Thus, the explanation of quantum theory in terms of a sub-quantum classical mechanics would require that particles be subjected to infinite friction while suffering zero dissipation at the same time. Such a strange sub-quantum mechanics could hardly be called 'classical'.

The energy functional $E[\rho,\phi]$ is chosen to be the expectation of a *local* "energy" function $\varepsilon(x,t)$, that is,

$$E[\rho,\phi] = \int d^3x\, \rho(x,t)\, \varepsilon(x,t) \ , \tag{22}$$

where $\varepsilon(x,t)$ depends on $\rho(x,t)$ and $\phi(x,t)$ and their derivatives.[4] The local energy appropriate to the non-relativistic regime is

$$\varepsilon(x,t) = \frac{1}{2}mv^2 + \frac{1}{2}mu^2 + V(x) \ , \tag{23}$$

where the scalar function $V(x)$ represents an additional "potential" energy. The justification of ε is to be found in deeper-level physics but we can note that ε is tightly constrained by requiring invariance under time reversal ($\vec{v} \to -\vec{v}$ and $\vec{u} \to \vec{u}$) and the low velocity regime [32][7].

Using eqs.(20) and (21) the energy E is

$$E = \int d^3x\, \rho \left(\frac{\hbar^2}{2m}(\vec{\nabla}\Phi)^2 + \frac{\hbar^2}{2m}(\vec{\nabla}\log\rho^{1/2})^2 + V \right) \tag{24}$$

so that, after some algebra [7],

$$\frac{dE}{dt} = \int d^3x\, \dot{\rho} \left(\hbar\dot{\Phi} + \frac{\hbar^2}{2m}(\vec{\nabla}\Phi)^2 + V - \frac{\hbar^2}{2m}\frac{\nabla^2\rho^{1/2}}{\rho^{1/2}} \right) \tag{25}$$

We impose that $\dot{E} = 0$ for spatially arbitrary choices of the initial conditions ρ and Φ, that is, at the initial t_0 we ought to be able to change ρ and Φ independently at different locations and still

[4] In an energy eigenstate the local energy $\varepsilon(x,t)$ is uniform in space and constant in time.

get $\dot{E} = 0$. This implies the integrand should vanish at the initial t_0. But any arbitrary time t can be taken as the initial time for evolution into the future. Therefore for all t we require that

$$\hbar\dot{\Phi} + \frac{\hbar^2}{2m}(\vec{\nabla}\Phi)^2 + V - \frac{\hbar^2}{2m}\frac{\nabla^2\rho^{1/2}}{\rho^{1/2}} = 0 \; , \tag{26}$$

which is the quantum version of the Hamilton-Jacobi equation. Equations (26) and the FP equation,

$$\dot{\rho} = -\vec{\nabla}\cdot(\rho\vec{v}) = -\frac{\hbar}{m}\vec{\nabla}\cdot\left(\rho\vec{\nabla}\Phi\right) \tag{27}$$

are the coupled dynamical equations we seek.

These two real equations can be written as a single complex equation by combining ρ and Φ into a complex function $\Psi = \rho^{1/2}\exp(i\Phi)$. Computing the time derivative $\dot{\Psi}$ and using eqs.(26) and (27) leads to the Schrödinger equation,

$$i\hbar\frac{\partial\Psi}{\partial t} = -\frac{\hbar^2}{2m}\nabla^2\Psi + V\Psi \; . \tag{28}$$

Earlier we had introduced m as a particle-specific constant that measures fluctuations, $\hbar$ as a constant that fixes units, and the entropic time t as a parameter designed to keep track of the accumulation of changes. Their relation to familiar physical quantities was a matter of conjecture. But now that we can see what role they play in the Schrödinger equation we can identify m with the particle mass, $\hbar$ with Planck's constant, and the entropic time t with physical time.[5]

Other attempts to derive quantum theory start from an underlying, perhaps stochastic, classical mechanics. The ED approach is different in that it does not assume an underlying classical substrate; ED provides a derivation of *both Schrödinger's equation and also Newton's* $F = ma$. Classical mechanics is recovered in the usual limits of $\hbar \to 0$ or $m \to \infty$. Indeed, writing $S = \hbar\Phi$ in eq.(26) and letting $m \to \infty$ with S/m fixed leads to the classical Hamilton-Jacobi equation

$$\dot{S} + \frac{1}{2m}(\vec{\nabla}S)^2 + V = 0 \; , \tag{29}$$

while eqs.(17), (20), and (21) give $m\vec{v} = \vec{\nabla}S$ and $\vec{u} = 0$ with vanishing fluctuations $\langle\Delta w^a \Delta w^b\rangle = \frac{\hbar}{m}\Delta t\,\delta^{ab} \to 0$.

6. Measurement in ED

In practice the measurement of position can be technically challenging because it requires the amplification of microscopic details to a macroscopically observable scale. However, no intrinsically quantum effects are involved: the position of a particle has a definite, albeit unknown, value x and its probability distribution is, by construction, given by the Born rule, $\rho(x) = |\Psi(x)|^2$. We can therefore assume that suitable position detectors are in principle available. First we consider observables other than position: how they are defined and how they are measured. Then we conclude with a few remarks on amplification and Bayes theorem.

6.1. Observables other than position

The fact that the Schrödinger equation (28) is linear and unitary makes the language of Hilbert spaces particularly convenient so from now we adopt Dirac's bra-ket notation and write $\Psi(x) = \langle x|\Psi\rangle$. For convenience we consider the case of a particle that lives on a discrete lattice.

[5] Where by 'physical' we mean that it is the time t that appears in the laws of physics.

Journal of Physics: Conference Series **504** (2014) 012009 doi:10.1088/1742-6596/504/1/012009

The generalization to a continuous space is straightforward. The probabilities of the previously continuous positions

$$\rho(x)\,d^3x = |\langle x|\Psi\rangle|^2\,d^3x \quad \text{become} \quad p_i = |\langle x_i|\Psi\rangle|^2 \; , \tag{30}$$

and if the state is

$$|\Psi\rangle = \sum_i c_i|x_i\rangle \quad \text{then} \quad p_i = |\langle x_i|\Psi\rangle|^2 = |c_i|^2 \; . \tag{31}$$

Since position is the only objectively real quantity there is no reason to define other observables except that they turn out to be convenient when considering more complex experiments. Consider a setup in which right before reaching the position detector the particle is subjected to additional interactions, say magnetic fields or diffraction gratings. Suppose the interactions in such a complex setup $\mathcal{A}$ are described by the Schrödinger eq.(28), that is, by a particular unitary evolution $\hat{U}_A$. The particle will be detected with certainty at position $|x_i\rangle$ provided it was initially in a state $|s_i\rangle$ such that

$$\hat{U}_A|s_i\rangle = |x_i\rangle \; . \tag{32}$$

Since the set $\{|x_i\rangle\}$ is orthonormal and complete, the corresponding set $\{|s_i\rangle\}$ is also orthonormal and complete,

$$\langle s_i|s_j\rangle = \delta_{ij} \quad \text{and} \quad \sum_i |s_i\rangle\langle s_i| = \hat{I} \; . \tag{33}$$

Now consider the effect of this setup $\mathcal{A}$ on some generic initial state vector $|\Psi\rangle$ which can always be expanded as

$$|\Psi\rangle = \sum_i c_i|s_i\rangle \; , \tag{34}$$

where $c_i = \langle s_i|\Psi\rangle$ are complex coefficients. The state $|\Psi\rangle$ will evolve according to $\hat{U}_A$ so that as it approaches the position detectors the new state is

$$\hat{U}_A|\Psi\rangle = \sum_i c_i \hat{U}_A|s_i\rangle = \sum_i c_i|x_i\rangle \; . \tag{35}$$

which, invoking the Born rule for position measurements, implies that the probability of finding the particle at the position x_i is

$$p_i = |c_i|^2 = |\langle s_i|\Psi\rangle|^2 \; . \tag{36}$$

Thus, the probability that the particle in the initial state $|\Psi\rangle$ after going through the setup $\mathcal{A}$ is found at position x_i is $|c_i|^2$.

The same experiment can be described from a point of view in which the setup $\mathcal{A}$ is a black box, a complex detector the inner workings of which are not emphasized. The particle is detected at $|x_i\rangle$ *as if* it had earlier been in the state $|s_i\rangle$. We can adopt a new language and say, perhaps inappropriately, that the particle has effectively been "detected" in the state $|s_i\rangle$, and therefore, the probability that the particle in state $|\Psi\rangle$ is "detected" in state $|s_i\rangle$ is $|\langle s_i|\Psi\rangle|^2$ — which reproduces Born's rule for a generic measurement device. The shift in language is not particularly fundamental — it is merely a matter of convenience but we can pursue it further and assert that the setup $\mathcal{A}$ is a complex detector that "measures" all operators of the form

$$\hat{A} = \sum_i \lambda_i |s_i\rangle\langle s_i| \tag{37}$$

where the eigenvalues λ_i are arbitrary scalars.

Some remarks are in order. Note that when we say we have detected the particle at x_i *as if* it had earlier been in state $|s_i\rangle$ with eigenvalue λ_i we are not implying that the particle was in the particular state $|s_i\rangle$ — this is just a figure of speech. It is in this sense that the corresponding value λ_i of the observable $\hat{A}$ has been "created by the act of measurement". To be more explicit:

if a sentence such as "a particle has momentum $\vec{p}$" is used only as a linguistic shortcut that conveys information about the wave function before the particle enters the complex detector then, strictly speaking, there is no such thing as the momentum of the particle. The momentum is not an attribute of the particle; it is an attribute of the epistemic state $\Psi(x)$.

Incidentally, note that it is not necessary that the eigenvalues of the operator $\hat{A}$ be real — they could be complex numbers. What is necessary is that its eigenvectors $|s_i\rangle$ be orthogonal. This means that $\hat{A}$ need not be Hermitian but its Hermitian and anti-Hermitian parts of $\hat{A}$ must be simultaneously diagonalizable — they must commute.

In the standard interpretation of quantum mechanics Born's rule (36) is a postulate; within ED it is the natural consequence of unitary time evolution and the fact that all measurements are ultimately position measurements. This raises the question of whether our scheme is sufficiently general to encompass all measurements of interest. While there is no general answer that will address all cases — who can, after all, even list all the measurements that future physicists might perform? — we can, nevertheless, ask whether our scheme includes a sufficiently large class of interesting measurements. How, for example, does one measure an observable for which there is no unitary transformation mapping its eigenstates to position eigenstates? Every case demands its own specific analysis. For example, how does one measure the energy of a free particle? A measurement device characterized by eigenvectors $\{|s\rangle\}$ measures all operators of the form $\hat{A} = \int ds\, \lambda(s)|s\rangle\langle s|$. Therefore the same device that measures the momentum $\hat{p}$ of a particle (*e.g.*, using a magnetic field or a diffraction grating followed by a position detector such as a photographic plate or a photoelectric cell) can also be used to infer the energy $\hat{H} = \hat{p}^2/2m$ of a free particle.

Here is another example: It is not so easy to place a probe inside the atom, so how does one measure the energy of an electron that is bound to an atom? In practice the energy of the bound particle is not measured directly; instead it is inferred from the energy of photons emitted in transitions between the bound states. Since photons are free particles measuring their energy is not in principle problematic. This is a special case of the general scheme in which the system of interest and the pointer variable of an apparatus become correlated in such a way that observation of the pointer allows one to infer a quantity of the system. The paradigmatic example is a Stern-Gerlach experiment in which the particle's position is the pointer variable that allows one to infer its spin.

The difficulty with the standard von Neumann interpretation is that it is not clear at what stage the pointer variable "collapses" and attains a definite value. This is precisely the difficulty of principle that is resolved in the entropic approach: the pointer variable is a position variable too and therefore always has a definite value.

7. Amplification

The technical problem of amplifying microscopic details so they can become macroscopically observable is usually handled with a detection device set up in an initial state of unstable equilibrium. The particle of interest activates the amplifying system by inducing a cascade reaction that leaves the amplifier in a definite macroscopic final state described by some pointer variable α.

An eigenstate $|s_i\rangle$ evolves to a position x_i and the goal of the amplification process is to infer the value x_i from the observed value α_r of the pointer variable. The design of the device is deemed successful when x_i and α_r are suitably correlated and this information is conveyed through a likelihood function $P(\alpha_r|x_i)$ — an ideal amplification device would be described by $P(\alpha_r|x_i) = \delta_{ri}$. Inferences about x_i follow from a standard application of Bayes' rule,

$$P(x_i|\alpha_r) = P(x_i)\frac{P(\alpha_r|x_i)}{P(\alpha_r)} \ . \tag{38}$$

The point of these considerations is to emphasize that there is nothing intrinsically quantum mechanical about the amplification process. The issue is one of appropriate selection of the information (in this case the data α_r) that happens to be relevant to a certain inference (in this case x_i). This is, of course, a matter of design: a skilled experimentalist will design the device so that no spurious correlations—whether quantum or otherwise—nor any other kind of interfering noise will stand in the way of inferring x_i.

Acknowledgments
I would like to thank D. Bartolomeo, C. Cafaro, N. Caticha, S. DiFranzo, A. Giffin, P. Goyal, D. T. Johnson, K. Knuth, S. Nawaz, M. Reginatto, C. Rodríguez, and J. Skilling for many discussions on entropy, inference and quantum mechanics.

References
[1] E. Nelson, *Quantum Fluctuations* (Princeton UP, Princeton 1985).
[2] L. Ballentine, Rev. Mod. Phys. **42**, 358 (1970).
[3] L. Ballentine, *Quantum Mechanics: A Modern Development* (World Scientific, Singapore 1998).
[4] C. Caves, C. Fuchs, and R. Schack, Studies in History and Philosophy of Modern Physics **38**, 244 (2007).
[5] N. Harrigan and R. Spekkens, Found. Phys. **40**,125 (2010).
[6] H. P. Stapp, Am. J. Phys. **40**, 1098 (1972).
[7] A. Caticha, J. Phys. **A 44**, 225303 (2011) (arXiv.org/abs/1005.2357).
[8] A. Caticha, *Entropic Inference and the Foundations of Physics* (USP Press, São Paulo, Brazil 2012); online at http://www.albany.edu/physics/ACaticha-EIFP-book.pdf.
[9] W. K. Wootters, Phys. Rev. **D**, 357 (1981).
[10] A. Caticha: Phys. Lett. **A244**, 13 (1998); Phys. Rev. **A57**, 1572 (1998); Found. Phys. **30**, 227 (2000).
[11] C. Brukner and A. Zeilinger, "Information and Fundamental Elements of the Structure of Quantum Theory," in *Time, Quantum, Information*, ed. L. Castell and O. Ischebeck (Springer, 2003) (arXiv:quant-ph/0212084).
[12] R. Spekkens, Phys. Rev. **A 75**, 032110 (2007).
[13] P. Goyal, K. Knuth, J. Skilling, Phys. Rev. **A 81**, 022109 (2010).
[14] L. Hardy, "Reformulating and Reconstructing Quantum Theory" (arXiv.org:1104.2066).
[15] M. Reginatto, "From information to quanta: a derivation of the geometric formulation of quantum theory from information geometry" (arxiv:1312.0429).
[16] L. de la Peña and A. M. Cetto, *The Quantum Dice, an Introduction to Stochastic Electrodynamics* (Kluwer, Dordrecht, Holland 1996).
[17] M. J. W. Hall and M. Reginatto, J. Phys. **A 35**, 3289 (2002); Fortschr. Phys. **50**, 646 (2002).
[18] G. 't Hooft, "Determinism beneath Quantum Mechanics" (arxiv:quant-ph/0212095); "Emergent quantum mechanics and emergent symmetries" (arxiv:hep-th/0707.4568).
[19] H. T. Elze and O. Schipper, Phys. Rev. **D66**, 044020 (2002) ; H. T. Elze, Phys. Lett. **A310**, 110 (2003).
[20] S. Adler, *Quantum Theory as an Emergent Phenomenon* (Cambridge UP, Cambridge 2004).
[21] G. Grössing, Phys. Lett. **A 372**, 4556 (2008) (arxiv:0711.4954); G. Grössing *et al*, "The Quantum as an emergent System" J. Phys.: Conf. Ser. **361** 012008 (2012).
[22] M. Schlösshauer, Rev. Mod. Phys. **76**, 1267 (2004).
[23] G. Jaeger, *Entanglement, Information, and the Interpretation of Quantum Mechanics* (Springer-Verlag, Berlin Heidelberg 2009).
[24] A. Caticha and A. Giffin, "Updating Probabilities," in *Bayesian Inference and Maximum Entropy Methods in Science and Engineering*, ed. by A. Mohammad-Djafari, AIP Conf. Proc. **872**, 31 (2006) (arXiv.org:physics/0608185).
[25] A. Caticha, *Found. Phys.* **30**, 227 (2000).
[26] D. T. Johnson and A. Caticha, "Entropic dynamics and the quantum measurement problem", *Bayesian Inference and Maximum Entropy Methods in Science and Engineering*, ed. by K. Knuth *et al.*, AIP Conf. Proc. **1443**, 104 (2012) (arXiv:1108.2550).
[27] S. Nawaz and A. Caticha, "Momentum and uncertainty relations in the entropic approach to quantum theory", *Bayesian Inference and Maximum Entropy Methods in Science and Engineering*, ed. by K. Knuth *et al.*, AIP Conf. Proc. **1443**, 112 (2012) (arXiv:1108.2629).
[28] D. T. Johnson, "Generalized Galilean Transformations and the Measurement Problem in the Entropic Dynamics Approach to Quantum Theory", Ph.D. thesis, University at Albany (2011) (arXiv:1105.1384).
[29] H. Price, *Time's Arrow and Archimedes' Point* (Oxford UP, New York 1996).

Journal of Physics: Conference Series **504** (2014) 012009 doi:10.1088/1742-6596/504/1/012009

[30] H. D. Zeh, *The Physical Basis of the Direction of Time* (Springer, 2002).
[31] E. Nelson, "Connection between Brownian motion and quantum mechanics," in *Einstein Symposium Berlin*, Lect. Notes Phys. **100**, p.168 (Springer-Verlag, Berlin 1979).
[32] L. Smolin, "Could quantum mechanics be an approximation to another theory?" (arXiv.org/abs/quant-ph/0609109).

Journal of Physics: Conference Series **504** (2014) 012010 doi:10.1088/1742-6596/504/1/012010

Spin and charge from space and time

Manfried Faber

Institute for Atomic and Subatomic Physics, Vienna University of Technology, Operngasse 9, A-1040 Wien, Austria

E-mail: faber@kph.tuwien.ac.at

Abstract. We follow the idea that particles are topological solitons, characterised by topological quantum numbers determining charge and spin. To describe such particles and their spin we use the three rotational degrees of freedom of local spatial dreibeins. Orbiting particles contribute by internal rotation to the total angular momentum. The mass of these particles is given by their field energy only.

1. Introduction

A pendulum model as depicted in Fig. 1 obeys a discretised version of the Sine–Gordon model for the rotational angle ϑ, a bosonic field in one time and and one space dimension, 1+1D [1]. Besides wave-like excitations this interesting model has stable localised excitations, kink solutions, solitons, behaving like particles. Their mass is field energy only. The wave-velocity c in the mechanical model is of the order of 1m/s. The Sine-Gordon model is a relativistic model. Therefore, solitons moving with velocity v are Lorentz contracted, see Fig. 1b, and their mass increases with the well–known γ factor, $\gamma = 1/\sqrt{1 - v^2/c^2}$, $m(v) = \gamma m(0)$.

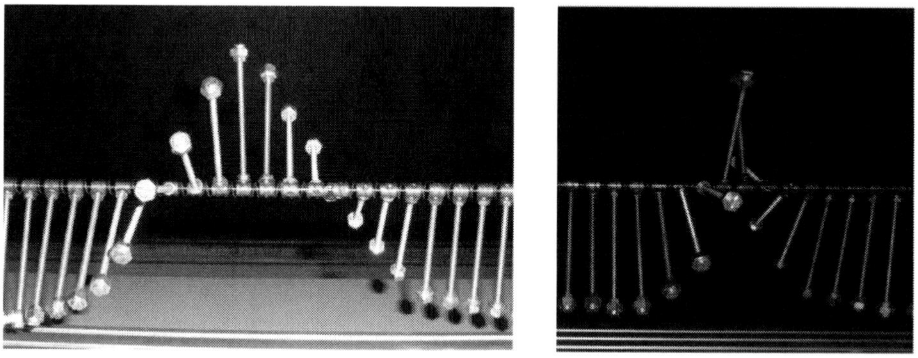

(a) (b)

Figure 1. A pendulum model obeying a discretised version of the Sine–Gordon model. A soliton at rest (a) has larger size than a moving soliton (b) which is Lorentz contracted. Mechanical model by Peter Pataki, photos by Gerald Pechoc.

There are kinks and antikinks, right-handed and left-handed solitons where the bosonic field ϑ varies from 0 to $\pm 2\pi$. Two–soliton configurations are time dependent since equal "charges" repel and opposite charges attract each other. Despite the fact that solitons have a finite size, in collisions they behave point-like. The corresponding time-depend two-soliton solutions are well known [ ]

$$\vartheta(x,t) = 4\,\mathrm{atan}\frac{\beta\sinh(\gamma x)}{\cosh(\beta\gamma t)}. \tag{1}$$

In Fig. 2 such solutions are depicted for some values of $\beta = v/c$, for $t = 0$, the time of shortest approach. Towards $\beta = 1$ the common profile gets steeper and steeper and converges to size zero, a step function.

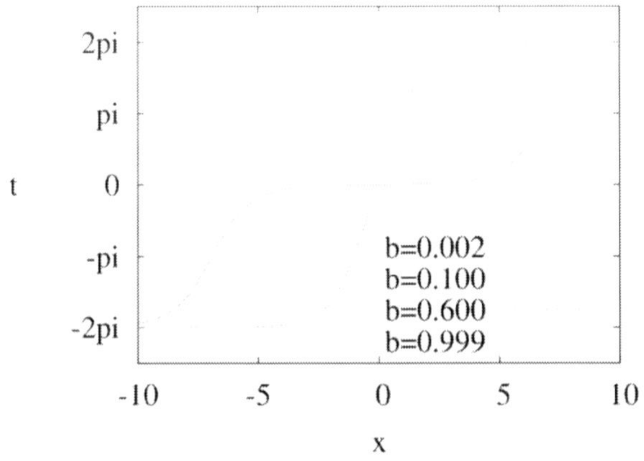

Figure 2. Field of two colliding solitons at $t = 0$, the time of shortest approach. For large values of β the two solitons are not really distinguishable. Approaching $\beta = 1$ the common profile gets arbitrary steep, closer and closer to a step function.

A generalisation of the 1+1D Sine–Gordon model to 3+1D was suggested in ref. [ ], the model of topological fermions (MTF). There the bosonic variable ϑ of the Sine–Gordon model is generalised to three Euler angles, to an SO(3) field. In this article we interpret this field of rotational matrices as orientations of local dreibeins in space which may rotate in time. As in the Sine–Gordon model the MTF has stable topological excitations and mass is field energy. Despite the fact that these solitons are extended, in high energy collisions they are compressed to nearly point-like particles. Charge and spin are topological quantum numbers.

Let us recall the most important properties of charge and spin. Charges feel Coulomb and Lorentz forces. Since Faraday's laws of electrolysis, J. J. Thomsons cathode ray experiments in 1897 and Millikan's oil drop experiments in 1909 we know that charge is quantised in units of the elementary electric charge $e_0 = 1.602 \cdot 10^{-19}$ C. Maxwell's electrodynamics and quantum mechanics give no explanation for this quantisation of electric charge.

There is no classical analogon of spin. From quantum mechanics we know that electrons have two spin-states, usually indicated as spin up and spin down. The spin is described by the SU(2) group, like the isospin. Whereas the isospin is an internal degree of freedom the spin has a relation to our 3–dimensional space: the spin $\mathbf{s}$ contributes like the orbital angular momentum $\mathbf{l}$ to the angular momentum $\mathbf{j}$ of particles, $\mathbf{j} = \mathbf{l} + \mathbf{s}$.

From special relativity we know that there is no absolute space and space and time are related. Gravitation theory teaches us that space–time is not rigid, it is deformable by matter and energy density. The length scales of space–time depend on the position, they are encoded in the metric tensor $g_{\mu\nu}$ in gravitational theory and described by local translations in Poincarè gauge theory.

2. The model

Now we follow the idea that charge is a property of space, a space which is Minkowskian at macroscopic distances and has rotational dislocations at microscopic distances as schematically shown in Fig. 3. We will show that dislocations which are described by a dreibein (triade) rotating in space by 2π with an appropriate Lagrangian is a nice model to describe elementary charges.

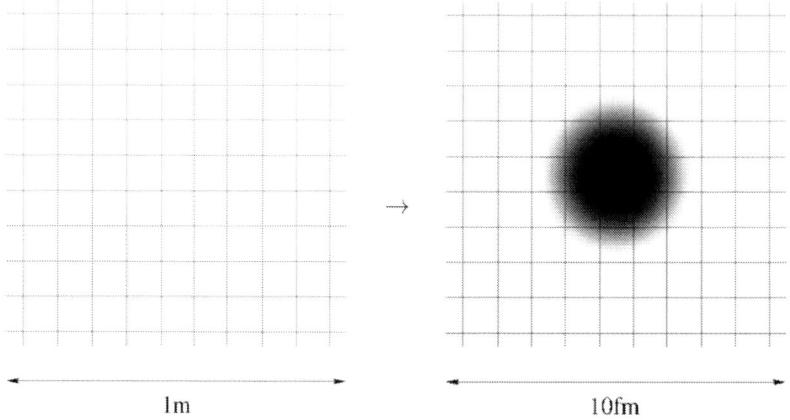

Figure 3. Schematic picture of space which behaves Minkowskian at macroscopic distances and is perturbed by dislocations at microscopic distances.

2.1. Degrees of freedom

The only degrees of freedom of this model are the above mentioned three rotational angles of spatial dreibeins (triades). The earliest description of rotations is due to Rodrigues (1840) by unit quaternions [ ]

$$Q = q_0 + q_1\mathbf{i} + q_2\mathbf{j} + q_3\mathbf{k}: \quad \mathbf{i}^2 = \mathbf{j}^2 = \mathbf{k}^2 = \mathbf{ijk} = -1, \quad q_0^2 + \vec{q}^2 = 1, \ \vec{q} = (q_1, q_2, q_3). \quad (2)$$

and was reinvented by Hamilton (1843). Today it is more common to use the group SO(3) or its double covering group SU(2) which is isomorphic to S^3, the sphere in 4D. We make a transition from the three imaginary quaternionic units to Pauli matrices

$$(\mathbf{i}, \mathbf{j}, \mathbf{k}) = -\mathrm{i}\vec{\sigma}, \quad x^\mu = (ct, \mathbf{r}), \quad q_0 = \cos\alpha, \ \vec{q} = \vec{n}\sin\alpha, \ \vec{n}^2 = 1, \quad (3)$$

and describe the dreibein field by an SU(2) field

$$Q(x) = \cos\alpha(x) - \mathrm{i}\vec{\sigma}\vec{n}(x)\sin\alpha(x) = \mathrm{e}^{-\mathrm{i}\alpha(x)\vec{\sigma}\vec{n}(x)}. \quad (4)$$

Models with such degrees of freedom are usually called non-linear sigma models. We have to keep in mind that due to the double covering, the SO(3) field configuration corresponding to $\pm Q(x)$ are identical.

2.2. Geometry of S^3

By the SU(2) field $Q(x)$, curves in space-time, parametrised by s or t: $x^\mu(s), x^\mu(t)$ are mapped to SU(2) and its parameter space, the sphere S^3, and reflect its geometry. To every point of such curves we can attribute a vector field, the connection one-form,

$$(\partial_s Q)\, Q^\dagger = -\mathrm{i}\vec{\sigma}\vec{\Gamma}_s \quad (5)$$

vectors with values in the tangential space of S^3. With a few lines one can see that in the parametrisation (4) the three-component vector $\vec{\Gamma}_s$ reads

$$\vec{\Gamma}_s = \partial_s\alpha\vec{n} + \sin\alpha\cos\alpha\,\partial_s\vec{n} + \sin^2\alpha\,\vec{n}\times\partial_s\vec{n}. \tag{6}$$

The arrows are used to avoid too many indices and indicate the vector character in the tangential space.

Further we can attribute to every area in space-time, e.g. $\mathrm{d}x\mathrm{d}y$, an area on S^3 and get an area density, a connection two-form or curvature by

$$\vec{R}_{\mu\nu} = \vec{\Gamma}_\mu \times \vec{\Gamma}_\nu, \tag{7}$$

as schematically shown in Fig. 4. Due to the equality of mixed partials, applied to Eq. (5), we

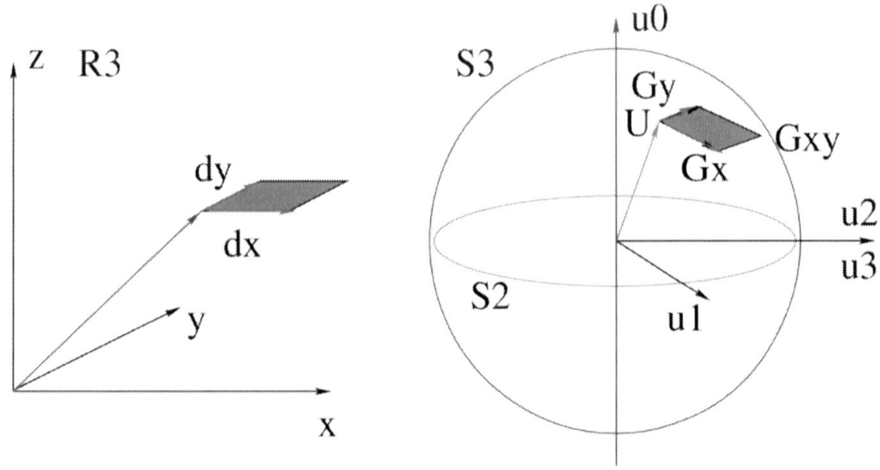

Figure 4. Schematic picture of maps of points, line and area elements from Minkowski space to S^3 : $\mathrm{d}x \wedge \mathrm{d}y \mapsto \vec{\Gamma}_x \times \vec{\Gamma}_y \mathrm{d}x\mathrm{d}y$

get the Maurer-Cartan structural equation

$$\vec{\Gamma}_\mu \times \vec{\Gamma}_\nu = \frac{1}{2}[\partial_\mu\vec{\Gamma}_\nu - \partial_\nu\vec{\Gamma}_\mu] \tag{8}$$

leading to the well-known curvature expression

$$\vec{R}_{\mu\nu} = \partial_\mu\vec{\Gamma}_\nu - \partial_\nu\vec{\Gamma}_\mu - \vec{\Gamma}_\mu \times \vec{\Gamma}_\nu, \tag{9}$$

which is covariant against basis rotations.

2.3. Relate geometry to physics

Defining the field strength tensor we start to relate the geometry to physics. With units appropriate to get Gauß's law for the elementary electric charge e_0 the dual field strength tensor reads

$$^\star\vec{F}_{\mu\nu} = \frac{e_0}{4\pi\epsilon_0}\vec{R}_{\mu\nu}. \tag{10}$$

Electric fields are described by spatial components of the curvature tensor and magnetic fields by space-time components. The arrow in $^\star\vec{F}_{\mu\nu}$ indicates that this model has three contributions for every field strength component. Like in QCD we call these internal components "color". Correspondingly we can introduce a dual vector field in appropriate units

$$\vec{C}_\mu = -\frac{e_0}{4\pi\varepsilon_0 c}\vec{\Gamma}_\mu. \tag{11}$$

2.4. Lagrangian

In relativistic field theories the most effective method to define the dynamics is by defining a Lagrangian. Our aim is to describe charges and their interaction with the electro-magnetic field. There is a most successful theory for this aim and this is Maxwell's electrodynamics. But it describes charges and their fields by different quantities, by fermion fields $\psi(x)$ and by gauge fields $A^\mu(x)$. From the successes of the Sine-Gordon model we get the idea to describe charges and their fields on the same footing, by a single field $Q(x)$. We aim to investigate how far we can get with this idea. We expect a non-linear theory. Maxwell's electrodynamics should be a very clever linearisation of this model by the above mentioned separation of charges and field.

From the success of gauge field theories we expect a Lagrangian quadratic in the field strength tensor. From the comparison with the Sine-Gordon model we know that stable solitonic configurations of finite energy can only be obtained with two competing contributions to the action, one which tries to smooth solitons and one shrinking solitons. Extended solitons, reacting at large energies like point-like particles need to be regular solutions without any substructure. Such a field configuration symmetric in three-dimensional space is a configuration of hedge-hog type. It should be convenient for point-like charges.

The only known possibility to get non-trivial stable field configurations with long-range interactions is by the Lagrangian

$$\mathcal{L} = \mathcal{L}_e - \mathcal{H}_p = -\frac{\alpha_f \hbar c}{4\pi}\left(\frac{1}{4}\vec{R}_{\mu\nu}\vec{R}^{\mu\nu} + \Lambda\right), \quad \frac{\alpha_f \hbar c}{4\pi}\frac{1}{4}\vec{R}_{\mu\nu}\vec{R}^{\mu\nu} = -\frac{1}{4\mu_0}\vec{F}_{\mu\nu}\vec{F}^{\mu\nu}. \tag{12}$$

The prefactor $\frac{\alpha_f \hbar c}{4\pi}$ follows from the expression (10) for the field strength tensor, it is necessary to get the dimension of an action density in the International System of Units (SI). The term quadratic in the field strength is a kinetic term describing in electrodynamics the dynamics of the photon fields. It agrees with the Skyrme term of the Skyrme model [ ]. It contains four derivatives of the soliton field $Q(x)$, maximally two of them are time derivatives. The Skyrme model includes another term, quadratic in derivatives, corresponding to the usual kinetic term for the connection field $\vec{\Gamma}_\mu$. As the Skyrme model shows, this term leads also to solitonic solutions, but to solitons with short range interactions. The Skyrme model is therefore accepted as an approximation to QCD. As we want to describe long range Coulombic interactions we have to avoid this kinetic term. According to the Hobart-Derrick theorem we have to include a term with less than three derivatives and the only possibility left is a term without derivatives, a potential term Λ as in Eq. (12). Solitons with properties of charges have fields extending up to infinity. At infinity the potential term should contribute with its vacuum value only. The only term left at infinity is therefore the kinetic term, quadratic in the field strengths. To get non-vanishing derivatives of fields and a constant potential term we have to have degenerate vacua. A good choice to allow for hedge-hog type solitons is a term $\Lambda(q_0)$ with even powers in q_0, schematically shown in Fig. 5, with potential minima at the equatorial sphere $\mathrm{S}^2_{\mathrm{equ}}$,

$$\Lambda(q_0) = \frac{q_0^{2m}}{r_0^4}, \quad m = 1, 2, 3, \cdots. \tag{13}$$

By dimensional reasons, to equilibrate the four derivatives in $\vec{R}_{\mu\nu}\vec{R}^{\mu\nu}$, this term has to include the forth power of a scale parameter r_0 which finally determines the size of solitons.

Like the cosmological constant in general relativity the term (13) has no derivatives. But in distinction to the cosmological constant it is coordinate dependent, it contributes in the center of solitons and does not contribute in the vacuum. One could call $\Lambda(q_0)$ a cosmological function. If the universe would start in a state with $Q = 1$ and $\alpha = 0$ one would expect an inflationary transition to a vacuum state with $\alpha = \pi/2$.

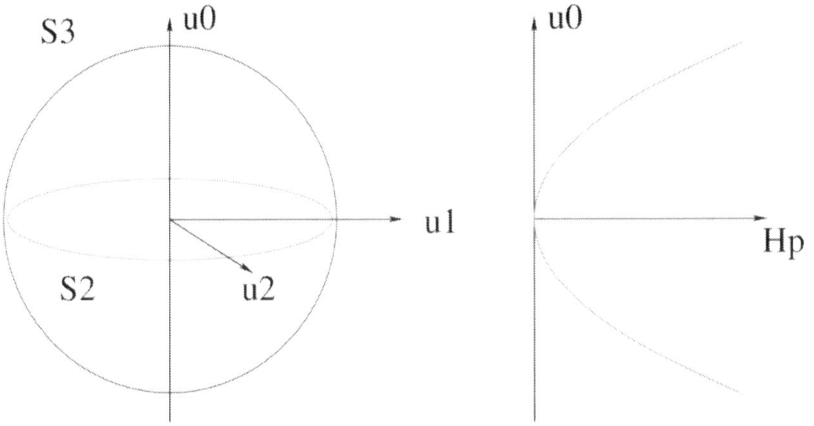

Figure 5. Schematical drawing of the potential term $\Lambda(q_0)$ with even powers in q_0 leading to a two-fold degeneracy of the vacuum states at the equatorial $S^2_{\rm equ}$.

3. Single solitons and charge quantisation

After defining the Lagrangian (12) and specifying the potential (13) we are ready to derive the equations of motion and to check whether there are long-range solitonic solutions. With an hedge-hog ansatz we construct spherical symmetric solutions,

$$\vec{n}(x) = \frac{\vec{r}}{r}, \quad \alpha(x) = \alpha(\rho), \quad \rho = r/r_0. \tag{14}$$

For vacuum configurations $\alpha = \pi/2$, Eq. (6) is reducing to the vector field of a dual Dirac monopole in the Wu-Yang representation with

$$\vec{\Gamma}_\mu = \vec{n} \times \partial_\mu \vec{n}. \tag{15}$$

The original Dirac monopole [,] has two types of singularities, the famous Dirac-string and the singularity in the center. In the Wu-Yang representation of the Dirac monopole [,] the Dirac string is removed, but the singularity in the center is still present. With the ansatz (14) we are able to remove also the singularity in the center.

Since there is no time dependence in the ansatz (14), we can use spherical coordinates r, ϑ, φ and get from Eq. (6) for the connection field

$$\vec{\Gamma}_r = \partial_\rho \alpha(\rho)\,\vec{n}, \;\; \vec{\Gamma}_\vartheta = \sin\alpha\left[\cos\alpha\,\vec{e}_\theta + \sin\alpha\,\vec{e}_\phi\right], \;\; \vec{\Gamma}_\varphi = \sin\theta\sin\alpha\left[\cos\alpha\,\vec{e}_\phi - \sin\alpha\,\vec{e}_\theta\right], \tag{16}$$

with $\vec{e}_\theta, \vec{e}_\phi$ unit vectors in internal color space. With the diagonal, spherical metric $g_{\mu\nu} = \mathrm{diag}(1, l_r^2, l_\vartheta^2, l_\varphi^2)$ and $(l_r, l_\vartheta, l_\varphi) = (1, r, r\sin\vartheta)$ we get the dual vector field

$$\vec{C}_r = -\frac{e_0}{4\pi\varepsilon_0 c}\frac{\vec{\Gamma}_r}{l_r}, \quad \vec{C}_\vartheta = -\frac{e_0}{4\pi\varepsilon_0 c}\frac{\vec{\Gamma}_\vartheta}{l_\vartheta}, \quad \vec{C}_\varphi = -\frac{e_0}{4\pi\varepsilon_0 c}\frac{\vec{\Gamma}_\varphi}{l_\varphi}. \tag{17}$$

The non-vanishing field strength components read

$$\begin{aligned}
\vec{E}_r &= -\frac{e_0}{4\pi\varepsilon_0}\frac{\vec{\Gamma}_\vartheta \times \vec{\Gamma}_\varphi}{r^2\sin\vartheta} = -\frac{e_0}{4\pi\varepsilon_0}\frac{\sin^2\alpha}{r^2}\vec{n}, \\
\vec{E}_\vartheta &= -\frac{e_0}{4\pi\varepsilon_0}\frac{\vec{\Gamma}_\varphi \times \vec{\Gamma}_r}{r\sin\vartheta} = -\frac{e_0}{4\pi\varepsilon_0}\frac{\alpha'(r)\sin\alpha}{r}(\cos\alpha\,\vec{e}_\theta + \sin\alpha\,\vec{e}_\phi), \\
\vec{E}_\varphi &= -\frac{e_0}{4\pi\varepsilon_0}\frac{\vec{\Gamma}_r \times \vec{\Gamma}_\vartheta}{r} = -\frac{e_0}{4\pi\varepsilon_0}\frac{\alpha'(r)\sin\alpha}{r}(\cos\alpha\,\vec{e}_\phi - \sin\alpha\,\vec{e}_\theta).
\end{aligned} \tag{18}$$

From the Lagrangian (12) follows the energy functional

$$H = H_e + H_p = \frac{\alpha_f \hbar c}{r_0} \int_0^\infty \mathrm{d}\rho \left[\frac{\sin^4 \alpha}{2\rho^2} + (\partial_\rho \cos\alpha)^2 + \rho^2 \cos^{2m} \alpha \right] \tag{19}$$

and by its variation the non-linear differential equation

$$\partial_\rho^2 \cos\alpha + \frac{(1-\cos^2\alpha)\cos\alpha}{\rho^2} - m\rho^2 \cos^{2m-1}\alpha = 0 \tag{20}$$

for $\cos\alpha(\rho)$. For this equation analytic solutions are known for $m = 2$ and $m = 3$ [,]. From an aesthetic point of view the $m = 3$ solution looks nicer, it reads

$$\alpha(\rho) = \arctan\rho \quad \text{with} \quad \rho = \frac{r}{r_0}. \tag{21}$$

The radial energy density, the integrand in the energy functional (19), has three contributions,

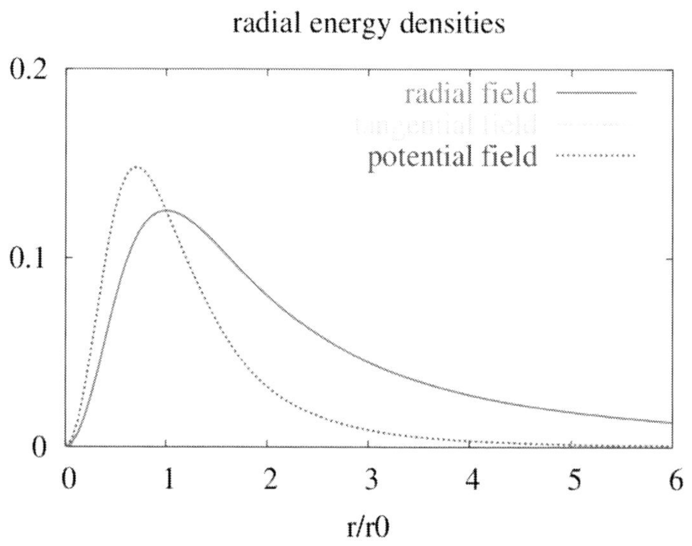

Figure 6. Radial energy densities (22) for the solution (21) of the differential equation (20) for $m = 3$.

$$h = \frac{\alpha_f \hbar c}{r_0} \Big[\frac{\rho^2}{2(1+\rho^2)^2} + \frac{\rho^2}{(1+\rho^2)^3} + \frac{\rho^2}{(1+\rho^2)^3} \Big]. \tag{22}$$

Up to the constant factor in front of the bracket in Eq. (22) they are shown in Fig. (6). For large distances the radial field decays with $1/r^4$ as expected for the energy density of the electric field of a point charge. Due to the regularity of the solution the singularity of the electric field of a point-like charge is removed. There is no singularity left. Two further terms contribute to the energy density, a tangential term originating in the components $\vec{E}_\vartheta$ and $\vec{E}_\varphi$ of the electric field and a contribution from the potential energy H_p. For the solution (21) the latter two contributions agree.

The total energy of this single soliton solution reads

$$E_1 = \frac{\alpha_f \hbar c}{r_0}\frac{\pi}{4} \quad \text{with} \quad \alpha_f \hbar c = 1.44 \text{ MeV fm}. \tag{23}$$

Comparing this result with the energy of the lightest charge particle, the electron with a rest energy of 0.511 MeV, we get $r_0 = 2.21$ fm. This value is of the same order of magnitude as the classical electron radius $\frac{\alpha_f \hbar c}{m_e c^2} = \alpha_f \frac{\hbar}{m_e c} = 2.82$ fm.

In the SI the scales for space, time, mass and charge are based on the requirement for a coherent system of units. In a certain sense these scales are man-made and can't be predicted by a model, they can only be adjusted by a comparison to experimental values. The only parameters of the model, r_0, c, $\hbar$ and α_f, correspond to these four scales. They were adjusted to the charge and the mass of the electron, the velocity of light and the dielectric constant of the vacuum. Therefore, the elementary electric charge can't be predicted. But, in distinction to Maxwell's electrodynamics this model predicts the quantisation of electric charge. Due to the geometric restrictions fractional multiples of e_0 can't be realised by regular field distribution of the SU(2)-field $Q(x)$. There is no contradiction to the existence of quarks, since they never appear as isolated particles.

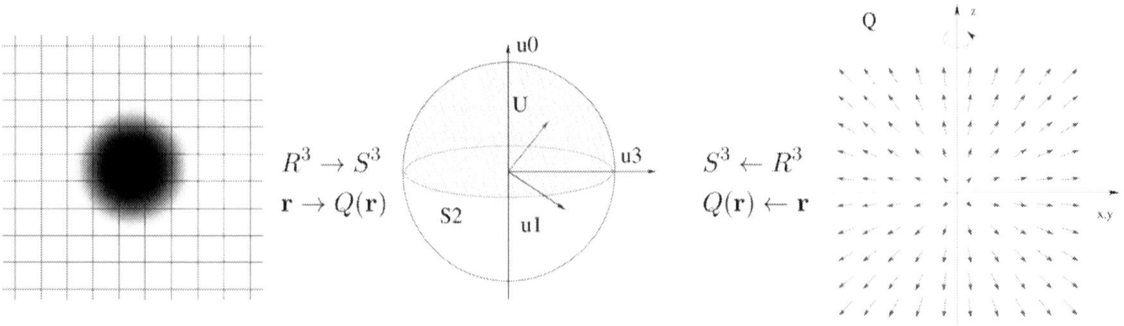

Figure 7. The field of rotational matrices $Q(\mathbf{r})$ of spatial dreibeins maps $R^3 \to S^3$. The center of the soliton is mapped to the origin and spatial infinity to the equatorial S^2_{equ}. The right diagram indicates this map by arrows of length $\sin\alpha(\mathbf{r})$ and direction $\vec{n}(\mathbf{r})$. A soliton covers only half of S^3. Points on the lower half of S^3 are furthermore characterised by dashed arrows.

We succeeded to follow the idea of Fig. 3 to describe a rotational dislocation in R^3 by assigning to every coordinate triple $\mathbf{r}$ a rotational matrix $Q(\mathbf{r})$, as shown in Fig. 7. A unit charge covers a hemisphere of S^3 only. To get a more precise picture of the field, $Q(\mathbf{r})$ is symbolised in the right diagram by arrows of length $\vec{q}(\mathbf{r}) = \sin\alpha(\mathbf{r})\vec{n}(\mathbf{r})$. The center of the soliton is mapped to the unit matrix and therefore a vector of length zero. Traversing the center of the soliton along a line through the origin, $Q(\mathbf{r})$ defines a continuous 2π rotation with positive chirality. The rotational axis agrees with the direction of the line. In this sense the configuration is a three-dimensional generalisation of the Sine-Gordon model with solitons defined by 2π rotations between degenerate ground states. The ground states build a discrete set in the Sine-Gordon model and a two-dimensional manifold in this model.

Releasing the condition $\vec{n}(x) = \frac{\vec{r}}{r}$ and allowing for global rotations of soliton configurations we can realise a continuum of soliton solutions for the Lagrangian (12). All these configurations belong to the same homotopy class, they can be "continuously deformed" into each other. There are four homotopy classes of SU(2)-field configurations. They differ by transformations with the non-trivial center element $z = -\mathbb{1}_2$ of SU(2), $Q \to -Q$, or by parity transformation Π_n in internal space, $\vec{n} \to -\vec{n}$. The far field of a soliton determines its interaction with other solitons

Transf.	1	z	Π_n	$z\Pi_n$
$\vec{n}$	$\vec{r}/r$	$-\vec{r}/r$	$-\vec{r}/r$	$\vec{r}/r$
q_0	≥ 0	≤ 0	≥ 0	≤ 0
Z	1	-1	-1	1
$2\mathcal{Q}$	1	1	-1	-1
diagram	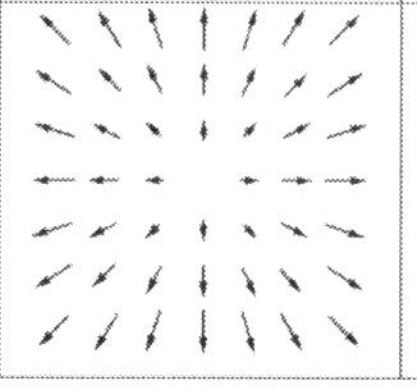	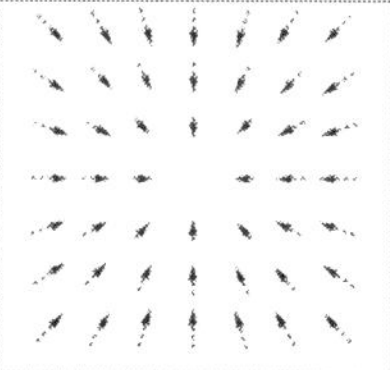	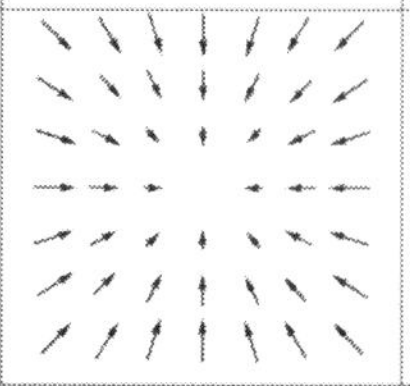	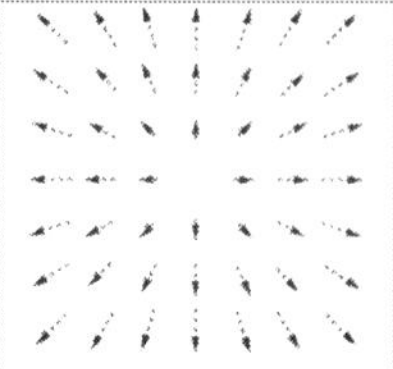

Table 1. There are four homotopy classes for soliton configurations. They differ by parity transformations in colour space $\Pi_n : \vec{n} \to -\vec{n}$ and transformations with the non-trivial center element $z = -\mathbb{1}_2 : Q \to zQ$. They can be characterised by the charge number Z and the topological charge $\mathcal{Q}$, defined in Eqs. (24) and (25). Length and direction of arrows depict the value of the imaginary part $\vec{q}(x) = \vec{n} \sin\alpha$ of the soliton field $Q(x)$. The sign of q_0 is indicated by the line type of the arrow, full lines for $q_0 > 0$ and dashed lines for $q_0 < 0$.

and therefore its charge. For the far field we have $\cos\alpha = 0$, arrows have unit length and arrows with full and dashed stem symbolise the same SU(2)-element on the equatorial S^2_{equ}. There are two possible coverings of S^2_{equ}. As discussed in detail in refs. [2, 9, 10, 11] we can distinguish them by choosing a closed surface $\mathcal{S}$ parametrised by coordinate lines u and v around the soliton center and determining

$$Z(\mathcal{S}) := -\frac{1}{4\pi} \oint_{\mathcal{S}} du dv \, [\partial_u \vec{n} \times \partial_v \vec{n}] \vec{n}. \tag{24}$$

$Z(\mathcal{S})$ is the charge number of the soliton. It is obvious that Π_n transformations change the sign of Z. Configurations which are topological equivalent to a configuration with arrows pointing outwards we have defined as the charge of an electron, i.e. a negative elementary charge, $Z = -1$, arrows pointing inwards as positive, $Z = +1$. Now we are going to discuss in more detail the spin properties which are related to the hemispheres of S^3.

4. Spin properties

4.1. Spin as topological quantum number

From quantum mechanics we know that spin-1/2 particles appear in two spin states, usually called spin-up and spin-down. Only one component of the spin-vector is measurable and this component can have two values only, $\pm\frac{\hbar}{2}$. By the interaction of a spin-1/2 particle with a magnetic field the expectation value of the spin can be rotated in arbitrary direction. If we would attribute in our model spin-up and spin-down states to the upper and lower hemisphere of S^3, this would contradict to the experiment, since field configurations on different hemispheres are topological different, there is no continuous transition between them. This is the reason why, as mentioned above, we have to attribute SO(3) and not SU(2) field configurations to the solitons. Due to the double covering of SO(3) by SU(2), in SO(3) there is no upper and lower hemisphere, there is no difference between $\pm Q(x)$ configurations. In this picture a single soliton in the universe does not have a spin. Only by the interaction of solitons, the spin direction appears. There are two possibilities to connect two spatially separated SO(3) configurations continuously, a trivial connection and a non-trivial connection, see Fig. (8). The trivial connection reads in the SU(2) language, that the two solitons occupy the same hemisphere of S^3 and the non-trivial configuration, that the two solitons occupy different hemispheres. Schematic arrow diagrams for these two configurations for opposite charges are depicted in Fig. 8. The number of coverings of

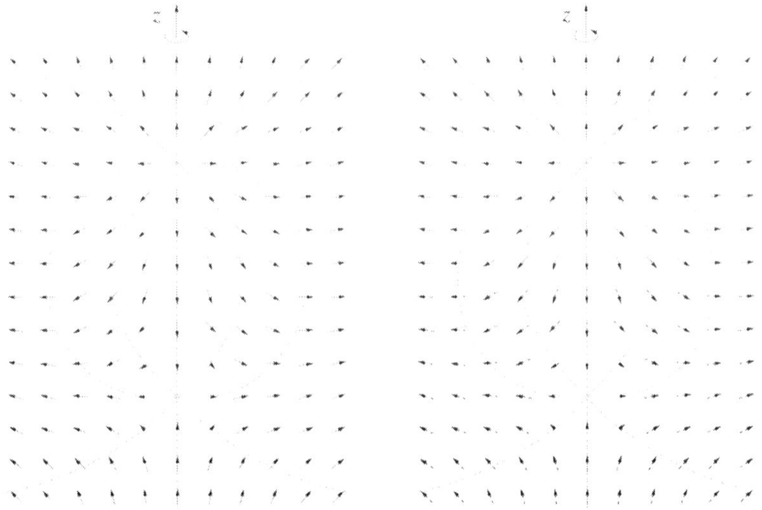

Figure 8. Schematic arrow diagram for the field $Q(\mathbf{r})$ and the electric field by blue (dashed) lines for two opposite unit charges. The arrows with full stem represent values $\vec{q} = \vec{n}\sin\alpha$ with positive values of $q_0 = \cos\alpha$, the dashed stems negative values of q_0, i.e. $\frac{\pi}{2} \leq \alpha \leq \pi$. The left configuration corresponds to a spin-zero and the right configuration to a spin-one state. The electric field lines follow lines of constant $\vec{n}$.

$\mathbb{S}^3$ is a conserved topological quantum number. This is one of the reasons we try now to describe the spin quantum number s by this topological quantum number, the topological charge $\mathcal{Q}$,

$$\mathcal{Q} = \frac{1}{V(\mathbb{S}^3)} \int_0^\infty \mathrm{d}r \int_0^\pi \mathrm{d}\vartheta \int_0^{2\pi} \mathrm{d}\varphi \vec{\Gamma}_r(\vec{\Gamma}_\vartheta \times \vec{\Gamma}_\varphi) \text{ with } V(\mathbb{S}^3) = \int_{S^2} \mathrm{d}^2 n \int_0^\pi \mathrm{d}\alpha \sin^2\alpha = 2\pi^2 \quad (25)$$

and define

$$s = |\mathcal{Q}|\,. \quad (26)$$

The field configurations of single solitons, see Table 1, cover half of $\mathbb{S}^3$ and are therefore characterised by $s = 1/2$. Such configurations are connected to the surrounding by the lines of constant $\vec{n}$-field. A 4π-rotation of the center of a soliton can be described by a big circle on $\mathbb{S}^3$. Since this lasso around $\mathbb{S}^3$ can be contracted to a point, one can understand that by appropriate movements of the constant $\vec{n}$-lines the 4π-rotation is topological equivalent to no rotation. This reflects the famous 4π-rotational invariance of spin-1/2 states.

Looking more carefully to the attractive two-soliton configurations of Fig. 8, one can imagine that at spatial infinity all unit-arrows turn finally in positive z-direction for both configurations. This indicates that the configurations are uncharged, as mentioned above. The spin properties we can read off from the central structure of the configuration. We can attribute a chirality χ to the soliton centers according to the sign of the topological charge and get

$$\mathcal{Q} =: \chi s. \quad (27)$$

Especially significant is the Q-field along the z-axis. We realise that in the left diagram with $s = 0$ the soliton-field describes a left-handed rotation of the spatial dreibein for negative z-values and a right-handed rotation for positive z-values. These opposite chiralities of the two solitons we would describe in corresponding quantum mechanical configurations by opposite magnetic quantum numbers m_s. In the right diagram of Fig. 8 we observe a complete 4π-rotation along the z-axis as we expect for two parallel soliton spins adding up to $s = 1$. In the left, $s = 0$, diagram the soliton field $Q(\mathbf{r})$ at the center of mass $\mathbf{r} = 0$ of the two-soliton system approaches $\alpha = \pi/2$ for large distances of the solitons only. Therefore this configuration does not cover the full upper hemisphere twice. The energy of this configuration is thus lower than the energy of the $s = 1$-state in the left diagram where we get exactly $\alpha(\mathbf{r} = 0) = \pi/2$ and the configuration covers both hemispheres of S^3 completely.

4.2. Spin as angular momentum

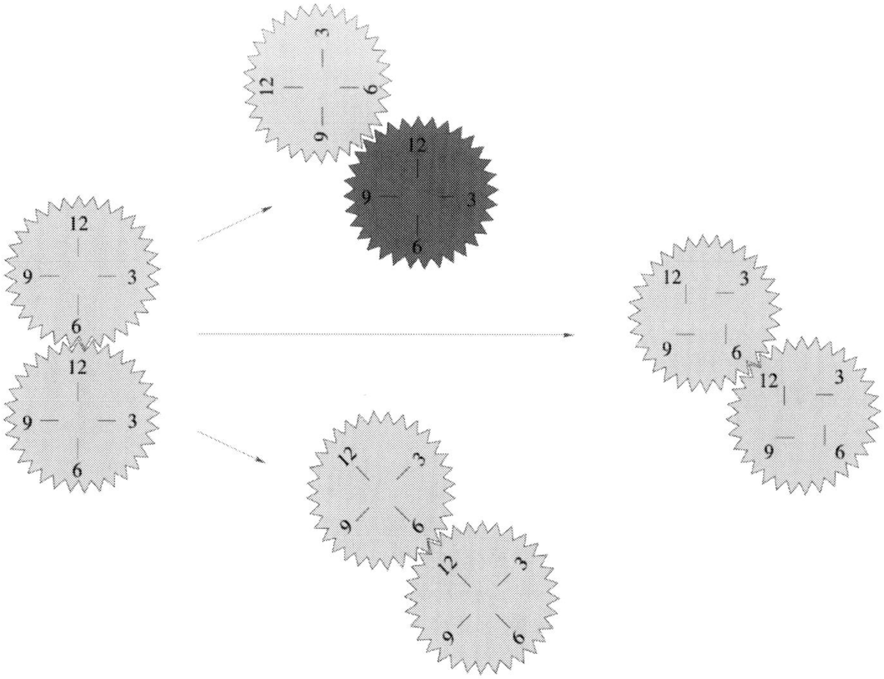

Figure 9. The rotation of gears of equal radius is compared to rotations of solitons. If a light gear rotates around a heavy (static) one, from the left to the upper diagram, the circulating gear performs a 4π-rotation. Another possibility for two gears of equal weight would be a rigid rotation, from the left to the lower diagram. Solitons circulate around as shown from left to right.

Whereas in quantum mechanics mainly the su(2)-algebra and the SU(2)-representations are important, the above discussion shows that in this model the manifold of the SU(2)-group and its map to R^3 is related to the spin properties. An essential property of spin we still didn't discuss, its contribution to angular momentum. In the static configurations of Fig. 8 nothing contributes to the angular momentum. But as soon as the solitons start to move around each other they have to rotate, they are connected to each other by the lines of constant $\vec{n}$-field, like gears. We can try to compare two unit charges to a pair of gears of equal size. If a heavy gear is at rest, another light gear circulating around performs a 4π-rotation, see Fig. 9 from the left to the upper diagram. For two gears of equal mass one would rather expect a rigid rotation, from left to bottom. But gears are rigid bodies and solitons are not rigid. That solitons perform a rigid rotation was a wrong assumption in ref. [9]. We are now going to discuss that solitons circulate rather as shown in Fig. 9 from left to right.

Figure 10. Two solitons with opposite charge and spin quantum number $s = 1$ rotating about a common center of mass. Four rotational angles $0, \frac{\pi}{4}, \frac{\pi}{2}$ and $\frac{3\pi}{4}$ are shown from left to right and top to bottom. This is not a rigid rotation. The line connecting the two soliton centers keeps its $\vec{n}$-value, $\vec{n} = (0, 0, -1)$ during the rotation.

For soliton configurations we have to take into account the boundary conditions defined by the broken vacuum. In Fig. 8 we assumed a vacuum configuration with $Q = -\mathrm{i}\vec{\sigma}\vec{n}$ and $\vec{n} = (0, 0, 1)$. This vacuum state has to remain unchanged during the rotation of the soliton pair. Therefore, solitons can't rotate like rigid bodies. In Fig. 10 a slow rotation of two solitons with opposite charges and $s = 1$ is shown for rotational angles of $0, \frac{\pi}{4}, \frac{\pi}{2}$ and $\frac{3\pi}{4}$. Along the line connecting the two soliton centers the vector $\vec{n}$ keeps its direction $\vec{n} = (0, 0, -1)$ during the rotation. It follows that there is an internal motion contributing to the total angular momentum. This two soliton state returns after a 2π-rotation to the original state as it should be for integer spin states.

Journal of Physics: Conference Series **504** (2014) 012010 doi:10.1088/1742-6596/504/1/012010

4.3. Angular momentum components

We can treat $\alpha_k = \alpha n_k$ as field variables of the SU(2)-field $Q(x)$ in Eq. (4). The relations

$$\frac{\partial \mathcal{L}}{\partial \partial_\mu \alpha_k} \partial_\nu \alpha_k = \frac{\partial \mathcal{L}}{\partial \Gamma_{\mu k}} \Gamma_{\nu k} \tag{28}$$

allow to consider $\vec{\Gamma}_\mu$ as generalised velocities and to derive from the Lagrange density (12) generalised momenta

$$\pi^\mu = \frac{\partial \mathcal{L}}{\partial \vec{\Gamma}_\mu} = -\frac{\alpha_f \hbar c}{4\pi} \vec{\Gamma}_\nu \times \vec{R}^{\mu\nu}. \tag{29}$$

The canonical energy-momentum tensor

$$\Theta^\mu{}_\nu(x) = \frac{\partial \mathcal{L}(x)}{\partial(\partial_\mu \alpha_k)} \partial_\nu \alpha_k - \mathcal{L}(x)\,\delta^\mu_\nu = \frac{\partial \mathcal{L}(x)}{\partial \vec{\Gamma}_\mu} \vec{\Gamma}_\nu - \mathcal{L}(x)\,\delta^\mu_\nu \tag{30}$$

turns out to be symmetric

$$\Theta^\mu{}_\nu = -\frac{\alpha_f \hbar c}{4\pi} \vec{R}^{\mu\sigma} \vec{R}_{\nu\sigma} - \mathcal{L}(x)\,\delta^\mu_\nu, \tag{31}$$

a property important for the derivation of the angular momentum tensor. This is different to classical electrodynamics where the canonical tensor is gauge dependent and asymmetric.

Global rotations of the soliton field modify the coordinates x_i and the orientations α_i of the local dreibeins. This leads to expressions for the orbital angular momentum

$$L_i = \frac{1}{c} \int \mathrm{d}^3x\, \varepsilon_{ijk} x^j \Theta^{0k}. \tag{32}$$

and the spin

$$\begin{aligned} S_i &= -\frac{1}{c} \int \mathrm{d}^3x \left(\frac{\partial \mathcal{L}}{\partial \partial_0 \alpha_l} \varepsilon_{ilk} \alpha_k \right) = \\ &= -\frac{1}{c} \int \mathrm{d}^3x\, \vec{\pi}^0 \sin\alpha [\cos\alpha\, \vec{n} \times \vec{e}_i + \sin\alpha\, \vec{n} \times (\vec{n} \times \vec{e}_i)]. \end{aligned} \tag{33}$$

Since the analytic solutions of two-soliton systems are not known, the determination of the angular momentum contributions for classical paths can only be determined approximately or numerically. This remains for future work.

5. Summary and outlook

Inspired by the success of the Sine-Gordon model to describe particles by topological solitons of a bosonic field, we tried a generalisation from 1+1D to 3+1D from 1 rotational angle to 3 rotational angles. Since three rotational angles describe rotations in a three-dimensional space, there is the possibility to interpret these rotations as rotations of spatial dreibeins. This leads to the tempting picture that both long-range interactions, gravitation and electromagnetism can be explained by properties of space-time, i.e. using exclusively space and time coordinates. It is an interesting question how far we can drive this idea and where discrepancies to experimental facts get obvious.

Let me enumerate some successes, limitations and open questions.

We describe charges by local dreibeins rotating by 2π along lines crossing the charge centers. Fields of single solitons are connected to the surrounding by lines of constant $\vec{n}$-field. Rotating soliton cores can therefore recover the original configurations only after a 4π-rotations. This indicates the fermionic character of such solitons. By an appropriate Lagrangian solitons are topological stable field configurations which survive scattering processes. Due to the topological character charge is quantised. Such a quantisation is already known from Dirac monopoles. Actually the solitons of this model are dual Dirac monopoles where the two types of singularities of Dirac monopoles were removed. The Dirac string is removed by a transition from the vector potential A_μ to the a three dimensional unit vector field $\vec{n}$ and the singularity in the monopole center is avoided by the transition to an SU(2) field Q. To the author it appears as a nice aspect that in the dual formulation static charges are described by the purely spatial components of the dual field strength tensor (10) whereas "moving" currents lead to the space-time components of the ${}^\star\vec{F}_{\mu\nu}$ and the curvature tensor $\vec{R}_{\mu\nu}$. In classical electrodynamics we find just the opposite correspondence, moving currents are described by static fields and static charges by time components of the fields.

The term quadratic in the field-strength of the Lagrangian (12) seems more natural than the potential term Λ. Nevertheless, such a Λ-term is known from gravitational theory as the cosmological constant. In this model Λ is a cosmological function. One could try to relate the transition from a state with $Q = 1$ and constant $\Lambda = 1/r_0^4$ to $\Lambda = 0$ to inflation known from cosmology and gets a release of energy density of $\alpha_f \hbar c(4\pi r_0^4) = 4.8$ keV/fm^3. The potential term leads to a two-fold degeneracy of the vacuum states. Due to the broken symmetry there are two massless Goldstone bosons described by the $\vec{n}$-component of dreibein rotations. This gives the possibility to interpret photons as Goldstone bosons of rotational symmetry breaking of the vacuum. Lines of constant $\vec{n}$-field can be seen as fibres of a Hopf map from S^3 (R^3 with broken vacuum) to S^2. One could try to follow the conjecture that the Hopf index of this map agrees with the photon number.

The far field of a single soliton, the field $\vec{n}$ covers S^2 once. The two possible topological different coverings determine the sign of the unit charge $\pm e_0$. In this far field limit there is left a rotational invariance around the direction of the $\vec{n}$-field, leading to a U(1) gauge invariance of the dual vector potential C_μ and the dual field strength tensor ${}^\star F_{\mu\nu}$; in this limit both are reduced to one-component fields in internal space as expected for a U(1)-gauge theory. In this model charges and fields are described by the same field $Q(x)$. An artificial separation of solitons and their fields allows to derive the appearance of Coulomb and Lorentz forces. The internal structure of a soliton can be further characterised by the sign of the topological charge $\mathcal{Q} = \pm 1/2$, we called it chirality. The absolute value of $\mathcal{Q}$ can be interpreted as spin quantum number s. The angular momentum property of spin appears in this model only for solitons in orbital motion.

The solitons show all the properties expected for charged particles in a fully relativistic model. The mass of the particles is field energy. Particles experience Lorentz contraction and relativistic mass increase. Moving particles lead to magnetic fields as expected from Lorentz transformations of the electric field. Solitons at rest have a finite size comparable to the classical electron radius. Since they have no real substructure, in central collisions in high energy scattering processes they are expected to be compressed to nearly point-like objects like the Sine-Gordon solitons.

This article investigates on a classical level solitonic excitations of space-time and compares basic properties, like mass, charge and spin of such solitons with particle properties. Quantum fluctuations were not included in these investigations. An appealing method to go beyond the classical behaviour could possibly be indicated by the silicon oil drop experiments of Yves Couder and his group [12]. In these experiments a classical particle interacts with a wave-field which results from a superposition of an external field and the field created by the particle. Under appropriate conditions this resulting pilot-wave field leads to interference [13], tunnelling [14] and spontaneous quantisation of closed orbits [15]. Transferring this experience to solitons would mean that particles on their classical path are disturbed by, and interacting with, a subquantum medium. Such a medium must not transfer energy and momentum by friction. Deceleration by friction processes can only be avoided by a medium propagating with the speed of light which does not prefer special reference systems. Further, to get interference phenomena with classical solitons resonance processes between the subquantum medium and solitons could be essential. Since solitons are extended objects one could imagine that they select and interact with certain frequencies from a wave-field.

Due to the low number of degrees of freedom electromagnetic fields are more restricted than in classical electrodynamics. On the other hand besides solutions of the Maxwell equations there is a possibility for non-vanishing non-quantised magnetic currents [,]. Further one can expect α-waves propagating through the vacuum [ ]. These presently unknown excitations one can try to shift to the subquantum medium.

Finally one can ask about strong and weak interactions. It is obvious that to include these interactions in the model an extension by further field degrees of freedom is necessary. If one adheres to the idea that the fields are properties of space-time this would lead to further dimensions which would be hidden inside of strongly interacting particles.

References

[1] Remoissenet M 2003 *Waves Called Solitons: Concepts and Experiments* Advanced Texts in Physics (Springer) ISBN 783540659198

[2] Faber M 2001 A Model for topological fermions *Few Body Syst.* **30** 149–186 (*Preprint* arXiv:hep-th/9910221 [hep-th])

[3] Rodrigues O 1840 Des lois géométriques qui régissent les déplacements d'un système solide dans l'espace, et de la variation des coordonnées provenant de ces déplacements considérés indépendamment des causes qui peuvent les produire *Journal de Mathématiques Pures et Appliquées* **5** 380–440

[4] Skyrme T H R 1961 A Nonlinear field theory *Proc. Roy. Soc. Lond.* **A260** 127–138

[5] Dirac P A M 1931 Quantised singularities in the electromagnetic field *Proc. Roy. Soc. Lond.* **A133** 60–72

[6] Dirac P A M 1948 The theory of magnetic poles *Phys. Rev.* **74** 817–830

[7] Wu T T and Yang C N 1975 Concept of nonintegrable phase factors and global formulation of gauge fields *Phys. Rev.* **D12** 3845–3857

[8] Wu T T and Yang C N 1975 Some remarks about unquantized nonabelian gauge fields *Phys. Rev.* **D12** 3843–3844

[9] Faber M 2012 Particles as stable topological solitons *J. Phys.: Conf. Ser.* **361**(1) 012022

[10] Faber M and Kobushkin A P 2004 Electromagnetic limit in a model for charged solitons *Phys. Rev.* **D69** 116002 (*Preprint* arXiv:hep-th/0207167)

[11] Borisyuk D, Faber M and Kobushkin A 2007 Electro-Magnetic Waves within a Model for Charged Solitons *J. Phys.* **A40** 525–531 (*Preprint* arXiv:arXiv:0708.3173 [hep-th])

[12] Couder Y, Fort E, Gautier C H and Boudaoud A 2005 From bouncing to floating: Noncoalescence of drops on a fluid bath *Phys. Rev. Lett.* **94**(17) 177801

[13] Couder Y and Fort E 2006 Single-particle diffraction and interference at a macroscopic scale *Phys. Rev. Lett.* **97**(15) 154101

[14] Eddi A, Fort E, Moisy F and Couder Y 2009 Unpredictable tunneling of a classical wave-particle association *Phys. Rev. Lett.* **102**(24) 240401

[15] Fort E, Eddi A, Boudaoud A, Moukhtar J and Couder Y 2013 Path-memory induced quantization of classical orbits *ArXiv e-prints* (*Preprint* arXiv:1307.6051 [cond-mat.soft])

[16] Faber M, Kobushkin A and Pitschmann M 2008 Shape vibrations of topological fermions *Adv. Stud. Theor. Phys.* **2** 11–22 (*Preprint* arXiv:0812.4225 [math-ph])

Which came first, spacetime or clocks?

G N Ord

Department of Mathematics, Ryerson University, Toronto, Ontario, Canada

E-mail: gord@ryerson.ca

Abstract. Emergent quantum mechanics seeks a deeper level theory, anticipating that such a theory will provide a clearer picture of the relation between the quantum and classical worlds. In this work we show that the quantum-classical divide is a manifestation of the transition from Newton's absolute time to relativity's path-dependent time. The prior theory in this case is that *particles are intrinsic clocks*. The emergence of separate classical and quantum behaviour is seen by considering different continuum limits in a single digital clock model. A continuum limit that constructs a continuous worldline provides a simple basis for Minkowski spacetime. An alternative limit in which the clock itself contains boost information leads to the Dirac equation.

Introduction

The four major theories of the twentieth century, special relativity (SR), quantum theory (QT), general relativity and quantum field theory are all considered more fundamental than their predecessor, Newtonian mechanics (NM). For our purposes, general relativity and quantum field theory shall be considered extensions of SR and QT respectively, and the transition from Newtonian mechanics to QM and SR will be our main concern.

Quantum mechanics, while empirically accurate, is currently a prescription rather than a mathematical expression of physical principles. Since there are a myriad of interpretations of QM, it has been difficult to determine through which interpretation we are to find a logically prior theory. If quantum theory is emergent, the question remains 'emergent from what?'.

By comparison, special relativity seems straightforward. From Einstein's two physically motivated postulates we arrive at Minkowski spacetime without the encumbrance of multiple interpretations. However, having arrived at spacetime by virtue of the light speed postulate, it is clear that we have *missed* quantum mechanics in the process. A sensible question is then; what is *wrong* with the conventional route to Minkowski space that it misses quantum mechanics?

One prime suspect for quantum incompatibility is the worldline. In Newtonian mechanics particles have worldlines and this is carried over to SR. However, worldlines are not a feature of quantum mechanics outside of the Bohm picture. If we are to question the existence of worldlines when in fact objects appear to *have* worldlines on macroscopic scales, it seems that the continuum limit involved in time-keeping is in question, and it is to this that we turn.

In section 1 we review a simple model of a digital clock that illustrates the transition to Newtonian absolute time. We see the logical steps necessary to go from a digital clock to an analog clock to absolute time. The continuum limit here uncovers unimodular complex numbers as model clocks.

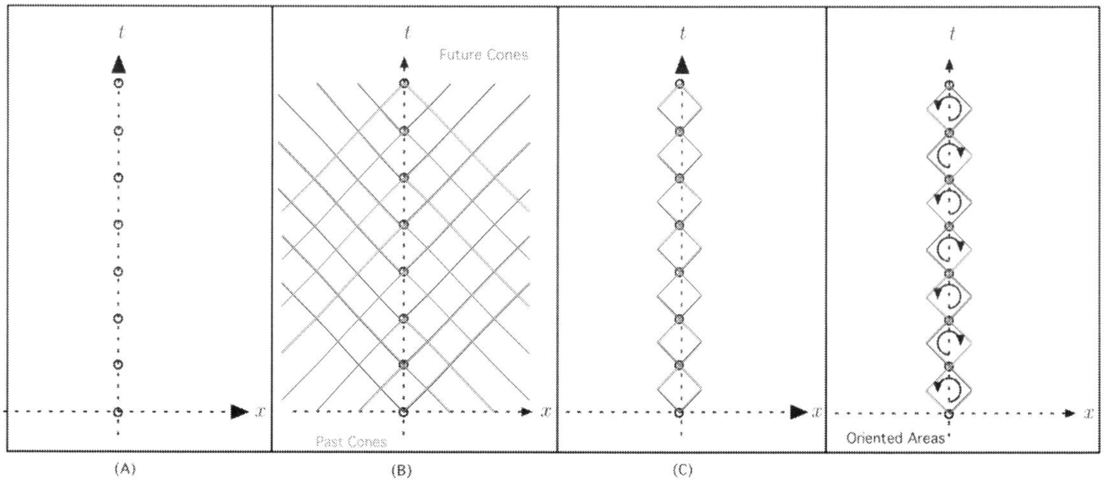

Figure 1. (A) Periodic events in a two dimensional space-time. (B) Past and future light-cones for the events. Units are chosen so that $c = 1$. (C) The causal areas between events. (D) The area boundary is chosen to provide a single traversal. This results in an alternating orientation of causal areas.

In section 2 we show how the clock must be modified for consistency with the Lorentz transformation. The continuum limit in this case uncovers Clifford algebra in place of complex numbers and absolute time is replaced by Minkowski spacetime.

In section 3 we take a closer look at the clock model that generated Minkowski spacetime and show that if we require a clock to *contain* the boost information of the Lorentz transformation, the transfer matrix of Minkowski spacetime becomes a generating function that satisfies the Dirac equation. Section 4 discusses the implication of this result.

1. Newtonian Timekeeping

Following[1] consider a periodic process in a stationary frame that provides events at fixed intervals. To make contact with special relativity in the next section we shall assume a speed of light $c = 1$ much greater than any other speed in the model. The Galilean transformation shall be assumed.

Fig. 1(A) represents a periodic sequence of events, stationary at the origin in an inertial frame. (B) shows the past and future light cones of these events and (C) the 'causal' areas that lie between the sequential events. Here the term causal is used because, regardless of any actual clock 'mechanism', the areas represent the maximum time-like domain of influence that successive ticks have in common. In (D) we colour the area boundaries to show that we can imagine the boundaries to be drawn as a single continuous curve from t=0 to the latest event and thence back to the origin. Drawing the boundary this way gives the causal areas an orientation and places the events at intersections of smooth segments of the two curves. If we think of the chain of areas in terms of the boundary curves there are two kinds of events, the outside corners and the path-crossing points where the curves intersect. We call the latter on-worldline events and the former off-worldline events. Our clock, from a path perspective, is equivalent to a couple of featureless photons confined between two walls that generate a 'tick' at crossing points.

Our periodic process is just a digital clock that ticks at the events. To construct an analog clock from this we require a continuum limit in which we allow for ever higher frequency ticks. We can do this using the same clock model, as illustrated in Fig. 2. To take the continuum limit for this sequence of clocks we represent the clock algebraically via the encoding illustrated in

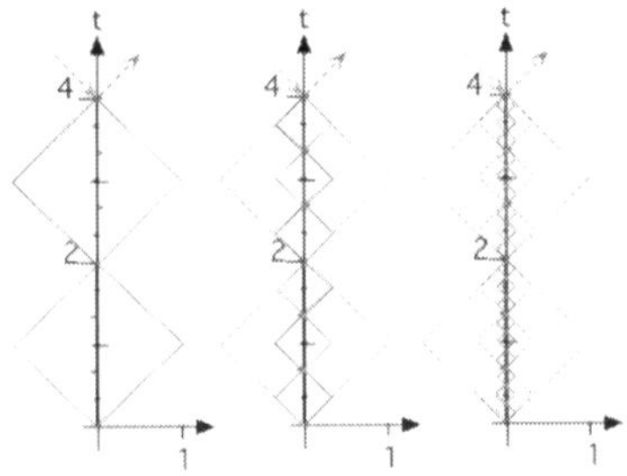

Figure 2. A full clock cycle for the original clock with similar clocks running at three and nine times the original frequency. As the frequency is increased, the density of events on the t-axis increases and the causal area between on-worldline events decreases. In the continuum limit the causal area shrinks to zero and the event sequence becomes a worldline.

Fig. 3. In this figure, a two-component 'state vector' records projections onto the left and right light cones and because the area boundaries are themselves null segments, successive states are orthogonal. The states consist of a period 4 sequence

$$s_k \in S_0 = \left\{ \begin{pmatrix} 1 \\ 0 \end{pmatrix}, \begin{pmatrix} 0 \\ 1 \end{pmatrix}, \begin{pmatrix} -1 \\ 0 \end{pmatrix}, \begin{pmatrix} 0 \\ -1 \end{pmatrix} \right\} \tag{1}$$

and the actual sequence is a clock *signal* that can be inverted to establish the time up to integer accuracy. If we think of the column vectors as the active display of a clock, algebraically the clock mechanism corresponds to left multiplication by the transfer matrix

$$T_0 = \begin{pmatrix} 0 & -1 \\ 1 & 0 \end{pmatrix} \tag{2}$$

where we note that $T_0^2 = -I_2$, a square root of minus the identity and a fourth root of the identity. In terms of the Pauli matrices T_0 is just $-i\sigma_y$. If the clock is in state s_k at time k then the subsequent state, after the next event is

$$s_{k+1} = T_0 s_k. \tag{3}$$

Since then

$$s_k = T_0^k s_0 \tag{4}$$

the power of the transfer matrix corresponds to the discrete displacement in time.

To make this clock analog, we have to construct a process that agrees with the clock at the integer events where the clock ticks, but provides extra events between ticks. The simplest way to do this is to construct a clock, similar to the original, but running at a higher frequency. Fig. 2 shows an example of the original clock with similar clocks running at three and nine times the original frequency.

In terms of the transfer matrix T_0, a clock running at n times the frequency will have a transfer matrix T_n such that $T_n^n = T_0$ so that we can take T_n as the n-th root of T_0 via an eigenvalue expansion:

$$T_n = \begin{pmatrix} \cos\left(\frac{\pi}{2n}\right) & -\sin\left(\frac{\pi}{2n}\right) \\ \sin\left(\frac{\pi}{2n}\right) & \cos\left(\frac{\pi}{2n}\right) \end{pmatrix} \tag{5}$$

We can use this to take the limit as $n \to \infty$, in which case we get:

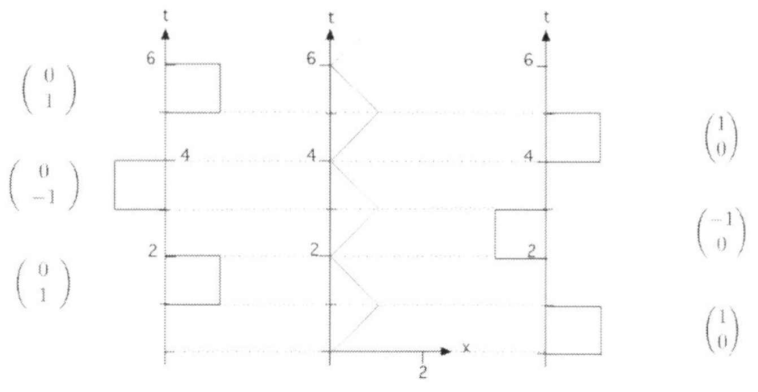

Figure 3. The right boundary of the chain of causal areas can be in any one of four states. The boundary densities and column vectors are illustrated. On-worldline events occur at the even integers. The continuum limit (6) suggest that the left and right densities taken together can be replaced by a 'clock face' that interpolates between the four states imitating a unimodular complex number in polar form.

$$\begin{aligned} T(t) = \lim_{n\to\infty} T_n^{nt} &= \begin{pmatrix} \cos\left(\frac{\pi t}{2}\right) & -\sin\left(\frac{\pi t}{2}\right) \\ \sin\left(\frac{\pi t}{2}\right) & \cos\left(\frac{\pi t}{2}\right) \end{pmatrix} \\ &= \mathrm{I}_2 \cos\left(\frac{\pi \mathrm{t}}{2}\right) + \mathrm{T}_0 \sin\left(\frac{\pi \mathrm{t}}{2}\right) \end{aligned} \tag{6}$$

where I_2 is the 2×2 identity matrix. Notice that $T(n) = T_0^n$ for integer n so that our 'analog clock' agrees with the digital clock at the integers where the original digital clock has events.

From equation (6) we can see that the analog transfer matrix $T(t)$ is just a rotation matrix and that the continuum limit replaces the set of four state vectors S_0 with the set of all column unit vectors $S = \left\{(u,v)^T \mid u^2+v^2 = 1, u, v \in \Re\right\}$. From the form of (6) we see that the transfer matrix looks like a complex number in polar form where T_0 is just an implementation of the unit imaginary. To see the connection to complex numbers more directly we can consider a change of variables $(x,y)^T = \frac{1}{\sqrt{2}}(I_2+T_0)(u,v)^T$. This diagonalizes the original transfer matrix T_0 which becomes $i\sigma_z$ and equation(6) becomes

$$\begin{aligned} T(t) &\rightarrow \begin{pmatrix} \exp[i\left(\frac{\pi t}{2}\right)] & 0 \\ 0 & \exp[-i\left(\frac{\pi t}{2}\right)] \end{pmatrix} \\ &= \exp[\left(\frac{i\pi\sigma_z t}{2}\right)] \end{aligned} \tag{7}$$

From equation (7) we see that the diagonal form of the transfer matrix has two unimodular complex numbers on its diagonal, both synchronized but rotating in opposite directions as t increases. As a result, both components of the resulting state vector are unimodular complex numbers rotating in opposite directions. Either component could be inverted over half periods to extract t as a function of the component itself. Notice in Fig. 4, the continuum limit smoothly interpolates a four state process creating a spiral that is conveniently described by a unimodular complex number.

Figure 4. The initial digital clock has only four states, however the continuum limit builds an extra dimension that appears algebraically through the arrival of complex numbers.

In a Newtonian world where the Galilean transformation holds, the above 'clock' measures time intervals without being affected by relative motion. The continuum limit would give the same result, regardless of the inertial frame. Since the clock *signal* is the same unimodular complex number for all such clocks, regardless of inertial frame, the clock argument t can be taken as absolute and assigned its own dimension in a two dimensional space-time. In a Newtonian world we can ascribe a separate reality to t as a universal time while assuming that real clocks emit *signals* such as $e^{i\omega t}$ *as particular implementations of time measurement.*

The above model has just illustrated the fact that the transition from a discrete clock to a continuum one is conveniently provided through the use of complex numbers. The unimodular complex number is the mathematical counterpart of an analog wristwatch where the counting of digital events is replaced by a continuous increase in phase. By assuming the Gallilean transformation, we have ensured that an ensemble of inertial clocks moving at different speeds keep the same time. The universality of our clock model, the fact that all discrete sequences of events can be interpolated this way, provides a justification of the *priority of time over clocks.*

In the next section, we modify the Galilean invariance of the above **G**-clock to be consistent with the light-speed postulate. In the context of our model, we shall see that this illustrates the transition from complex numbers to Clifford algebras.

2. Minkowski space

In the original digital clock the transition to successive states is accomplished algebraically through left multiplication by T_0. Even and odd states are not distinguished by T_0 as the causal areas are symmetric rectangles, Fig. 5. However, relativistically, a clock moving at constant speed v in the x direction has to stay in states 1 and 3 longer than states 2 and 4, Fig. 6. We need a new transfer matrix T_{m+n} that has to function as an $(m+n)$-th root of T_0^2 while weighting the occupation of state 1 more heavily than state 2 $(m > n)$. This means increasing the magnitude of the 1-1 element and decreasing the magnitude of the 2-2 element. To see how this may be done expand $T(t)$ to first order in $t = 1/n$:

$$T(1/n) \simeq \begin{pmatrix} 1 & -\frac{\pi}{2n} \\ \frac{\pi}{2n} & 1 \end{pmatrix}$$

where we assume $n \gg 1$. The ones on the diagonal equally weight 1-1 and 2-2 transitions. If we consider the replacement $T(1/n) \rightarrow T_v(1/n) = T(1/n) + w\frac{\pi}{2n}\sigma_z$ this will effect an increased residence time in the odd states $(w > 0)$ and a decreased residence time in the even states as suggested by Fig. 6. We choose w so that the 'moving clock runs slow' in accordance with the

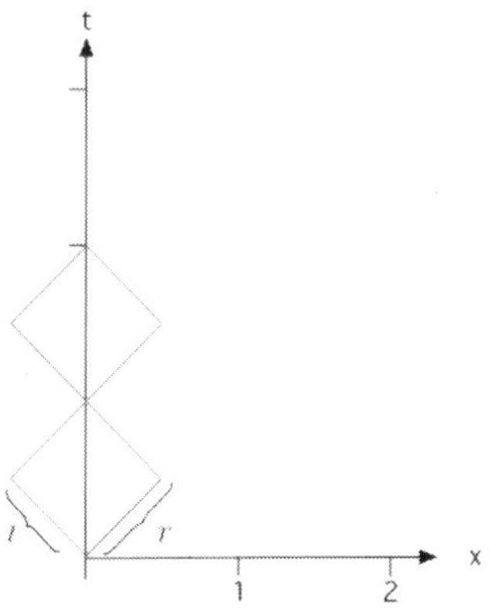

Figure 5. A clock cycle in a co-moving frame. The null boundaries cross at successive on-worldline events. Successive causal regions have the same area but opposite orientations.

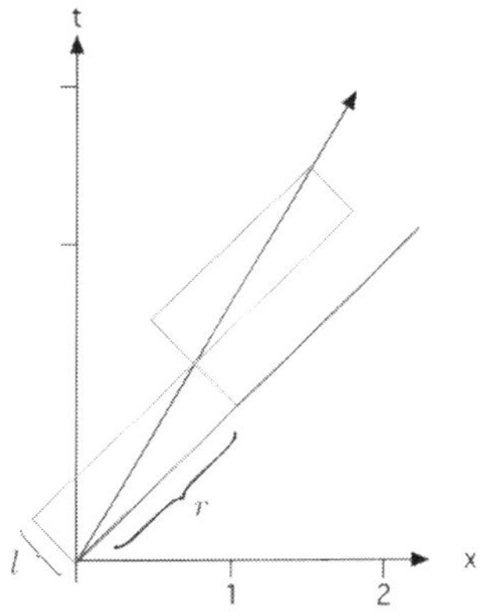

Figure 6. A clock cycle in a moving frame. The null boundaries remain null but the left and right projections change length. The signs and magnitudes of the causal areas are invariant.

Lorentz transformation and the area preservation of the light clock[2]. To be in accordance with Fig. 6 the first on-worldline event should take place at $t = 2\gamma$ where $\gamma = 1/\sqrt{1-v^2}$ is the time dilation factor. So we must have:

$$T_v(2\gamma) = \lim_{n\to\infty} T_v(1/n)^{2n\gamma} \quad = \quad -I_2. \tag{8}$$

For equation (8) to hold, the eigenvalues of T_v raised to the power $2\gamma n$ must be -1 in the limit as $n \to \infty$. The eigenvalues of T_M to lowest order in $1/n$ are $\lambda = 1 \pm \frac{i\pi\sqrt{1-w^2}}{2n}$ giving

$$\lim_{n\to\infty} \lambda^{2\gamma n} = e^{\mp i\pi\sqrt{1-w^2}\gamma}. \tag{9}$$

For this to equal -1 we have $w = \pm v$. Choosing the positive value of v for w we can calculate the transfer matrix for a clock moving at speed v with respect to the lab frame whose transfer matrix was given by (6). The matrix is

$$T_v(t) \quad = \quad \mathrm{I}_2 \cos\left(\frac{\pi \mathrm{t}}{2\gamma}\right) + \gamma(\mathrm{v}\sigma_\mathrm{z} - \mathrm{i}\sigma_\mathrm{y}) \sin\left(\frac{\pi \mathrm{t}}{2\gamma}\right). \tag{10}$$

Equation (10) is the appropriate transfer matrix for the moving clock and, as in the Newtonian case, the continuum limit increases the density of on-worldline events based on the assumption that clocks have worldlines and can emit continuous signals. The mathematical convenience and physical justification for Minkowski spacetime can be seen by viewing the transfer matrix as a set of instructions for constructing an analog time-signal along a worldline. Notice that the transfer matrix has a structure that can be partitioned into 'kinematic' and 'dynamic' information. Setting the base frequency of the above **M**-clock at ω we have

$$T_v(t) = \mathrm{I}_2 \cos\left(\frac{\omega \mathrm{t}}{\gamma}\right) + \underbrace{\gamma(v\sigma_z - i\sigma_y)}_{\text{kinematics}}\overbrace{\sin\left(\frac{\omega t}{\gamma}\right)}^{\text{dynamics}}. \tag{11}$$

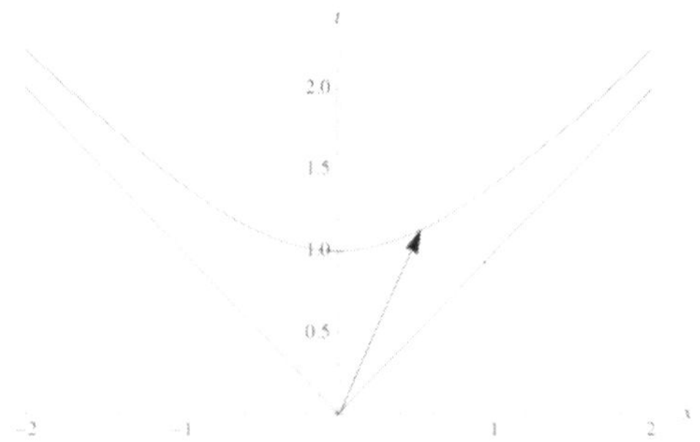

Figure 7. The odd part of the transfer matrix is a vector. The unit vector starts at the origin and ends on the first on-worldline event. The Euclidean length of the vector is direction dependent, however the causal area it represents is a Lorentz invariant.

The dynamic information is clock-specific and contains the frequency and phase information. Comparing this to the transfer matrix for the **G**-clock equation (6) we see that the matrix implementation of the unit imaginary T_0 has been replaced by the (bi)vector $\mathbf{J_v} = \gamma(v\sigma_z - i\sigma_y)$ that has similar properties. Notice that $\mathbf{J_v}^2 = -\mathrm{I}_2$ and that time evolution generated by T_v exhibits the time dilation appropriate for the Lorentz transformation. Unlike the unit imaginary, the vector $\mathbf{J_v}$ has a directional aspect, providing as it does, the direction of the proper time axis of the moving clock, Fig. 7. The intrinsic normalization of $\mathbf{J_v}$ is through spacetime *area* as opposed to length. It is this feature that leads to the mixed signature of spacetime. As we shall see in the next section, the 'wave-particle duality' of quantum mechanics is an alternative manifestation of the spacetime area-invariance of the Lorentz transformation.

In terms of information, equation (10) is a direct algebraic encoding of the first half of Wheeler's famous maxim[3]:

"Spacetime tells matter how to move; matter tells spacetime how to curve."

Equation (10) is literally the instruction for the clock to generate a spiral signal as it moves along its worldline at constant velocity. The coefficient of the vector part of (10) is plotted in Fig. 8 as a colour map showing the phase of the clock along the worldline. In the continuum limit, the transfer matrix has filled in the worldline of the clock with an increasing density of events that is mapped onto the phase of the trig function. The vector $\mathbf{J_v}$ ensures the phase is properly distributed according to the Lorentz transformation for each velocity. Fig. 9] shows the ensemble of clock phases for all velocities.

Since for all clocks, the kinematic structure of $T_v(t)$ is independent of the clock's base frequency, this suggests that we ascribe this feature of clocks to an ambient *spacetime* using frame vectors σ_x and $-i\sigma_y$ for the x and t coordinates respectively. With this convention, boosts are handled algebraically like rotations in a Euclidean space. This is a mathematical convenience, but one that has precedence in Newton's clock. In that case we concluded that because all discrete periodic clocks could be smoothly interpolated along their worldline to give a unimodular complex number with time linearly related to phase, time itself became 'fundamental' and clock signals became the physical approximations of 'reality'.

In this relativistic case, although we can no longer divorce time from space via the Galilean transformation, the algebraic analog is clear. We simply ascribe the kinematic information of clocks to 'spacetime'. In doing so, we are keeping the worldline feature of macroscopic clocks intact and inventing spacetime as a global attribute of Nature that controls boosts. Here spacetime becomes the reality that all physical clocks approximate. Individual clock *signals* are then the derivative concept, not the spacetime that represents all clocks.

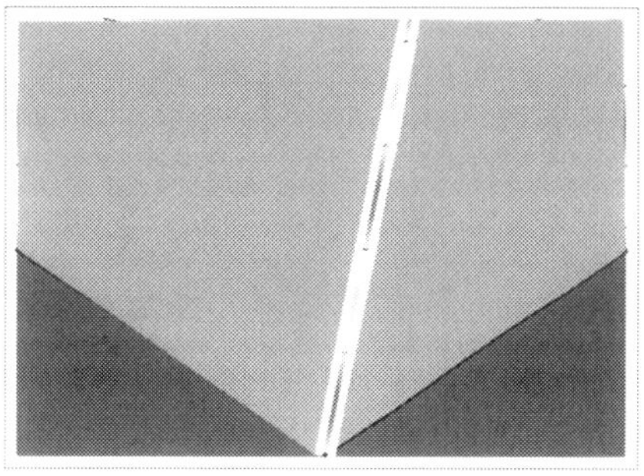

Figure 8. The coefficient of the vector term in equation (10) plotted as a colour map along the direction of $\mathbf{J_v}$ inside the light cone.

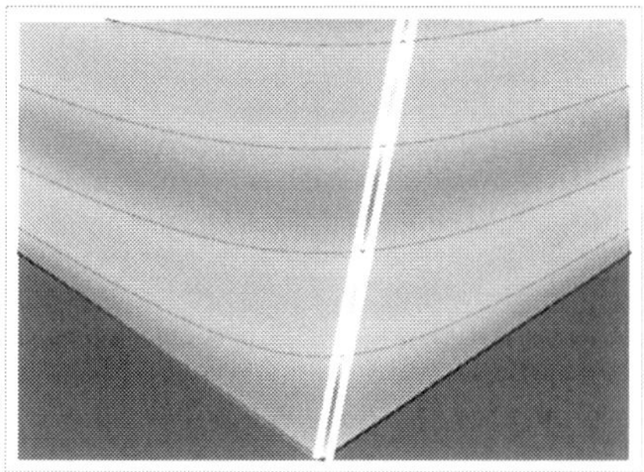

Figure 9. The coefficient of the vector in equation (10) plotted against a background of velocities $\{-1 < v < 1\}$ inside the light cone.

The argument that Newton's absolute time and Minkowski's spacetime are logically prior concepts to clock signals is rooted in the idea that real objects, including clocks, have worldlines. While macroscopically this is true, microscopic verification is simply not possible below some physical scale. Furthermore, the uncertainty principle calls the concept of smooth worldlines into question, suggesting fractal rather than smooth paths[4] as a basis for quantum propagation. This being the case, the above argument for a logically prior spacetime is weakened by its assumption of a continuum limit that 'fills in' the worldline to arbitrarily fine scales. In the next section we relinquish worldlines in favour of a continuum limit that packages the boost information with the clock itself.

3. The Dirac clock

In the previous two sections we associated discrete periodic clocks with chains of oriented areas, Fig. 1. We argued that all such periodic clocks give rise to either Newton's clock or Minkowski's clock depending on how boosts affect timekeeping. The universality of the transcription to absolute time or Minkowski spacetime encouraged a perspective in which time or spacetime comes first and clock signals are derivative concepts. The continuum limit is however in question since the worldlines of quantum particles are not verifiable. As a result, placing the continuum limit along the worldline, as we have done in the previous sections, is suspect. By placing the continuum limit along worldlines we risk building the quantum Zeno effect into the clock model through this limit, thus excluding quantum propagation!

A less restrictive continuum limit can be obtained by 'repackaging' the information that we commonly ascribe to spacetime. Consider an isolated periodic clock in empty space in which we are a 'special'[1]observer. We observe a single discrete event of the clock and we use that to choose the origin of our spacetime coordinate system. With a single observation we do not know the speed or direction of the clock with respect to our chosen frame so the orientation of our time axis will not in general coincide with that of the clock, neither will the 'instruction set' for the clock be written in terms of our arbitrary reference frame. In fact, if we do not believe that Nature uses an informative spacetime to 'tell the clock how to move', the clock itself must have prepackaged information that we as observers can *access* from whatever spacetime coordinate system we construct. The situation in comparison to previous clocks is visually depicted in figures [10] and [11]. In Fig. 10 the continuum limit is taken along the worldline with spacetime providing instructions that keep the timekeeping appropriate to the clock's inertial frame. In

[1] Special in the sense that we can observe the process without disturbing it in any way.

Journal of Physics: Conference Series **504** (2014) 012011 doi:10.1088/1742-6596/504/1/012011

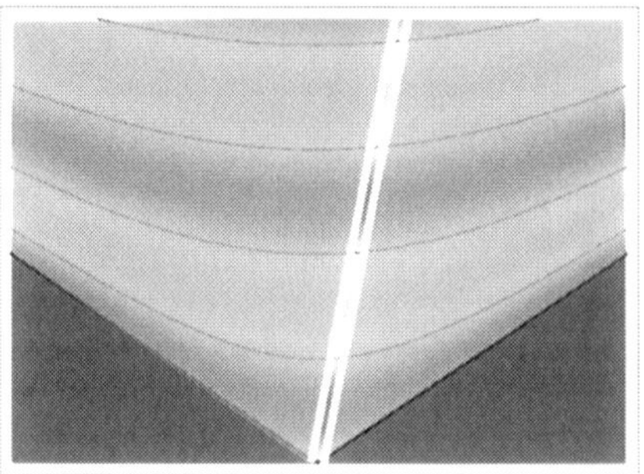

Figure 10. The coefficient of the vector term in equation (10) plotted as a colour map along the worldline of the clock. The continuum limit for Minkowski space is taken along the worldline of the clock. The transfer matrix 'tells the clock how to keep time' with the clock moving at fixed v.

Figure 11. The continuum limit for the generating function for boosts is taken at fixed t. It encodes the information the clock must access if it can be read from an arbitrary inertial frame. The resulting generating function satisfies the Dirac equation.

Fig. 11 we assume the clock itself contains the information needed for the whole ensemble of relative velocities from which our chosen spacetime is drawn. Shifting this information from spacetime to the clock itself corresponds to associating the clock with the *generating function* for the transfer matrices that we associate with Minkowski's clock. In this case the continuum limit is no longer along the worldline, it is at whatever value of t we choose to observe the clock in our own frame. This changes the role of the continuum limit from implementing a continuous evolution along the worldline to one that encodes the temporal history along the spatial direction.

To implement the transfer matrix as a generating function we use complex exponentials to count ticks of the clock[5, 6]. Algebraically we replace I_2 by $\exp[-ip\sigma_z\epsilon]$ in the $t = \epsilon << 1$ expansion of the transfer matrix, take the continuum limit, and then find the state of the clock at a specific *spacetime* point by inverting what has become a Fourier transform. Thus we consider replacing the short-time transfer matrix by

$$\begin{pmatrix} 1 & -m\epsilon \\ m\epsilon & 1 \end{pmatrix} \rightarrow \begin{pmatrix} e^{ip\epsilon} & -m\epsilon \\ m\epsilon & e^{-ip\epsilon} \end{pmatrix}.$$

The continuum limit of this generating function gives

$$\begin{aligned} T(p,t) &= \lim_{\epsilon\rightarrow 0} \begin{pmatrix} \exp[ip\epsilon] & -m\epsilon \\ m\epsilon & \exp[-ip\epsilon] \end{pmatrix}^{t/\epsilon} \\ &= \cos(Et)I_2 - \frac{1}{E}(ip\sigma_z + im\sigma_y)\sin(Et) \end{aligned} \quad (12)$$

where $E = \sqrt{p^2 + m^2}$ and for generality we have replaced $\pi/2$ by m and $1/n$ by ϵ. Notice here the generating function form of the transfer matrix still has a multivector structure that is a combination of a scalar and a unit vector since

$$(\frac{-1}{E}(ip\sigma_z + im\sigma_y))^2 = \frac{1}{E^2}(-p^2 - m^2)I_2 = -I_2. \quad (13)$$

The generating function has singled out one-dimensional subspaces of a 2-dimensional domain via a unit vector. To return to 'configuration' space notice that the generating function is also a Fourier transform and can be inverted as such. Namely

$$\begin{aligned} T(x,t) &= \lim_{\epsilon\to 0}\frac{1}{2\pi\epsilon}\int_{-\pi/\epsilon}^{\pi/\epsilon} e^{ix(p\epsilon)}\begin{pmatrix}\exp[ip\epsilon] & -m\epsilon\\ m\epsilon & \exp[-ip\epsilon]\end{pmatrix}^{t/\epsilon} d(p\epsilon)\\ &= \frac{1}{2\pi}\int_{-\infty}^{\infty} e^{-ipx}\left(I_2\cos(Et)+\frac{1}{E}(ip\sigma_z+im\sigma_y)\sin(Et)\right)dp, \qquad |x|<t. \quad (14)\end{aligned}$$

the inverse Fourier transform of $T(p,t)$. Here the division of the integral by ϵ in the continuum limit gives $T(x,t)$ as a density. Notice the relation between Eqn. (14) and Fig. 11. In Eqn. (14), t and m are fixed and the integral over p samples all possible boosts. This corresponds to all possible values of v in Minkowski's clock in Fig. 11. The continuum limit in the generating function assigns a phase to each point in Minkowski's spacetime based on the fundamental frequency m of the clock. We call this model Dirac's clock because (14) satisfies the free-particle Dirac equation[1, 7, 8]. The only difference between the **M**-clock and this clock is the role of the continuum. The limit for the **D**-clock does not form worldlines and the causal areas between events is not reduced to zero in the limit. This means that the paths associated with the transfer matrix products, squashed into worldlines in the **M**-clock, expand into the future cone of the origin for the **D**-clock Fig. 12.

4. Discussion

The Dirac equation is commonly regarded as a unique harbinger of relativistic quantum mechanics, yet the generating function (14) satisfies this enigmatic equation without any appearance of a quantum context. How can this be?

The answer lies at the heart of the quest for an emergent quantum mechanics. In their conventional formulations, QM and SR are based on very different paradigms. The former is about process and propagation, while the latter is about events and objects. The conventional merger of the two at the Dirac equation gives the appearance of a theory more fundamental than either constituent. However, obtaining a fundamental equation by merging two approximations is an inspired kludge, not a derivation.

The above clock models show us how spacetime and quantum propagation emerge from periodic processes in a relativistic world and why, once emerged, they appear to represent conflicting paradigms. Minkowski spacetime arises through a continuum limit that creates a worldline for a clock, the causal area between events going to zero in the continuum limit (Fig. 2). Timekeeping information is ultimately transmitted along rays of constant velocity in a two dimensional spacetime that acquires responsibility for the Lorentz transformation. That is, a fair statement of the significance of the **M**-clock is:

> *Any* periodic sequence of events, stationary in an inertial frame, may be smoothly interpolated to form a worldline described by a transfer matrix of the form (11).

This may be construed to mean that it is a property of an ambient spacetime that all digital clocks behave in this way, and that both spacetime and the worldline are physically real in functionality, if not observability. This interpretation, clearly useful for macroscopic objects, is questionable in its assumption of scale invariance.

As an analogy, in the case of diffusion, the assumption of scale invariance is the condition that the mean free time of a diffusing particle be zero in the continuum limit. In the derivation of the diffusion equation from kinetic theory the vanishing of the mean free time is analogous to the disappearance of the causal areas between events in the **M**-clock. While it is easy to recognize the 'non-physical' aspects of this in the diffusion equation[2], it takes a model clock to

[2] For example, shrinking the mean free time to zero creates Wiener paths of infinite length with fractal dimensions of 2. It also gives the diffusion equation an infinite signal velocity. These features are non-physical, but have approximate validity over scales much larger than the actual mean free time.

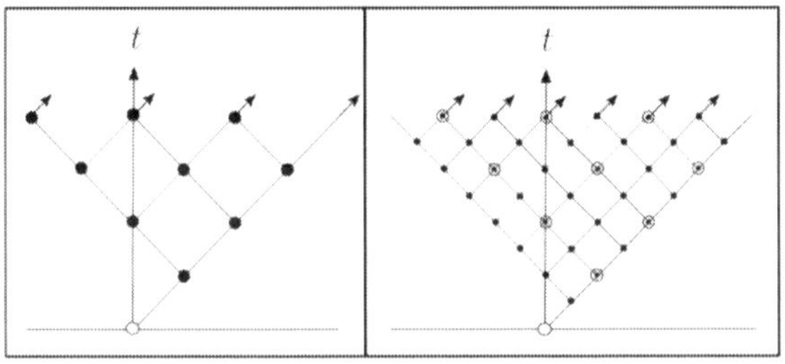

Figure 12. Refining the lattice for the generating function (Dirac's clock Eqn. (14)). Plotted are four steps in time for the (1, 1) element of T_0 and its first refinement at half the original step size. Comparing this to the refinement of the step size for Minkowski's clock (Fig. 2), notice the proliferation of paths rather than the constriction of a causal chain.

demonstrate that the spacetime of Minkowski contains assumptions of scaling that are ultimately unphysical.

Dirac's clock addresses the implicit scale invariance of the 'Spacetime telling matter how to move' assumption in Minkowski's clock. It simply shifts the boost information from a containing spacetime to the clock itself, preserving the dynamical information that Minkowski's clock omits. This is done mathematically by replacing the transfer matrix by a generating function(Fig. 12). It reverses the roles of time and clock-signal that the **M**-clock inherits from its Newtonian predecessor. Whereas clock-signals are functions of time for Minkowski, spacetime is a function of clock-signals for Dirac. This latter perspective has the advantage that mass is inextricably linked to spacetime, as is the Fourier uncertainty principle. Conventional treatments have to reintroduce these to Minkowski's spacetime (via dynamics and quantization respectively) since they are both removed in the **M**-clock's continuum limit.

The parallel emergence of Minkowski spacetime and the Dirac equation shown here(Fig. 13) exposes a weakness in conventional interpretations of both constructs. The clock model shows that spacetime is an abstraction of properties that are shared by all clocks *from the perspective of a continuum limit in which the clock period is always much smaller than* **any** *other timescale under consideration.* Spacetime is clearly relevant for macroscopic objects but the scale invariance eliminates the possibility of intrinsic timekeeping for microscopic objects.

Similarly the conventional picture of the Dirac equation arising as a transplant of the quantum mechanics of Schrödinger into the relativistic domain creates the illusion that the origin of quantum propagation is not itself relativistic. The **D**-clock shows that this is highly unlikely. The path-dependent phase of the **D**-clock is a direct inheritance of the time dilation present in the **M**-clock *but absent in the* **G**-*clock.* In this view, Schrödinger's equation inherits its path-dependent phase from relativity. Its apparent independence from relativity is an artifact similar to the apparent independence of kinetic energy from 'c' in non-relativistic mechanics.

Returning to the question in the title: 'Which came first, spacetime or clocks?' the three clock models illustrate that the answer illuminates the boundary between classical and quantum physics. If the spacetime of Minkowski is a reality that is independent of individual clocks and holds to arbitrarily small scales, then all clocks appear as classical objects with worldlines. Quantum propagation is consequently 'squeezed out' and must ultimately have a separate origin. On the other hand, if spacetime is a scale-dependent ensemble concept and there exist clocks that contain all the boost information they need to be consistent with an ambient spacetime, Dirac propagation is an intrinsic property of such clocks. In this case, as is increasingly suggested by both theory and experiment,[9, 10, 7, 11, 12], "particles are clocks".

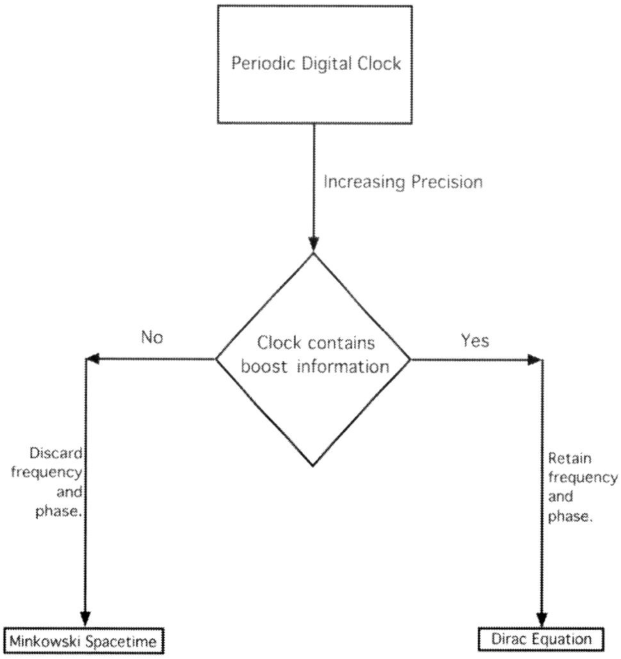

Figure 13. The relation between boost information and clocks in the continuum limit determines the choice of quantum or classical behaviour. If boost information is relegated to spacetime via worldlines in the continuum limit, clocks measure an ambient spacetime and the physics is classical. If clocks include the information for boosts, the resulting generating function obeys the Dirac equation. This suggests that the physics is no longer classical.

Acknowledgments
The author would like to thank AINS and the Fetzer Franklin Fund for the opportunity to participate in EmQM13, and the Channon Foundation for financial support of this work.

References
[1] Ord G 2012 *J. Phys.: Conf. Ser.* **361** 012007
[2] Ord G and Mann R 2011 *Int. J. Theor. Phys.* **51** 652–666 DOI 10.1007/s10773-011-0945-1
[3] Ford K and Wheeler J A 2010 *Geons, Black Holes, and Quantum Foam: A Life in Physics* (W. W. Norton and Company)
[4] Ord G N 1983 *J. Phys. A* **16** 1869–1884
[5] Thompson C J 1979 *Mathematical Statistical Mechanics* (Princeton University Press)
[6] Ord G N 1992 *J. Stat. Phys.* **66** 647–659
[7] Ord G N 2009 *Ann. Phys.* **324** 1211–1218 DOI 10.1016/j.aop.2009.03.007
[8] Ord G N and Mann R B 2003 *Phys. Rev. A* **67** 022105
[9] Penrose R 2013 *Ideas Roadshow* ed Burton H (Open Agenda Publishing) URL http://www.ideasroadshow.com/issues/roger-penrose-2013-07-12
[10] Lan S Y, Kuan P C, Estey B, English D, Brown J M, Hohensee M A and Müller H 2013 *Science* **339** 554–557 (*Preprint* http://www.sciencemag.org/content/339/6119/554.full.pdf)
[11] Poirier B 2013 *talk at this conference*
[12] Jizba P 2013 *this volume*

Cooperative dynamical processes: the emergence of relativistic quantum theory

Petr Jizba
FNSPE, Czech Technical University in Prague, Břehová 7, 115 19 Praha 1, Czech Republic
E-mail: p.jizba@fjfi.cvut.cz

Fabio Scardigli
Dipartimento di Matematica, Politecnico di Milano, Piazza Leonardo da Vinci 32, 20133 Milano, Italy
E-mail: fabio@phys.ntu.edu.tw

Abstract. There is a theoretical evidence that relativistically invariant quantum dynamics at (enough) large space-time scales can result from a cooperative process of two inter-correlated *non-relativistic* stochastic dynamics, operating at different energy scales. We show that the Euclidean transition amplitude for a relativistic particle is identical to the transition probability of a Brownian particle propagating in a granular space. We discuss the issue of the robustness of the special-relativistic quantum mechanics thus obtained under small changes in the granular-space distribution. Experimental implications for early Universe cosmology are also briefly outlined.

1. Introduction

Scenarios that strive to describe quantum theory as an emergent, non-primitive concept typically run into difficulties when trying to address a relativistic generalization. In this paper we discuss a possible way out of this situation by showing that the observed relativistic behavior in the quantum world might well be just a statistically emergent phenomenon out of deeper non-relativistic level of quantum dynamics. Key in our discussion is the concept of *emergence*, i.e., an idea appearing in various areas of contemporary science. The emergence paradigm has a long tradition also in physics. For instance, in gravity, starting from the insights of Sacharov [1], the notion of emergent gravitational interaction has shaped large parts of theoretical cosmology during the last 40 years. In particle and condensed matter physics one can observe a wealth of phenomena exhibiting several inter-correlated time or energy scales with cooperative behaviors. The ensuing emergent dynamics is then, as a rule, not reducible to the behaviors of its parts. In these systems, the observed macroscopic-scale dynamics and related degrees of freedom differ drastically from the actual underlying microscopic-scale physics [2]. The concept of *Superstatistics* provides an explicit realization of this paradigm: It shows that the emergent behavior can be often regarded as a *superposition* of several statistical systems that operate at different spatio-temporal scales [3, 4]. To date, many successful applications are known, in hydrodynamic turbulence [5], turbulence in quantum liquids [6], pattern forming systems [7] or scattering processes in high-energy physics [8].

In order to work, the superstatistics requires the existence of sufficient separations between the spatio-temporal scales of the relevant dynamics, so that the system has enough time to relax to a local equilibrium state and to stay within it for some time. Usually we are concerned with two scales. As in Ref. [3], we consider an intensive parameter ζ that fluctuates on a much larger time scale than the typical relaxation time of the local dynamics. The random variable ζ can be in practice identified, e.g., with an inverse temperature [3, 4], a coupling constant [3], a friction constant [9], volatility [10] or einbein [11]. The *adiabatic Ansatz* allow us to understand the superstatistics on an intuitive ground. Namely, the system under consideration, during its evolution, travels within its state space X (described by a state variable $A \in X$) which is partitioned into small cells characterized by a sharp value of ζ. Within each cell, the system is described by the conditional distribution $p(A|\zeta)$. As ζ varies adiabatically from cell to cell, the joint distribution of finding the system with a sharp value of ζ in the state A is $p(A,\zeta) = p(A|\zeta)p(\zeta)$ (Bayes theorem). The resulting macro-scale (or emergent) statistical distribution $p(A)$ for finding a system in the state A is obtained by marginalizing over the distribution of the variable being discarded, i.e.,

$$p(A) = \int \mathrm{d}\zeta\, p(A|\zeta)p(\zeta)\,. \tag{1}$$

In mathematical language $p(A)$ is known as the marginal distribution. From a technical point of view, a sufficient time scale separation between two relevant dynamics of the studied system allows to qualify superstatistics as a form of slow modulation [12].

In this paper, we recast the Feynman transition amplitude of a relativistic scalar particle into a form which, after being analytically continued to imaginary times, coincides with the superstatistics marginal probability (1). The derivation is based on the Lévy–Khinchine theorem for infinitely divisible distributions [13, 14]. For illustration we consider the dynamics and the propagator of a Klein–Gordon (i.e., neutral spin−0) particle. Our reasonings can be also extended to charged spin−0, spin−$\frac{1}{2}$, Proca's spin−1 particles and to higher-spin particles phrased via the Bargmann–Wigner wave equation [11]. Further generalization to external electromagnetic potential has been reported in Refs. [11, 15]. The above formulation can be visualized as if the particle would randomly propagate (in the sense of Brownian motion) through an inhomogeneous or granular medium ("vacuum") [15].

Our argument is based upon a recent observation [10, 11, 15] that the Euclidean path integral (PI) for relativistic particles may be interpreted as describing a doubly-stochastic process that operates at two separate spatio-temporal scales. The short spatial scale, which is much smaller than particle's Compton length $\lambda_C = 1/mc$ ($\hbar = 1$ is assumed throughout this paper), describes a Wiener (i.e., Galilean relativity) process with a sharp Newtonian (galilean-invariant) mass. The large spatial scale, which is of order λ_C, corresponds to distances over which the fluctuating Newtonian mass changes appreciably. At scales much larger than λ_C the particle evolves according to a genuine relativistic dynamics, with a sharp value of the mass coinciding with the Einstein rest mass. Particularly striking is the fact that when we average the particle's velocity over the structural correlation distance (i.e., over particle's λ_C) we obtain the velocity of light c. So the picture that emerges from this analysis is that the particle (with a non-zero mass!) propagates over the correlation distance λ_C with an average velocity c, while at larger distance scales (i.e., when a more coarse grained view is taken) the particle propagates as a relativistic particle with a sharp mass and an average velocity that is subluminal. Quite remarkably, one can observe an identical behavior in the well-known Feynman's checkerboard PI [17, 18] to which the transition amplitude (1) reduces in the case of a relativistic Dirac fermion in $1+1$ dimensions [11, 15].

An expanded presentation, including the issue of reparametrization invariance, bibliography, and proofs of the main statements and formulas can be found in Ref. [15].

2. Path integrals and combined statistics

When a conditional probability density function (PDF) is formulated through a PI, then it satisfies the semigroup equation for continuous Markovian processes — the Einstein–Smoluchowski equation (ESE), namely [16]

$$p(y,t''|x,t) = \int_{-\infty}^{\infty} \mathrm{d}z\, p(y,t''|z,t')p(z,t'|x,t)\,, \tag{2}$$

with t' being any time between t'' and t. Converse is also true, namely any transition probability satisfying ESE possesses a genuine PI representation [17]. In complex dynamical systems one often encounters probabilities formulated as a superposition of PI's,

$$\wp(x',t'|x,t) = \int_0^{\infty} \mathrm{d}\zeta\; \omega(\zeta,T) \int_{x(t)=x}^{x(t')=x'} [\mathrm{d}x\,\mathrm{d}p]\; e^{\int_t^{t'} \mathrm{d}\tau(ip\dot{x}-\zeta H(p,x))}\,. \tag{3}$$

Here $\omega(\zeta,T)$ with $T = t'-t$ is a normalized PDF defined on $\mathbb{R}^+\times\,\mathbb{R}^+$. The form (3) typically appears in non-perturbative approximations to statistical partition functions, in polymer physics, in financial markets or in systems with reparametrization invariance. The random variable ζ is then related to the inverse temperature, coupling constant, volatility or vielbein, respectively.

The existence of different time scales and the flow of the information from slow to fast degrees of freedom typically creates the irreversibility on the macroscopical level of the description. The corresponding information thus is not lost, but passes in a form incompatible with the Markovian description. The most general class of distributions $\omega(\zeta,T)$ on $\mathbb{R}^+\times\,\mathbb{R}^+$ for which the superposition of Markovian processes remain Markovian, i.e., when also $\wp(x',t'|x,t)$ satisfies the ESE (2), was found in Ref. [10]. The key is to note that in order to have (2) satisfied by $\wp$, the rescaled PDF $w(\zeta,T) \equiv \omega(\zeta/T,T)/T$ should satisfy the ESE for homogeneous Markovian process

$$w(\zeta,t_1+t_2) = \int_0^{\zeta} \mathrm{d}\zeta'\, w(\zeta',t_1)w(\zeta-\zeta',t_2)\,. \tag{4}$$

Because of the convolution appearing of the right-hand-side of (4), the Laplace image $\mathcal{L}(w)(p_\zeta,T) \equiv \tilde{w}(p_\zeta,T)$ fulfills the simple multiplicative functional equation. By assuming continuity in T, it follows that the multiplicative semigroup $\tilde{w}(p_\zeta,T)_{T\geq 0}$ satisfies $\tilde{w}(p_\zeta,T) = \{\tilde{w}(p_\zeta,1)\}^T$. From this we can deduce that the distribution of ζ at T is completely determined by the distribution of ζ at $T=1$. In addition, because $\tilde{w}(p_\zeta,1) = \{\tilde{w}(p_\zeta,1/n)\}^n$ for any $n\in\mathbb{N}^+$, $w(\zeta,1)$ is infinitely divisible. The Lévy–Khinchine theorem [14, 13] then ensures that $\log\tilde{w}(p_\zeta,T) \equiv -TF(p_\zeta)$ must have the generic form

$$\log\tilde{w}(p_\zeta,T) = -T\left(\alpha p_\zeta + \int_0^{\infty}(1-e^{-p_\zeta u})\nu(\mathrm{d}u)\right), \tag{5}$$

where $\alpha \geq 0$ is a drift constant and ν is some non-negative measure on $(0,\infty)$ satisfying $\int_{\mathbb{R}^+}\min(1,u)\nu(\mathrm{d}u) < \infty$. Finally the Laplace inverse of $\tilde{w}(p_\zeta,T)$ yields $\omega(\zeta,T)$. Once $\omega(\zeta,T)$ is found, then $\wp(x',t'|x,t)$ possesses a PI representation on its own. What is the form of the new Hamiltonian? To this end we rewrite (3) in Dirac operator form as [10]

$$\wp(x',t'|x,t) = \langle x'|\int_0^{\infty}\mathrm{d}\zeta\, w(\zeta,T)e^{-\zeta\hat{H}}|x\rangle = \langle x'|\{\tilde{w}(\hat{H},1)\}^T|x\rangle = \langle x'|e^{-TF(\hat{H})}|x\rangle\,. \tag{6}$$

Hence, the identification $\mathcal{H}(\boldsymbol{p},\boldsymbol{x}) = F(H(\boldsymbol{p},\boldsymbol{x}))$ can be made. Here one might worry about the operator-ordering problem. For our purpose it suffices to note that when H is $\boldsymbol{x}$-independent, the former relation is exact. In more general instances the Weyl ordering is a natural choice because in this case the required mid-point rule follows automatically and one does not need to invoke the gauge invariance [10, 26]. In situations when other non-trivial configuration space symmetries (such as non-holonomic symmetry) are required, other orderings might be more physical [10].

3. Emergence of relativistic quantum theory

Our attention now turns to the Feynman transition amplitudes or better its Euclidean version — transition probabilities. We will now show that the latter naturally fits into the structure of superstatistics PI's. To this end we note that the choice $\alpha = 0$ and $\nu(du) = 1/(2\sqrt{\pi}u^{3/2})du$ leads to $F(p_\zeta) = \sqrt{p_\zeta}$. This identifies $w(\zeta, T)$ with the Lévy distribution with the scale parameter $T^2/2$. Moreover, when $H(\boldsymbol{p}, \boldsymbol{x}) = \boldsymbol{p}^2c^2 + m^2c^4$ then (3) can be cast into the form (see also Refs. [10, 11, 15])

$$\int_{\boldsymbol{x}(0)=\boldsymbol{x}}^{\boldsymbol{x}(T)=\boldsymbol{x}'} [\mathrm{d}\boldsymbol{x}\,\mathrm{d}\boldsymbol{p}] \exp\left\{\int_0^T \mathrm{d}\tau \left[\mathrm{i}\boldsymbol{p}\cdot\dot{\boldsymbol{x}} - c\sqrt{\boldsymbol{p}^2 + m^2c^2}\right]\right\}$$

$$= \int_0^\infty \mathrm{d}\mathfrak{m}\,\mathcal{F}_{\frac{1}{2}}\left(\mathfrak{m}, Tc^2, Tc^2m^2\right) \int_{\boldsymbol{x}(0)=\boldsymbol{x}}^{\boldsymbol{x}(T)=\boldsymbol{x}'} [\mathrm{d}\boldsymbol{x}\,\mathrm{d}\boldsymbol{p}] \exp\left\{\int_0^T \mathrm{d}\tau \left[\mathrm{i}\boldsymbol{p}\cdot\dot{\boldsymbol{x}} - \frac{\boldsymbol{p}^2}{2\mathfrak{m}} - mc^2\right]\right\} \quad (7)$$

where $T = t' - t$, and

$$\mathcal{F}_p(z, a, b) = \frac{(a/b)^{p/2}}{2K_p(\sqrt{ab})}\, z^{p-1}\, \mathrm{e}^{-(az+b/z)/2}, \quad (8)$$

is the generalized inverse Gaussian distribution [14] (K_p is the modified Bessel function of the second kind with index p). The LHS of (7) represents the PI for the free spinless relativistic particle in the Newton–Wigner representation [19]. The full Klein–Gordon (KG) kernel which also contains the negative-energy spectrum can be obtained from (7) by considering the Feshbach–Villars representation of the KG equation and making the substitution [11]

$$\mathcal{F}_{\frac{1}{2}}\left(\mathfrak{m}, tc^2, tc^2m^2\right) \mapsto \frac{1+\mathrm{sgn}(t)\sigma_3}{2}\, \mathcal{F}_{\frac{1}{2}}\left(\mathfrak{m}, |t|c^2, |t|c^2m^2\right). \quad (9)$$

The matrix nature of the smearing distribution (σ_3 is the Pauli matrix) naturally includes the Feynman–Stückelberg causal boundary condition and thus treats both particles and antiparticles in a symmetric way [11, 20]. When the partition function is going to be calculated, the trace will get rid of the sgn(t) term and 1/2 is turned to 1.

The explicit form of the identity (7) indicates that $\mathfrak{m}$ can be interpreted as a Galilean-invariant Newton-like mass which takes on continuous values distributed according to $\mathcal{F}_{\frac{1}{2}}\left(\mathfrak{m}, Tc^2, Tc^2m^2\right)$ with $\langle\tilde{m}\rangle = m + 1/Tc^2$ and $\mathrm{var}(\mathfrak{m}) = m/Tc^2 + 2/T^2c^4$. Fluctuations of the Newtonian mass can be then depicted as originating from particle's evolution in an inhomogeneous or granular medium. Granularity, as known, for example, from solid-state systems, typically leads to corrections in the local dispersion relation [21] and hence to alterations in the local *effective mass*. The following picture thus emerges: on the short-distance scale, a non-relativistic particle can be envisaged as propagating via classical Brownian motion through a single grain with a local mass $\mathfrak{m}$. This fast-time process has a time scale $\sim 1/\mathfrak{m}c^2$. An averaged value of the local time scale represents a transient temporal scale $\langle 1/\mathfrak{m}c^2\rangle = 1/mc^2$ which coincides with particle's Compton time T_C — i.e., the time for light to cross the particle's Compton wavelength. At time scales much longer than T_C (large-distance scale), the probability that the particle encounters a grain which endows it with a mass $\mathfrak{m}$ is $\mathcal{F}_{\frac{1}{2}}(\mathfrak{m}, Tc^2, Tc^2m^2)$. As a result one may view a single-particle relativistic theory as a single-particle non-relativistic theory where the particle's Newtonian mass $\mathfrak{m}$ represents a fluctuating parameter which approaches on average the Einstein rest mass m in the large t limit. We stress that t should be understood as the *observation time*, a time after which the observation (position measurement) is made. In particular, during the period t the system remains unperturbed. One can thus justly expect that in the long run all mass fluctuations will be washed out and only a sharp time-independent effective mass will be perceived. The form of $\langle\mathfrak{m}\rangle$ identifies the time scale at which this happens with $t \sim 1/mc^2$, i.e. with the Compton time T_C. It should be stressed that above mass fluctuations have nothing to do with the *Zitterbewegung* which is caused by interference between positive- and negative-energy wave components. In our formulation both regimes are decoupled.

We may also observe that by coarse-graining the velocity over the correlation time T_C we have

$$\langle|\boldsymbol{v}|\rangle_{T_C} = \left.\frac{\langle|\boldsymbol{p}|\rangle}{\langle\mathfrak{m}\rangle}\right|_{T_C} = c\,. \tag{10}$$

So on a short-time scale of order λ_C the spinless relativistic particle propagates with an averaged velocity which is the speed of light c. But if one checks the particle's position at widely separated intervals (much larger than λ_C), then many directional reversals along a typical PI trajectory will take place, and the particle's net velocity will be then less than c — as it should be for a massive particle. The particle then acquires a sharp mass equal to Einstein's mass, and the process (not being hindered by fluctuating masses) is purely Brownian. This conclusion is in line with the well-known Feynman checkerboard picture [15, 18] to which it reduces in the case of $(1+1)$D relativistic Dirac particle.

4. Robustness of the emergent relativistic physics

Understanding the robustness of the emergent Special Relativity under small variations in the mass-smearing distribution function $\mathcal{F}_{\frac{1}{2}}$ can guide the study of the relation between Einsteinian SR and other deformed variants of SR, such as Magueijo–Smolin and Amelino-Camelia's doubly special relativity [22, 23], or (quantum) κ-Poincaré deformation of relativistic kinematics [24]. In DRS models a further invariant scale ℓ is introduced, besides the usual speed of light c, and ℓ is typically considered to be of order of the Planck length. A small variation $\delta\mathcal{F}_{\frac{1}{2}}$ of the smearing function originates the new Hamiltonian

$$\bar{H} = \frac{\epsilon_1}{4} + \left(1 + \frac{\epsilon_0}{2}\right)\sqrt{\boldsymbol{p}^2c^2 + m^2c^4 + \frac{\epsilon_2}{4}}\,, \tag{11}$$

with $\epsilon_1 = -2\,(1+\epsilon_0/2)\,\sqrt{\epsilon_2}$ (see Ref. [15] for details). By setting

$$\epsilon_1 = 2\left(\sqrt{\frac{1}{1-c^2m^2\ell^2}} - 1\right), \quad \epsilon_2 = \frac{4c^6m^4\ell^2}{1-c^2m^2\ell^2}\,,$$

the new Hamiltonian $\bar{H}$ can be easily identified with

$$\bar{H} = c\,\frac{-m^2c^2\ell \mp \sqrt{\boldsymbol{p}^2(1-m^2c^2\ell^2) + m^2c^2}}{1-m^2c^2\ell^2}\,, \tag{12}$$

which coincides with the Magueijo–Smolin's doubly special relativistic Hamiltonian, in, say, its version [25].

It should be stressed that the Hamiltonian (11) (when also negative energy states are included) violates CPT symmetry. This is a typical byproduct of the Lorentz symmetry violation in many deformed SR systems.

For the Hamiltonian (12) a relation analog to (7) holds, where now the smearing function has the form $\mathcal{F}_{\frac{1}{2}}(\mathfrak{m}, Tc^2\lambda, Tc^2m^2\lambda)$ with $\lambda = 1/(1-m^2c^2\ell^2)$. The correlation distance $1/mc\lambda$ can be naturally assumed as the minimal size L_{GRAIN} of the "grain of space" of the polycrystalline medium, which is linked to the new invariant scale ℓ by

$$L_{\text{GRAIN}} := \frac{1}{mc\lambda} = \lambda_C(1 - m^2c^2\ell^2)\,. \tag{13}$$

By tuning the size L_{GRAIN} of these "grains of space" it is possible to pass continuously from Lorenz symmetry to other different symmetries, as those enjoyed by DSR models. We can in principle speculate that each large scale symmetry could originate from a specific kind of space(time) granularity.

5. Cosmological implications

When spacetime is curved, a metric tensor enters in both PI's in (7) in a different way, yielding different "counterterms" [16, 26]. For instance, in Bastianelli–van Nieuwenhuizen's time slicing regularization scheme [26] one has (when $\hbar$ is reintroduced)

$$\begin{aligned}\frac{\boldsymbol{p}^2}{2\mathfrak{m}} \quad &\mapsto \quad \frac{g^{ij}p_ip_j}{2\mathfrak{m}} + \frac{\hbar^2}{8\mathfrak{m}}(R + g^{ij}\Gamma^m_{il}\Gamma^l_{jm})\,,\\ \sqrt{\boldsymbol{p}^2 + m^2c^2} \quad &\mapsto \quad \sqrt{g^{ij}p_ip_j + \frac{\hbar^2}{4}(R + g^{ij}\Gamma^m_{il}\Gamma^l_{jm}) + m^2c^2} \; + \; \hbar^4\Phi(R, \partial R, \partial^2 R) \; + \; \mathcal{O}(\hbar^6)\,,\end{aligned} \tag{14}$$

where g^{ij}, R, Γ^j_{kl} and $\Phi(\ldots)$ are the (space-like) pull-back metric tensor, the scalar curvature, the Christoffel symbol, and a non-vanishing function of R and its first and second derivatives, respectively. This causes the superstatistics identity (7) to break down, as can be explicitly checked to the lowest order in $\hbar$. The respective two cases will thus lead to different physics. Because Einstein *equivalence principle* requires that the *local* spacetime structure must be identified with a Minkowski spacetime possessing Lorentz symmetry, one might assume the validity of (7) at least locally. However, in different space-time points one has, in general, a different typical length scale of the local inertial frames, depending on the gravitational field. The characteristic size of the local inertial (i.e. Minkowski) frame is of order $1/|K|^{1/4}$ where $K = R_{\alpha\beta\gamma\delta}R^{\alpha\beta\gamma\delta}$ is the Kretschmann invariant and $R_{\alpha\beta\gamma\delta}$ is the Riemann curvature. Relation (7) tells us that the special relativistic description breaks down in regions of size smaller than λ_C. For curvatures large enough, namely for strong gravitational fields, the size of the local inertial frame can become smaller than λ_C, that is $1/|K|^{1/4} \lesssim \lambda_C$. In such regions the special relativistic description is no more valid, and according to (7) must be replaced by a Newtonian description of the events. For instance, in Schwarzschild geometry we have $K = 12\,r_s^2/r^6$, and the breakdown should be expected at radial distances $r \lesssim (\lambda_C^2 r_s)^{1/3}$ (r_s is the Schwarzschild radius) which are — apart from the hypothetical case of micro-black holes (where $\lambda_C \simeq r_s$) — always deeply buried below the Schwarzschild event horizon. In the cosmologically relevant Friedmann-Lemaître-Robertson-Walker (FLRW) geometry, we have $K = 12\,(\dot{a}^4 + a^2\ddot{a}^2)/(ac)^4$, and the breakdown happens when $(\dot{a}^4 + a^2\ddot{a}^2) \gtrsim (ac/\lambda_C)^4$, where $a(t)$ is the FLRW scale factor of the Universe and $\dot{a} = da/dt$. Applying the well-known Vilenkin–Ford model [27] for inflationary cosmology, where $a(t)$ is given by: $a(t) = A\sqrt{\sinh(Bt)}$ with $B = 2c\sqrt{\Lambda/3}$ (Λ is the cosmological constant), we obtain a temporal bound on the validity of local Lorentz invariance, which, expressed in FLRW time, is

$$t \lesssim \frac{1}{B} \operatorname{arcsinh}\left[\frac{B\lambda_{C}}{(8c^4 - (B\lambda_{C})^4)^{1/4}}\right] \equiv \bar{t}\,. \tag{15}$$

By using the presently known [28] value $\Lambda \simeq 10^{-52}\text{m}^{-2}$ and the τ-lepton Compton's wavelength $\lambda^{\tau}_{C} \simeq 6.7 \times 10^{-16}$m (yielding the tightest upper bound on t), we obtain $\bar{t} \simeq 4 \times 10^{-24}$s. Note that, since $B\lambda_{C} \ll c$, then $\bar{t} \simeq \lambda_{C}/c = t_{C}$. Such a violation of the local Lorentz invariance naturally breaks the particle-antiparticle symmetry since there is no unified theory of particles and antiparticles in the non-relativistic physics — formally one has two separate theories. If the resulting matter-antimatter asymmetry provides a large enough CP asymmetry then this might have essential consequences in the early Universe, e.g., for leptogenesis. In this respect, $\bar{t}$ is compatible with the *nonthermal* leptogenesis period that typically dates between 10^{-26}–10^{-12}s after the Big Bang.

Further insights are triggered by the view presented above. If we assume a breakdown of Lorentz invariance in high curvature regions whose size falls below λ_C, then we should notice that the (very) early Universe was exactly in this situation. To describe physics in such regions the Newtonian framework seems to be the right one. If this is indeed the case, then the speed of light is no more constrained to be universal, neither to have the present day numerical value. In particular, light (and with it, information) could travel much faster than now. Fully thermalization of different regions of the very early Universe wouldn't have been an issue at all, in that epoch. Hence, one of the main problems demanding the existence of an inflationary period wouldn't have been there. From this side, our approach has wide contacts with the variable-speed-of-light cosmology [29], proposed exactly to explain the horizon problem and to construct an alternative to cosmic inflation.

This approach stimulates therefore the research of cosmological models where we have violation of Lorentz invariance in small regions, a situation likely happened at the beginning of the Universe. Later, along with the expansion, a full Lorentz symmetry is recovered. Analogies can be certainly considered between our approach and the so called Hořava–Lifschitz gravity [30], a model where relativistic invariance is violated at short scales. Unfortunately present day cosmological models based on such theory are plagued by a number of adjustable phenomenological parameters, which largely screen their actual predictive power. Nevertheless, the construction of Lorentz invariance violating models able to predict measurable consequences on the CMB spectrum remains a main target of this research line.

6. Conclusions and perspectives

The presented picture of emergent special relativity, where the observed space-time symmetries have a superstatistics basis, suggests an alternative reading of relativistic signals phrased in terms of quantum-mechanical probabilities. It is also clear that the Special Relativistic invariance is encoded in a *specific* grain smearing distribution. Notably, the exact Lorentz Symmetry of a space-time has no fundamental significance in our analysis, as it is only an accidental symmetry of the coarse-grained configuration space in which the particle executes a standard Wiener process. In the passage from grain to grain particles's Newtonian mass fluctuates according to an inverse Gaussian distribution. The observed inertial mass of the particle is thus not a fundamental constant, but it reflects the particle's interaction with the granular vacuum (cosmic field). This, in a sense, supports Mach's view of the phenomenon of inertia.

Interactions can be included in our framework in two different ways. The interaction with a background field (such as electromagnetic field) can be directly treated with the superstatistics prescription (7), see [11]. On the other hand, the multi-particle interactions can be consistently formulated by "embedding" the relativistic PI in QFT via the worldline quantization. Such an embedding may help to study several cosmological implications of systems with granular space. If any of such systems quickly flows to the infrared fixed point, any direct effect due to the space discreteness, and related SR violation, might be insignificant on cosmological scales (where Lorentz and diffeomorphism invariance are restored), while it might be crucial in the early Universe, e.g., for leptogenesis and the ensuing baryogenesis. Consequences on the detailed structure of the Cosmic Microwave Background spectrum will be explored in future work.

The presented approach implies a preferred frame. In this connection it is worth of noting that, despite the fact that (7) is not manifestly LS invariant, one may use the Stückelberg trick and introduce a new fictitious variable into the PI (7), in such a way that the new action will have the reparametrization symmetry, but will still be dynamically equivalent to the original action. For relevant details see Ref. [15]. By not knowing the source, one may then view this artificial gauge invariance as being a fundamental or even a defining property of the relativistic theory. One might, however, equally well, proclaim the "polycrystalline" picture as being a basic (or primitive) edifice of SR and view the reparametrization symmetry as a mere artefact of an artificial redundancy that is allowed in our description. It is this second view that we favored here.

The presented scenario cannot directly accommodate the massless particles such as photons (identity (7) holds true only for $m \neq 0$). One possibility would be to use the PI representation of Polyakov–Wheeler for massless particles and try to construct a similar superstatistics duality between Einsteinian and Galilean relativity as in the case of massive particles. This procedure is, however, not without technical difficulties and currently is under investigation. Conceptually is far more simpler to assume that the photon has a small mass. At present, there are a number of experimental limits to the mass of the photons [31]. For instance, tests based on Coulomb's law and the galactic vector potential set an upper limit of $m_\gamma \lesssim 10^{-18}$ eV/c$^2 \simeq 10^{-57}$g. This gives the domain correlation distance for the photon $\simeq 1/m_\gamma c^2 \simeq 10^{43}$m which is bigger than the radius of observable Universe ($\simeq 10^{26}$ m) and so in this picture the photon mass does not fluctuate — it is a quasi-invariant.

Finally, it is hoped that our approach should reinforce the links between the superstatistics paradigm and the approach to quantum gravity based on stochastic quantization [32]. In particular, the outlined granular space could be a natural model for the noise terms in a Parisi–Wu stochastic-like quantization approach to gravity.

Acknowledgments

The authors are grateful to H. Kleinert, Z. Haba, M. Sasaki, M. Sakellariadou and L.S. Schulman for discussions and useful feedbacks. P.J. was supported by GAČR Grant No. P402/12/J077.

References

[1] Sacharov A, 2000 *Gen. Rel. Grav.* **32** 365, reprinted from *Doklady Akademii Nauk SSSR* **177** 70, November 1967.
[2] see, e.g., Anderson P 1972 *Science* **177** 393
Laughlin R B 2005 *A Different Universe: Reinventing Physics from the Bottom Down* (Basic Books: London)
Licata I and Sakaji A 2008 *Physics of Emergence and Organization* (WS: London)
[3] Beck C 2001 *Phys. Rev. Lett.* **87** 180601
Beck C and Cohen E G D 2003 *Physica* **A322** 267
Wilk G and Wlodarczyk Z 2000 *Phys. Rev. Lett.* **85** 2770
[4] Hanel R, Thurner S and Gell-Mann M 2011 *Proc. Natl. Acad. Sci. (USA)* **108** 6390
[5] Reynolds A 2003 *Phys. Rev. Lett.* **91**, 084503
[6] Beck C and Miah S 2012 (*Preprint* arXiv:1207.4062)
[7] Daniels K E, Beck C and Bodenschatz E 2004 *Physica* D **193** 208
[8] Wilk G and Wlodarczyk Z 2009 *Eur. Phys. J.* A **40** 299
Beck C 2009 *Eur. Phys. J.* A **40** 267
[9] Beck C 2007 in Radons G *et al* (eds): *Anomalous Transport: Foundations and Applications*

(John Wiley: London)
[10] Jizba P and Kleinert H 2008 *Phys. Rev.* E **78** 031122
[11] Jizba P and Kleinert H 2010 *Phys. Rev.* D **82** 085016
[12] Allegrini P, Barbi F, Grigolini P and Paradisi P 2006 *Phys. Rev.* E **73** 046136
[13] Bertoin J 1996 *Lévy Processes* (CUP: Cambridge)
[14] Feller W 1966 *An Introduction to Probability Theory and its Applications, Vol. II* (John Wiley: London)
[15] Jizba P and Scardigli F 2012 *Phys. Rev.* D **86** 025029
Jizba P and Scardigli F 2012 *Int. J. Mod. Phys.* B **26** 1241003
Jizba P and Scardigli F 2012 *J. Phys.: Conf. Ser.* **361** 012026
Jizba P and Scardigli F 2012 *J. Phys.: Conf. Ser.* **343** 012048
Jizba P and Scardigli F 2013 *Eur. Phys. J.* C **73** 2491. [arXiv:1301.4091]
[16] Kleinert H 2009 *Path Integrals in Quantum Mechanics, Statistics, Polymer Physics and Financial Markets* (WS: Singapore 2009).
[17] Feynman R P and Hibbs A R 1965 *Quantum Mechanics and Path Integrals* (McGraw-Hill: New York)
[18] Gaveau B, Jacobson T, Kac M and Schulman L S 1984 *Phys. Rev. Lett.* **53** 419
Jacobson T and Schulman L S 1984 *J. Phys.* A **17** 375
[19] Newton T and Wigner E 1949 *Rev. Mod. Phys.* **21** 400
[20] Feshbach H and Villars F 1958 *Rev. Mod. Phys* **30** 24
[21] see, e.g., Johnson E A and MacKinnon A 1993 *J. Phys.: Condens. Matter* **5** 5859
[22] Amelino-Camelia G 2002 *Int. J. Mod. Phys.* D **11** 35
Magueijo J and Smolin L 2002 *Phys. Rev. Lett.* **88** 190403
Amelino-Camelia G 2002 *Nature* **418** 34
[23] Kowalski-Glikman J 2005 *Introduction to Doubly Special Relativity* (Springer: Berlin)
[24] Daszkiewicz M, Lukierski J and Woronowicz M 2008 *Phys. Rev.* D **77** 105007
[25] Magueijo J and Smolin L 2003 *Phys. Rev.* D **67** 044017
[26] Bastianelli F and van Nieuwenhuizen P 2005 *Path Integrals and Anomalies in Curved Space* (CUP: Cambridge)
[27] Vilenkin A and Ford L H 1982 *Phys. Rev.* D **26** 1231
Scardigli F, Gruber C and Chen P 2011 *Phys. Rev.* D **83** 063507
[28] see, e.g., Tegmark M et al. 2006 *Phys. Rev.* D **74** 123507
[29] Barrow J D 1998 *Phys. Rev.* D **59** 4. [arXiv:astro-ph/9811022]
Albrecht A, Magueijo J 1999 *Phys. Rev.* D **59** 043516. [arXiv:astro-ph/9811018]
[30] Horava P 2009 *Phys. Rev.* D **79** 084008. [arXiv:0901.3775]
[31] Chibisov G V 1976 *Soviet Physics Uspekhi* **19** 624
Lakes R 1998 *Phys. Rev. Lett.* **80** 1826
[32] Rumpf H 1986 *Phys. Rev.* D **33** 185
Rumpf H 1993 *Prog. Theor. Phys. Suppl.* **111** 63
de Aguiar T C, Menezes G and Svaiter N F 2009 *Class. Quantum Grav.* **26** 075003
Ambjorn J, Loll R, Westra W and Zohren S 2009 *Phys. Lett.* B **680** 359

Stochastic mechanics of relativistic fields

Edward Nelson
Department of Mathematics, Princeton University, Princeton NJ 08544, USA
E-mail: nelson@math.princeton.edu

Abstract. After a brief review of stochastic mechanics of nonrelativistic particle systems, the paper discusses in a qualitative way issues concerning the application of stochastic mechanics to relativistic fields and their relation to quantum field theory. It is suggested that gauge theories will be essential to this program.

1. Stochastic mechanics of nonrelativistic particle systems

A *basic Lagrangian* is

$$L = \frac{1}{2}m_{ij}v^i v^j - \varphi + A_i v^i \tag{1}$$

Here the mass tensor m_{ij} is a Riemannian metric, v^i is the velocity, φ is the scalar potential, and A_i is the covector potential; the summation convention is used. This is the general form of a Lagrangian for which the force is a dynamical variable (depending only on position and velocity). For simplicity, we discuss only the case for which the configuration space is $\mathbb{R}^{sN}$ for N particles in s space dimensions. But tensor notation is still useful; the masses of the various particles can be included in the diagonal constant mass tensor m_{ij}. Also for simplicity we assume here that $A_i = 0$ and denote the scalar potential φ by V.

Hamilton's principal function is

$$S(x,t) = -\int_t^{t_1} L(X(s,x,t), \dot{X}(s,x,t), s)\, ds \tag{2}$$

where $X(s,x,t)$ is the configuration at time s starting at x at time t and $\dot{X}$ is its time derivative. The requirement that S be stationary when the flow is perturbed by a time-dependent vector field leads to the Hamilton-Jacobi equation

$$\frac{\partial S}{\partial t} + \frac{1}{2}\nabla^i S \nabla_i S + V = 0 \tag{3}$$

Let w be the Wiener process on $\mathbb{R}^{sN}$, the stochastic process of mean 0 characterized by

$$dw^i dw_i = \hbar dt + o(dt) \tag{4}$$

Stochastic mechanics postulates that the motion of the configuration is a Markov process governed by the stochastic differential equation

$$dX^i = b^i(X(t),t)dt + dw^i \tag{5}$$

where b^i is the mean forward velocity. Call this *basic stochasticization*; this terminology is better than "stochastic quantization" since the physics remains classical. The fluctuations are of order $dt^{\frac{1}{2}}$ in time dt, and with a coefficient larger than $\hbar$ in (4) this postulate could be falsified by experiment, without violating the Heisenberg uncertainty principle.

Now try to substitute the Markov process X into (2) and require that the conditional expectation $\mathbb{E}_t$ of the action, given the configuration at time t, be stationary with respect to variations of the forward velocity b^i. The trajectories of the process X are not differentiable, so replace the derivatives in $\dot{X}$ by difference quotients, and replace the integral by a Riemann sum. There is a singular term whose conditional expectation is 0, so it drops out, and there is a singular term that is a constant, independent of the trajectory, so it drops out from the variation. Then pass to the limit when the Riemann sum becomes an integral. The result is the *stochastic principal function*

$$S(x,t) = -\mathbb{E}_{x,t}\int_t^{t_1}\left(\frac{1}{2}b^ib_i + \frac{\hbar}{2}\nabla_i b^i - V\right)(X(s),s)ds \tag{6}$$

where $\mathbb{E}_{x,t}$ is the expectation conditioned by $X(t) = x$.

Let

$$R = \frac{\hbar}{2}\log\rho \tag{7}$$

where ρ is the time-dependent probability distribution of the configuration. Computation shows that

$$\begin{aligned}&\frac{\partial S}{\partial t} + \frac{1}{2}\nabla^i S\nabla_i S + V - \frac{1}{2}\nabla^i R\nabla_i R - \frac{\hbar}{2}\Delta R = 0\\&\frac{\partial R}{\partial t} + \nabla_i R\nabla^i S + \frac{\hbar}{2}\Delta S = 0\end{aligned} \tag{8}$$

The first equation is the *stochastic Hamilton-Jacobi equation*. There is no deterministic analogue of the second equation since $R = 0$ when $\hbar = 0$. These two coupled nonlinear partial differential equations determine the process X.

With

$$\psi = e^{R+iS} \tag{9}$$

these equations are equivalent to the Schrödinger equation

$$\frac{\partial\psi}{\partial t} = -\frac{i}{\hbar}\left(-\frac{1}{2}\Delta + V\right)\psi \tag{10}$$

As I reported to this conference two years ago [1], despite many successes there are problems with stochastic mechanics as a candidate for a physically realistic theory. There can be instantaneous signaling between widely separated correlated but dynamically uncoupled systems. Although the probability distribution of position measurements at a single time are the same for quantum mechanics and stochastic mechanics (because $|\psi|^2 = \rho$ is the probability density of the configuration), measurements at two different times, even when mutually compatible, can differ in the two theories.

The problem is that stochastic mechanics describes a Markov process in multidimensional configuration space, not in physical space.

2. Stochastic mechanics of fields

There are two motivations for applying stochastic mechanics to fields. One is the hope that since fields live on physical spacetime nonlocality problems may be avoided. The other is that it may provide useful technical tools in constructive quantum field theory. The strategy is to apply basic stochasticization to a basic field Lagrangian.

Consider a real scalar field φ on d-dimensional spacetime. I shall discuss the free field of mass $\mu > 0$ and the field with a φ^4 interaction. Choose a spacelike hyperplane $\mathbb{R}^s$, where s, the number of space dimensions, is $d-1$. The configuration space is a set of scalar functions φ on $\mathbb{R}^s$. Impose a spatial cutoff and a momentum cutoff. That is, put the system in a box of side λ and represent the free field as a set of harmonic oscillators (a device going back to Jeans) with momentum bounded by κ. Then we have a system with finitely many degrees of freedom.

For the free field, of course, the limit $\lambda \to \infty, \kappa \to \infty$ presents no difficulty. But it is unwise to attempt to formulate an interacting field on the same space since by Haag's theorem it must have an inequivalent representation of the canonical commutation relations.

Formally, the φ^4 theory has the interaction Hamiltonian

$$\int_{\mathbb{R}^s} \varphi^4(x)\, dx \tag{11}$$

With our cutoffs, the integral becomes a sum. But this expression is hopelessly singular as $\kappa \to \infty$, and the interaction Hamiltonian is replaced by

$$\int_{\mathbb{R}^s} :\varphi^4(x):\, dx \tag{12}$$

The colons denote Wick ordering, which can be expressed simply by summing over quadruples of *distinct* oscillators.

For $d = 2$ it was shown that the Hamiltonian is bounded below by a constant independent of κ, giving the limit $\kappa \to \infty$. This was the origin of hypercontractivity, which has grown into a rich topic in analysis. The construction of the limit $\lambda \to \infty$ followed—in fact, with φ^4 replaced by any polynomial of even degree.

For $d = 3$ the φ^4 theory was rigorously established by Glimm and Jaffe, with mass and coupling constant renormalization, by heroic work with cluster expansions.

For $d > 4$ Aizenman established a no-go theorem: the limiting theory with cutoffs removed is non-interacting.

For $d = 4$ there are partial no-go results, and most workers in constructive quantum field theory believe that the limiting theory with cutoffs removed is non-interacting also for $d = 4$, but there is no complete proof—the problem is open.

In this work in constructive quantum field theory, Minkowski space was replaced by Euclidean space, giving a problem in probability theory: the Euclidean fields commute. Then the relativistic theory was reconstructed from the Euclidean theory. This is a powerful analytic technique, but it does not give a candidate for quantum theory as emergent from an underlying classical theory.

What does stochastic mechanics have to say? We have a basic Lagrangian. But the separation of the energy for the free field into kinetic energy and potential energy is not a relativistically invariant procedure. Probability theory and relativity do not mix well, due to the indefinite nature of the Minkowski metric.

For the ground state of the free field, the corresponding random field of stochastic mechanics was constructed by Guerra and Ruggiero [2]. It is a Gaussian field with Euclidean-invariant covariance function. But it plays a different role than in Euclidean field theory; relativistic effects persist. Suppose that the field is coupled to an external potential localized in a bounded spacetime region Ω, bounded in the past by t_1 and in the future by t_2. Consider the Markov process, not necessarily for the ground state, of the free field at time t_1. It interacts with the external potential and has a certain value at t_2. By the basic theorem of stochastic mechanics, the probability distribution of the process at time t_2 is the same as that given by quantum mechanics. Since the quantum field is relativistic, the influence of the external potential affects only the points at time t_2 that lie in the future cone of some point in Ω.

But we cannot conclude a similar result for the Markov process itself. Supraluminal influences persist, alas, and the hope that stochastic mechanics applied to field theory gives a reasonable candidate for quantum field theory as emergent from a classical theory does not seem to be valid.

3. Epiphany at the airport

This was the discouraged note on which I ended my talk at the conference. But returning home I had a long wait at the Zurich airport, and a ray of hope appeared.

Consider the electromagnetic field and represent the vector potentials A_μ as harmonic oscillators with basic stochasticization. Then, although they are subject to supraluminal signaling, the physically real electric field and magnetic field should not be, and they transform covariantly under Lorentz transformations. Gauge theories can be expected to play the role of harmonizing the requirements of probability theory and of relativity theory. If spinor fields can be treated stochastically, with their associated currents represented as random fields, and if the divergence problems endemic to relativistic field theory can be handled (two big ifs), the way may be open to a theory treating the phenomena at present described by quantum filed theory as emerging from classical physics with intrinsic randomness.

References

[1] Nelson E 2012 *J. Phys.: Conf. Series* **361**
[2] Guerra F and Ruggiero P 1973 *Phys. Rev. Lett.* **31**

Elements of physics for the 21st century

Werner A Hofer
Department of Physics, University of Liverpool, L69 3BX Liverpool, Britain
E-mail: whofer@liverpool.ac.uk

Abstract. Given the experimental precision in condensed matter physics – positions are measured with errors of less than 0.1pm, energies with about 0.1meV, and temperature levels are below 20mK – it can be inferred that standard quantum mechanics, with its inherent uncertainties, is a model at the end of its natural lifetime. In this presentation I explore the elements of a future deterministic framework based on the synthesis of wave mechanics and density functional theory at the single-electron level.

1. Introduction

The paper describes research presented at the EmQM 13 conference. It gives an overview of work on quantum mechanics through about fifteen years, from the first paper on extended electrons and photons published in 1998 [1], to the last paper on quantum nonlocality and Bell-type experiments in 2012 [2]. A final section contains the first steps towards a density functional theory of atomic nuclei, presented for the first time at the conference in Vienna. My publications on quantum mechanics possess a gap from about 2002 to 2010. This was due to the realization that I could not account for a simple fact: I could not explain, how the electron changes its wavelength, when it changes its velocity. I felt at the time that not understanding this simple fact probably meant that I could not understand the electron. Hence I only continued the development of this framework after, prompted by a student of mine, I had found a solution which seemed to make sense. For that I have to thank this particular student. I also have to thank Gerhard Grössing and Jan Walleczek for organizing this great conference, and the Fetzer Franklin Fund for very generous financial support.

I think we can say today that we actually do understand quantum mechanics. Maybe not in the last details, and maybe not in its full depth, but in the basic physics which it describes, and the deep flaws buried within its seemingly indisputable axioms and theorems. In that, we differ from Richard Feynman, who famously thought that nobody could actually understand it. However, this was said before two of the most important inventions for science in the twentieth century became available to researchers: high-performance computers, and scanning probe microscopes. Computers changed the way science is conducted, since they allow for exquisite experimental control and an extensive numerical analysis of all experiments. This, in turn, means that successful theory and successful quantitative predictions, based on local quantities, make it increasingly implausible, that processes exist, which are operating outside space and time. Then, the solution to the often paradoxical theoretical predictions and sometimes incomprehensible experimental outcomes can only lie in the rebuilding of a model in microphysics which is both, rooted in space and time, and which allows for a description of single events at the atomic scale. This paper aims at delivering the first building blocks of such a comprehensive model.

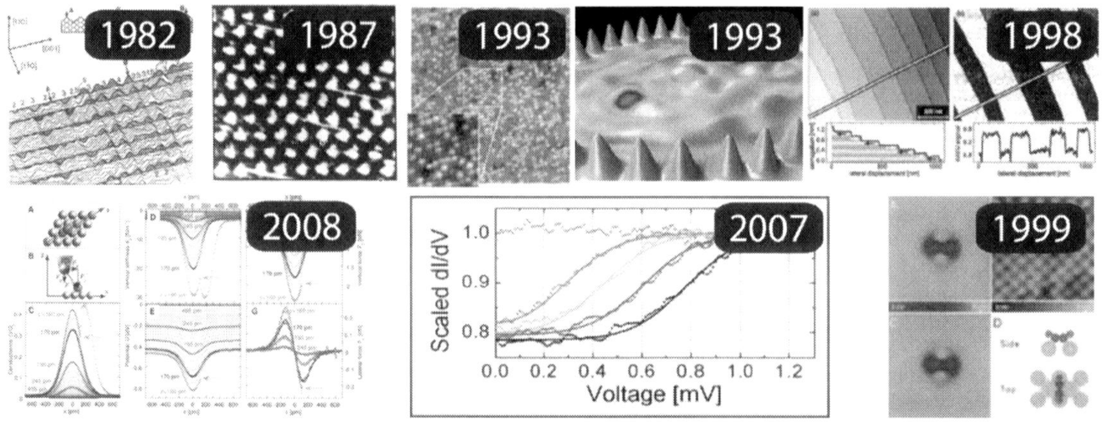

Figure 1. Top frames, clockwise, development of scanning probe microscopy over the last 30 years. Au-terraces 1982[4], Au-atoms 1987[5], Atomic species 1993[6], electron waves 1993[7], electron-spin 1998[8], atomic vibrations 1999[9], spin-flip excitations 2007[11], forces on single atoms 2008[3].

Computers and scanning probe instruments make it possible to measure and analyze single processes on individual atoms and even electrons. Today we can measure and simulate the forces necessary to push one atom across a smooth metal surface [3], vibrations created by single electrons impinging on molecules [9], torques on molecules created by ruptures of single molecular bonds [10], or single spin-flip excitations on individual atoms [11]. See Fig. 1 for the development of experiments over the last thirty years. Distances, in these experiments, can be measured with an accuracy of 0.05 pm [12], which is about 1/4000th of an atomic diameter; energies with a precision of 0.1meV, which is about 1/20000th of the energy of a chemical bond [11]. Given these successes and this accuracy, of which physicists could only dream at the time of Einstein, Schrödinger, Heisenberg, or Dirac, it would be intellectually deeply unsatisfying if we were today still limited to the somewhat crude theoretical framework of standard quantum mechanics, developed at the beginning of the last century.

2. The main problem

The main problem faced by theorists today is the precision of experiments at the atomic scale, because it exceeds by far the limit encoded in the uncertainty relations. This has been the subject of debate for some time now, following the publication of Ozawa's paper in 1988 [13], which demonstrated that the limit can be broken by certain measurements. An even larger violation can be observed in measurements of scanning probe instruments [14]. If the instrument measures, via its tunneling current, the variation of the electron density across a surface, then a statistical analysis of such a measurement is straightforward. In the conventional model electrons are assumed to be point particles. The same assumption is made in quantum mechanics, when the formalism is introduced via Hamiltonian mechanics and Poisson brackets. It is also the conventional wisdom in high energy collision experiments, where one finds that the radius of the electron should be less than 10^{-18}m. If this is correct, then the density is a statistical quantity derived from the probability of a point-like electron to be found at a certain location. This has two consequences: (i) A measurement of a certain distance with a certain precision for a particular point on the surface can only be distinguished from the measurement at a neighboring point if the standard deviation is lower than a certain value. (ii) A certain energy limit allows only a certain lower limit for the standard deviation in these measurements. One can now show quite easily [14] that the standard deviation at realistic energy limits (in case of a silver surface the band energy) is about two orders of magnitude larger than the possible value for state-of-the-art measurements today. The allowed limit for the standard deviation in the experiments is about 3pm, while the standard deviation from the band energy limit is about 300pm. The consequence for the standard framework of quantum mechanics is quite devastating: the uncertainty principle, and by association the whole framework of operator mechanics, becomes untenable, because it is contradicted by experiments. It is precisely this contradiction, which has been claimed by theoretical physicists to be impossible. It also has one consequence, which can be seen as the one principle of the following: The density of electron charge is a real physical quantity. It has the same ontological status as electromagnetic fields or macroscopic mass or charge distributions. The only difference, and the origin of many of the complications arising in atomic scale physics is that the density not only interacts with external fields, but it also interacts with an electron's internal spin density.

The theoretical framework combines two separate models. Both of them are due to physicists born in Vienna, so the location of a workshop on emergent quantum mechanics, from my personal perspective, could not have been better chosen. The first of these physicists is Erwin Schrödinger, born in Vienna in 1887, the second one is Walter Kohn, born in Vienna in 1923. The fundamental statements, underlying these two separate models, are the following: (i) A system is fully described by its wavefunction (Schrödinger). (ii) A system is fully described by its density of electron charge (Kohn). I have been asked, at this workshop, whether the violation of the uncertainty relations could be accounted for by a reduced limit of the constant. Such a solution disregards the ultimate origin of the uncertainty relations. They are based, conceptually, on the assumption that electrons are point particles (this is the link to classical mechanics and Poisson brackets), and the obligation to account for wave properties of electrons. If wave properties are real, then there will be no theoretical limit to the precision in their description.

3. Wavefunctions and charge density

If the density of electron charge is a real physical property, then a common framework must be developed, which allows to map the density onto wavefunctions in the Schrödinger theory. Wavefunctions famously do not have physical reality in the conventional model. However, their square does, according to the Born rule. Here, we want to demonstrate that this is correct to some extent also within the new model, but with one important limitation: even though wavefunctions do not have the same reality as mass or spin densities, they can be assembled from these two - physically real - properties.

3.1. Density and energy

Electrons play a key role in modern physics. Indeed, one could argue that all of physical sciences at the atomic and molecular level, Physics, Chemistry, and Biology is concerned with only one topic: the behavior and properties of electrons. This is also reflected in the celebrated theorem of Walter Kohn: all properties of a physical system, composed of atoms, are defined once the distribution of electron charge within the system is determined [15]. The solution to the problem of electron density distribution is formulated in density functional theory in a Schrödinger-type equation. The spin density is, in this framework, denoted as an isotropic spin-up or spin-down component of the total charge density, the energy related to this spin-density is computed with the help of Pauli-type equations.

However, the framework does not provide physical insights into either spin-densities at the single electron level, or how spin-densities will change in external magnetic fields. What was missing, so far, was a clear connection between the density of electron charge, on the one hand, and the spin density on the other hand. A connection, which should explain the physical origins of wavefunctions in the standard model. It should also explain, how density distributions may change as a consequence of changes to the electron velocity, thus underpinning the wave properties of electrons, found in all experiments. This poses a problem for both, standard quantum theory and density functional theory, because in both cases free electrons are described by plane waves:

$$\psi(\mathbf{r}) = \frac{1}{\sqrt{V}} \exp(i\mathbf{kr}) \tag{1}$$

In this case the Born rule gives a constant value for the probability density, for the mass density and for the charge density: free electrons, then, do not have any distinctive property related to their velocity.

It turns out to be surprisingly simple to construct a more physical model. Once it is accepted that electrons must be extended, the wave features must be part of the density distributions of free electrons themselves. In this case the density of charge must also be wavelike. Assigning a wave-like behavior to the density of electron mass moving in z-direction by:

$$\rho(z,t) = \frac{\rho_0}{2}\left[1 + \cos\left(\frac{4\pi}{\lambda}z - 4\pi\nu t\right)\right] \tag{2}$$

where ρ_0 is the inertial mass density of the electron, and λ and ν depend on the momentum and frequency according to the de Broglie and Planck rules. At zero frequency infinite wavelength, describing an electron at rest, the mass density is equal to the inertial mass density. However, if the electron moves, then the density is periodic in z and t. This requires the existence of an additional energy reservoir to account for the variation in kinetic energy density. We next introduce the spin density as the geometric product of two field vectors, $\mathbf{E}$ and $\mathbf{H}$, which are perpendicular to the direction of motion. These fields are:

$$\mathbf{E}(z,t) = \mathbf{e}_1 E_0 \sin\left(\frac{2\pi}{\lambda}z - 2\pi\nu t\right) \qquad \mathbf{H}(z,t) = \mathbf{e}_2 H_0 \sin\left(\frac{2\pi}{\lambda}z - 2\pi\nu t\right) \tag{3}$$

Spin, in this picture, is the geometric product of the two vector components. It is thus a chiral (and for the free electron imaginary) field vector, which is either parallel or anti-parallel to the direction of motion. The total energy density is constant and equal to the inertial energy density if we impose a condition on the spin amplitude [16]:

$$\begin{aligned}
E_{kin}(z,t) &= \frac{1}{2}\rho_0 v_{el}^2 \cos^2\left(\frac{2\pi}{\lambda}z - 2\pi\nu t\right) \\
E_{spin}(z,t) &= \left(\frac{1}{2}\epsilon_0 E_0^2 + \frac{1}{2}\mu_0 H_0^2\right)\sin^2\left(\frac{2\pi}{\lambda}z - 2\pi\nu t\right) \\
&\left(\frac{1}{2}\epsilon_0 E_0^2 + \frac{1}{2}\mu_0 H_0^2\right) =: \frac{1}{2}\rho_0 v_{el}^2 \\
\Rightarrow \quad E_{tot} &= E_{kin}(z,t) + E_{spin}(z,t) = \frac{1}{2}\rho_0 v_{el}^2 = \text{constant}
\end{aligned} \tag{4}$$

It should be noted that not only the frequency, but also the intensity of the spin component depends on the velocity of the electron. This is in marked contrast to classical electrodynamics, where the energy of a field only depends on the intensity but not on the frequency. Here, it establishes a link between the quantum behavior of electrons and the quantum behavior of electromagnetic fields. This link gives a much more precise explanation for the validity of Planck's derivation of black body radiation. If every electromagnetic field, due to emission or absorption of energy by electrons, must follow the same characteristic, then every energy exchange must also be proportional to the frequency of the field. Then Planck's assumption, that $E = h\nu$ is nothing but a statement of this fact. However, that also the intensity follows the same rule, has been unknown so far. In our view this could be the fundamental principle for a general framework of a non-relativistic quantum electrodynamics to be developed in the future. Electrostatic repulsion of such an extended electron has to be accounted for, as it is in density functional theory (DFT), by a negative cohesive energy of the electron of -8.16eV. In DFT this energy component is known as the self-interaction correction.

3.2. Wavefunctions

It is straightforward to assemble wavefunctions from mass and spin density components, following this route. Wavefunctions are in our framework multivectors containing the even elements of geometric algebra in three dimensional space [17]. The even elements are real numbers and bivectors (product of two vectors), the 4π symmetry, which is the basis of Fermi statistics in the conventional framework, follows from the symmetry properties of multivectors under rotations in space. The real part ψ_m of a general wavefunction can be written as a scalar part, equal to the square root of the number density:

$$\psi_m = \rho^{1/2} = \rho_0^{1/2} \cos\left(\frac{2\pi}{\lambda} z - 2\pi\nu t\right) \tag{5}$$

In geometric algebra, this is the scalar component of a general multivector. The bivector component ψ_s is the square root of the spin component, times the unit vector in the direction of electron propagation, times the imaginary unit. It is thus:

$$\psi_s = i\mathbf{e}_3 S^{1/2} = i\mathbf{e}_3 S_0^{1/2} \sin\left(\frac{2\pi}{\lambda} z - 2\pi\nu t\right) \tag{6}$$

The scalar component and the bivector component for an electron are equal to the inertial number density:

$$\rho_0 = S_0 \qquad \Rightarrow \qquad \rho + S = \rho_0 = \text{constant} \tag{7}$$

The same result can be reached by applying the Born rule, for the wavefunction defined as:

$$\begin{aligned} \psi &= \rho^{1/2} + i\mathbf{e}_3 S^{1/2} \qquad \psi^\dagger = \rho^{1/2} - i\mathbf{e}_3 S^{1/2} \\ \psi^\dagger \psi &= \rho + S = \rho_0 = \text{constant} \end{aligned} \tag{8}$$

The difference to the conventional formulation is that the wavefunction is a multivector, not a complex scalar. It also makes the spin component a chiral vector, which is important for the understanding of spin measurements.

Formally, we can recover the standard equations of wave mechanics, if we define the Schrödinger wavefunction as a complex scalar, retaining the direction of the spin component as a hidden variable. The wavefunction for a free electron then reads:

$$\psi_S = \rho^{1/2} + iS^{1/2} = \rho_0^{1/2} \exp\left[i\left(\frac{2\pi}{\lambda} z - 2\pi\nu t\right)\right] \tag{9}$$

In the conventional framework the dependency of the wavefunction and the Schrödinger equation on external scalar or vector potentials is usually justified with arguments from classical mechanics and energy conservation. In our approach, the justification is the changed frequency and wavevector of electrons if they are subject to external fields. If we assume that the frequency of the electron varies from the value inferred from the de Broglie and Planck relations:

$$i\hbar\frac{\partial \psi_S}{\partial t} = h\nu\psi \neq -\frac{\hbar^2}{2m}\nabla^2\psi_S = \frac{p^2}{2m}\psi_S, \tag{10}$$

then the difference, which is observed in the photoelectric effect, can be accounted for by an additional term in the equation which is linear with the measured scalar potential. Then:

$$i\hbar \frac{\partial \psi_S}{\partial t} = -\frac{\hbar^2}{2m} \nabla^2 \psi_S + V \psi_S \tag{11}$$

The second situation, where this can be the case, observed for example in Aharonov-Bohm effects, is when the wavelength does not comply with the wavelength inferred from the frequency and the Planck and de Broglie relations. In this case one can account for the observation by including the vector potential in the differential term of the equation to arrive at the general equation [16]:

$$i\hbar \frac{\partial \psi_S}{\partial t} = \frac{1}{2m} \left(i\hbar\nabla - e\mathbf{A}\right)^2 \psi_S + V \psi_S \tag{12}$$

The important difference to standard wave mechanics is that all these effects occur at a local level and can therefore be analyzed locally.

3.3. *Many-electron systems*

In a many-electron system motion of electrons is correlated throughout the system and mediated by crystal fields within the material. If the spin component in general is a bivector, and if it is subject to interactions with other electrons in the system, then the general, scalar Schrödinger equation will not describe the whole physics of the system. Simply accounting for all interactions by a scalar effective potential v_{eff} would recover the Kohn-Sham equations of DFT, if exchange and correlation were included. It would do so, however, for both, density components and spin components, since:

$$\begin{gathered} \left(-\frac{\hbar^2}{2m} \nabla^2 + v_{eff}\right) \left(\rho^{1/2} + i\mathbf{e}_3 S^{1/2}\right) = \mu \left(\rho^{1/2} + i\mathbf{e}_3 S^{1/2}\right) \\ \left(-\frac{\hbar^2}{2m} \nabla^2 + v_{eff} - \mu\right) \rho^{1/2} = \left(-\frac{\hbar^2}{2m} \nabla^2 + v_{eff} - \mu\right) S^{1/2} = 0 \end{gathered} \tag{13}$$

In this case the solutions of the equation, single Kohn-Sham states, would not lend themselves to a local analysis of physical events, and would not include an independent spin component. We therefore propose a different framework for a many electron system, which scales linearly with the number of electrons and remains local. Such a model can be achieved by including a bivector potential into a generalized Schrödinger equation in the following way:

$$\left(-\frac{\hbar^2}{2m} \nabla^2 + v_{eff} + i\mathbf{e}_v v_b\right) \left(\rho^{1/2} + i\mathbf{e}_s S^{1/2}\right) = \mu \left(\rho^{1/2} + i\mathbf{e}_s S^{1/2}\right) \tag{14}$$

where we have changed the spin component to describe a general spin direction $\mathbf{e}_s$. The geometric product of two vectors is the sum of a real scalar and an imaginary vector: $\mathbf{e}_v\mathbf{e}_s = \mathbf{e}_v \cdot \mathbf{e}_s - i\mathbf{e}_v \times \mathbf{e}_s$. The equation of motion for a general many-electron system then reads:

$$\begin{gathered} \left(-\frac{\hbar^2}{2m} \nabla^2 + v_{eff} - \mu\right) \rho^{1/2} = \mathbf{e}_v \cdot \mathbf{e}_s v_b S^{1/2} \\ \left(-\frac{\hbar^2}{2m} \nabla^2 + v_{eff} - \mu\right) \mathbf{e}_s S^{1/2} + \mathbf{e}_v v_b \rho^{1/2} = -\mathbf{e}_v \times \mathbf{e}_s v_b S^{1/2} \end{gathered} \tag{15}$$

If $\mathbf{e}_v = 0$, we recover Eq. (13). As inspection shows, the coupled equations only have a solution if the direction of the bivector potential is equal to the direction of spin ($\mathbf{e}_v = \mathbf{e}_s$), which reduces the problem to:

$$\left(-\frac{\hbar^2}{2m}\nabla^2 + v_{eff} - \mu\right)\left(\rho^{1/2} - S^{1/2}\right) = v_b\left(\rho^{1/2} + S^{1/2}\right) \tag{16}$$

For $v_b = 0$ and $\tilde{\rho}^{1/2} = \rho^{1/2} - S^{1/2}$ this equation is identical to the Levy-Perdew-Sahni equation derived for orbital free DFT in the 1980s [18].

$$\left(-\frac{\hbar^2}{2m}\nabla^2 + v_{eff} - \mu\right)\tilde{\rho}^{1/2} = 0 \tag{17}$$

One can reduce the expression to the conventional Schrödinger equation for the hydrogen atom by setting $v_{eff} = v_n$, the Coulomb potential of the central nucleus. The equation then has two groundstate solutions, both radially symmetric:

$$\rho^{1/2} = \pm\frac{C}{2}e^{-\alpha r} \qquad S^{1/2} = \pm\frac{C}{2}e^{-\alpha r} \tag{18}$$

where C is a constant, and α is the inverse Bohr radius. The vector $\mathbf{e}_s$ is the radial unit vector $\mathbf{e}_r$ and the two spin directions are inward and outward. The same solution will apply to all s-like charge distributions, also, therefore, to the valence electron of silver (see the discussion of Stern-Gerlach experiments below).

The great advantage of the formulation is the simplicity and the reduced number of variables. Both, ρ and S are scalar variables. In addition, we have to find the directions of the unit vectors, $\mathbf{e}_s = \mathbf{e}_v$ for every point of the system. This reduces the time independent problem to a problem of finding five scalar components in real space. Compared to the standard formulation of many-body physics, where one has to find a wavefunction of $3N$ variables, where N is the number of electrons, or to standard DFT, which scales cubic with N, the approach is much simpler.

However, the effective potential v_{eff} and the bivector potential v_b in this model are generally not known and have to be determined for every system. Judging from the development of standard DFT this process will probably take at least ten years of development before reliable methods can be routinely used in simulations. But we think, that this method and this approach to many-body physics will also be an element of physics in the 21st century.

4. Experiments

As stated in the introduction, we consider the fact that quantum mechanics does not allow for a detailed analysis of single events a major drawback of the theory. However, a theoretically more advanced model will have to pass the test that it can actually deliver these insights. This value statement, i.e. that a theoretical framework is superior not because it obtains higher precision in the numerical predictions, but it is superior because it provides causal insight into physical processes, is somewhat alien to the current debate about quantum mechanics. The tacit agreement seems to be that no theory can provide such an insight. This is one of the fundamental assumptions of the Copenhagen school. There, it is stated *that no theoretical model can be more than a coherent framework for obtaining numbers in experimental trials.* But we do actually not know that this is true, because the assumption that it is true contains an assumption about reality. The assumption that reality cannot *in principle* be subjected to an analysis in terms of cause and effect in physical processes. The argument thus is not even logically consistent with its own believe system. Here, we want to show that the analysis of single events in terms of cause and effect is possible also at the atomic scale. This, we think, demonstrates more than anything else the problems of the standard framework.

4.1. Acceleration of electrons

We are quite used to the fact that the wavelength of an electron is inverse proportional to its momentum. It is thus also quite normal to write a wavefunction of a particular free electron, which contains a variation of its amplitude according to this momentum. However, when an electron is accelerated, then standard theory is referring us to the Ehrenfest theorem [19]. Incidentally, also Paul Ehrenfest was born in Vienna, in 1880. But his theorem only describes the change of an expectation value in a system. It does not allow us to understand, how the wavefunction changes its wavelength, or how the frequency of the wave increases when it interacts with an accelerating potential. Within the present model, this is exactly described at the local level by a new equation, which we call the *local* Ehrenfest theorem. Its mathematical expression is:

$$\mathbf{f} = -\nabla\phi = \rho_0 \frac{d\mathbf{v}}{dt} \tag{19}$$

It states that the force (density) at a particular point of the electron is exactly equal to the gradient of an external potential ϕ, and that it is described by its classical formulation, the acceleration of its inertial mass. The reason that it is described by this equation is that the number density or the mass density (here we use the two notions interchangeable) is complemented by the spin density to yield a constant:

$$\rho + S = \rho_0 = \text{constant} \tag{20}$$

The same applies to the square of the wavefunction, which is:

$$\psi^*\psi = \rho + S = \rho_0 = \text{constant} \tag{21}$$

The time differential of momentum density at a particular point is therefore:

$$\frac{d}{dt}\left(m\psi^*\psi\mathbf{v}\right) = m\left(\psi^*\psi\right)\frac{d\mathbf{v}}{dt} \tag{22}$$

However, what is hidden in the classical expression is the shift of energy from the mass component to the spin component as the electron accelerates:

$$\dot{S} = -\dot{\rho} \tag{23}$$

Here we find the reason for the change of wavelength in an acceleration process: the spin component increases in amplitude, and as gradually more energy is shifted into this component the wavelength becomes shorter and the frequency increases. A process, which so far has remained buried underneath the mathematical formalism and is now open to analysis.

4.2. Stern-Gerlach experiments

An inhomogeneous magnetic field leads to deflection of atoms, if they possess a magnetic moment. This effect was used, in the ground breaking experiments on silver by Gerlach and Stern in 1922 [20], to demonstrate that the classically expected result, i.e. a statistical distribution around a central impact, is not in line with experimental outcomes. The assumption that the orbital moment would cause the deflection was also untenable, because in this case one would observe an odd number of impact locations. Within the new model the effect is easy to understand. Above, we derived the solution for the electron mass and spin density of a hydrogen-like atom. Assuming that the valence electron of silver can be described in a similar model, we find two different spin directions: one, parallel to the radial vector and directed outward, the other, parallel to the radial vector and directed inward (see Fig. 2, left images). The induced spin densities S_i (see Fig. 2, centre), as the atoms enter the field, are due to the changes of the

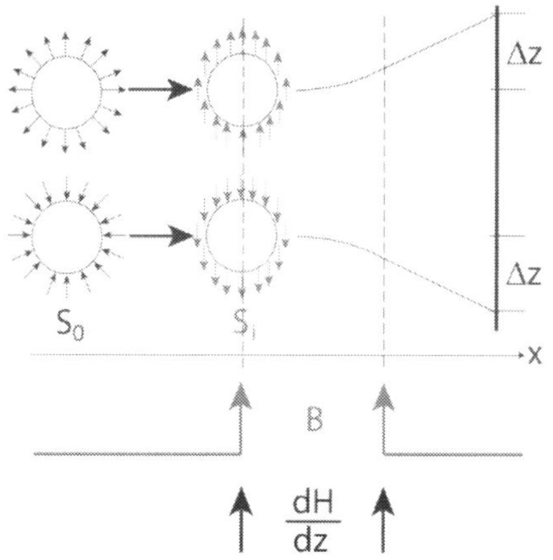

Figure 2. Spin measurement of a hydrogen-like atom. Left: the spin densities S_0 are parallel to the radial vector. Center: the direction of the induced spin densities S_i is parallel or antiparallel to the magnetic field. Right: due to the inhomogeneous field the atoms are deflected upward or downward.

spin orientation in a time-dependent magnetic field, which comply with a Landau-Lifshitz like equation [16]:

$$\mathbf{S} = \mathbf{e}_S \cdot S \qquad \frac{d\mathbf{e}_S}{dt} = \text{constant} \cdot \mathbf{e}_S \times \left(\mathbf{v} \times \frac{d\mathbf{B}}{dt}\right) \tag{24}$$

Then the induced spin densities will lead to a precession around the magnetic field **B** in two directions, which will give rise to induced magnetic moments parallel, or anti parallel to the field. In an inhomogeneous field the force of deflection is then directed either parallel or antiparallel to the field gradient, leading to two deflection spots on the screen, exactly as seen in the experiments (Fig. 2, right). While therefore in the standard model, which assumes that: (i) Spin is isotropic; (ii) A measurement breaks the symmetry of the spin, no process exists, which could actually explain the symmetry breaking of the initially isotropic spin, the situation is completely different in the new model. Here the process is described by:

(i) Spin is isotropic.
(ii) The measurement induces spins aligned with the magnetic field.
(iii) The induced spins lead to positive or negative deflections in a field gradient.

The description is fully deterministic, since the initial direction of spin densities determines the experimental outcome. Statistics only enter the picture, if the initial spin densities are unknown, which they are in practice.

4.3. Interference experiments

Double slit experiments are so difficult to understand in the framework of standard quantum mechanics that Richard Feynman called them "a phenomenon which is impossible, absolutely impossible, to explain in any classical way, and which has in it the heart of quantum mechanics. In reality, it contains the only mystery" [21]. The work done recently is already quite convincing: whether it is with Bohm-type trajectories, fluctuating fields [22], or whether by weak measurements [23], the result always seems to be that one particle passes through one particular opening. In the standard framework the effects are calculated e.g. with the help of Feynman path integrals. The process described in this mathematical framework is shown

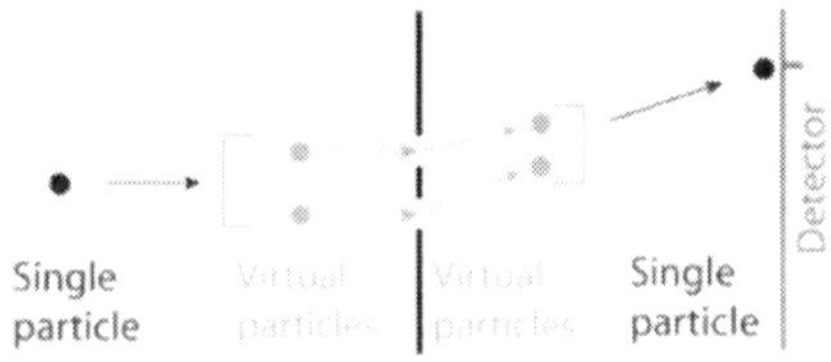

Figure 3. Double slit interferometry, Feynman path integrals. A single particle is assumed to split into virtual particles prior to the interferometer. After the interferometer all particles recombine, the acquired phases along their path determining the interference amplitude. A single particle is detected at the detector screen.

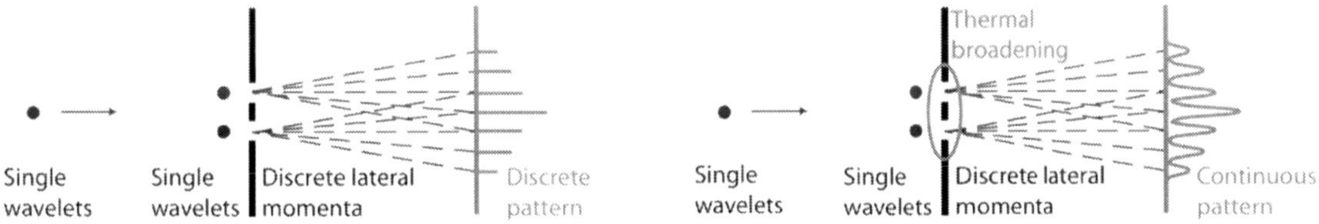

Figure 4. Double slit interferometry, real picture. Left: A single particle passing through an opening of the interferometer acquires a discrete lateral momentum due to interactions with the discrete interaction spectrum of atomic scale systems. The interference pattern is a series of sharp impact regions. Right: due to the thermal energy of the slit environment and interactions with molecules in air the impact regions broaden with a Gaussian until they resemble the wave-like interference pattern in an optical interferometer.

in Fig. 3. A single particle, upon entering the vicinity of the interferometer, is assumed to split into a number of virtual particles. Each virtual particle passes exactly one opening of the interferometer, where it acquires a characteristic phase. After the interferometer, all particles are again recombined interfering in a particular way due to their acquired phases. A single impact is observed on the detector screen.

It is quite clear that this has nothing to do with real events. However, this insight does not solve the problem, what actually happens so that single entities (electrons or photons), will acquire certain deflections in an interferometer, and why these deflections have an uncanny resemblance to interference patterns of light in an interferometer. In our view, this problem could have been solved a long time ago by Duane [24]; a solution which was later taken up by Lande [25]. The key observation for their model is that every atomic scale system has a discrete interaction spectrum. This means that every interaction of such a system with a single photon or electron can only cause observable changes in the particle's dynamics, if a discrete amount of energy is exchanged, typically corresponding to the excitation of single lattice vibrations. Given this fact, it is impossible that the particle acquires a continuous lateral momentum. Consequently, it also cannot be detected in intermediate regions, unless its trajectory is additionally determined by thermal broadening of the actual interaction. This model of the process can be experimentally verified. The key to such a verification is the separation of the individual effects changing the particle's trajectory (see Fig. 4). In a liquid helium environment the thermal motion of atoms is frozen. In addition, in an ultrahigh vacuum environment, no interactions with molecules are possible. In a low temperature interferometer the impacts will be sharply defined images of the particle beam deflected by interactions with the atomic environment, while a gradual increase of the temperature of the interferometer should lead to a gradual broadening of the impact regions. This broadening, moreover, should reflect the thermal energy range of the slit environment. Performing such a controlled experiment seems entirely feasible today, and in our view it will establish that indeed the interaction with the atomic environment, and not some fictitious splitting and recombination process is at the bottom of this - hundred years old - mystery.

4.4. Interference of large molecules

It has been claimed, in a number of high-impact publications since 1999, that large molecules can be made to interfere on gold gratings, and that these experiments show both, the coherence of the molecules over macroscopic trajectories (range of cm), and that the wavelength of these molecules is equal to the de Broglie wavelength of their inertial mass. However, in line with the Duane-Lande model of interference, a different explanation for the experimental data seems more likely.

As the exemplar of the standard interpretations we use in the following the first experiments on C_{60} molecules, which were published in the journal *Nature* [26]. Due to the interest of the Chemistry community in these molecules, their properties have been extremely well researched in the past. Theorists routinely calculate their electronic properties, their phonon spectrum, and their light absorption and emission spectrum. They have been adsorbed on surfaces and their charge density distribution has been compared to the results of STM experiments, which verified the theoretical results in great detail. As every Chemist will know, phonon or vibrational modes of organic molecules are varied and range from a few meV (breathing modes, torsion) to a few hundred meV (stretch modes). This particular molecule contains 60 carbon atoms, it thus has 180 modes of vibration which cover the whole energy range.

In the experiments the molecules are heated with laser light, reaching velocities of a few hundred meters per second, and then passed through a grating with a width of about 50 nm, and a depth of 100 nm. All molecules presented in this type of experiments so far are polarizable; they can possess an electric dipole moment. After the grating it is observed that the molecular impacts show a variation, which is taken as proof that the molecules possess a de Broglie wavelength and interfere as coherent waves.

This seems problematic for various reasons. First, it is well known that the electronic density is fully characterizing a many-electron system. A de Broglie wavelength can in principle not be defined on the basis of the electronic density of these molecules. Second, it is also well known that internal degrees of freedom of molecular systems start mixing after very short timescales, in the range of picoseconds. That a molecule is heated with a laser - most likely leading to excitation of electronic transitions - and then spends microseconds preserving a state vector related to its translational motion, while shaking rapidly due to vibrational excitations is not obvious. Third, it is even less obvious that such a molecule, with its time dependent dipole moment, will not induce dipole moments in the slit system itself, which then interact with the molecule's dipole to alter its trajectory. And fourth, an interference of state vectors of this molecule would mean that the molecule will split into several individual molecules. Given that such a creation of additional molecules violates the energy principle by several MeV, this is highly unlikely.

So how does it really work? Most likely the polarizable molecule is excited by laser light so that most of its low lying vibrational excitations are activated. This molecule enters the interferometer with a time-dependent dipole moment. As the molecule interacts with the atomic environment of the interferometer, it induces electric dipoles into the slit system. These time-dependent dipole moments interact with the molecular dipole moments until the molecule has passed the interferometer. Due to the interaction the molecules acquire a distinct lateral momentum. The momentum leads to a deflection on the detector screen. The deflection is interpreted as the result of a de Broglie wave, because the distance from the point of no deflection to the point of impact is inverse proportional to the velocity of the molecule. It is

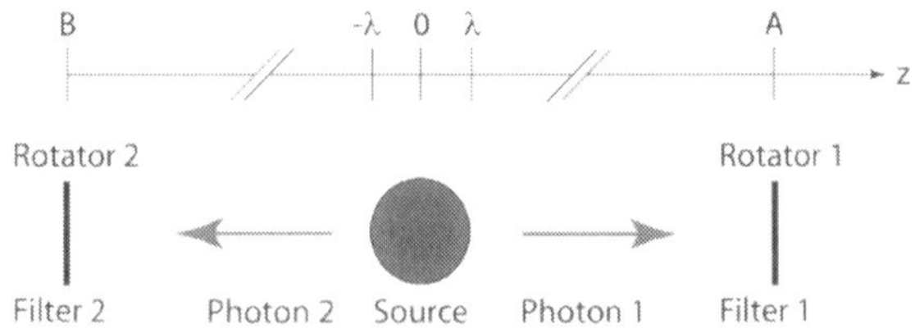

Figure 5. Aspect-type experiment. Two photons are emitted from a common source with an initial unknown polarization angle φ_0. Their polarization is then measured at points A and B. (From Ref. [2]).

inverse proportional to the velocity of the molecule, because the time constant of the interaction duration depends on the time the molecule spent in the slit environment. Then a faster molecule will spend less time, therefore acquire less lateral momentum, therefore end up closer to the point of no deflection. This, however, would not be the sign of a de Broglie wavelength, but the effect of the constant distance from the entry to exit of the interferometer (100nm). This whole scenario should be relatively easy to simulate with modern electronic structure methods. One could also pin down the effect by using non-polarizable molecules in the experiments.

4.5. Aspect-type experiments

These experiments have been puzzling physicists for at least thirty years. The height of the confusion was probably reached with Aspect's review paper in the journal *Nature* in 1999, where he stated: "The violation of Bell's inequality, with strict relativistic separation between the chosen measurements, means that it is impossible to maintain the image a la Einstein where correlations are explained by common properties determined at the common source and subsequently carried along by each photon. We must conclude that an entangled EPR photon pair is a non-separable object; that is, it is impossible to assign individual local properties (local physical reality) to each photon. In some sense, both photons keep in contact through space and time" [27].

We shall show in the following that exactly such a model a la Einstein can explain all experimental data and that the confusion arises from a fundamental technical error in Bell's derivations. To explain the experiments in detail at the single photon level, let us start with setting up a system composed of a source of photons at the point $z = 0$, and two polarization measurements at arbitrary points $z = A$ and $z = -B$. We assume that the polarization measurements contain rotations in the plane perpendicular to z. We also assume that the two photons are emitted from the source with an arbitrary angle of polarization φ_0. It is irrelevant for the following, whether the field vectors of the two photons rotate during propagation. If they do, this will show up only as an additional angle Δ between their polarization measurements at A or $-B$. The setup of the experiment is shown in Figure 5. A single measurement at A consists of two separate processes: First, the polarization angle is altered by an angle φ_A. Mathematically, this is a rotation in three dimensional space and in the plane perpendicular to the direction of motion, which can be described by the geometric product of a rotator in this plane (a geometric product) $\varphi_A \mathbf{e}_1\mathbf{e}_2$ acting on the photon's field vector $\mathbf{S}$, which is parallel to $\mathbf{e}_3$. To take care of normalization, we describe such a rotation as:

$$R(A) = \exp\left[(\varphi_A + \varphi_0)\,\mathbf{e}_1\mathbf{e}_2(\mathbf{e}_3)\right] = e^{i(\varphi_A+\varphi_0)} \tag{25}$$

Then, the photon is detected, if the probability p which depends on the angle of rotation and the initial angle of polarization, is larger than a certain threshold:

$$p[R(A)] = [\Re(R(A))]^2 = \cos^2(\varphi_A + \varphi_0) \tag{26}$$

Depending on how we define our threshold, which is a function of the measurement equipment, an impact at a certain angle of measurement φ_A and a certain initial angle φ_0 is fully determined by the knowledge of these two angles. The single event is thus fully accounted for. However, in the actual experiments the angle φ_0 is unknown, and it is randomly distributed over the whole interval $[0,2\pi]$. A set of N experiments will thus lead to a random value for the probability, covering the whole interval [0,1]. The single measurement is thus random. The same is true for a measurement at point $-B$. Also here the polarization measurement is described by a rotation, with a different and fully independent angle φ_B. The probability of detection is, along the same lines:

$$p[R(B)] = [\Re(R(B))]^2 = \cos^2(\varphi_B + \varphi_0) \tag{27}$$

Also in this case the single event is fully accounted for if the initial angle φ_0 and the angle of polarization φ_B are known. Again, a set of N experiments will lead to a random value for the probability, covering the whole interval [0,1].

One could now assume that the correlation probability is the product of the two measurement probabilities at points A and $-B$, respectively. This is exactly what Bell assumed in the derivation of his inequalities, when he wrote [28]:

$$P(\mathbf{a}, \mathbf{b}) = \int d\lambda \rho(\lambda) A(\mathbf{a}, \lambda) B(\mathbf{b}, \lambda) \tag{28}$$

Here, λ has the same meaning as the initial angle φ_0, and the crucial error lies in the assumption that the correlation probability is the product of individual probabilities. This is manifestly incorrect, because it disregards the mathematical properties of rotations. Two separate rotations at A and $-B$ have to be accounted for by a product of individual rotations, thus:

$$\begin{aligned} R(A) \cdot R(B) &= \exp\left[(\varphi_A + \varphi_0)\, \mathbf{e}_1\mathbf{e}_2(\mathbf{e}_3)\right] \cdot \exp\left[(-\varphi_B - \varphi_0)\, \mathbf{e}_1\mathbf{e}_2(\mathbf{e}_3)\right] \\ &= \exp\left[i\,(\varphi_A - \varphi_B)\right] \end{aligned} \tag{29}$$

It is impossible, from these two rotations, to derive a probability which is the product of two positive numbers. Furthermore, the hidden variable φ_0, which is present in the probability of individual polarization measurements, is canceled out in the correlation derived from two separate rotations. The correct form of the probability for the correlation derived from the two rotations will be:

$$p[R(A), R(B)] = [\Re(R(A) \cdot R(B))]^2 = \cos^2(\varphi_A - \varphi_B) \tag{30}$$

These probabilities are equal to the correlation probabilities derived in the Clauser-Horne-Shimony-Holt formalism [29]:

$$C^{++} = C^{--} = \cos^2(\varphi_A - \varphi_B) \qquad C^{+-} = C^{-+} = 1 - \cos^2(\varphi_A - \varphi_B) \tag{31}$$

They lead to the standard expectation values measured in Aspect-type experiments:

$$E(\varphi_A, \varphi_B) = \cos\left[2(\varphi_A - \varphi_B)\right] \tag{32}$$

And they violate the Bell inequalities in the exact same way as found in the experiments:

$$\begin{aligned} S(\varphi_A, \varphi_A', \varphi_B, \varphi_B') &= E(\varphi_A, \varphi_B) - E(\varphi_A, \varphi_B') + \\ &+ E(\varphi_A', \varphi_B) + E(\varphi_A', \varphi_B') = 2\sqrt{2} \end{aligned} \tag{33}$$

if $\varphi_A = 0, \varphi_A' = 45, \varphi_B = 22.5, \varphi_B' = 67.5$. To repeat the findings: a model based on polarizations and rotations in space recovers all experimental results. It allows for a cause-effect description of every single measurement. It also violates the Bell inequalities. Not, because it is a non-local model, but because Bell made a fundamental error in the derivation of his inequalities.

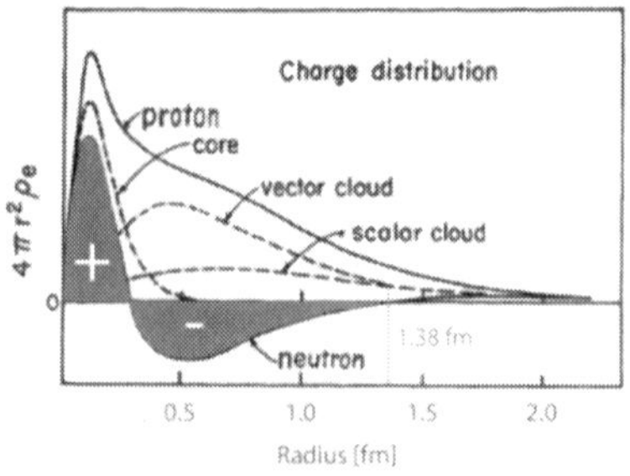

Figure 6. Neutron scattering experiments by Littauer et al. [30]. A neutron consists of a positive core and a negative shell. The radius of the neutron is about 1.38fm.

5. Towards a density model of atomic nuclei

5.1. Electrons and neutrons

It is quite natural, if one considers the electron an extended particle, to ask, what shape and form it might have apart from the atomic environment. We know from DFT that its density, consequently its shape, will depend on the potential environment. After all, we find much higher densities of electron charge in heavier atoms with a higher number of central charges than we find in hydrogen. So one may also ask, in what shape and form an electron exists, for example, in a neutron. We know that the neutron decays outside an atomic nucleus in about 880 seconds to a proton and an electron, with an excess energy of 785 keV, which is mostly converted into X-ray radiation.

$$n^0 \rightarrow p^+ + e^- + 785\text{keV} \tag{34}$$

We also know, from scattering experiments (see Fig. 6), that a neutron contains a core of positive charge, which one could tentatively call the "proton" and a shell of negative charge, which one could to first instance identify as the "electron". If the electron exists in such a high density phase, then one could also seek its eigenstates with the help of a Schrödinger equation adapted to the much smaller lengthscales and much higher energy scales. However, for such an assumption to make sense it first has to be determined, where the additional mass of the neutron compared to isolated protons and electrons comes from. Here, it has to be remembered that the radius of a neutron is much smaller than the radius of a hydrogen atom. Therefore, the electrostatic field of an electron outside hydrogen has a very low energy of about 11 eV, while this field has a large energy of close to 1 MeV for an electron with a radius of 1.38 fm:

$$W_0^e = \frac{1}{2}\int_\infty^{r_e} \epsilon_0 |\mathbf{E}|^2 dV = \frac{1}{4\pi\epsilon_0}\frac{e^2}{r_e} \approx 11\text{eV} \qquad W_n^e = \frac{1}{4\pi\epsilon_0}\frac{e^2}{r_n} \approx 1040\text{keV} \tag{35}$$

Here, one finds that the electrostatic energy alone, considering mass equivalents, can account for the excess mass. Next, it is necessary to analyze nuclear units. We know from atomic physics that atomic units are defined from fundamental constants and determine the solution of the hydrogen problem with the Schrödinger equation. Let me just remind the reader that an exponentially decaying wavefunction $\psi(r) = \rho_0^{1/2}\exp(-\alpha r)$ leads to the following characteristic equation and the solution for α:

$$\left(-\frac{\hbar^2\alpha^2}{2m} + \frac{2\hbar^2\alpha}{2mr} - \frac{e^2}{4\pi\epsilon_0 r}\right)\psi(r) = E\psi(r) \qquad \frac{2\hbar^2\alpha}{2mr} - \frac{e^2}{4\pi\epsilon_0 r} = 0 \quad \rightarrow \quad \alpha = \frac{me^2}{4\pi\epsilon_0\hbar^2} \tag{36}$$

If a similar solution exists for the neutron, then the decay constant must be different. We account for this hypothesis by rescaling the Planck constant in a nuclear environment so that:

$$\hbar_n = x\hbar \qquad \alpha_n = \frac{1.89 \times 10^{10} m^{-1}}{x^2} \qquad E_n = \frac{E_H}{x^2} \tag{37}$$

The Schrödinger equation in a nuclear environment then reads:

$$\left(-\frac{1}{2}\nabla^2 - \frac{1}{r}\right)\psi_n(r) = E_n\psi_n(r) \tag{38}$$

The total energy is the sum of the positive energy of the electrostatic field and the negative energy of the eigenvalue, it is known to be 785 keV. It depends, ultimately, on only two values: the radius of the neutron, which is known from scattering experiments, and the scale x. With a_0 the Bohr radius we get:

$$W_n = \frac{e^2}{4\pi\epsilon_0 a_0}\left(\frac{a_0}{r_n} - \frac{1}{2x^2}\right) = \left(\frac{a_0}{r_n} - \frac{1}{2x^2}\right) \times 27.211[eV] \tag{39}$$

The scale x can therefore be calculated from experimental values. With $r_n = 1.38(415)$ fm and $W_n = 785$ keV we get for the scale x and the energy scale E_n:

$$x = \frac{1}{18779^{1/2}} = \alpha_f \qquad E_n = 511\text{keV} = m_e c^2 \tag{40}$$

Both of these values are very fundamental. In the standard model the fine structure constant α_f describes the difference in coupling between nuclear forces and electrostatic forces, while the rest energy of the electron E_n is one of the fundamental constants in high energy physics. At present, we do not have a clear indication of the significance of this finding. It is quite improbable, that this result should be a mere coincidence. After all, the identity relies on two experimental values, the radius of the neutron and the mass of the neutron. Had these values been different, the fine structure constant or the rest energy of the electron would not have been the result of this derivation. We expect that a nuclear model on the basis of high-density electrons, which we also tentatively assume to be an element of physics in the 21st century, will be able to answer this important question.

5.2. Magic nuclei

It is known that certain numbers of nucleons, assumed to be protons and neutrons in the conventional model, lead to increased stability of atomic nuclei. If high-density electrons are the glue that holds protons together, then protons in a nucleus will be in a regular arrangement. In this case the problem of nuclear organization becomes to first instance a problem of three dimensional geometry. Starting from a single proton, and adding one proton after the other, always under the condition that the distances between protons are constant, will automatically lead to a shell model of atomic nuclei, where a certain number of protons corresponds to closed shells. In Fig. 7 we show the first seven closed shells. In particular the first four, with 4, 16, 28, and 40 protons, correspond to magic nuclei in nuclear physics. Larger shells do not necessarily, but it has to be considered that we do not yet have a comprehensive model of interactions within an atomic nucleus, which could account for the observed nuclear masses. Compared to DFT the additional complication within a nucleus are the relatively large volume of protons, which probably cannot be taken into account with a model of point charges, and the unknown role of nuclear forces. Also, it is quite unclear at present if the electrostatic interactions within the nucleus have the same intensity as in a vacuum, how screening works, and what role the energy

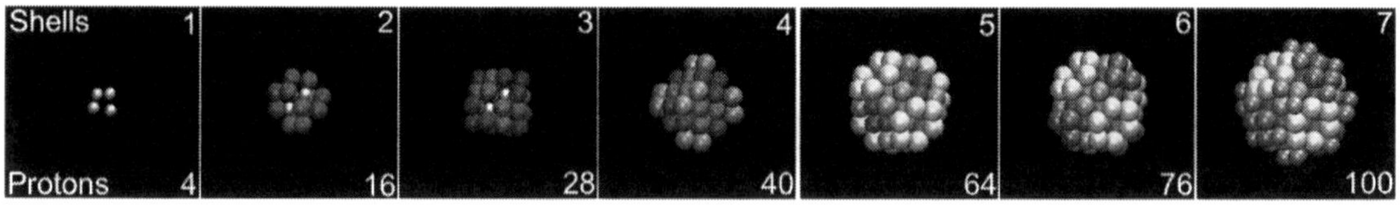

Figure 7. Closed shells of atomic nuclei for up to 100 protons. The shell model is only based on geometry and does not include detailed interactions at this point.

of electrostatic fields will play in the overall picture. The first steps towards such a model are therefore highly tentative and it is to be expected that a fully quantitative model of atomic nuclei is still a long time in the future. However, such a model could provide a unified basis for discussions in nuclear physics, which connects it seamlessly to other fields of physics: something, which is manifestly not the case at present.

Acknowledgments

Helpful discussions with Krisztian Palotas are gratefully acknowledged. The work was supported by the Royal Society London and the Canadian Institute for Advanced Research (CIFAR).

References

[1] W A Hofer, Physica A, 178-196 (1998).
[2] W A Hofer, Front. Phys. 7, 504-508 (2012).
[3] M Ternes et al., Science 319, 1066-1069 (2008).
[4] G Binnig and H Rohrer, Surf. Sci. 126, 236-244 (1983).
[5] V M Hallmark et al., Phys. Rev. Lett. 59, 2879-2882 (1987).
[6] P T Wouda et al., Surf. Sci. 359, 17-22 (1996).
[7] M F Crommie, CP Lutz, DM Eigler, Science 262, 218-220 (1993).
[8] M Bode, Rep. Progr. Phys. 66, 523- (2003).
[9] B C Stipe, MA Rezaei, W Ho, Phys. Rev. Lett. 82, 1724-1727 (1999).
[10] K R Harikumar et al., Nat. Chem. 3, 400-408 (2011).
[11] C F Hirjibehedin et al., Science 317, 1199-1202 (2007).
[12] H Gawronski, M Mehldorn and K Morgenstern, Science 319, 930-933 (2008).
[13] M Ozawa, Phys. Rev. Lett. 60, 385-388 (1988).
[14] W A Hofer, Front. Phys. 7, 218-222 (2012).
[15] P Hohenberg and W Kohn, Phys. Rev. B136, 864-871 (1964).
[16] W A Hofer, Found. Phys. 41, 754-791 (2011).
[17] C Doran and A Lasenby, Geometric Algebra for Physicists, Cambridge University Press, Cambridge (2002).
[18] M Levy, J P Perdew and V Sahni, Phys. Rev. A 30, 2745-2748 (1984).
[19] P Ehrenfest, Z. Phys. 45, 455-457 (1927).
[20] W Gerlach and O Stern, Z. Phys. 9, 353-355 (1927).
[21] R P Feynman, The Feynman Lectures Vol. III, pp. 1-1, Addison Wesley, Reading, Mass. (1965).
[22] G Groessing, S Fussy, J Mesa Pascasio and H. Schwabl, Physica A 389, 4473-4484 (2010).
[23] S Kocsis et al., Science 322, 1170-1173 (2011).
[24] W Duane, Proc. Nat. Acad. Sci. 9, 158 (1923).
[25] A Lande, From Dualism to Unity in Quantum Physics, Cambridge University Press, Cambridge UK (1960)
[26] M Arndt et al., Nature 401, 680-683 (1999).
[27] A Aspect, Nature 398, 189-190 (1999).
[28] J S Bell, Physics 1, 195-199 (1964).
[29] F Clauser, M A Horne and A Shimony, RA Holt, Phys. Rev. Lett. 23, 880-883 (1969).
[30] R M Littauer et al., Phys. Rev. Lett. 7, 144-147 (1961).

Causality and local determinism versus quantum nonlocality

M Kupczynski

Département de l'Informatique, UQO, Case postale 1250, succursale Hull, Gatineau, QC, Canada J8X 3X 7

E-mail: marian.kupczynski@uqo.ca

Abstract. The entanglement and the violation of Bell and CHSH inequalities in spin polarization correlation experiments (SPCE) is considered to be one of the biggest mysteries of Nature and is called *quantum nonlocality*. In this paper we show once again that this conclusion is based on imprecise terminology and on the lack of understanding of probabilistic models used in various proofs of Bell and CHSH theorems. These models are inconsistent with experimental protocols used in SPCE. This is the only reason why Bell and CHSH inequalities are violated. A probabilistic non-signalling description of SPCE, consistent with quantum predictions, is possible and it depends explicitly on the context of each experiment. It is also deterministic in the sense that the outcome is determined by supplementary local parameters describing both physical signals and measuring instruments. The existence of such description gives additional arguments that quantum theory is emergent from some more detailed theory respecting causality and local determinism. If quantum theory is emergent then there exist perhaps some fine structures in time-series of experimental data which were not predicted by quantum theory. In this paper we explain how a systematic search for such fine structures can be done. If such reproducible fine structures were found it would show that quantum theory is not predictably complete, which would be a major discovery.

1. Introduction

Quantum theory (QT), giving accurate predictions for statistical distributions of outcomes obtained in physical experiments, is unable to tell which outcome will be observed and when. In spite of this it has been often claimed that QT provides the most complete description of individual physical systems and that *outcomes of quantum measurements are produced in an irreducibly random* way.

According to a statistical interpretation first advocated by Einstein [1] QT does not provide a complete description of individual physical systems and may be emergent from some more detailed theory of physical phenomena. If contexts of experiments are correctly taken into account this interpretation is free of paradoxes [2].

In particular, the long range correlations predicted by QT for EPR type experiments and confirmed in spin polarization correlation experiments (SPCE) seem to *"cry for explanation"*.

Searching for an intuitive explanation of these correlations Bell analyzed local realistic and local stochastic hidden variable models. These models led to so called Bell and CHSH inequalities (CHSH) which were violated by some QT predictions [3, 4].

In *local realistic* models values of physical observables are predetermined by a source and recorded passively by a measuring apparatus. A notion of *local realism* is understood here in a narrow and specific sense. In stochastic models experimental outcomes are obtained locally in an irreducibly random way.

A general belief was that no other local models were possible, which is why the experimentally confirmed violation of CHSH [5, 6] has been incorrectly interpreted as a mysterious nonlocality of Nature with correlations coming from outside of space-time and paraphrased sometimes as: *"two perfectly random dice tossed in two far away locations produce perfectly correlated outcomes"*.

Of course a photon is not a die, correlations are not perfect and the confusion came from imprecise terminology and from the lack of understanding of probabilistic models used in various proofs of CHSH. All these models used a unique probability space and a joint probability distribution to describe all the results of different experiments performed in incompatible experimental settings. Non-probabilistic models used predetermination of the results and/or *counterfactual* definiteness.

Several authors pointed out that these assumptions were very restrictive and if not used CHSH simply could not be proven [7-34]. Local probabilistic models for SPCE consistent with QT were constructed [8-10, 19, 21, 25-27, 32-33]. Experiments in various domains of science were found with correlations violating CHSH [18]. Several experiments in quantum optics and in neutron interferometry could be simulated event by event in a local and causal way [35, 36]. The references we are giving in this paper are by no means exhaustive. Each cited paper contains many additional references.

These results seem to be unknown, not understood or ignored by part of the physical community.

A possibility of a rational explanation of strong correlations of the outcomes of SPCE gives arguments in favor of QT being an emergent theory. We will show that this idea has many other experimentally testable implications.

In particular, if QT is emergent then all pure quantum ensembles are mixed statistical ensembles and its various sub-ensembles may have slightly different properties [19, 20, 23, 25, 26]. Such differences can be detected by using purity tests which we discussed extensively many years ago in a different context [37-39].

Moreover some reproducible fine structures may exist in time series of data which are destroyed when a standard statistical analysis is done. To find these fine structures one has to study experimental data using more sophisticated statistical tools [39-41]. The existence of such structures would give not only convincing arguments that QT is emergent but it would also show that QT is not predictably complete [25, 39].

This paper is organized as follows.

In section 2 we show that if we insist on *irreducible randomness* of local experiments the most general probabilistic model is equivalent to the model found in a stochastic hidden variables approach. Therefore the violation of CHSH shows only that the assumption of *the irreducible randomness* in QT is incorrect giving additional arguments that QT is emergent from some microscopic theory respecting *causality* and *local determinism*.

In section 3 we construct a local contextual probabilistic model for which it is impossible to prove CHSH. This model is able to reproduce all quantum predictions for SPCE. We also show that contrary to the general belief QT does not predict strict anti-correlations of outcomes in SPCE. At the end of this section we talk about a wonderful migration of monarch butterflies which would be impossible without some kind of causality and determinism in Nature.

In section 4 we discuss other experimental implications of the assumption that QT is an emergent theory. In particular we show some statistical tools which can be used to detect fine structures in time series of data not predicted by QT.

The last section contains conclusions.

2. Probabilities and irreducible randomness

A probability is a complex notion [14] therefore we have to tell how we define it.

Let us consider a simple experiment consisting on flipping a coin c_1 by a flipping device d_1. We are repeating this experiment several times and we notice that knowing all preceding outcomes we are unable to predict what will be the next outcome. Besides we notice that if we increase the number of repetitions and calculate the relative frequencies of appearances of heads (H) or tails (T) then these frequencies approach 0.5. We say that *outcomes are produced in a random way with probability 0.5* and our experiment is called a *random experiment*. If we label H as 1 and T as -1 we introduce a random variable X_1 obeying a Bernoulli probability distribution $B(0.5)$ defined by probabilities: $P(X_1 = -1) = P(X_1 = 1) = 0.5$.

We were dealing here with a fair coin and with a fair device. If we took a coin c_2 which was crooked or/and a not fair flipping device d_2 we could observe that the probability p of observing H was not equal to 0.5. In this case we would introduce a random variable X_2 obeying a probability distribution $B(p)$ defined by the probabilities: $P(X_2 = -1) = p$ and $P(X_2 = 1) = 1 - p$.

Therefore a probability is neither a property of a coin nor a property of a flipping device. A probability is only a property of a random experiment [21-23]. Therefore we are dealing always with conditional probabilities [17-18] depending on the experimental context and we will write it explicitly when it is necessary. Since QT predicts the probabilities of various random experiments it is a contextual theory.

We concluded that the outcomes of coin flipping experiments were produced in a random way. However, we know that a motion of a coin obeys the laws of classical physics so if we knew all important parameters having impact on its motion we could in principle predict with certainty each final outcome. Of course we don't know and we don't control these parameters but this type of randomness is often considered less perfect and is called a classical or *reducible randomness*.

The fathers of QT postulated without proof that the randomness observed in quantum measurements and phenomena is more fundamental than in classical physics being intrinsic and irreducible. We will show that the violation of CHSH gives indication that this is not a case.

If both quantum and classical randomness are reducible we can only say that given experimental outcomes are produced in a random way if the time series of these outcomes passes with success all tests of randomness invented by statisticians. We regularly produce such series of "random" outcomes using generators of pseudo-random numbers but a person knowing the algorithm knows exactly what will be the next number of a series. Similarly consecutive digits of the decimal development of π pass with success all tests of randomness.

Let us consider now two far away laboratories performing experiments x and y respectively on two physical signals S_1 and S_2 produced by some source S. Let us assume that outcomes of the experiments (x, y) are (a, b) where $a = \pm 1$ and $b = \pm 1$. The outcomes of these experiments gathered during some period of time form two samples of data $a_1, a_2, \ldots a_n \ldots$ and $b_1, b_2, \ldots b_n \ldots$. In mathematical statistics these outcomes are observations of two time series of random variables $A_1, A_2 \ldots A_n \ldots$ and $B_1, B_2 \ldots B_n \ldots$ called sampling distributions.

If all A_i are independent and identically distributed (i.i.d) as some random variable A and all B_i are i.i.d as some random variable B then outcomes of the experiments x and y are completely described by conditional joint probability distributions $P(A = a, B = b|x, y, S_1, S_2)$. Since the signals S_1 and S_2 can be correlated at the source:

$$P(A = a, B = b|x, y, S_1, S_2) \neq P(A = a|x, S_1)P(B = b|y, S_2) \tag{1}$$

However, *if S is sending a pure statistical ensemble of signals,* for example, if each beam is composed of some *identical* physical systems *and if the outcomes are obtained in irreducibly random way* then *the random variables A and B are independent* and

$$P(A = a, B = b|x, y, S_1, S_2) = P(A = a|x, S_1)P(B = b|y, S_2). \tag{2}$$

From (2) we obtain immediately the factorization of expected values $E(AB) = E(A)E(B)$.

If we insist on *irreducible randomness* of the act of measurement we may only obtain non trivial correlations if we assume that S is sending a mixed ensemble of signals in which each couple $(S_1(\lambda_1), S_2(\lambda_2))$ is included with a probability $P(\lambda)$ where $\lambda = (\lambda_1, \lambda_2)$ are labels of different couples. We are talking about signals because our reasoning holds even if a signal is not a beam of some discrete physical systems. In this case instead of (2) we obtain:

$$P(a, b|x, y) = \sum_{\lambda \in \Lambda} P(\lambda)P(a|x, \lambda_1)P(b|y, \lambda_2) \tag{3}$$

and

$$E(AB) = E(AB|x, y) = \sum_{\lambda \in \Lambda} P(\lambda)E(A|\lambda_1, x)E(B|\lambda_2, y). \tag{4}$$

Using (4) and choosing in an appropriate way $P(\lambda)$ we may prove that $Cov(A, B) \neq 0$. A probabilistic model defined by (3) describes the experiment (x, y) as a mixed ensemble of pairs of *independent* random experiments labelled by λ. Correlations which may be created in this way are quite limited. Besides, to estimate the probabilities in (3) we should follow a completely different experimental protocol than the experimental protocol used in SPCE. A more detailed discussion of this point is given in [27, 28].

To make this point clearer let us consider a simple experiment consistent with (3). Charlie is sending to Alice and Bob particularly chosen pairs of dice and let x and y be simply *rolling a die* experiments. Then in order to estimate the probabilities of various outcomes using (3) the following protocol may be used. Alice and Bob receive a first pair of dice and they roll their die many times to estimate corresponding probabilities. They proceed in the same way with all correlated pairs sent by Charlie and at the end they average all the obtained estimates. It does not resemble the SPCE protocol.

If we perform four different experiments (x, y), (x', y), (x, y') and (x', y') using the same signals prepared by S and if all $|E(A|\lambda_1)| \leq 1$ and $|E(A|\lambda_2)| \leq 1$ then using (4) we can easily prove CHSH inequalities:

$$|E(AB) - E(AB')| + |E(A'B) + E(A'B')| \leq 2 \tag{5}$$

which are violated in SPCE and in many experiments from other domains of science [18].

Equation (4) was used in *stochastic hidden variable models* and called *locality condition*. This terminology is the source of unjustified speculations about the non-locality of Nature.

Equation (5) can also be proven in a different way using *local realistic hidden variable models*. In these models outcomes of all experiments are predetermined and there exists a unique probability space and a joint probability distribution for all possible experiments (x, y) [11, 14, 22, 23, 29]. The outcomes obtained in experiments x and y are evaluations of two functions $A(\lambda_1) = \pm 1$ and $B(\lambda_2) = \pm 1$. Therefore following Bell [2] we obtain:

$$E(AB) = \sum_{\lambda \in \Lambda} P(\lambda)A(\lambda_1)B(\lambda_2). \tag{6}$$

Using (6) we can first prove Bell inequalities and deduce (5) as a corollary [3-4]. A very simple proof is given for example in the appendix II of [9].

In experiments x and x' one is measuring incompatible physical observables. Therefore, all experiments (x, y), (x', y), (x, y') and (x', y') require different experimental settings. It was already known to Boole and Kolmogorov that a unique probability space and a joint probability distribution able to describe such experiments do not need to exist [13-16, 22-26, 33-34].

Since (5) is violated a correct probabilistic description of experiments has to be contextual imitating the contextual description provided by QT. In SPCE outcomes of measurements of physical observables A and B depend on a triplet (S, x, y) defining a particular experimental context. In QT this experimental context is also described by a triplet.

Physical systems prepared by S are represented by a wave function $\psi \in H_1 \otimes H_2$ or by a density operator ρ. The local physical observables A and B are represented by hermitian operators $\hat{A} \otimes I$ and $I \otimes \hat{B}$. Using this notation, conditional expectation values $E(AB|\rho) = Tr\rho\hat{A}\hat{B}$ are calculated. *There is no reason for the existence of a functional relation between expected values obtained in incompatible experimental settings.*

One obtains (5) only if a physical state ρ is a convex sum of separable states: $\rho = \sum_{i=1}^{k} p_i \rho_i \otimes \tilde{\rho}_i$ where $0 < p_i < 1$ because in this case [28]:

$$E(AB|\rho) = \sum_{i=1}^{k} p_i E(A|\rho_i) E(A|\tilde{\rho}_i). \tag{7}$$

Of course, the entangled spin singlet state in SPCE is not a convex sum of separable states, which is why probabilistic models leading to (5) cannot be used.

In the next section we will present a contextual local probabilistic model able to reproduce all QT predictions for SPCE.

3. Memory and local determinism in Nature

In order to preserve correlations between signals created at the source some memory of these correlations has to be preserved when the signals reach measuring instruments and when the measurements are completed. To achieve it, a measurement process, instead of being irreducibly random, has to be strictly deterministic and has to depend both on a signal and on a state of the instrument in a moment of measurement.

An outcome is not predetermined at the source but it is only known when a measurement is completed. Therefore a measured value of the physical observable is not an attribute of individual physical systems but it is a contextual property created during the act of measurement [21, 26, 42].

As we told in the preceding section, for experiments requiring incompatible experimental settings a unique probability space and a joint probability distribution able to describe these experiments do not exist, and each couple of experiments (x, y) may be only described by a specific probability space Λ_{xy}.

Such type of contextual probabilistic description [15-17, 21-22, 25, 26] reconciles in some sense Einstein and Bohr. Einstein [1] claimed that QT might not provide a complete description of individual physical systems. Bohr insisted that any description of quantum phenomena is contextual depending on the whole experimental setting [42].

In a contextual probabilistic model (3) is replaced by:

$$P(a, b|x, y) = \sum_{\lambda \in \Lambda_{xy}} P(\lambda) P(a|x, \lambda_1, \lambda_x) P(b|y, \lambda_2, \lambda_y) \tag{8}$$

where $P(\lambda) = P(\lambda_1, \lambda_2) P_x(\lambda_x) P_y(\lambda_y)$, $\lambda = (\lambda_1, \lambda_2, \lambda_x, \lambda_y)$ and Λ_{xy} is different for each pair of experiments. The correlations between signals created by S are coded in $P(\lambda_1, \lambda_2)$. To preserve the partial memory of these correlations we have to assume a local determinism thus $P(a|x, \lambda_1, \lambda_x)$ and $P(b|y, \lambda_2, \lambda_y)$ are 0 or 1.

Probabilistic models defined by (8) are consistent with the experimental protocol of SPCE and they can reproduce any correlations predicted by QT and violating CHSH. *Local choices of experimental settings made by experimenters are completely independent* and the order in which they are made has no influence on the correlations observed.

One may also define a different contextual model violating CHSH by replacing in (6) Λ by Λ_{xy} and adding parameters describing instruments [27].

Summation in equation (8) or in modified equation (6) can be replaced by integration over variables λ without changing the conclusions [26, 27, 30-31]. There exist various proofs of (5) not using explicitly (4) or (6). However, in all these proofs the contextual character of physical observables is neglected and a *counterfactual reasoning* is used [9, 14, 21, 22, 24, 29].

Since models defined by (8) contain probability distributions of $P_x(\lambda_x)$ and $P_y(\lambda_y)$ therefore they do not predict perfect anti-correlations for the outcomes of SPCE. They do not differ in this respect from QT. According to statistical interpretation the spin singlet wave function commonly used to describe the preparation of physical systems in SPCE does not give any information about individual outcomes but only it is a mathematical tool which together with appropriate operators can be used to predict statistical regularities in experimental data.

In SPCE observables A and B are the spin projections on two directions characterized by angles θ_A and θ_B respectively. Calculation for a singlet state gives the expectation value $E(AB|\psi) = -\cos(\theta_A - \theta_B)$. However, directions of spin polarization analyzers are never sharp and are only defined by some small intervals I_A and I_B containing angles close to θ_A and θ_B respectively. Therefore even if detection efficiencies were perfect and if we dealt with a perfect singlet spin state the prediction for the measured expectation values is only:

$$E(AB|\psi) = -\int_{I_A}\int_{I_B} \cos(\theta_1 - \theta_2)\, d\rho_A(\theta_1)\, d\rho_B(\theta_2) \tag{9}$$

From (9) it is clear that, contrary to a general belief, QT does not predict strict anti-correlation of clicks in experiments measuring A and B because in a case of continuous variables the probabilities it provides should be treated as probability densities and integrated respectively. It was pointed out many years ago [19, 21]. More detailed discussion of this problem may be found in [24-26].

The existence of probabilistic models (8) able to reproduce predictions of QT seems to indicate that QT is an emergent theory and that there is no irreducible randomness in Nature.

One also may find many arguments in favour of determinism in biology. As an example we describe below *a wonderful migration* of monarch butterflies.

The monarch butterflies make massive, over 3000 km long, southward and northward yearly migrations from Ontario province in Canada to the sanctuaries of the Mariposa Biosphere Reserve in Mexico. In difference to birds, no single individual makes the entire round trip.

The length of these journeys exceeds *a normal lifespan of monarchs which is around six weeks.* A normal life cycle of monarchs has 4 stages. From eggs they hatch into caterpillars, and then wrap up in cocoons where by *metamorphosis* they become butterflies. They get out from the cocoons, mate, lay eggs and die. The last generation of the summer, *with life span exceeding 7 months*, flies to one of many overwintering sites.

The overwintering generation does not reproduce until it leaves for the northward migration sometime in February and March and arrives as far north as Texas. Only the second, third or fourth generations return to their northern locations in the United States and Canada in late spring.

A true mystery is *how monarchs can return to the same overwintering spots over a gap of several generations.* The only possibility is that the *migration patterns are deterministically coded in their genes.* But even if these migration patterns are inherited it is still amazing how a small butterfly navigates using the sun, sky and earth's magnetic field to arrive exactly where it should. More information can be found for example on Wikipedia.

There are plenty other phenomena in biology which also indicate that: "*Nature does not play dice*" and that *irreducible randomness* is a false concept.

4. Is quantum theory predictably complete?

If QT is an emergent theory this fact has many model independent implications which in principle can be discovered by a detailed study of time series of experimental data.

In particular any model of sub-phenomena describes a pure quantum state as a mixed statistical ensemble with respect to some additional parameters. Any sub-ensemble of a pure ensemble is indistinguishable from the initial ensemble. The sub-ensembles of a mixed ensemble can differ between them. Since we do not control the distribution of supplementary parameters during the experiment time series of data may differ from run to run of the same experiment [19, 20, 43].

Therefore in an experiment in which it is believed that some beams are prepared in a pure quantum state one can make systematic changes of intensity and of geometry of the beams trying to change the properties of the initial statistical ensemble. If the initial ensemble was mixed there is a chance to discover it.

To detect statistically significant differences between different experimental runs one has to use non parametric statistical tests which we studied many years ago and called *purity tests* [37, 38]. Since not many physicists know these tests we explain below two important tests: *run* and *rank tests*.

A *run* is a sequence of like elements that are preceded and followed by different elements or no elements at all. For example, in a sequence 00101100011011 we have eight runs and in a sequence 11111100000111 we have three runs. If the appearance of 1 and 0 is purely random the number of runs can neither be too small nor too big. If a sample size is $n = n_1 + n_2$,where n_i are the numbers of 0 and 1 respectively, the test statistics to test the randomness of the series is $R = \#$ of runs. The expected value $E(R)$ and its variance $Var(R) = \sigma_R^2$ are functions of n_1 and n_2 and the probability distribution is known. For large values of n a statistics $Z = (R - E(R))/\sigma_R$ is normally distributed.

In *rank tests* we rank the observations in increasing order and then we use the ranks for actual observations. For example in Mann–Whitney U test we verify a null hypothesis that there is no significant difference between two populations. We combine two random samples $S(1)$ and $S(2)$ from these populations having sizes n_1 and n_2 respectively and we rank all the observations. To all tied observations the same averaged rank is assigned. The Mann–Whitney U statistic is a function of n_1, n_2 and R_1 which is the sum of ranks scored by the observations from the sample $S(1)$. For large samples one can use the standard normal distribution and tables exist for smaller samples.

A detailed discussion of these and other purity tests may be found in [37-39]. Using purity tests we may detect that data obtained in our experiment are a sample from some mixed statistical ensemble.

In order to answer the question of whether QT is predictably complete we must study more carefully time series of experimental data in order to find some fine stochastic structures if they exist [39, 40].

Let us go back to our discussion of two samples of experimental data which we made on page 3 of this paper and consider data coming only from one experiment x. The outcomes of this experiment gathered during some period of time form a sample $a_1, a_2 \ldots a_n \ldots$ where individual outcomes are obtained in some discrete moments of time. In mathematical statistics these outcomes are observations a time series of random variables $A_1, A_2 \ldots A_n \ldots$.

If all A_t are i.i.d as some random variable A then outcomes of the experiment x are completely described by a conditional probability distribution $P(A = a|x, S_1)$. *This is what is believed to be true for quantum measurements.* Therefore histograms and empirical probability distributions are compared with the predictions of QT and a fine structure of time series of data is not studied carefully enough.

If A_t are not i.i.d random variables then experimental data are only completely described be the properties of the whole time series A_t.

A study of a time series is a delicate and a difficult task [39-41]. To detect a fine structure in a time series one has to use various models and try to match available data.

In particular for stationary time series a large class of autoregressive integrated moving average models ARIMA was studied extensively [41]. We define below a simplest autoregressive model $AR(p)$ using a standard notation.

A time series it is a family of random variables Z_t where $t = 0, 1 \ldots$ for simplicity. Time series is stationary if $E(Z_t) = \mu$, $Var(Z_t) = \sigma^2$ and covariance function $\gamma(k)$ at lag k does not depend on t where $\gamma(k) = Cov(Z_t, Z_{t+k}) = E((Z_t - \mu)(Z_{t+k} - \mu))$.

A white noise it is a time series a_t where a_t are normal i.i.d random variables with zero mean. The autocorrelation function $\rho(k)$ at lag k (ACF) is defined as $\rho(k) = \gamma(k)/\gamma(0)$ and it is easy to see that $\rho(0) = 1$, $\rho(k) = \rho(-k)$ and $|r(k)| \leq 1$.

Autoregressive time series $AR(p)$ with $\mu = 0$ satisfies the following equation [40]:

$$Z_t - \Phi_1 Z_{t-1} - \Phi_2 Z_{t-2} - \ldots - \Phi_p Z_{t-p} = a_t \tag{10}$$

where Φ_i are real numbers.

Using (10) one can prove that ACF is a quickly decaying function when k increases. Another important function having well known properties is so called partial autocorrelation function (PAC) which measures the importance of the lag k in $AR(p)$. These two autocorrelation functions are used to find out whether a studied time series can be modeled as $ARIMA$ process and in particular as $AR(p)$ process.

We cannot assume that our time series of experimental outcomes can be described using $AR(p)$ model. Therefore we have to analyze carefully our data in order to find out which model if any is a correct one. In particular we have to study:

- Descriptive statistics.
- Histogram and normal scores plot.
- Simple time series plot (z_t, t).
- Lagged scatter plots $((z_t, z_{t+k})$.
- Empirical ACF and PAC plots.

As we mentioned above a descriptive statistics and a histogram average out any fine structure of a time series. To illustrate this point we simulated a sample of size 500 of $AR(2)$:

$$Z_t - 0.25 Z_{t-1} - 0.5 Z_{t-2} = a_t \tag{11}$$

where a_t were normal i.i.d. with a unit variance.

A standard descriptive analysis: summary, histogram, and normal scores showed that our simulated data could be viewed as a sample drawn from some normally distributed population.

Using the statistical package S+ we found that empirical ACF was decaying and empirical PAC had a clear ≪cut-off≫ at lag 2 allowing us to conclude the sample was drawn from a stationary $AR(2)$ time series. Moreover the estimated values of the coefficients in equation (10) were: 0.243 and 0.487 very close to the true values 0.25 and 0.5 respectively [40]. It shows a power of statistical tools.

5. Conclusions

We have shown that one can explain entanglement and long range correlations without invoking *nonlocality* and *quantum magic*.

The violation of CHSH inequalities seems to indicate that *irreducible randomness* is a false notion and that *"Nature does not play dice"* supporting views of Einstein, 't Hooft [45] and many participants of this conference.

If QT is emergent from some detailed theory of physical phenomena then we can probably find confirmation of this idea by analyzing more carefully time series of experimental data.

Careful testing of CHSH inequalities was important because it allowed to check QT predictions and to eliminate probabilistic models which were unable to reproduce these predictions.

Since using local and contextual probabilistic models one cannot prove CHSH inequalities therefore further costly testing of these inequalities seems unnecessary. It would be better instead to search for reproducible fine structures in time series of data and answer the question whether QT is predictably complete [39, 40].

The contextual statistical interpretation of QT not only avoids various paradoxes but it allows also a consistent description of quantum measurement process [45].

According to this interpretation a wave function is not an attribute of a single physical state and it cannot be changed instantaneously. The reduced wave packet describes only the preparation of the sub-ensemble of the initial ensemble of physical systems for subsequent measurements.

Therefore there are no strict anti-correlations in SPCE and a registration of a particular click on Alice's detector does not give any *deterministic* prediction what happens or what will happen at Bob's detector. These facts are often misunderstood by members of quantum information community.

QT is extremely successful theory and perhaps it is the best description of physical phenomena we were able to find. Therefore even if QT is an emergent theory it does not mean that we will succeed to find more detailed consistent description of all observed physical phenomena.

However a possibility of explaining or simulating several quantum phenomena without referring to *quantum magic* is already an important step forward.

In spite of successes of QT there are still many worrying facts. In quantum field theory we introduce infinities by considering point-like charges and masses and we have to remove these infinities by using a complicated process of renormalization.

If one takes seriously the idea that hadrons are extended particles one may define a unitary S-matrix such that one cannot prove the *optical theorem* [46]. If the violation of the optical theorem was confirmed in LHC experiments it would require a reformulation of the standard model (SM).

Testing SM is a difficult task requiring many free parameters, various phenomenological inputs and Monte Carlo simulation of events. Even in quantum mechanics we have to use approximations and semi-empirical models containing free parameters in order to explain experimental data. It is worrying because in some sense QT becomes un-falsifiable [25, 43].

Let us conclude with two citations. One due to Bell [2]: *"What is proved by impossibility proofs is lack of imagination"* and one due to Bohr [42]: *"knowledge presents itself within a conceptual framework adapted to previous experience and ... any such frame may prove too narrow to comprehend new experiences"*.

Acknowledgments

I would like to thank Gerhard Grössing, Andrei Khrennikov, Theo Nieuwenhuizen, Kristel Michielsen and Hans de Raedt for interesting discussions.

References

[1] Einstein A 1949 *Albert Einstein: Philosopher-Scientist* ed P A Schilpp (NY: Harper and Row)
[2] Ballentine L E 1998 *Quantum Mechanics: A Modern Development* (Singapore: World Scientific)
[3] Bell J S 2004 *Speakable and Unspeakable in Quantum Mechanics* (Cambridge: Cambridge UP)
[4] Clauser J F, Horne M A, Shimony A, and Holt R A 1969 *Phys. Rev. Lett.* **23** 880
[5] Aspect A, Dalibard J and Roger G 1982 *Phys. Rev. Lett.* **49** 1804
[6] Weihs G, Jennewein T, Simon C, Weinfurther H and Zeilinger A 1998 *Phys. Rev. Lett.* **81** 5039
[7] Accardi L 1981 Topics in quantum probability *Physics Reports* **77** 169
[8] Accardi L and M Regoli M 2002 Locality and Bell's inequality *QP-XIII, Foundations of Probability and Physics* ed A Khrennikov (Singapore: World Scientific) p 1
[9] Accardi L 2005 Some loopholes to save quantum nonlocality *Foundations of Probability and Physics-3* (AIP Conf. Proc. **750**) ed A Khrennikov (New York: Melville) p 1
[10] Aerts D 1986 A possible explanation of the probabilities of quantum mechanics *Journal of Mathematical Physics* **27** 202
[11] Fine A 1982 Hidden variables, joint probability and the Bell inequalities *Phys. Rev. Lett.* **48** 291
[12] Hess K and Philipp W 2001 A possible loophole in the Bell's theorem and the problem of decidability between the views of Einstein and Bohr *Proc. Natl. Acad. Sci.* USA **98** 14228
[13] Hess K and Philipp W 2005 Bell's theorem: critique of proofs with and without inequalities *AIP Conf. Proc.* **750** 150
[14] Khrennikov A Yu 1999 *Interpretation of Probability* (Utrecht: VSP)
[15] Khrennikov A Yu 2007 Bell's inequality: nonlocality, "death of reality", or incompatibility of random variables *Quantum Theory Reconsideration of Foundations-4* (*AIP Conf. Proc.* **962**) ed G Adenier et al (New York: Melville) p 121
[16] Khrennikov A Yu 2009 Violation of Bell's Inequality and nonKolmogorovness *Foundations of Probability and Physics-5* (*AIP Conference Proceedings* **1101**) ed L Accardi et al (New York: Melville) p 86
[17] Khrennikov A Yu 2009 *Contextual Approach to Quantum Formalism* (Dortrecht:Springer)
[18] Khrennikov A Yu 2010 *Ubiquitous Quantum Structure* (Berlin: Springer)
[19] Kupczynski M 1984 New test of completeness of quantum mechanics *Preprint* IC/84/242.
[20] Kupczynski M 1986 On some new tests of completeness of quantum mechanics *Phys. Lett.* A **116** 417
[21] Kupczynski M 1987 Pitovsky model and complementarity *Phys. Lett.* A **121** 51
[22] Kupczynski M 1987 Bertrand's paradox and Bell's inequalities *Phys. Lett.* A **121** 205
[23] Kupczynski M 2002 On the completeness of quantum mechanics *arXiv*:quant-ph/028061
[24] Kupczynski M 2005 Entanglement and Bell inequalities *J. Russ. Laser Res.* **26** 514
[25] Kupczynski M 2006 Seventy years of the EPR paradox *Albert Einstein Century International Conference* (*AIP Conf. Proc.* vol **861**) ed J-M Alimi and A Funzfa (New York: Melville) p 516 (arXiv: 0710.3397 [quant-ph])
[26] Kupczynski M 2007 EPR paradox ,locality and completeness of quantum *AIP Conf. Proc.* vol **962** 274 (arXiv: 0710.3510 [quant-ph])
[27] Kupczynski M 2012 Entanglement and quantum nonlocality demystified *AIP Conf. Proc.* vol **1508** 253 (arXiv: 0710.3550 [quant-ph])
[28] Kupczynski M Operational approach to the entanglement *Preprint* arXiv:1208.6532 [quant-ph]
[29] De Muynck V M , De Baere W and Martens H 2001 Interpretations of quantum mechanics, joint measurement of incompatible observables and counterfactual definiteness *Found. Phys.* **24** 1589
[30] Nieuwenhuizen T M 2009 Where Bell went wrong *AIP Conf. Proc.* **1101** 127
[31] Nieuwenhuizen T M 2011 Is the contextuality loophole fatal for the derivation of Bell inequalities *Found. Phys.* **41** 580
[32] Pitovsky I 1983 Deterministic model of spin statistics *Phys. Rev.* D **27** p 2316
[33] Pitovsky I 1994 George Boole's conditions of possible experience and the quantum puzzle Brit. J. Phil. Sci. **45** 95
[34] De Raedt H, Hess K and Michielsen K 2011 Extended Boole-Bell inequalities applicable to Quantum Theory *J. Comp. Theor. Nanosci.* **8** 10119
[35] Michielsen K, Jin F, and De Raedt H 2011 Event-based corpuscular model for quantum optics experiments *J. Comp. Theor. Nanosci.* **8** 1052
[36] H. De Raedt, F. Jin, and Michielsen K 2012 Discrete-event simulation of neutron interferometry experiments *AIP Conf. Proc.* **1508** 172
[37] Kupczynski M 1977 On some important statistical tests *Riv. Nuovo Cimento* **7** 215
[38] Gajewski J and Kupczynski M 1979 Purity tests for π-d charge multiplicity distributions *Lett. Nuovo Cimento* **26** 81
[39] Kupczynski M 2009 Is the quantum theory predictably complete *Phys. Scr.* **T135** 014005 (*arXiv*:0810.1259

[quant-ph])

[40] Kupczynski M 2010 Time series, stochastic processes and completeness of quantum theory *AIP Conf. Proc.* **1327** 394 (*arXiv*:1009.3002 [quant-ph])

[41] Box G E P, Jenkins G M and Reinsel G C 2008 *Time Series Analysis Forecasting and Control* (Hoboken: Wiley)

[42] Bohr N 1963 *Essays 1958-1962 on Atomic Physics and Human Knowledge* (NewYork: Wiley)

[43] Kupczynski M 1973 Is Hilbert space language too rich *Int. J. Theor. Phys.* **79** 319

[44] `t Hooft G 2007 The mathematical basis for deterministic quantum mechanics *Beyond the Quantum* ed T M Nieuwenhuizen *et al.* (Singapore: World Scientific) 3

[45] Allahverdyan A E, Balian R and Nieuwenhuizen T M 2013 Understanding quantum measurement from the solution of dynamical models *Physics Reports* **525** 1

[46] Kupczynski M 2013 On a possible violation of the optical theorem in LHC experiments *arXiv*:1308.1447 [hep-ph]

Weak measurement and its experimental realisation

R Flack[1] **and B J Hiley**[1,2]

[1]HEP, University College, Gower Street, London
[2]TPRU, Birkbeck, University of London, Malet Street, London WC1E 7HX.

E-mail: r.flack@ucl.ac.uk; b.hiley@bbk.ac.uk

Abstract.
The relationship between the real part of the weak value of the momentum operator at a post selected position is discussed and the meaning of the experimentally determined stream-lines in the Toronto experiment of Kocsis *et al* is re-examined. We argue against interpreting the energy flow lines as photon trajectories. The possibility of performing an analogous experiment using atoms is proposed in order that a direct comparison can be made with the trajectories calculated by Philippidis, Dewdney and Hiley using the Bohm approach.

1. Introduction

The notion of a weak measurement has opened up new ways to explore quantum processes [3, 15, 25, 31, 35]. In contrast to the usual von Neumann or strong measurement, which gives information about the eigenvalues of a dynamical operator, weak measurements enable us to obtain information about small induced phase changes. This process allows us to experimentally investigate new features of a quantum process. Weak values of the momentum operator give us access to the components of the energy-momentum tensor which, in turn, have a direct significance for the Bohm approach.

In fact the real parts of these weak values enable us to measure components of the energy-momentum tensor of not only the electromagnetic field, but also the Schrödinger, Pauli and Dirac fields [23, 26, 36]. It is here that we make contact with the Bohm approach because these components are directly related to Bohm momentum, $\boldsymbol{p}_B = \nabla S$, a parameter that plays a fundamental role in this approach.

This connection has already been recognised and used in the experiments of Kocsis *et al* [25]. Here weak measurements have been made to obtain values for the Poynting vector and from these measurements, sets of energy flow lines have been constructed. This specific experiment used a weak electromagnetic source so that only one photon entered the apparatus at a time. The information about the weak value of the momentum operator was then found using standard photon counting techniques. As single photons are involved, the question as to the meaning of Poynting's vector for a single photon is raised, a question that has remained unanswered since the inception of the notion of a photon [32]. Can the flow lines in this experiment shown in Figure 1 be regarded as photon trajectories?

The one model that could possibly give an answer to this question is that introduced by Bohm [7]. Recall that in that model, the close connection between the classical Hamilton-Jacobi equation and the real part of the Schrödinger equation under polar decomposition of the wave function, enables a meaning to be given to the notion of a trajectory of a particle even in quantum

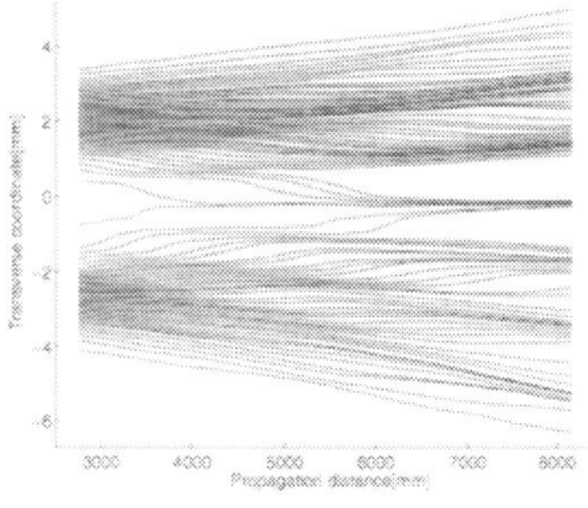

Figure 1. Experimentally constructed electromagnetic energy flow lines. From Kocsis *et al.* [25]

mechanical interference situations such as the two-slit experiment [7–9, 12]. The trajectories for a Schrödinger particle are shown in Figure 2. Not only do these results hold for the Schrödinger particle, they also hold for the Pauli and Dirac particles [14, 22]. In all these cases a meaning can be given to a trajectory because in a suitable non-relativistic limit, the trajectory actually becomes identical to the classical trajectory.

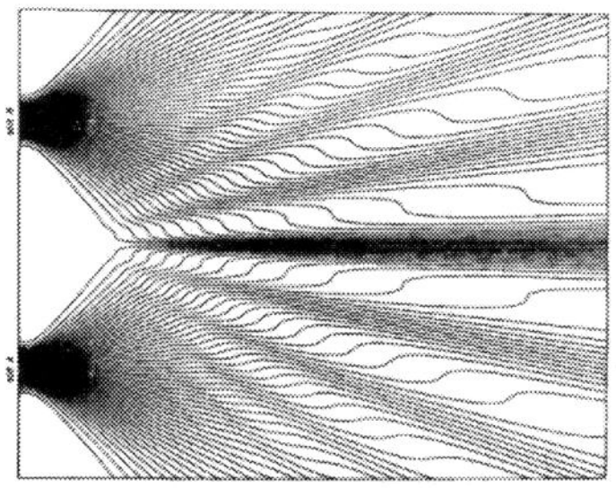

Figure 2. Two-slit trajectories calculated in Bohm approach [30]

Although the Bohm approach allows for the notion of a trajectory for Schrödinger particles, it does not give a meaning to the trajectory of a photon. This feature has long been known and has already been discussed by Bohm [8], while a more comprehensive treatment can be found in Bohm, Hiley and Kaloyerou [10] and Kaloyerou [24].

In this paper we will discuss the difficulties that arise in attempting to give a meaning to the notion of a photon trajectory. This will lead us to suggest that atoms should be used instead of photons. We will also discuss the principles behind the design of an experiment that uses atoms, rather than photons, in a typical two-slit situation. This will require us to make a weak measurement on the atoms after they have passed through the two-slits. From these results we will be able to construct trajectories using techniques that were pioneered by Kocsis *et al* [25]. In this way we will be able to compare these experimentally constructed trajectories with those calculated by Philippidis *et al* [30]. In this case we will be comparing like-with-like.

2. What is a weak value?

2.1. Definition of weak value

We start with a formal definition of a weak value given in Aharonov *et al* [3]. The weak value of an operator $\hat{A}$ is

$$\langle \hat{A} \rangle_{W_{\phi,\psi}} = \frac{\langle \phi | \hat{A} | \psi \rangle}{\langle \phi | \psi \rangle}.$$

The first thing we notice is that it is a complex number so it actually gives rise to two values. Notice further that as the real part of $\langle\phi|\psi\rangle$ approaches zero, such values become very large, which must already throw up doubts as to the usefulness of the concept of a weak value. However such doubts should be put aside and the idea pursued further. Let us now turn to see what extra information the weak value of the momentum operator, $\hat{P}$, gives us.

2.2. *Weak value of the momentum operator*

Clearly the weak value depends upon the post selected state ϕ, so let us consider the weak value of the momentum operator at a post selected position so that we can write $|\phi\rangle = |x\rangle$, then

$$\langle x|\hat{P}|\psi(t)\rangle = -i\nabla_x\psi(x,t).$$

If we now write $\psi(x,t) = R(x,t)e^{iS(x,t)}$ we find

$$\langle\hat{P}\rangle_{W_{\psi(x,t)}} = \nabla S(x,t) - i\nabla_x\rho(x,t)/2\rho(x,t).$$

One immediately recognises that the real part is the Bohm momentum, $\boldsymbol{P}_B = \boldsymbol{\nabla}S$, used in the Bohm approach to quantum mechanics, while the imaginary part is the osmotic momentum, with which we will not be concerned in this paper. (For details of the osmotic momentum see Hiley [23], Bohm and Hiley [11] and references therein.)

This result may seem model specific and has little to add to standard quantum mechanics, but it should be noted that the Bohm momentum is intimately related to the $(0,j)$ component of the energy-momentum tensor, $T^{\mu\nu}$, when one treats the wave function as a field [19,33]. In fact we find

$$\rho P^i_B(x,t) = T^{0i}(x,t) \quad \text{and} \quad \rho E_B(x,t) = T^{00}(x,t). \tag{1}$$

Here $\boldsymbol{P}_B$ and E_B are the Bohm momentum and Bohm energy respectively.

There are two points to emphasise here. Firstly the weak value of the momentum is the T^{0j} component of the energy-momentum tensor. Secondly, it gives the possibility of attaching a value to the Bohm momentum, thus making the approach open to experimental investigation. Furthermore it should be noted that $\boldsymbol{P}_B$ is not the eigenvalue momentum revealed in a strong or von Neumann measurement of the momentum operator. This difference has been discussed at length in Belinfante [4]. These results are not confined to the Schrödinger particle but also hold for the Pauli and Dirac particles [19, 20]. Thus we can now obtain additional empirical information about quantum processes.

Note further that the Bohm momentum has nothing to do with classical physics as seems to have taken root from the way Bohm originally introduced his ideas [7,8]. It is actually a feature of the quantum formalism itself. It should also be pointed out that exactly the same results can be obtained for the Schrödinger and Pauli particles using the von Neumann-Moyal algebraic approach [23].

With this particular example we see that the weak value of the four-momentum operator can give us information about the energy-momentum content of the quantum process. The next question is to ask how we can access these weak values experimentally.

2.3. *Weak measurement*

The general method for extracting information in a weak measurement using the Stern-Gerlach approach has been discussed in detail by Duck *et al* [15] and by Flack and Hiley [16]. In that discussion it is the weak value of the of the spin operator that is measured. In this paper we will be interested in the weak value of the momentum operator, so we will not discuss this example further here.

To measure the weak value of the momentum operator we have to find some way of coupling the momentum operator to some other observable which can then be used to extract the weak value $\langle P\rangle_W$. This can be done in the final stage of the measurement where a strong von Neumann measurement is made of this operator. This will involve a collapse of a part of the wave function without destroying the weak value. Let us call this operator, $\hat{A}$, the nature of which will be specified later. Then the interaction Hamiltonian can be written as $H_I = g(t)\hat{P}.\hat{A}$.

To see how this works, let us start with a particle in an initial state $|\Psi(t_0)\rangle = |\psi(t_0)\rangle|\xi\rangle$ where the momentum operator $\hat{P}$ acts on $|\psi(t_0)\rangle$ and the operator $\hat{A}$ acts on $|\xi\rangle$. Now introduce the time development operator $U(t,t_0) = \exp\left[-i\int_{t_0}^t H_I(t')dt'\right]$ so that the solution of the Schrödinger equation can be written in the form

$$|\Psi(t)\rangle|\xi\rangle = \exp\left[-iD\Delta t\hat{P}.\hat{A}\right]|\Psi(t_0)\rangle|\xi\rangle$$

where we have writen

$$\eta = \int_{t_0}^t g(t')dt' = D\Delta t.$$

Here D is the strength of the interaction and Δt is the time the interaction is active. Let us denote the complete orthonormal set of eigenfunctions of $\hat{A}$ by $|\xi_n\rangle$, so that

$$\hat{A}|\xi_n\rangle = a_n|\xi_n\rangle.$$

Since we will be interested in the weak value of the momentum at a specific point x, we will multiply from the left by $\langle x|$ and expand the n-th component of the exponential to form

$$\langle x|e^{-i\eta a_n\hat{P}}|\psi(t_0)\rangle|\xi_n\rangle = \langle x|\psi(t_0)\rangle\sum_{m=0}^{m=\infty}\frac{(-i\eta a_n)^m}{m!}\frac{\langle x|\hat{P}^m|\psi(t_0)\rangle}{\psi(x,t_0)}|\xi_n\rangle.$$

This can be rewritten as

$$\begin{aligned}&\langle x|e^{-i\eta a_n\hat{P}}|\psi(t_0)\rangle|\xi_n\rangle = \\ &= \psi(x,t_0)\left[e^{-i\eta a_n}\langle P(x)\rangle_W + \sum_{m=2}^{\infty}\frac{(-i\eta a_n)^m}{m!}\left[\langle P^m(x)\rangle_W - \langle P(x)\rangle_W^m\right]\right]|\xi_n\rangle.\end{aligned}$$

When the term involving the sum $\sum_{m=2}$ is small so that it can be neglected, we can write the final state of the particle in the form

$$\psi(x,t)|\xi\rangle = \sum_n \psi(x,t_0)c_n e^{-i\eta a_n\langle P(x)\rangle_W}|\xi_n\rangle. \tag{2}$$

We have chosen $|\xi\rangle = \sum_n c_n|\xi_n\rangle$. In this way we see that the weak measurement has changed the state $|\xi\rangle$ so that it becomes

$$|\xi_f\rangle = \sum_n d_n|\xi_n\rangle \quad \text{with} \quad d_n = c_n e^{-i\eta a_n\langle P\rangle_W}.$$

Thus the weak value $\langle P(x)\rangle_W$ can be found from the coefficients d_n. This means that if we now make a suitable strong measurement on the final state $|\xi_f\rangle$, we can pick out one of the eigenstates $|\xi_n\rangle$, say, and then from d_n we can find the weak value $\langle P(x)\rangle_W$.

If we try to determine d_n simply from the probability of finding $|\xi_n\rangle$, we will not find $\langle P(x)\rangle_W$ because it appears in the imaginary exponent. To abstract the information from these coefficients, we need to make a strong measurement on an operator complementary to $\hat{A}$, i.e. one that does not commute with with $\hat{A}$. Let us choose $\hat{B}|\mu_n\rangle = m_n|\mu_n\rangle$ with $[\hat{A}, \hat{B}] \neq 0$. Now suppose

$$|\xi_j\rangle = \sum_i b_{ji}|\mu_i\rangle,$$

then

$$|\xi_f\rangle = \sum_{n,i} d_n b_{in}|\mu_i\rangle$$

so that

$$\hat{B}|\xi_f\rangle = \sum_{n,k} m_k d_n b_{kn}|\mu_k\rangle.$$

We now make a strong measurement of $\hat{B}$, and obtain the probability of finding one particular state, say $|\mu_r\rangle$. This probability is given by $|\sum_n d_n b_{rn}|^2$, from which we can abstract the weak value $\langle P(x)\rangle_W$ as required.

2.4. What can couple to the momentum of an atom?
In order to find what couples to the momentum in an atomic beam, we need to remember that as an electrically neutral atom with a magnetic dipole moment $\boldsymbol{\mu}$ moves, it appears to have an effective *electric* dipole moment [1],

$$\boldsymbol{d} = (\boldsymbol{v} \times \boldsymbol{\mu})/c.$$

The energy of this dipole in an electric field will be

$$U = -\boldsymbol{d}.\boldsymbol{E} = \boldsymbol{v}.(\boldsymbol{E} \times \boldsymbol{\mu})/2$$

so the term that we couple to the momentum operator is $(\boldsymbol{E} \times \boldsymbol{\mu})$. If we replace the magnetic moment by its operator equivalent, $\hat{\mu} = \frac{e\hbar}{2mc}\hat{\boldsymbol{\sigma}}$, our interaction Hamiltonian becomes

$$H_I = -\frac{e\hbar}{4m^2c}(\hat{\boldsymbol{P}}.(\boldsymbol{E} \times \hat{\boldsymbol{\sigma}})). \tag{3}$$

This interaction Hamiltonian has been used to demonstrate the Aharonov-Casher effect [2, 13].

We now have a specific operator, $(\boldsymbol{E} \times \hat{\boldsymbol{\sigma}})$ to replace the operator $\hat{A}$ used in Section 2.2 so that it is now possible to measure, among other things, the Bohm momentum, $\boldsymbol{P}_B(x)$ in various situations. In this paper we are interested in exploring the two-slit experiment using Schrödinger particles, namely atoms.

3. Measurements to find trajectories in a two-slit interferometer

3.1. Bohm trajectories
In order to proceed we first discuss exactly how the notion of a particle trajectory arises the Bohm approach. The simplest way to see this is to follow Bohm [7, 12] and split the Schrödinger

equation into its real and imaginary parts under polar decomposition, $\psi = R\exp(iS)$[1]. It is then straight forward to show that the real part of the Schrödinger equation becomes

$$\partial S/\partial t + (\nabla S)^2/2m + Q + V = 0. \quad (4)$$

This equation is known as the quantum Hamilton-Jacobi equation. We emphasise that this equation follows exactly from the Schrödinger equation and no new mathematical structure is added.

The equation has a remarkable similarity to classical Hamilton-Jacobi equation where S, the phase of the wave function, replaces the classical action and a new quality of energy, the quantum potential energy, $Q = -\nabla^2 R/2mR$, appears. If we assume the canonical relation $\boldsymbol{P}_B = \boldsymbol{\nabla S}$, is still valid in the quantum case, then we have defined at every point in space, a momentum $\boldsymbol{P}_B(x)$.

In contrast to the observable momentum eigenvalue, Bohm assumed that $\boldsymbol{P_B}$ was the actual momentum 'possessed' by the particle, the "beable", a term introduced by Bell [5]. In this way the particle is assumed to actually possess simultaneously a position and a momentum, where, to repeat, this momentum is not the eigenvalue of the momentum operator when we are in the x-representation. Note that if we are working in the p-representation, then $\boldsymbol{P_B} = \boldsymbol{p}$.

One of the original objections to the Bohm approach was that since $\boldsymbol{P_B}$ was not an observable, it was not a meaningful quantity. Since there seemed to be no way to attribute an experimental value to $\boldsymbol{P_B}$, it, along with the whole approach, could easily be dismissed. However several factors have arisen over the last ten years that has clarified its meaning. These factors taken together with the fact that the momentum corresponds to a weak value, it is now possible to explore its predictions experimentally.

The first of these factors, as has already been pointed out, is that there is a connection between the terms that are used in the Bohm model and the components of the energy momentum tensor when the Schrödinger wave is treated as a quantum field in its own right. In fact this relationship has already been pointed out by Takabayasi [34] many years ago. Since then this relationship has been extended to apply to the Pauli and Dirac particles by Hiley and Callaghan [19]. It is from this work the the relationship shown in equation (1) was established. Takabayasi [34] has also shown that the quantum potential appears in the T^{jj} components of the energy-momentum tensor so even the quantum potential is open to experimental determination.

Secondly the Bohm momentum is identical to a momentum defined by Moyal [27], the Moyal momentum. Moyal shows that the transport equation for this momentum is the real part of the Schrödinger equation, which, of course, includes the quantum potential. All of this might be dismissed as a compelling factor because "it is simply replacing one semi-classical theory for another". However this conclusion is not correct. Hiley [23] has shown that the Moyal algebra is isomorphic to the algebra introduced by von Neumann in his classic 1931 paper [29]. This algebra leads to the uniqueness of the Schrödinger representation and is therefore at the heart of the quantum formalism. These results show that the Moyal theory and, in consequence the Bohm approach, are central to standard quantum formalism.

What the Moyal algebra describes is a non-commutative statistical theory with the Moyal (Bohm) momentum emerging as the conditional expectation value of the momentum derived using the distribution function $F_\psi(X, P)$. This distribution function is, in fact, a two-point density matrix, $\rho(X, P)$, where (X, P) are the mean position and mean momentum of an extended region of phase space, the quantum blob in the language of de Gosson [17].

Thirdly, there is the relatively recent discovery to which we have given prominence in this paper, namely, that the Bohm momentum is simply the real part of the weak value of the momentum operator evaluated at a post-selected position [23, 26, 36]. Thus if we can measure

[1] We put $\hbar = 1$ throughout.

$\langle \boldsymbol{P} \rangle_W$ in, say, a two-slit experiment, we will have a value of $\boldsymbol{P}_B(x)$ defined at all points in space from which we can construct the trajectories shown in Figure 2. Furthermore if we can also find the weak value corresponding to the Bohm kinetic energy, viz. $\langle P^2 \rangle_W$, then we can also put a value to the quantum potential energy Q, so that *all* the dynamic variables used in the Bohm approach can be investigated experimentally.

3.2. *Experimental realisation using photons*

A preliminary investigation of the possibility of experimentally determining the form of trajectories in a two-slit type interference apparatus has already been carried out using photons [25], but there are some difficulties in interpreting their energy flow lines as photon trajectories. What this important experiment actually shows is that it is possible to construct energy flow lines in this case.

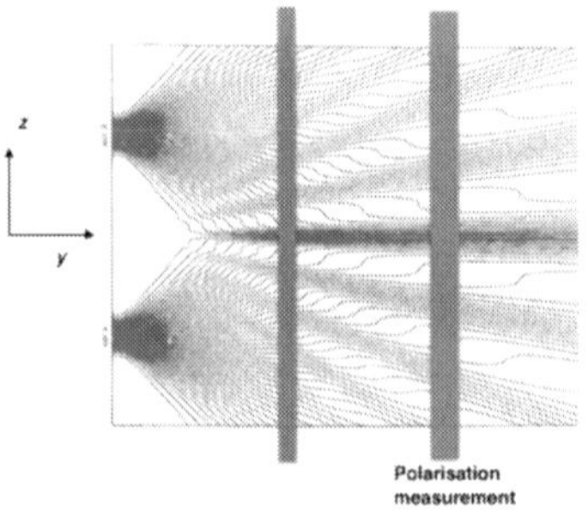

Figure 3. Schematic weak measurement apparatus for $\langle P_z \rangle_W$

In the actual experiment single photons are fed into a 50-50 beam splitter before being re-combined using two collimated fibre couplers that act as the two slits. The recombined beam is then passed through a thin birefringent calcite crystal, which introduces a weak effect by inducing a phase change in the overall wave function. This phase change is proportional to $\langle P \rangle_W$ as shown in equation (2) and can then be revealed in a strong measurement on the polarisation vector, following the principles outlined in section 2.2.

Figure 3 shows a schematic view of the apparatus required to measure $\langle P_z \rangle_W$, the transverse momentum at various points between the slit system and the final screen on which the fringes can be displayed. The initial beam is polarised in some definite direction before being passed through the thin birefringent crystal. The phase change introduced by the crystal leaves the spatial wave function unchanged, but the polarisation vector is rotated by a small amount.

If the interaction Hamiltonian coupling the weak device to the beam is of the form $H_I = g(t)\hat{P}.\hat{S}$, where $\hat{S}$ is the polarisation operator, then the final wave function will be of the form given in equation (2). To extract the information we need, the beam is passed through another polariser to separate out the left handed and right handed components and the number of photon counts per unit time in each component is determined. From these counts the transverse momentum at a set of positions, x, can be found from

$$\langle P_z(x) \rangle_W = G \sin^{-1}\left(\frac{N_R - N_L}{N_R + N_L}\right)$$

where G is a factor that depends on properties of the birefringent crystal. (For full details see Kocsis*et al.* [25]).

To construct the energy flow lines, $P_{B_Z}(x)$, is combined with the known $P_{B_y}(x)$ to construct the tangent vector to the flow line at the point x. This is repeated for a set of points between the thin crystal and the final screen. The points are chosen so that it is possible to join up the tangent vectors to produce the flow lines. (For the details of the extrapolation method used to produce these flow lines, see the paper of Kocsis *et al.* [25].)

These flow lines have some resemblance to those calculated by Philippidis *et al* [30] from the Schrödinger equation using the Bohm approach. However direct comparison should not be expected because, in the photon case, Maxwell's equations must be satisfied which gives rise to properties that are different from those arising from the Schrödinger equation (see Bohm [6].) Notice we prefer to talk about energy flow lines because, as we have seen from the relation between $P^i_B(x,t)$ and the energy-momentum tensor components $T^{0i}(x,t)$, these weak values are momentum fluxes. Thus in the case of light, we are actually measuring the Poynting vector of the electromagnetic field. To interpret these flow lines as "photon trajectories" raises some deep questions which we will discuss in the next section.

3.3. Meaning of the energy flow lines

As we have seen, the weak values measured in the Kocsis *et al.* [25] experiment enables the Poynting vector to be calculated at a number of points. The important feature of their experiment is that only one photon enters the apparatus at a given time so this raises a couple of questions. "What is the meaning of the Poynting vector for a single photon?" and "Can each energy flow line be regarded as a trajectory for the photon?"

The notion of a trajectory implies that we can give meaning to a photon existing momentarily at a space-time point, but remember the photon has a sharply defined energy and momentum. Thus it could be argued, that because of the uncertainty principle, a photon cannot exist at a point in space-time.

However such a conclusion is based on the assumption that all physical properties are eigenvalues of some operator or other. But as we have already pointed out in the Bohm approach, the uncertainty principle does not apply to the beables, $\boldsymbol{x}$ and $\boldsymbol{P}_B(x)$. The uncertainty principle only tells us is that we cannot *measure* position and momentum simultaneously using a *strong* measurement. In other words, we cannot associate the particle with simultaneous eigenvalues of the operators $\hat{\boldsymbol{X}}$ and $\hat{\boldsymbol{P}}$. However as we have seen $\boldsymbol{P}_B$ is the real part of the weak value of the operator $\hat{\boldsymbol{P}}$ and therefore is not ruled out on the grounds of the uncertainty principle. As $\boldsymbol{x}$ is the position eigenvalue in the case we are considering, the particle does not have a strong value of $\hat{\boldsymbol{P}}$, but it has a weak value of the same operator given by $\langle P(x)\rangle_W$. All of this follows from the properties of the Schrödinger formalism. Thus the Bohm interpretation gives simultaneously $\boldsymbol{x}$ and $\boldsymbol{P}_B(x)$ to the particle that obeys the Schrödinger equation.

The photon, on the other hand, is not a Schrödinger particle and therefore we must proceed with caution. Indeed Bohm himself has already pointed out that we must treat the photon in a very different way [8]. The reasons for coming to this conclusion are discussed in detail in Bohm and Hiley, chapter 11, [12]. Here we will simply point out that, unlike the Schrödinger particle with its finite rest mass, the photon has no position state $|x\rangle$. Or as Muthukrishnan, Scully and Zubairy [28] put it, there is no particle creation operator that creates a photon at an exact point in space.

An even simpler way of looking at it is to realise there is no way a photon can be 'slowed down' to reveal a non-relativistic, classical behaviour as is the case for a Schrödinger particle. In the latter, we simply have to find situations where the quantum potential is negligible to find the resulting behaviour is classical [18].

In the classical limit of quantum electrodynamics, we are left simply with a continuous field, which satisfies Maxwell's equations. This implies we should start with the electromagnetic field and develop a way of describing a quantised field by treating the classical field as a 'beable', and treating the fields in an analogous way to the way we treat the position and momentum of the Schrödinger particle.

Following up that line of reasoning, it might be tempting to take the $\boldsymbol{E}$ and $\boldsymbol{B}$ as the beables and then consider the energy density $T^{00}(x) = (E^2(x) + B^2(x))/2$ and Poynting vector $T^{0j}(x) = \boldsymbol{E}(x) \times \boldsymbol{B}(x)$ as giving information about the local energy and momentum. Recall the Poynting vector give us the momentum of an electromagnetic beam

$$\boldsymbol{p} = \frac{1}{4\pi c}\int(\boldsymbol{E} \times \boldsymbol{B})d\tau.$$

Clearly we can regard this momentum as being shared between the individual photons in the beam. Indeed dividing the Poynting vector by the number of photons in the beam gives us the average momentum carried by a photon. The crucial question as far as the experiment of Kocsis *et al* is concerned is whether we can use this momentum to calculate the 'trajectory' of a photon.

Unfortunately in this relativistic case we do not get a consistent set of trajectories in all frames. This difficulty has already been discussed in Bohm, Hiley and Kaloyerou [10] so we will not repeat the arguments here. Instead lets us give an example to illustrate the type of problem that arises. Consider an electrostatic field $\boldsymbol{E}$ uniform in the x-direction. The Poynting vector is zero, and this would correspond to the photons at rest. Consider a Lorentz transformation, again in the x-direction. $\boldsymbol{E}$ is unchanged and there is still no $\boldsymbol{B}$ field, so the Poynting vector remains zero and the photons are still at rest, but clearly the photons cannot be at rest in both frames.

These and other arguments show that we cannot obtain a consistent way of attaching the results obtained from these fields to construct meaningful photon trajectories. This criticism about trajectories is in no way a criticism of the energy flow results obtained in the experiment of Kocsis *et al.* [25]. The experiment is still important because it demonstrates for the first time how weak measurements can be used to measure components of the energy-momentum tensor in the quantum domain.

4. Weak measurements with atomic beams

The objections against photon trajectories do not apply to Schrödinger particles. In the Bohm approach the $T^{0\mu}$ components are given a meaning in terms of the quantum Hamilton-Jacobi equation where they appear as Bohm momentum $\boldsymbol{P_B}$ and the Bohm energy $E = -\partial S/\partial t$. The quantum Hamilton-Jacobi equation (4), like its classical counterpart, is simply an expression for energy conservation. We know that the classical Hamilton-Jacobi gives an ensemble of possible trajectories. The presence of Q in equation (4) does not deny the possibility of an ensemble of trajectories, it merely alters the form of the trajectories as shown in Figure 2.

Furthermore if Q becomes negligible then the resulting trajectories become the classical trajectories as was shown in Hiley and Mufti [18]. Since we know in the classical limit, a particle follows one of these trajectories and as these trajectories deform into their corresponding quantum counterparts, it is natural to assume that in the quantum case of a Schrödinger particle, we can retain the notion of a localised particle. Any response of this particle to the global environment is 'communicated' through the quantum potential Q.

To calculate these trajectories in a given experiment, say a two-slit interferometer, we simply integrate the expression for $\boldsymbol{P}_B(x) = m\, d\boldsymbol{r}_B/dt$. which gives us the set of trajectories shown in Figure 2. In order to establish the experimental legitimacy of these trajectories we propose to carry out an experiment along the lines of Kocsis *et al.* [25], but in this case using atoms. The atoms must be neutral but possess a magnetic moment so that we can use the interaction Hamiltonian given in equation (3). This means that we must replace the thin birefringent crystal with some form of uniform electric field. This will ensure that the momentum operator is coupled to the spin of the atom in an appropriate way.

We are at the moment in the planning stage of developing a two-slit interference experiment using a beam of atoms. In this experiment we plan to measure the weak value $\langle P_z \rangle_W$ which will allow us to calculate the transverse Bohm momentum. This value, together with the momentum in the direction of the beam, which in this case is along the y-axis (see Figure 3) will allow us to determine tangent momentum vectors at a series of points in the interference region. From these an ensemble of atomic trajectories can be calculated along the lines indicated in the experiment of Kocsis *et al.* [25]. Thus we will be able to compare the experimental results with those predicted by the Bohm approach.

5. Conclusion

Following on from the important experimental work of Kocsis *et al.* [25], who measured the energy flow lines shown in Figure 1, we have in this paper discussed the possibility of using similar techniques to measure the two-slit trajectories of atoms. The advantage of our proposed experiment is that in the case of atoms, a well defined notion of a trajectory is provided by the Bohm approach to the quantum formalism. This approach uses the real part of the Schrödinger equation to define what we mean by a trajectory. Thus we have a much clearer meaning of the notion of a trajectory for an atom than can be given to a photon. In the case of the atom, its classical behaviour is described by the Hamilton-Jacobi equation, which is clearly the limit of equation (4), whereas in the case of the photon, the classical behaviour is that of a field.

Since the Bohm approach based on the Schrödinger particle predicts well defined trajectories which have actually been calculated by Philippidis *et al.* [30] (See Figure 2), our proposals will show whether the trajectory calculations agree with experiment.

Acknowledgments

The authors would like to thank Markus Arndt of the University of Vienna, Peter Barker of UCL, and Jim Clark of the Cockcroft Institute for their help and advice in the design stages of the proposed atomic experiment.

6. Appendix

As we have already indicated, it is not possible to treat photons in the same way as Schrödinger particles in the Bohm approach. Since in the classical limit we have to deal with fields, we choose the field and its conjugate momentum to be the beables of the theory [8, 10, 12]. Since we are dealing with fields, the key question then is to ask how the concept of a photon appears in this approach.

We start with a Fourier analysis of the vector potential, namely,

$$\boldsymbol{A}(\boldsymbol{x}) = (4\pi/V)^{1/2} \sum_{k,\mu} \epsilon_{k,\mu} q_{k,\mu} e^{i\boldsymbol{k}.\boldsymbol{x}}$$

with $q^*_{k,\mu} = q_{k,\mu}$. We are working in the gauge $\boldsymbol{\nabla}.\boldsymbol{A} = 0$ and restricting our consideration to the transverse parts on the electromagnetic field.

Introducing the conjugate momentum $\boldsymbol{\Pi}_{k,\mu} = \partial q^*_{k,\mu}/\partial t$, the transverse part of the electric field is

$$\boldsymbol{E}(\boldsymbol{x}) = -\frac{1}{c}\frac{\partial \boldsymbol{A}(\boldsymbol{x})}{\partial t} = -\left(\frac{4\pi}{Vc^2}\right)^{1/2} \sum_{k,\mu} \epsilon_{k,\mu} \boldsymbol{\Pi}^*_{k,\mu} e^{i\boldsymbol{k}.\boldsymbol{x}}$$

and the transverse part of the magnetic field is

$$\boldsymbol{B}(\boldsymbol{x}) = \nabla \times \boldsymbol{A} = -(4\pi/V)^{1/2} \sum_{k,\mu} (\boldsymbol{k} \times \boldsymbol{\epsilon}_{k,\mu}) q_{k,\mu} e^{i\boldsymbol{k}.\boldsymbol{x}}.$$

The beables of the theory are then taken to be $q_{k,\mu}$ and $\boldsymbol{\Pi}_{k,\mu}$. The quantum Hamilton-Jacobi equation becomes

$$\frac{\partial s}{\partial t} + \sum_{k,\mu} \frac{\partial s}{\partial q_{k,\mu}} \frac{\partial s}{\partial q^*_{k,\mu}} + \sum_{k,\mu} (kc)^2 q_{k,\mu} q^*_{k,\mu} - \frac{\hbar^2}{2R} \sum_{k,\mu} \frac{\partial^2 R(\ldots q_{k,\mu} \ldots)}{q_{k,\mu} q^*_{k,\mu}} = 0.$$

The equation of motion of the $q_{k,\mu}$ derived from the Hamiltonian

$$H = \sum_{k,\mu} (\boldsymbol{\Pi}_{k,\mu} \boldsymbol{\Pi}^*_{k,\mu} + k^2 c^2 q_{k,\mu} q^*_{k,\mu})$$

becomes

$$\ddot{q}_{k,\mu} + k^2 c^2 q_{k,\mu} = \frac{\partial}{\partial q^*_{k,\mu}} \left(\frac{\hbar^2}{2R} \sum_{k',\mu'} \frac{\partial^2}{q_{k',\mu'} q^*_{k',\mu'}} \right).$$

Since Maxwell's equations for empty space follow when the right-hand term is zero, we see that the presence of this term profoundly modifies the behaviour of the electromagnetic field. This modification means that the oscillator $q_{k,\mu}$ can transfer a large quantity of energy and momentum even when $q_{k,\mu}$ is very small, because when $q_{k,\mu}$ is small the right-hand term may become very large. It is this rapid non-local and non-linear term that enables us to explain the notion of a photon.

In practice we cannot produce a beam of photons with all the photons having a sharp δ-function distribution in momentum. Usually there is a small distribution in momenta $f(\boldsymbol{k} - \boldsymbol{k}_0)$ so that the wave functional takes the form

$$\Psi = \sum_{k,\mu} f_\mu(\boldsymbol{k} - \boldsymbol{k}_0) q_{k,\mu} e^{-kct} \Psi_0$$

where Ψ_0 is the ground state of the electromagnetic field which is given by

$$\Psi_0 = \exp[-\sum_{k,\mu} (kc q_{k,\mu} q^*_{k,\mu} + ikct/2)]. \tag{5}$$

What Bohm showed was that the excess energy over the ground state is only appreciable within a region in which the wave packet $g(\boldsymbol{x})$ is appreciable, where

$$g(\boldsymbol{x}) = \sum_{k,\mu} f_\mu(\boldsymbol{k} - \boldsymbol{k}_0) e^{i\boldsymbol{k}.\boldsymbol{x}} \epsilon_{k,\mu}. \tag{6}$$

In order to understand the photoelectric effect, we must introduce the interaction Hamiltonian

$$H = \frac{1}{2m} [\boldsymbol{p} - (e/c)\boldsymbol{A}(\boldsymbol{x})]^2.$$

The photoelectric effect corresponds to the transition of a radiation oscillator from an excited state to the ground state, while the atomic electron is ejected with an energy $E = h\nu - I$, where I is the ionisation potential of the atom. The initial wave functional of atom plus the field containing only one quantum is

$$\Psi_i = \psi_0 \exp(-iE_0t/\hbar) \sum_{k,\mu} f_\mu(\boldsymbol{k} - \boldsymbol{k}_0) q_{k,\mu} e^{-kct} \Psi_0(\ldots q_{k,\mu} \ldots).$$

Now solving the Schrödinger equation, we obtain an asymptotic wave functional

$$\Psi_f = \Psi_0(\ldots q_{k,\mu} \ldots) \sum_{k,\mu} f_\mu(\boldsymbol{k} - \boldsymbol{k}_0) \times \frac{\exp[i\boldsymbol{k'}.\boldsymbol{r} - i\hbar(k'^2/2m)t]}{r} g_\mu(\theta, \phi, k'),$$

where the energy of the outgoing electron is $E = \hbar^2 k'^2/2m = \hbar kc + E_0$. The function $g_\mu(\theta, \phi, k')$ is the amplitude associated with the ψ-field of the outgoing electron. The outgoing electron wave packet, centred on $r = (\hbar k'/m)t$ will eventually become separated from its initial wave function ψ_0. If the electron happens to enter the outgoing packet, the initial wave function can subsequently be ignored. The system then acts for all practical purposes as if its wave field were given by equation (5) from which we conclude that the radiation field is in the ground state, while the electron has been liberated. It is readily shown that, as in the usual interpretation, the probability that the electron appears in the direction θ, ϕ can be calculated from $|g_\mu(\theta, \phi, k')|^2$. Thus the energy swept from the quantum field is just equivalent to one photon energy.

The picture is that the one quantum of energy is spread over the extent of the wave packet described by equation (6). Then during the interaction between the field and the electron, one quantum of energy is swept from the field and this is used to release the electron. Thus in this picture, the photon does not exist at a point like a Schrödinger particle. Bohm shows that the same picture can be made to explain the Compton effect. In fact other examples exhibiting this behaviour will be found in Bohm, Hiley and Kaloyerou [10] and Kaloyerou [24].

References

[1] Al-Jaber, S. M., Xingshu Zhu and Walter C. Henneberger. Interaction of a moving magnetic dipole with a static electric field. *Euro. J. Phys.*, **12** (1991) 268-70.
[2] Aharonov, Y. and Casher, A., Topological Quantum Effects for neutral Particles, *Phys. Rev. Letts*, **53**, (1984) 319-321.
[3] Aharonov, Y., Albert, D. Z. and Vaidman, L. How the Result of a Measurement of a Component of the Spin of a Spin-1/2 Particle Can Turn Out to be 100, *Phys. Rev. Lett.*, **60** (1988) 1351-4.
[4] Belinfante, F. J., *A Survey of Hidden-Variables Theories*, Pergamon Press, Oxford, 1973.
[5] Bell, J. S. *Speakable and Unspeakable in Quantum Mechanics*, Cambridge University Press, Cambridge, 1987.
[6] Bohm, D., *Quantum Theory*, p. 97-8, Prentice-Hall, Englewood Cliffs, N.J., 1951.
[7] Bohm, D., A Suggested Interpretation of the Quantum Theory in Terms of Hidden Variables I, *Phys. Rev.*, **85**,(1952) 166-179.
[8] Bohm, D., A Suggested Interpretation of the Quantum Theory in Terms of Hidden Variables II, *Phys. Rev,*, **85**, (1952), 180-193.
[9] Bohm, D. and Hiley, B. J., An Ontological Basis for Quantum Theory: Non-relativistic Particle Systems, *Phys. Reports*, **144**, (1987) 323-348.
[10] Bohm, D., Hiley, B. J. and Kaloyerou, P.N., An Ontological Basis for the Quantum Theory II ; a Causal Interpretation of Quantum Fields,*Phys. Reports*, **144**, (1987) 349-375.
[11] Bohm, D. and Hiley, B. J., Non-locality and Locality in the Stochastic Interpretation of Quantum Mechanics, *Phys. Reports*, **172**, (1989) 92-122.
[12] Bohm, D. and Hiley, B. J., *The Undivided Universe: an Ontological Interpretation of Quantum Theory*, Routledge, London 1993.

[13] Cimmino, A., Opat, G. I., Klein, A. G., Kaiser, H., Werner, S. A., Arif, M. and Clothier, R., Observation of the topological Aharonov-Casher phase shift by neutron interferometry, *Phys.Rev. Lett.* **63** (1989) 380-383.

[14] Dewdney, C., Holland, P. R., Kyprianidis, A. and Vigier, J-P., Spin and non-locality in quantum mechanics, *Nature*, **336** (1988) 536-44.

[15] Duck, I. M., Stevenson, P. M. and Sudarshan, E. C. G., The sense in which a "weak measurement"of a spin-1/2 particle's spin component yields a value 100, *Phys. Rev.*, **40** (1989) 2112-17.

[16] Flack, R and Hiley, B. J., Weak measurements, the Energy-momentum Tensor and the Bohm Approach, accepted to appear in Gao, S. *20th Anniversary of the Novel Concept of Protective Measurements*, 2014.

[17] de Gosson, M., Quantum Blobs, *Found. of Phys.* **43** (2012) 1-18.

[18] Hiley, B. J. and Mufti, A. H., The Ontological Interpretation of Quantum Field Theory Applied in a Cosmological Context, in *Fundamental Problems in Quantum Physics*, ed. Ferrero, M. and van der Merwe, A., 141-156,. Kluwer, Dordrecht, 1995.

[19] Hiley, B. J. and Callaghan, R. E., The Clifford Algebra approach to Quantum Mechanics A: The Schrödinger and Pauli Particles. arXiv:Maths-ph: 1011.4031.

[20] Hiley, B. J. and Callaghan, R. E., The Clifford Algebra Approach to Quantum Mechanics B: The Dirac Particle and its relation to the Bohm Approach. arXiv: Maths-ph:1011.4033.

[21] Hiley, B. J., Weak Values:Approach through the Clifford and Moyal Algebras, *J. Phys.: Conference Series*, **361** (2012) 012014.

[22] Hiley, B. J., Callaghan, R. E., Clifford Algebras and the Dirac-Bohm Quantum Hamilton-Jacobi Equation, *Found. Phys.*, **42** (2012) 192-208.

[23] Hiley, B. J., On the Relationship between the Moyal Algebra and the Quantum Operator Algebra of von Neumann, arXiv 1211.2098.

[24] Kaloyerou, P. N., The Causal Interpretation of the Electromagnetic Field, *Phys. Reports.* **244** (1994) 287-385.

[25] Kocis, S., Braverman, B., Ravets, S., Stevens, M. J., Mirin, R. P., Shalm, L.K., Steinberg, M. A., Observing the Average Trajectories of Single Photons in a Two-Slit Interferometer, *Science.* bf 332 (2011) 1170-73.

[26] Leavens, C. R., Weak Measurements from the point of view of Bohmian Mechanics, *Found. Phys.*, **35** (2005) 469-91.

[27] Moyal, J. E., Quantum Mechanics as a Statistical Theory, *Proc. Camb. Phil. Soc.* **45**, (1949), 99-123.

[28] Muthukrishnan, A., Scully,M. O. and Zubairy, M. S., The concept of the photon revisited, in [32] pp. S-18-27, 2003.

[29] von Neumann, J., Die Eindeutigkeit der Schrödingerschen Operatoren, *Math. Ann.*, **104** (1931) 570-87.

[30] Philippidis, C., Dewdney, C. and Hiley, B. J., Quantum Interference and the Quantum Potential, *Nuovo Cimento*, **52B**, (1979) 15-28.

[31] Ritchie, N.W., Story, J. G. and Hulet, R. G., Realization of a Measurement of a "Weak Value", *Phys. Rev. Lett.* **66** (1991) 1107-10.

[32] Roychoudhuri, C.,Kracklauer, Al F. and Creath, K., eds. *The nature of light: what is a photon?*, CRC Press, 2008.

[33] Takabayasi, T., Remarks on the Formulation of Quantum Mechanics with Classical Pictures and on Relations between Linear Scalar Fields and Hydrodynamical Fields, *Prog. Theo. Phys.*, **9** (1953) 187-222

[34] Takabayasi, T., On the Formulation of Quantum Mechanics associated with Classical Pictures, *Prog. Theo. Phys.*, **8** (1952) 143-182.

[35] Steinberg, A, Feizpour, A, Rozema, M. and Hayat, A., In praise of weakness, *Phys. World* March, (2013) 35-40.

[36] Wiseman, H. M., Grounding Bohmian mechanics in weak values and Bayesianism, *New J. Phys.*, **9** (2007) 165-77.

EmQM13: Emergent Quantum Mechanics 2013 IOP Publishing
Journal of Physics: Conference Series **504** (2014) 012017 doi:10.1088/1742-6596/504/1/012017

Non-locality and destructive interference of matter waves

Helmut Rauch
Atominstitut, Vienna University of Technology, 1020 Wien, Austria

Abstract. Quantum mechanics with massive particles becomes an important tool for fundamental research and applied science since many previously named "Gedanken" experiments become feasible. Neutrons are massive particles which couple to gravitational, nuclear and electro-magnetic interactions and they are sensitive to topological effects as well. Therefore they are proper tools for testing quantum mechanics where several previously named "hidden" parameters become measurable. Widely separated coherent beams can be produced by means of perfect crystal interferometers and they can be influenced individually. Spinor symmetry, spin superposition and quantum beat effect experiments have been performed and topological phases have been observed. Recent experiments related to the decoherence problem have shown that interference effects can be revived even when the overall interference pattern seems to be incoherent. All retrieval processes involve inherently unavoidable losses which stem partly from the theory itself and partly from an imperfect environment. Related post-selection experiments shed a new light on questions of quantum non-locality and support the request for more complete quantum measurements in the future. A more rational explanation of non-locality effects may be obtained when the plane wave components outside the wave packets are included in the discussion. This can also help to discuss entanglement and contextuality effects in a new light. In all quantum experiments more information can be extracted by more complete quantum experiments which will be important in the future to get a better understanding of quantum physics. An example may be the consideration of the Compton frequency and of proper time effects of matter waves.

1. Introduction

Neutrons exhibit wave and particle features and are a proper tool for testing quantum phenomena with massive particles (e.g. Rauch and Werner [1]). With perfect crystal interferometers widely separated coherent beams can be produced and influenced by nuclear, electroweak and gravitational interactions. Highly efficient polarizers, spin rotators and detectors are available. Most experiments use thermal neutrons with velocities in the order of 2000 m/s and wavelengths in the order of about 2 Å. The spin state of neutrons can be described by typical two-level systems where transitions can be induced by electromagnetic interaction. The two coherent beam paths through the interferometer can also be seen as a two-level system and related entanglements can be achieved and used for novel quantum measurements. High degree of coherence, expressed by the visibility of the interference pattern (up to 95%) and high order interferences (up to 200) can be observed. The coherence features are usually described in terms of the well-known Glauber [2] formalism. The coherence function is defined as the auto-correlation function of the wave function, which reads for a stationary situation as

$$\Gamma^{(1)}(\vec{\Delta}) = \left\langle \psi^*(\vec{r})\psi(\vec{r}+\vec{\Delta}) \right\rangle = \int g(\vec{k}) e^{i(\vec{k}\vec{\Delta})}\, d^3k, \tag{1}$$

where $\vec{\Delta}$ denotes the spatial shift of the two wave fields and $g(\vec{k})$ the momentum distribution in the related direction. The characteristic widths of this function define the coherence lengths Δ_i which are related to the widths δk_i of the related momentum distribution as: $\Delta_i \delta k_i \leq 1$; $i = x, y, z$. From measurements of the momentum distributions and the loss of contrast at high order interference, which is described by $|\Gamma^{(1)}(\vec{\Delta})|$, one obtains coherence lengths in the order of about 200-500 Å, which represents also the size of the related wave packets (Rauch et al. [3]) Pre- and post-selection experiments indicate that coherence exists even when the coherence pattern is smeared out due to inhomogeneities along the beam paths or statistical motions of the interferometer crystal during the time-of-flight the neutron spends within the interferometer. Since the beam cross section is usually much larger than the size of the wave packets one has to average Eq. (1) classically over the beam cross section which then also includes variations of the parameters of the apparatus. The complementarity and duality feature of quantum physics is an essential part for its understanding. In this context the Englert-Greenberger-Yasin relation can be considered as a further "two level" system where pure particle features exist at the north and south poles of the Bloch sphere and pure wave properties along the equator (Greenberger and Yasin [4], Englert [5]). This relation combines path predictabilities P with wave properties V, which equals the visibility of the interference pattern

$$P^2 + V^2 \geq 1. \tag{2}$$

This provides the basis for further entanglement experiments with other two level systems.

2. Momentum post-selection

In the course of several neutron interferometer experiments it has been established that smoothed out interference properties at high interference order can be restored even behind the interferometer when a proper spectral filtering is applied. This is shown schematically in Fig. 1 (Rauch [6]). The interference pattern reads as:

$$I_0(\chi) \propto 1 + |\Gamma(\chi)| \cos\chi \tag{3}$$

where $\chi = K \cdot \Delta = nk\Delta = -Nb_c\lambda D_{\text{eff}}$ is the phase shift given by the particle density N and the coherent scattering length b_c of the phase shifter, the wavelength of the neutrons λ and the effective path length within the phase shifter D_{eff}. For Gaussian wave packets the coherence function $|\Gamma(\chi)|$ has also a Gaussian shape which reduces the interference fringes at high order. The related momentum distribution becomes modulated as

$$I_0(\chi) \propto \exp\left[\frac{-(k-k_0)^2}{2\delta k^2}\right] \left\{1 + \cos\left(\frac{\chi k_0}{k}\right)\right\}. \tag{4}$$

Related measurements have been done by Jacobson et al. [7], who showed that interference fringes can be restored when a proper momentum filter is applied. Fig. 2 shows more recent measurements where the intensity wiggles have been observed for various phase shifts (Baron and Rauch [8]). In this case the imperfectness of the neutron beam (wavelength spread) causes the contrast reduction at high order.

The increasing modulation at high order is visible. This indicates a forceless beam modulation due to interference. The corresponding spatial distribution can be calculated for Gaussian beams

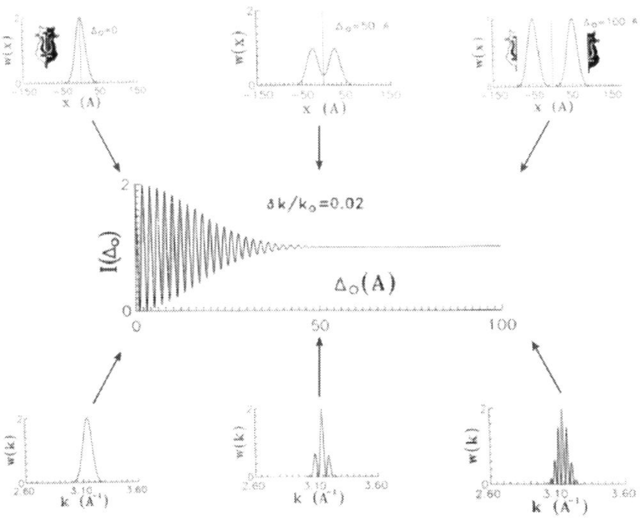

Figure 1. Wave packets and momentum distributions for various phase shifts at low and high interference order (Rauch [6])

as well and show that at high order (phase shifts larger than the coherence length) a double humped intensity distribution arises which can be written as ($\delta x \delta k = 1/2$; Fig. 1)

$$I_0(x) = |\psi(x) + \psi(x + \Delta)|^2 \propto e^{-x^2/2\delta x^2} + e^{-(x+\Delta)^2/2\delta x^2} + 2e^{-x^2/4\delta x^2} e^{-(x+\Delta)^2/4\delta x^2} \cos\chi \quad (5)$$

This means that the neutron occupies at the same time two separate regions of space, i.e. so-called Schrödinger cat-like states. Fig. 3 shows such a double humped distribution measured with a double loop interferometer (Baron and Rauch [8]). Δ_1 produces the Schrödinger cat-like state within the first loop and Δ_2 scans the spatial distribution within the second loop.

As a conclusion of this section we can say that a loss of visibility of the interference pattern at high order does not mean a loss of coherence but a shift of coherence features from ordinary space into momentum space. By a proper post-selection procedure the interference pattern can be retrieved. In all cases a complete retrieval is impossible since unavoidable losses occur at any interaction the system experiences (Rauch [9]).

3. Visibility loss due to noisy interaction

Magnetic fields of strength B and length L produce a phase shift of half of the Larmor rotation angle α

$$\chi = \frac{\alpha}{2} = \frac{\mu B L}{\hbar v} \quad (6)$$

where μ is the magnetic moment of the neutrons and v their velocity. Within a quasi-static approximation which may be valid when the time-of flight through the field $\tau_{\text{tof}} = L/v \leq 1/v_{\text{field}}$ is shorter than the typical time change of the field the loss of contrast for a Gaussian fluctuating field (ΔB) can be written as

$$\bar{I}_0 \propto 1 + e^{-\left(\frac{\mu L}{\hbar v}\right)(\Delta B)^2/2} \cos\chi. \quad (7)$$

Related measurements with unpolarized neutrons have shown the surprising effect that the interference pattern can be restored completely when the same noise field is applied to both beams at the same distance from the splitting (Baron and Rauch [8]; Fig. (4)). When the

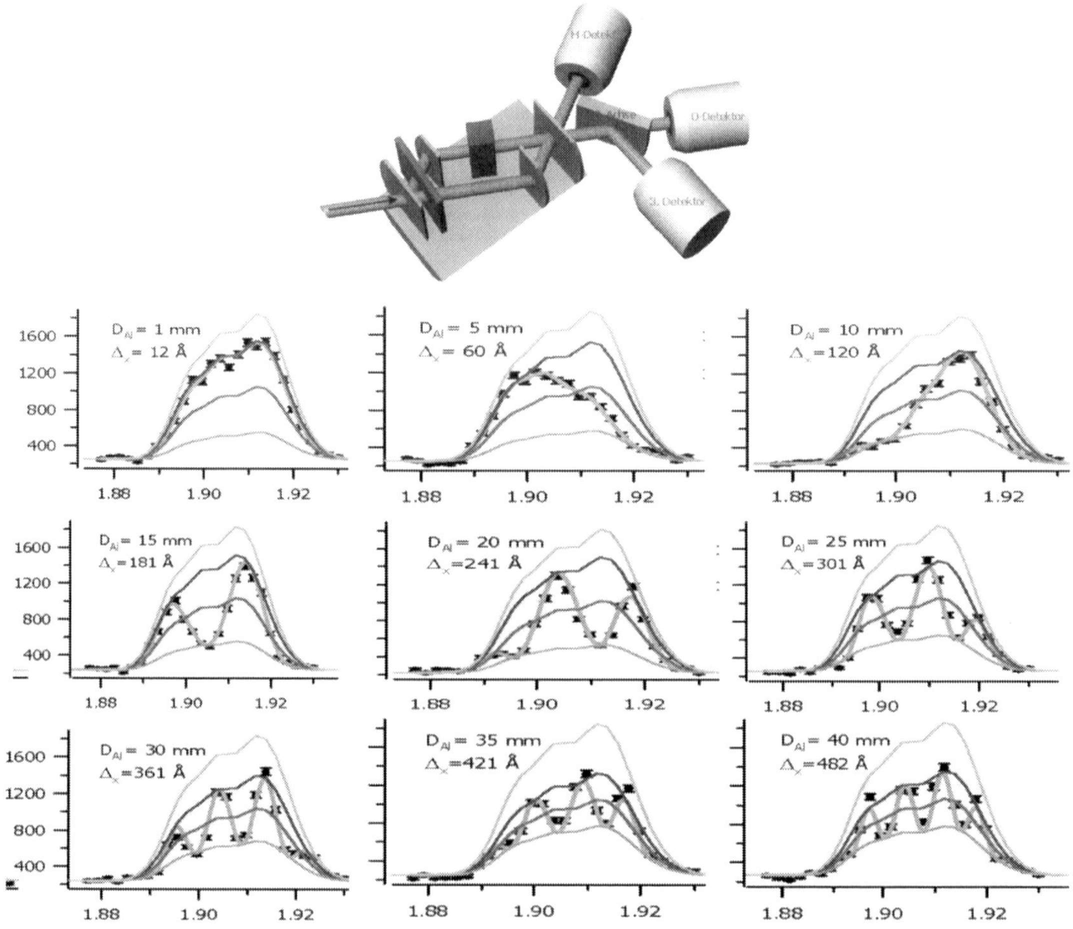

Figure 2. Experimental arrangement (top) with a third crystal as a momentum (wavelength) filter and typical momentum scans at different interference order (Baron and Rauch [8])

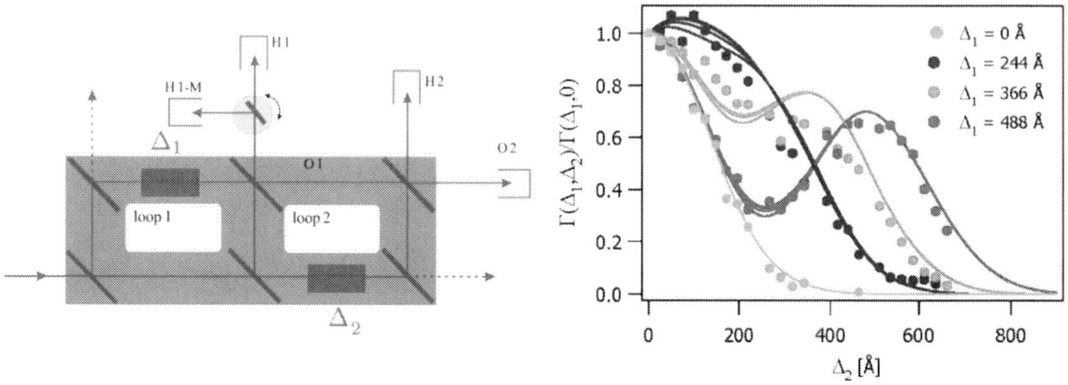

Figure 3. Double loop interferometer (left) and measured spatial distribution for different phase shifts in the first loop (Baron and Rauch [8])

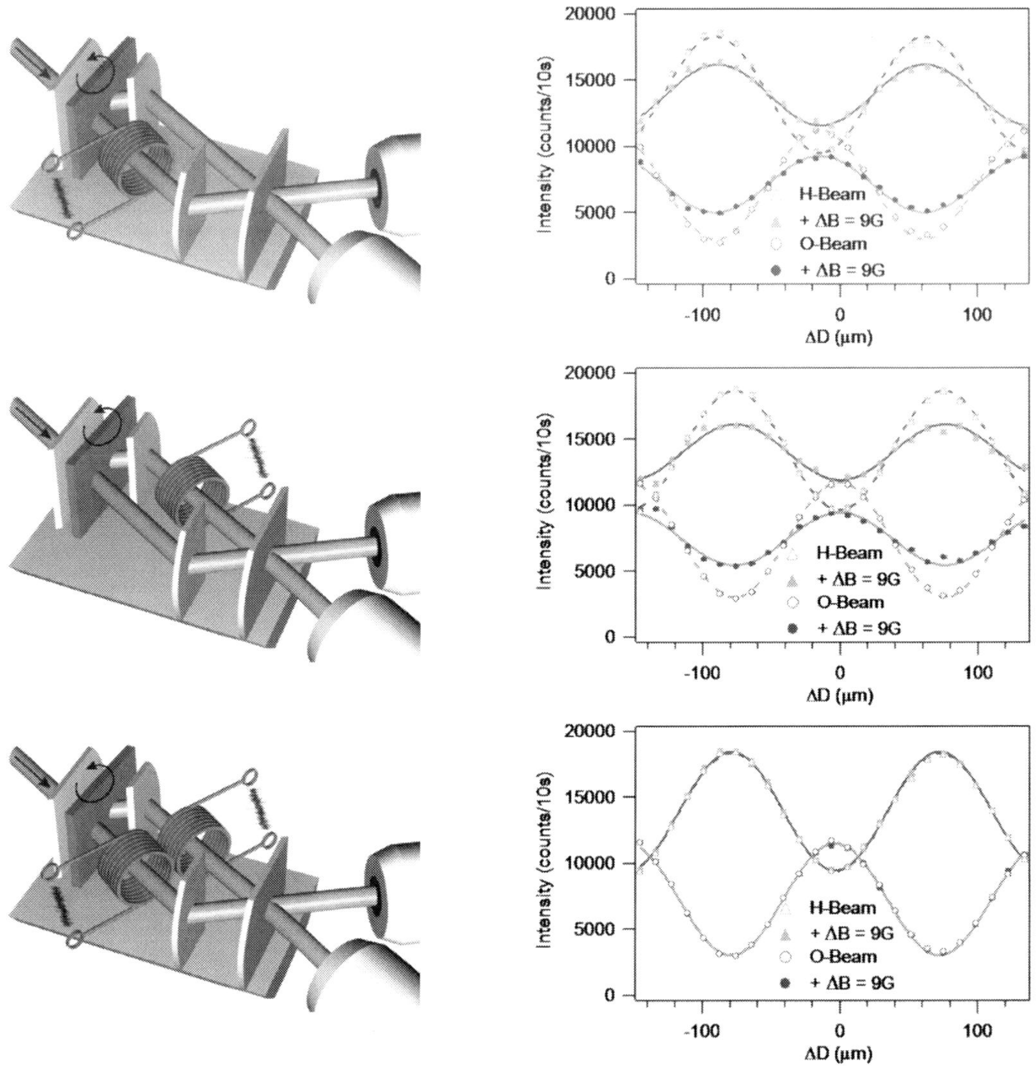

Figure 4. Interference pattern with (full lines) and without magnetic noise field (dashed lines) with frequencies between 0 and 20kHz and a mean amplitude of 9 G (Baron and Rauch [8]). The noise field has been applied to the beams separately (top and middle) and to both beams synchronously (bottom).

relative position of the coils is varied, i.e. a time delay $\Delta t = \Delta x/v$ is introduced, the autocorrelation function of the noise signal can be measured.

At high order when the contrast disappears due to the wavelength dispersion a smearing out effect on the modulated momentum distribution occurs, as shown in Fig. 5. In this case the dephasing factor of Eq. (7) adds to Eq. (4).

Within the experimental errors the loss of contrast is independent of the spatial separation of the Schrödinger cat-like states. The reversibility features when the same noisy field is applied to both beams indicate that we deal with a dephasing rather than a decoherence effect (Stern et al. [10]). The question of how "real" decoherence (irreversibility) can be achieved remains open. As long as the same disturbance is applied to both beams the coherence features

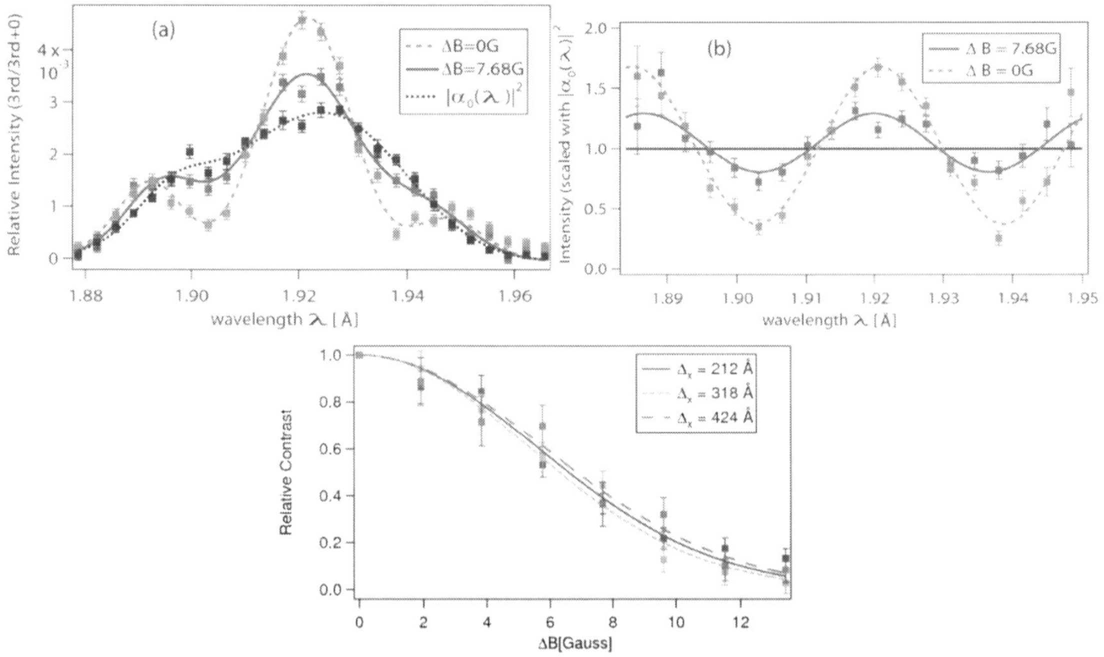

Figure 5. Reduction of the momentum beam modulation when noisy fields are applied, shown on an absolute scale (a) and in a background-corrected way (b). The general loss of contrast as a function of the field strength is shown below (Sulyok et al. [11])

are preserved and that holds even when absorption processes occur with the same probabilities in both beams (Summhammer et al. [12]).

4. Inelasticity of noisy fields

In time-dependent fields like those discussed in section 3 photon exchanges between the field and the neutron occur. In the case of a Rabi resonance flipper a one photon exchange occurs and the neutron polarization changes from spin-up to spin-down and vice versa (Badurek et al. [13]). For non-resonance magnetic fields multi photon exchange has been observed (Summhammer et al. [14]). Here we want to see whether these inelastic processes can yield decoherence.

One assumes oscillating fields within a region $0 < x < L$ which is described as

$$\vec{B}(\vec{r},t) = [B_0 + B(t) \cdot (\Theta(x) - \Theta(x - L))]\hat{z} \tag{8}$$

$$B(t) = \sum_{i=1}^{N} B_i \cos(\omega_i t + \varphi_i). \tag{9}$$

Taking into account that the kinetic energy of the neutrons ($\sim 20\,\text{meV}$) is much higher than the maximal potential barrier ($\mu B_i \sim 0.5\,\text{neV}$) one obtains after some analytical efforts the wave field behind the field region (Sulyok et al. [15])

$$\Psi_{III}(x,t) = \sum_{\vec{n}} J_{n_1}(\beta_1) \cdots J_{n_N}(\beta_N) \mathrm{e}^{-\vec{n}\vec{\eta}} \mathrm{e}^{-\mathrm{i}k_n x} \mathrm{e}^{-\mathrm{i}\omega_n t} \tag{10}$$

with

$$\begin{aligned}
\omega_{\vec{n}} &= \omega_0 + \vec{n}\vec{\omega} \qquad \vec{k}_{\vec{n}}^2 = k_0^2 - \frac{2m}{\hbar^2}\mu B_0 + \frac{2m}{\hbar}\vec{n}\vec{\omega} \\
\eta_i &= \varphi_i + \frac{\omega_i T + \pi}{2} \qquad \beta_i = 2\alpha_i \sin\frac{\omega_i T}{2} \qquad \alpha_i = \frac{\mu B_i}{\hbar\omega_i} \qquad T = \frac{L}{v_0} \\
\vec{n} &= (n_1, \ldots, n_N), \quad \vec{\varphi} = (\varphi_1, \ldots, \varphi_N), \quad \vec{\omega} = (\omega_1, \ldots, \omega_N), \quad \vec{\eta} = (\eta_1, \ldots, \eta_N),
\end{aligned}$$

and where $J_{n_i}(\beta_i)$ denote the Bessel functions of order n_i, which determine the transition amplitudes. From that one gets the interference pattern as:

$$I_0(x,t) = \frac{1}{2}\left|\Psi_I(x,t) + e^{i\chi}\Psi_{III}(x,t)\right|^2 = 1 + \mathrm{Re}\left\{e^{i\chi}\sum_{\vec{n}} J_{n_1}\cdots J_{n_N}\cdot e^{i\vec{n}(\xi_i+\vec{\omega}t)}\right\} \tag{11}$$

with

$$\xi_i = \eta_i - \frac{\omega_i x}{v_0}.$$

If the fundamental frequency of all frequencies is ω_f, the interference pattern can be expressed in a Fourier series

$$I_0(x,t) = \sum_{m=-\infty}^{\infty} c_m(x) e^{im\omega_f t} \tag{12}$$

where a comparison with Eq. (9) gives for unpolarized neutrons where only even m-terms remain:

$$c_m = \delta_{m0} + \sum_{\substack{\vec{n};\,\vec{n}\vec{\omega} = m\omega_f \\ \sum_i n_i \text{ even}}} J_{n_1}(\beta_1)\cdots J_{n_N}(\beta_N) e^{i\vec{n}\vec{\xi}} \cos\chi \tag{13}$$

which shows that the Fourier coefficient belonging to the frequency $m\omega_f$ contains the same product of Bessel functions as the transition amplitudes for an energy exchange $m\hbar\omega_f$. The arguments of the Bessel functions also contain a $\sin(\omega t/2)$-term defining a "resonance"-condition. If the time-of-flight $T = L/v$ through the field region fulfills $\omega_i T = 2\pi \cdot L$ $(L = 1, 2, 3, \ldots)$ no resulting energy exchange occurs.

Related experiments have been done with a time resolved analysis of the interference pattern, where the periodicity of the applied magnetic field ($\omega_f = 2\pi f_f$; $f_f = 1\,\mathrm{kHz}$) also manifests itself in the interference pattern (Sulyok et al. [15]). The related energy transfers lie in the order of 1 peV and energy gain and energy loss processes are equal, which is caused by the symmetric sinusoidal fields. The calculated results show good agreement with the measured values (Fig. 6). One notices that absorption and emission probabilities are equal which results from the symmetric interaction potential (Eq. (8)).

This procedure has been extended experimentally up to a 10 mode field and further by numerical methods up to a 1000 mode field. If random phases are assumed, and Gaussian variations of the amplitude and different frequency spectra, one simulates a noisy field. Fig. 7 shows the results for a rather low frequency and a higher frequency spectrum.

As a conclusion of this section it should be mentioned that a loss of visibility in an interference experiment does a priori not indicate a loss of coherence, but indicates that decoherence is rather difficult to achieve and probably as difficult as to preserve coherence. More and more complete quantum measurements may roll back decoherence to more sophisticated investigations, but some decoherence will stay as a residual of all interactions the system experienced, due to unavoidable losses (Rauch [9]).

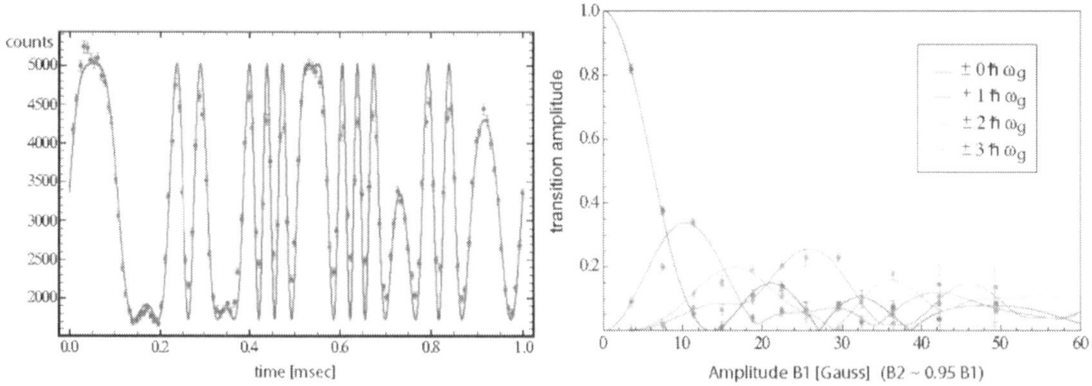

Figure 6. Example of a time-resolved interference pattern for a two mode field f1 = 3kHz and f2= 3kHZ and B1 = 30 G (left) and extracted multi- photon transition probabilities (right) (Sulyok et al. [15])

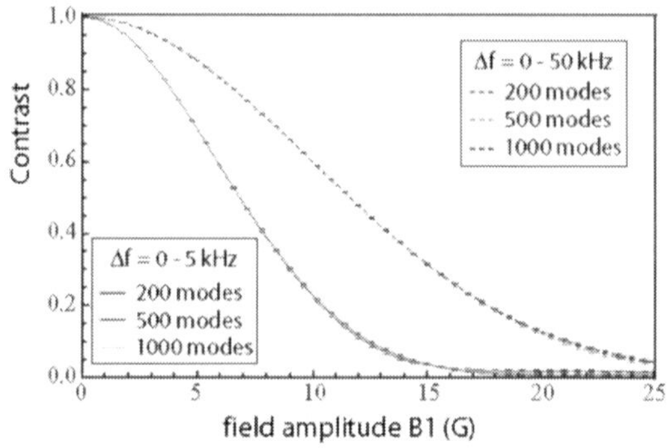

Figure 7. Calculated loss of contrast for a low frequency and a high frequency field as a function of the mean field amplitude (Sulyok et al. [15])

5. Discussion

Neutron interference experiments have shown that washed out interference pattern can be restored up to a high degree when proper post-selection procedures are applied. One may conclude that the particle-wave duality relation Eq. (2) has to be modified accordingly. Even in the case of a multi-mode noise field with multi-photon exchange the contrast can be retrieved when the same noise field is applied to the other beam or a time-resolved measurement is done. Whenever the reason for dephasing is known retrieval up to a certain degree is possible, but unavoidable quantum losses of coherence remain from any interaction the quantum system experienced (Rauch [9]). More complete measurements may show how much more information is inherent in any experiment and usually a small part is extracted. It also means that a complete isolation from the environment is impossible and a residual entanglement between the quantum object and the environment remains causing causal connections between the micro- and macro world (Zeh [16], Zurek [17]). This is not in contradiction with the idea that irreversibility is a fundamental property of nature and reversibility is only an approximation, a conclusion stated by several authors (e.g. Prigogine [18], Haag [19], Blanchard and Jadczyk [20]). At the same time it is closely connected to the quantum measurement problem, because a complete retrieval becomes impossible since in any interaction small loss mechanisms are unavoidable and they are caused by boundary and initial conditions (Giulini et al. [21], Namiki et al. [22]). The appearance of entropy associated with decoherencing effects reflects the presence of an arrow in time in quantum theory, i.e. a fundamental irreversibility in the formalism of the theory itself (Rauch [9]). This irreversibility comes into play only through initial and boundary conditions in our universe.

References

[1] Rauch H, Werner S A 2000 *Neutron Interferometry*, Clarendon Press, Oxford
[2] Glauber R J 1963 *Phys. Rev.* **130**, 2529; **131**, 2766
[3] Rauch H, Woelwitsch H, Kaiser H, Clothier R, Werner S A 1996 *Phys. Rev.* **A 53**, 902
[4] Greenberger D M, Yasin A 1988 *Phys. Lett.* **128**, 391
[5] Englert B G 1996 *Phys. Rev. Lett.* **77**, 2154
[6] Rauch H 1993 *Phys. Lett.* **A 173**, 240
[7] Jacobson D L, Werner S A, Rauch H 1994 *Phys. Rev.* **A 49**, 3196
[8] Baron M, Rauch H 2011 *AIP Conf. Proc.* **1327**, 89
[9] Rauch H 2006 *J. Physics: Conf. Series* **36**, 164
[10] Stern A, Aharonov Y, Imry Y 1990 *Phys. Rev.* **A 41**, 3436
[11] Sulyok G, Hasegawa Y, Klepp J, Lemmel H, Rauch H 2010 *Phys. Rev.* **A 81**, 053609
[12] Summhammer J, Rauch H, Tuppinger D 1987 *Phys. Rev.* **A 36**, 4447
[13] Badurek G, Rauch H, Summhammer J 1983 *Phys. Rev. Lett.* **51**, 1015
[14] Summhammer J, Hamacher K A, Kaiser H, Weinfurter H, Jacobson D L, Werner S A 1995 *Phys. Rev. Lett.* **75**, 3206
[15] Sulyok G, Lemmel H, Rauch H 2012 *Phys. Rev.* **A 85**, 033624
[16] Zeh H D 1970 *Found. Phys.* **1**, 69
[17] Zurek W D 1991 *Physics Today* Oct. **36**
[18] Prigogine I 1991 *Proc. Ecol. Phys. Chem.*, p.8
[19] Haag R 1990 *Nucl. Phys.* **B 18**, 135
[20] Blanchard P, Jadczyk A 1993 *Phys. Lett.* **A 175**, 157
[21] Giulini D, Joos E, Kiefer C, Kupsch J, Stamatescu I O, Zeh H D 1996 *Decoherence and the Appearance of a Classical World in Quantum Theory*, Springer, Berlin
[22] Namiki M, Pascazio S, Nakazato H 1997 *Decoherence and Quantum Measurements*, World Scientific, Singapore

Testing superdeterministic conspiracy

Sabine Hossenfelder
Nordita, KTH Royal Institute of Technology and Stockholm University
Roslagstullsbacken 23, SE-106 91 Stockholm, Sweden
E-mail: hossi@nordita.org

Abstract. Tests of Bell's theorem rule out local hidden variables theories. But any theorem is only as good as the assumptions that go into it, and one of these assumptions is that the experimenter can freely choose the detector settings. Without this assumption, one enters the realm of superdeterministic hidden variables theories and can no longer use Bell's theorem as a criterion. One can like or not like such superdeterministic hidden variables theories and their inevitable nonlocality, the real question is how one can test them. Here, we propose a possible experiment that could reveal superdeterminism.

1. Introduction

For any locally causal theory, the attempt to explain quantum effects by use of randomness induced through hidden variables can be shown to be in disagreement with experiment via Bell's theorem [1], or its generalization respectively [2]. Throughout the last decades, experiments have established that the hidden variables theories for which Bell's theorem applies are not realized in nature [3–9]. There are various loopholes in the conclusions that can be drawn from these experiments and not all of these loopholes have yet been satisfactorily closed (for a recent summary see e.g. [10]). But even so, local hidden variables are strongly disfavored.

However, Bell's theorem uses the assumption that one has the freedom to choose the detector's settings without modifying the prepared state that one wishes to measure. Without this freedom the conclusion of Bell's theorem does not apply. Superdeterminism requires a non-local correlation between the prepared state and the detector which seems unappealing. But the question whether the time evolution of our universe is fundamentally deterministic or not is too important to leave it to taste – it is an hypothesis that must be tested experimentally.

2. A Proposal for an Experiment

The essential difference between standard quantum mechanics and superdeterministic hidden variables theories is that in the former case two identically prepared states can give two different measurement outcomes, while in the latter case that is not possible. Unfortunately, 'identically prepared' includes the hidden variables, and it is difficult to identically prepare something one cannot measure. That is after all the reason why it looks indeterministic.

To circumvent this problem, instead of trying to prepare identical states one can make repeated measurements on the same state. For that, one takes two non-commuting variables (for example the spin or polarization in two different directions) and measures them alternately. In standard quantum mechanics the measurement outcomes will be non-correlated; the measurement of one observable prevents the state from being in an eigenstate of the other

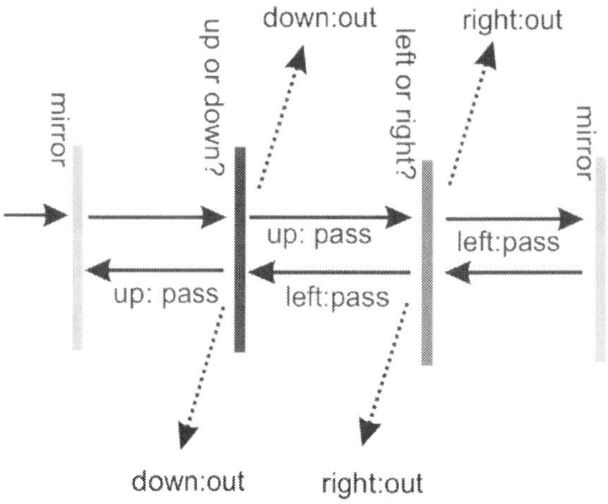

Figure 1. Sketch of proposed experiment. The prepared state is bounced between two mirrors. Two non-commuting observables (for example the up-down and left-right component of spin) are measured alternatingly. Only one combination passes. In standard quantum mechanics, the probability for the particle to pass decreases exponentially with time. In a superdeterministic theory with hidden variables, a particle that passes once will in principle continue to pass indefinitely. In practice, it will pass until noise destroys the autocorrelation of the time-series.

observable. In a superdeterministic hidden variables theory, the observables will be correlated - provided one can make a case that the hidden variables do not change in between the measurements. Figure 1 depicts an example for an experimental setup.

The provision that the hidden variables do not change is the reason why the test is only 'almost' model independent because it brings in an assumption about the origin of the hidden variables which is necessary to make the hypothesis testable at all. To arrive at concrete estimates that are explained in more detail in [11] , we thus assume that the hidden variables are due to the environment (the experimental setup) down to the relevant scales of the interactions taking place. That means we take on an effective field theory perspective and decouple all scales below the resolution of the experiment. As so often when it comes to the foundations of quantum mechanics, the noise is the main problem in this experimental setup. In practice, we will have to make the system small and cool and measure quickly enough so we have a chance to see the correlation between subsequent measurements.

To estimate whether it is possible to measure the correlations, let us denote one of the two non-commuting observables with O and label with κ the times at with O is measured. We are then interested in the autocorrelation time of the time-series of measurements of O. To simplify matters, we can look at the correlation of subsequent measurements with the first measurement:

$$\mathrm{Corr}_\kappa = \frac{E(\phi, O, t = 0 \wedge \phi, O, t = \kappa)}{E(\phi, O^2)} \ , \tag{1}$$

where E denotes the expectation value.

For the autocorrelation we make the usual ansatz of exponential decay

$$\mathrm{Corr}_\kappa = \exp\left(-\kappa/\tau\right) \ , \tag{2}$$

where τ is some timescale, the autocorrelation time, that encodes the rate of change of the experimental setup – the noise – and thus the environmentally induced hidden variables. Here, we implicitly assumed that the divergence of the system from a perfect repetition happens by incremental disturbances that are described by a homogeneous Poisson-process, i.e. the changes that lead to the decay of the autocorrelation are statistically independent and uniform in time. The larger the experiment and the shorter the typical timescales on which the relevant degrees of freedom change, the smaller τ and the faster the correlation will vanish. In the limit $\tau \to 0$, one reproduces standard quantum mechanics.

The question of whether one can test superdeterministic hidden variables theories then comes down to the question of whether we can find a setting in which the autocorrelation time is large enough to be measureable.

Consider as a concrete example a photodetector constituted of N atoms. Infalling photons excite an electron into the conduction band with an energy gap of ΔE. At a temperature T one single atom has a probability of $\exp(-\Delta E/T)$ to become excited by thermal motion, and it remains so for the average electron-hole recombination time τ_r which is typically of the order of a nanosecond. The time $\tilde{\tau}$ for N atoms to undergo a statistical change is then

$$\tilde{\tau} \approx \frac{\exp(\Delta E/T)}{N} \tau_r \ . \tag{3}$$

There are many other sorts of noise, but this is the one with the highest frequency, and thus the one relevant for the decay time.

A microscopically small detector with an extension of some μm would be constituted of approximately $N \approx 10^{15}$ atoms. A semi-conductor with a fairly large band gap of $\Delta E \approx 1$ eV at $T \sim 300$K then has $\tilde{\tau} \approx 10^{-6}$s. This autocorrelation time is feasible to measure and is also considerably larger than the time it would take a photon to bounce back between the mirrors in such a setup. We therefore have reason to be cautiously optimistic that such an experimental test can be realized with today's technological possibilities.

3. History

Interestingly, in a footnote of a 1970 paper [12], Eugene Wigner mentions Von Neumann discussing exactly this type of experiment:

> "Von Neumann often discussed the measurement of the spin component of a spin-1/2 particle in various directions. Clearly, the possibilities for the two possible outcomes of a single such measurement can be easily accounted for by hidden variables [...] However, Von Neumann felt that this is not the case for many consecutive measurements of the spin component in various different directions. The outcome of the rst such measurement restricts the range of values which the hidden parameters must have had before that rst measurement was undertaken. The restriction will be present also after the measurement so that the probability distribution of the hidden variables characterizing the spin will be different for particles for which the measurement gave a positive result from that of the particles for which the measurement gave a negative result. The range of the hidden variables will be further restricted in the particles for which a second measurement of the spin component, in a different direction, also gave a positive result..."

A longer discussion with Schrödinger followed this proposal. The conclusions could be summarized with saying that the experiment cannot test generic superdeterminism, but only certain types. If we think about it for a moment, this is unquestionably true because we can never rule out generic superdeterminism anyway: If time evolution was fundamentally indeterministic, one could always postulate some nonlocal hidden variables that remain unmeasurable. Such an hypothesis is however scientifically uninteresting because it cannot be falsified. To make superdeterminism scientifically interesting, one has to make additional assumptions about the origin of the hidden variables. Only then can one test it.

4. Conclusions

We have here proposed an experiment to test superdeterministic theories with hidden variables that are environmentally induced and estimated the prospects to realize this test. We have argued that it seems possible to test this type of theory with presently existing technology. Such an experiment could reveal that the apparent indeterminism of quantum mechanics is not fundamental and the collapse of the wave-function is not fundamentally probabilistic.

References

[1] J. S. Bell, *"On the Einstein-Podolsky-Rosen paradox,"* Physics **1**, 195 (1964).
[2] J. F. Clauser, M. A. Horne, A. Shimony, R. A. Holt, *"Proposed experiment to test local hidden variable theories,"* Phys. Rev. Lett. **23**, 880-884 (1969).
[3] A. Aspect, J. Dalibard, G. Roger, *"Experimental test of Bell's inequalities using time varying analyzers,"* Phys. Rev. Lett. **49**, 1804-1807 (1982).
[4] G. Weihs *et al.*, *"Violation of Bell's inequality under strict Einstein locality conditions,"* Phys. Rev. Lett. **81**, 5039-5043 (1998), [quant-ph/9810080].
[5] W. Tittel, *et al.*, *"Violation of Bell inequalities by photons more than 10 km apart,"* Phys. Rev. Lett. **81**, 3563-3566 (1998), [quant-ph/9806043]
[6] M. A. Rowe *et al*, *"Experimental violation of a Bell's inequality with efficient detection,"* Nature, **409**, 791 (2001).
[7] D. L. Moehring, M. J. Madsen, B. B. Blinov, C. Monroe, *"Experimental Bell Inequality Violation with an Atom and a Photon,"* Phys. Rev. Lett. 93, 090410 (2004).
[8] M. Ansmann *et al*, *"Violation of Bell's inequality in Josephson phase qubits,"* Nature **461**, 504 (2009).
[9] Y. Hasegawa *et al*, *"Violation of a Bell-like inequality in single-neutron interferometry,"* Nature **425**, 45 (2003).
[10] W. Rosenfeld *et al*, *"Towards a loophole-free test of Bell's inequality with entangled pairs of neutral atoms,"* Adv. Sci. Lett. **2**, 469 (2009), [arXiv:0906.0703 [quant-ph]].
[11] S. Hossenfelder, *"Testing super-deterministic hidden variables theories,"* Found. Phys. **41**, 1521 (2011) [arXiv:1105.4326 [quant-ph]].
[12] E. P. Wigner, *"On Hidden Variables and Quantum Mechanical Probabilities,"* Am. J. Phys. **38**, 1005 (1970).

Classical probability model for Bell inequality

Andrei Khrennikov
International Center for Mathematical Modeling
in Physics, Engineering, Economics, and Cognitive Science
School of Mathematics, Linnaeus University, Växjö, S-35195, Sweden
E-mail: `Andrei.Khrennikov@lnu.se`

Abstract. We show that by taking into account randomness of realization of experimental contexts it is possible to construct common Kolmogorov space for data collected for these contexts, although they can be incompatible. We call such a construction "Kolmogorovization" of contextuality. This construction of common probability space is applied to Bell's inequality. It is well known that its violation is a consequence of collecting statistical data in a few incompatible experiments. In experiments performed in quantum optics contexts are determined by selections of pairs of angles (θ_i, θ'_j) fixing orientations of polarization beam splitters. Opposite to the common opinion, we show that statistical data corresponding to measurements of polarizations of photons in the singlet state, e.g., in the form of correlations, can be described in the classical probabilistic framework. The crucial point is that in constructing the common probability space one has to take into account not only randomness of the source (as Bell did), but also randomness of context-realizations (in particular, realizations of pairs of angles (θ_i, θ'_j)). One may (but need not) say that randomness of "free will" has to be accounted for.

1. Introduction

The role of Bell's inequality [1] for quantum foundations and nowadays for quantum technologies was discussed in hundreds of papers, see, e.g., [2] for the extended bibliography. Recently crucial experimental tests closing the fair sampling loophole were performed [3]–[4]. Leading experimenters promise that the final (clean and loophole-free) test will be performed in a few years, may be even next year. Thus it seems that the "Bell inequality epic" [1]–[35] is near the end. Nevertheless, the probabilistic structure of the Bell argument has not yet been completely clarified. And the present paper is a step towards such clarification. In physics, violation of Bell's inequality is typically coupled to notions of *realism, locality, and free will* [1], [3]–[11]. In mathematically oriented literature, violation of Bell's inequality is considered as exhibition of *nonclassicality of quantum probability* - impossibility to use the Kolmogorov model of probability theory [36]. There can be found numerous publications on this topic, see, e.g., [2], [12]–[14], [17]–[30], [31]–[35]. In particular, in a series of works, e.g., [2], [17], [15], [33], the author of this paper presented the viewpoint that violation of Bell's inequality is a consequence of combining in one inequality of statistical data collected for a few incompatible experimental contexts. (In experiments performed in quantum optics contexts are determined by selections of pairs of angles (θ_i, θ'_j) fixing orientations of polarization beam splitters.) Data for each context can be described by the classical probability space, Kolmogorov space, but there is no common space for this data [2]. This viewpoint was strongly supported by the famous theorem of A. Fine [11] coupling Bell's inequality with existence of the joint probability distribution for all measured random variables.

What would N. Bohr say about such a viewpoint? He would definitely, cf. [37], say that there is no reason for violation of classicality even if data were collected for incompatible contexts. For him, measurement apparatus is a classical device and even if one collects outputs of a few classical devices, there is no reason to expect a violation of laws of classical probability. In this note we confirm such a position. We show that by taking into account the randomness of realization of experimental contexts it is possible to construct the common Kolmogorov space for data collected for these contexts, although they can be incompatible. We call such a construction "Kolmogorovization" of contextuality. This construction of common probability space is applied to Bell's inequality.

Opposite to the common opinion, we show that *statistical data corresponding to measurements of polarizations of photons in the singlet state, e.g., in the form of correlations, can be described in the classical probabilistic framework.* The crucial point is that in constructing the common probability space one has to take into account not only randomness of the source (as Bell did), but also randomness of context-realizations; in particular, realizations of the pairs of the angles (θ_i, θ'_j) determining orientations of polarization beam splitters (PBSs). Roughly speaking randomness of (pseudo)-random generators controlling selections of orientations of PBSs has to be added to randomness of the source. Although everybody recognizes that randomness of these generators plays an important role in the Bell framework, its presence is ignored in theoretical constructions representing the EPR-Bohm-Bell correlations. These generators are viewed as some "technicalities" which are interesting for experimenters, but not theoreticians. Recently the author started to pay serious attention to various experimental technicalities, to embed them into theoretical models, to lift them to the level of fundamental studies, e.g., [15]. The present paper can be considered as a further step in this direction.

In the Kolmogorov model constructed in this paper *quantum probabilities and correlations are represented as conditional probabilities and correlations* (under condition of fixed experimental settings). Such conditional probabilities and correlations can violate Bell's inequality, although unconditional ones have to satisfy it. We remark that, as was shown in numerous experiments, theoretical quantum probabilities and correlations coincide (of course, approximately) with probabilities and correlations found by experimenters for the fixed experimental settings, e.g., the angles (θ_i, θ'_j) in test of Bell type inequalities. Therefore our classical probabilistic model represents experimental probabilities and correlations as well.

Our model is *local.* However, it is *not objective (realistic)* in the sense of Einstein, Podolsky, Rosen and Bell: the values of observables are not determined by the state of a system emitted by a source, by, so to say, a hidden variable λ. However, from Bohr's viewpoint one cannot expect such type of objectivity, since it is defined solely in terms of the states of systems emitted by a source. And our model is *objective in a more general sense* (which might satisfy Bohr): the hidden variable λ and the random parameters in random generators selecting the experimental settings determine the values of observables (e.g., spin or polarization projections). We remark that even in classical physics randomness of measurement devices is routinely taken into account and if statistical data is collected by using a few different devices is has to be weighted. Kolmogorovization of Bell's argument is reduced to such a standard procedure. One may (but need not) say that randomness of "free will" has to be accounted for, but we postpone the discussion on this topic, see section 4.

The main problem handled in this note is the following[1]. There are a few experimental

[1] This note is based on the work [38]. Unfortunately, the latter suffered of a few presentational problems. First of them is that there were present two totally different classical Kolmogorov models for violation of Bell's inequality, one due to D. Avis and another due to the author (and it is reconsidered in the present note). And author's model was vaguely presented by mixing the von Mises frequency approach to probability (which is typically

contexts and statistical data (of any origin) collected for these contexts. *Is it possible to construct a common probability space representing these data?* We show that the answer is always positive!

This result changes the viewpoint on the role of quantum probabilistic calculus. As was mentioned, the common viewpoint is that we use quantum probabilistic rules, because classical probabilistic rules are violated, see, e.g., Feynman [39] and the author [40], [2]. (Besides violation of Bell's inequality, there is typically stressed that the formula of total probability is violated by statistical data collected in the two slit and other interference experiments.) In this note we show that such an "impossibility viewpoint" has to be changed. In the light of Kolmogorovization it is clear that the calculus of quantum probabilities is just a calculus of conditional probabilities defined in the classical probabilistic framework. It is not surprising that conditional probabilities, although generated in the common probability space, but with respect to different conditions, follow rules which are different from the rules which were derived for unconditional probabilities, such as, e.g., Bell type inequalities.

2. CHSH-inequality

We recall the rigorous mathematical formulation of the CHSH inequality:

Theorem 1. *Let $A^{(i)}(\omega)$ and $B^{(i)}(\omega), i = 1, 2,$ be random variables taking values in $[-1, 1]$ and defined on a single probability space $\mathcal{P}$. Then the following inequality holds:*

$$|< A^{(1)}, B^{(1)} > + < A^{(1)}, B^{(2)} > + < A^{(2)}, B^{(1)} > - < A^{(2)}, B^{(2)} > | \leq 2. \tag{1}$$

Correlation is defined as it is in classical probability theory:

$$< A^{(i)}, B^{(j)} >= \int_\Omega A^{(i)}(\omega)B^{(j)}(\omega)d\mathbf{P}(\omega).$$

Experimental tests of the CHSH-inequality are based on the following methodology. One should put statistical data collected for four pairs of PBSs settings:

$$\theta_{11} = (\theta_1, \theta_1'), \theta_{12} = (\theta_1, \theta_2'), \theta_{21} = (\theta_2, \theta_1'), \theta_{22} = (\theta_2, \theta_2'),$$

into it. Here $\theta = \theta_1, \theta_2$ and $\theta' = \theta_1', \theta_2'$ are selections of angles for orientations of respective PBSs.

Following Bell, the selection of the angle θ_i determines the random variable

$$a_{\theta_i}(\omega) = \pm 1.$$

There are two detectors coupled to the PBS with the θ-orientation: "up-spin" (or "up-polarization") detector and "down-spin" (or "down-polarization") detector. A click of the up-detector assigns to the random variable $a_\theta(\omega)$ the value +1 and a click of the down-detector assigns to it the value -1. In the same way selection of the angle θ' determines

$$b_{\theta_i'}(\omega) = \pm 1.$$

considered as mathematically non-rigorous) and the Kolmogorov measure-theoretic approach. This mixing of two approaches made the impression that the essence of the model is in usage of the von Mises approach. In the present note we proceed solely in the Kolmogorov framework. Another problem of [38] is that the upper bound for the CHSH-expression consisting of contextual correlations was 8 and not 4, i.e., two times more than the reasonably acceptable for four correlations (even conditional) of random variables bounded by one. In this note we show that this was an artifact of too straightforward usage of the CHSH-inequality for unconditional classical correlations. This inequality does not provide the best bound for these correlations. We show that in our Kolmorovization-model the CHSH-expression for unconditional correlations can be bounded by 1, instead of generally given 2. By taking into account randomness of realization of 4 experimental contexts we obtain the natural bound 4 for conditional correlations.

If one assumes that these observables can be represented as random variables on common Kolmogorov probability space, then their correlations have to satisfy the CHSH-inequality (1). However, the correlations calculated with the aid of the quantum formalism as well as experimental correlations violate this inequality. Therefore the above assumption about Kolmogorovness of data has to be rejected, see A. Fine [11], see A. Khrennikov [17], [2] for discussions. In this paper we do not question this conclusion. Our point is that from the classical probabilistic viewpoint there is no reason to assume that these correlations can be reproduced in common probability space. However, they can be embedded in "large Kolmogorov space" as conditional correlations.

2.1. *Random experiment taking into account random choice of settings*

a) There is a source of entangled photons.

b) There are two pairs of PBSs with the corresponding pairs of detectors. PBSs in pairs are oriented with angles numbered $i = 1, 2$ and $j = 1, 2$. The pairs of PBSs (with their detectors) are located in spatially separated labs, say Lab1 (PBSs $i = 1, 2$) and Lab2 (PBSs $j = 1, 2$). The source is connected (e.g., by optical fibers) to the Labs.

c) In each lab there is a distribution device; at each instance of time[2]: $t = 0, \tau, 2\tau, \ldots$, it opens the gate to only one of the two channels going to the corresponding PBSs. For simplicity, we suppose that in each Lab channels are opened with equal probabilities

$$\mathbf{P}(i) = \mathbf{P}(j) = 1/2.$$

Now in each of labs we define two observables corresponding to PBSs (with their detectors):

A1. $A^{(i)}(\omega) = \pm 1, i = 1, 2$ if in Lab1 the i-th channel is open and the corresponding (up or down) detector coupled to ith PBS fires;

A0. $A^{(i)}(\omega) = 0$ if in Lab1 the i-th channel is blocked.

B1. $B^{(j)}(\omega) = \pm 1, j = 1, 2$ if in Lab2 the j-th channel is open and the corresponding (up or down) detector coupled to jth PBS fires;

B0. $B^{(j)}(\omega) = 0$ if in Lab2 the j-th channel is blocked.

We remark that in the present experiments testing the CHSH-inequality experimenters do not use four PBSs, but only two PBSs (one at each side); not two pairs of detectors at each side, but just one pair. The random choice of orientations $\theta = \theta_1, \theta_2$ (on one side) and $\theta' = \theta'_1, \theta'_2$ (on another side) is modeled with the aid of additional devices preceding the corresponding PBSs. (Pseudo)-random generators specify parameters in these devices corresponding to selection of two different orientations for each of two PBSs. The main reason for the use of this scheme is that it is simpler in realization and it is essentially cheaper. The latter is very important, since photo-detectors with approximately 100% efficiency are extremely expensive. There are no doubts that everybody would agree that the two experimental schemes under discussion represent the same physical situation. However, the scheme with four PBSs is more natural from the probabilistic viewpoint. At each side there are two pairs of detectors determining two random variables. Roughly speaking there are two observers at each side, Alice1, Alice2 and Bob1, Bob2, each of them monitors her/his pair of detectors and, for each emitted system, she/he has to assign some value, if non of the detectors clicks she/he assigns 0. (We consider the ideal situation without losses).

[2] Timing can be experimentally realized with the aid of the cell method used in [4].

2.2. *Kolmogorovization of incompatible statistical data*

We now construct a proper Kolmogorov probability space for the EPR-Bohm-Bell experiment.

This is a general construction for combining of probabilities produced by a few incompatible experiments.

For the fixed pair of orientations (θ_i, θ'_j), there are given probabilities $p_{ij}(\epsilon, \epsilon'), \epsilon, \epsilon' = \pm 1$, to get the values (ϵ, ϵ'). These are either experimental probabilities (frequencies) or probabilities produced by the mathematical formalism of QM. For the singlet state and measurements of polarization, we have:

$$p_{ij}(\epsilon, \epsilon) = \frac{1}{2}\cos^2\frac{\theta_i - \theta'_j}{2}, p_{ij}(\epsilon, -\epsilon) = \frac{1}{2}\sin^2\frac{\theta_i - \theta'_j}{2}. \tag{2}$$

However, this special form of probabilities is not important for us. Our construction of unifying Kolmogorov probability space works well for any collection of numbers $p_{ij}(\epsilon, \epsilon')$ such that for any pair (i, j) :

$$0 \leq p_{ij}(\epsilon, \epsilon') \leq 1, \sum_{\epsilon,\epsilon'} p_{ij}(\epsilon, \epsilon') = 1.$$

Let us now consider the set of points Ω :

$$\omega = (\epsilon_1, 0, \epsilon'_1, 0), (\epsilon_1, 0, 0, \epsilon'_2), (0, \epsilon_2, \epsilon'_1, 0), (0, \epsilon_2, 0, \epsilon'_2),$$

where $\epsilon = \pm 1, \epsilon' = \pm 1$. These points correspond to the following events: e.g., $(\epsilon_1, 0, \epsilon'_1, 0)$ means: at the left hand side PBS N 1 is coupled and PBS N 2 is uncoupled and the same is at the right hand side, the result of measurement at the left hand side after PBS N 1 is given by ϵ_1 and at the right hand side by ϵ'_1.

We define the following probability measure on Ω :

$$\mathbf{P}(\epsilon_1, 0, \epsilon'_1, 0) = \frac{1}{4}p_{11}(\epsilon_1, \epsilon'_1), \mathbf{P}(\epsilon_1, 0, 0, \epsilon'_2) = \frac{1}{4}p_{12}(\epsilon_1, \epsilon'_2)$$

$$\mathbf{P}(0, \epsilon_2, \epsilon'_1, 0) = \frac{1}{4}p_{21}(\epsilon_2, \epsilon'_1), \mathbf{P}(0, \epsilon_2, 0, \epsilon'_2) = \frac{1}{4}p_{22}(\epsilon_2, \epsilon'_2).$$

We now define random variables $A^{(i)}(\omega), B^{(j)}(\omega)$:

$$A^{(1)}(\epsilon_1, 0, \epsilon'_1, 0) = A^{(1)}(\epsilon_1, 0, 0, \epsilon'_2) = \epsilon_1, A^{(2)}(0, \epsilon_2, \epsilon'_1, 0) = A^{(2)}(0, \epsilon_2, 0, \epsilon'_2) = \epsilon_2;$$

$$B^{(1)}(\epsilon_1, 0, \epsilon'_1, 0) = B^{(1)}(0, \epsilon_2, \epsilon'_1, 0) = \epsilon'_1, B^{(2)}(\epsilon_1, 0, 0, \epsilon'_2) = B^{(2)}(0, \epsilon_2, 0, \epsilon'_2) = \epsilon'_2,$$

and we put these variables equal to zero in other points; for example, $A^{(1)}(0, \epsilon_2, \epsilon'_1, 0) = A^{(1)}(0, \epsilon_2, 0, \epsilon'_2) = 0$. Thus if the channel going to PBS N1 at Alice1/Alice2 side is closed, then, since none of the detectors following it fires, Alice1 assigns the value 0 to her observable $A^{(1)}$.

These random variables are local in the sense that their values do not depend on experimental context and the results of measurements on the opposite side; for example, $A^{(1)}(\omega)$ depends only on the first coordinate of $\omega = (\omega_1, \omega_2, \omega_3, \omega_4)$.

We find two dimensional probability distributions; the nonzero ones are given by

$$\mathbf{P}(\omega \in \Omega : A^{(1)}(\omega) = \epsilon_1, B^{(1)}(\omega) = \epsilon'_1) = \mathbf{P}(\epsilon_1, 0, \epsilon'_1, 0) = \frac{1}{4}p_{11}(\epsilon_1, \epsilon'_1), \ldots,$$

$$\mathbf{P}(\omega \in \Omega : A^{(2)}(\omega) = \epsilon_2, B^{(2)}(\omega) = \epsilon'_2) = \frac{1}{4}p_{22}(\epsilon_2, \epsilon'_2).$$

Then, e.g.,

$$\mathbf{P}(\omega \in \Omega : A^{(1)}(\omega) = \epsilon_1, A^{(2)}(\omega) = \epsilon_2) = \mathbf{P}(\emptyset) = 0.$$

We also consider the random variables which monitor selection of channels: $\eta_a = i, i = 1, 2$ if the channel to the ith PBSs on the Alice1/Alice2 side is open and $\eta_b = j, j = 1, 2$ if the channel to the jth PBSs on the Bob1/Bob2 side is open. Thus

$$\eta_a(\epsilon_1, 0, 0, \epsilon_2') = 1, \eta_a(\epsilon_1, 0, \epsilon_1', 0) = 1, \ \eta_a(0, \epsilon_2, 0, \epsilon_2') = 2, \eta_a(0, \epsilon_2, \epsilon_1', 0) = 2,$$

$$\eta_b(\epsilon_1, 0, \epsilon_1', 0) = 1, \eta_b(0, \epsilon_2, \epsilon_1', 0) == 1, \ \eta_b(\epsilon_1, 0, 0, \epsilon_2') = 2, \eta(0, \epsilon_2, 0, \epsilon_2') = 2.$$

Here

$$\mathbf{P}(\omega \in \Omega : \eta_a(\omega) = 1) = \sum_{\epsilon_1, \epsilon_2'} \mathbf{P}(\epsilon_1, 0, 0, \epsilon_2') + \sum_{\epsilon_1, \epsilon_1'} \mathbf{P}(\epsilon_1, 0, \epsilon_1', 0)$$

$$= 1/4\Big[\sum_{\epsilon_1, \epsilon_2'} p_{12}(\epsilon_1, \epsilon_2') + \sum_{\epsilon_1, \epsilon_1'} p_{11}(\epsilon_1, \epsilon_1')\Big] = 1/2.$$

In the same way $\mathbf{P}(\omega \in \Omega : \eta_a(\omega) = 2) = 1/2; \mathbf{P}(\omega \in \Omega : \eta_b(\omega) = 1) = \mathbf{P}(\omega \in \Omega : \eta_b(\omega) = 2) = 1/2$. We remark that the random variables η_a and η_b are independent, e.g.,

$$\mathbf{P}(\omega \in \Omega : \eta_a(\omega) = 1, \eta_b(\omega) = 1) = \mathbf{P}(\epsilon_1, 0, \epsilon_1', 0) = 1/4 \sum_{\epsilon_1, \epsilon_1'} p_{11}(\epsilon_1, \epsilon_1') = 1/4$$

$$= \mathbf{P}(\omega \in \Omega : \eta_a(\omega) = 1)\mathbf{P}(\omega \in \Omega : \eta_b(\omega) = 1) = 1/4.$$

In the same way

$$\mathbf{P}(\omega \in \Omega : \eta_a(\omega) = 1, \eta_b(\omega) = 2) = \mathbf{P}(\epsilon_1, 0, 0, \epsilon_2')) = 1/4 \sum_{\epsilon_1, \epsilon_2'} p_{12}(\epsilon_1, \epsilon_1') = 1/4$$

$$= \mathbf{P}(\omega \in \Omega : \eta_a(\omega) = 1)\mathbf{P}(\omega \in \Omega : \eta_b(\omega) = 2) = 1/4$$

and so on. These random variables are local in the sense that their values do not depend on experimental context and the results of measurements on the opposite side; for example, $\eta_a(\omega)$ depends only on the first coordinate of $\omega = (\omega_1, \omega_2, \omega_3, \omega_4)$.

We now find conditional probabilities for the results of joint measurements of observables $A^{(i)}(\omega)$ and $B^{(j)}(\omega), i, j = 1, 2$, conditioned on opening of channels going to corresponding PBSs. The definition of the classical (Kolmogorov, 1933) conditional probability is based on Bayes formula, e.g.,

$$\mathbf{P}(\omega \in \Omega : A^{(1)}(\omega) = \epsilon_1, B^{(1)}(\omega) = \epsilon_2' | \eta_a = 1, \eta_b = 1)$$

$$= \frac{\mathbf{P}(A^{(1)}(\omega) = \epsilon_1, B^{(1)}(\omega) = \epsilon_2', \eta_a(\omega) = 1, \eta_b(\omega) = 1)}{\mathbf{P}(\omega \in \Omega : \eta_a(\omega) = 1, \eta_b(\omega) = 1)}$$

$$= \frac{\mathbf{P}(\epsilon_1, 0, \epsilon_1', 0)}{1/4} = p_{11}(\epsilon_1, \epsilon_1').$$

In general, we obtain:

$$\mathbf{P}(\omega \in \Omega : A^{(i)}(\omega) = \epsilon_i, B^{(j)}(\omega) = \epsilon_j' | \eta_a = i, \eta_b = j) = p_{ij}(\epsilon_i, \epsilon_j').$$

In particular, if initially the weights $p_{ij}(\epsilon_i, \epsilon_j')$ were calculated with the aid of the rules of quantum mechanics, for the singlet state and polarization observables, i.e., they have the meaning of quantum probabilities, then the classical conditional probabilities coincide with these quantum probabilities. In this way all quantum probabilities can be represented classically. However, they have to be treated as conditional probabilities. In principle, such a viewpoint on quantum probabilities is well established, see, e.g., [34], [2]. The novel contribution is that concrete classical probabilistic construction for embedding of quantum probabilities into classical Kolmogorov model was presented.

To complete the picture of conditioning, we also present probabilities for observations for closed channels, e.g.,

$$\mathbf{P}(\omega \in \Omega : A^{(1)}(\omega) = \epsilon_1, B^{(1)}(\omega) = \epsilon'_2 | \eta_a = 2, \eta_b = 1) =$$
$$= \frac{\mathbf{P}(A^{(1)}(\omega) = \epsilon_1, B^{(1)}(\omega) = \epsilon'_2, \eta_a(\omega) = 2, \eta_b(\omega) = 1}{\mathbf{P}(\omega \in \Omega : \eta_a(\omega) = 2, \eta_b(\omega) = 1)}$$
$$= \frac{\mathbf{P}(\emptyset)}{1/4} = 0.$$

3. Classical conditional probability viewpoint on violation of the CHSH-inequality

Since the CHSH-inequality is an inequality for correlations, we find them. We are interested in two types of correlations, so to say, absolute, $E(A^{(i)}B^{(j)})$, and conditional, $E(A^{(i)}B^{(j)}|\eta_a = i, \eta_b = j)$. We have:

$$E(A^{(i)}B^{(j)}) = \sum_{\epsilon_i, \epsilon'_i} \epsilon_i \epsilon'_j \mathbf{P}(A^{(i)} = \epsilon_i, B^{(j)} = \epsilon_j) = (1/4) \sum_{\epsilon_i, \epsilon'_i} \epsilon_i \epsilon'_j p_{ij}(\epsilon_i, \epsilon'_j)$$

and

$$E(A^{(i)}B^{(j)}|\eta_a = i, \eta_b = j) = \sum_{\epsilon_i, \epsilon'_i} \epsilon_i \epsilon'_j \mathbf{P}(A^{(i)} = \epsilon_i, B^{(j)} = \epsilon_j | \eta_a = i, \eta_b = j)$$
$$= \sum_{\epsilon_i, \epsilon'_i} \epsilon_i \epsilon'_j p_{ij}(\epsilon_i, \epsilon'_j).$$

Suppose that the weights p_{ij} are selected as quantum probabilities for the singlet state, denote the corresponding correlations C_{ij}, then

$$E(A^{(i)}B^{(j)}|\eta_a = i, \eta_b = j) = C_{ij}$$

and

$$E(A^{(i)}B^{(j)}) = C_{ij}/4.$$

The classical correlations satisfy to conditions of Theorem 1 and, hence, to the CHSH-inequality, i.e., the conditional correlations satisfy to the inequality:

$$|C_{11} + C_{12} + C_{21} - C_{22}| \leq 8. \tag{3}$$

Since the conditional correlations coincide with the quantum correlations, we obtain an inequality which has to be satisfied by the quantum correlations. However, in the right-hand side there is not 2 as in the CHSH, but 8 and, hence, no problem arises.

Roughly speaking the bound of 8 is too high to make any meaningful constraint on correlations. Thus, in this situation the CHSH-inequality gives too rough an estimate, since it is clear that even conditional correlations of random variables with values $\{-1, 0, +1\}$ are bounded by 1. Thus the worst straightforward estimate has to be

$$|C_{11} + C_{12} + C_{21} - C_{22}| \leq 4. \tag{4}$$

We now go another way around and from the inequality (4) for conditional correlations we obtain an inequality for unconditional ("absolute") correlations:

$$|E(A^{(1)}B^{(1)}) + E(A^{(1)}B^{(2)}) + E(A^{(2)}B^{(1)}) - E(A^{(2)}B^{(2)})| \leq 1. \tag{5}$$

4. Conclusion

We showed that one can easily violate the CHSH-inequality in the classical probabilistic framework (the Kolmogorov model [36]).

We emphasize that a violation of the CHSH-inequality for quantum correlations represented as classical conditional correlations is not surprising. In our approach the main problem is to explain Tsirelson bound.

We repeat our interpretation of the obtained result. Our model is local. It seems that "nonlocality" is an artificial issue without direct relation to violation of Bell's type inequalities. However, our model is not realistic in the sense of naive realism of Einstein, Podolsky, Rosen, and Bell, i.e., the measured values cannot be predetermined on the basis of the state of a system. Nevertheless, a kind of contextual realism (of Bohr's type) is preserved: by taking into account not only the state of a system, but also randomness of (pseudo)-random generators selecting orientations of PBSs, it is possible to violate the CHSH-inequality.

Appendix: Taking into account free will?

Recently, the so-called free will problem became an important topic in quantum foundational discussions. In particular, the recent spike of activity was generated by 't Hooft's position that quantum mechanics can be consistently explained by using the totally deterministic picture of nature [41]. Of course, anybody having elementary education in cognitive science and philosophy would consider the "free will" discussions in quantum foundations as very primitive, cf. with [42]. They have practically no relation to such discussions in cognitive science and philosophy. Therefore one may, in principle, ignore usage of this misleading terminology and just say that in the Bell type experiments randomness of a source of entangled systems is combined with random selection of experimental settings, e.g., in the form of angles, orientations of PBSs. The latter is realized with the aid of (pseudo)-random generators. Of course, experimenter has the freedom of choice of these random generators (as well as she has also the freedom to do or not the experiment at all). However, after generators were selected randomness involved in the Bell experiment (in fact, a few experiments) became of purely physical nature, i.e., the mental element is totally excluded. Nevertheless, one may argue that an experimenter can in principle interrupt the production of random numbers by changing at the arbitrary instant of time the parameters of random generators and here her free will again will play a role. In principle, we can accept this position and say that our model includes the mental randomness produced by brain's functioning. However, if we consider the brain as a physical system whose activity is based on spikes produced by neurons (or/and electromagnetic fields generated in the brain), then we again obtain the pure physical account of randomness in the Bell type experiments. Free will would play an exceptional role only if one accepts that human consciousness is not reduced to physical processes in the brain. Such a position is still valuable, e.g., [43]. By accepting it we would agree that Kolmogorovization is successful only as the result of accounting for combined physical-mental randomness.

G. 't Hooft rejects the existence of free will. We mention that his position is close to the position of the majority of cognitive scientists and philosophers working on the problem of free will [42]. The common opinion is that free will is one of the traces of the God-based picture of the world. The brain is a sort of deterministic device (which, of course, contains various noisy signals) and all acts of free will are dynamically predetermined by brain's state in previous instances of time. We state again that our model need not be based on such a position. We are fine by assuming the existence of a nonphysical pure mental elements or even God, in any event the classical probability model for Bell's experiment is well defined.

Acknowledgments
This work was partially supported by MPNS COST Action MP1006 (Fundamental Problems in Quantum Physics), by visiting fellowship to Institute for Quantum Optics and Quantum Information, Austrian Academy of Sciences, and by the grant of Linnaeus University, Mathematical Modeling of Complex Hierarchic Systems.

References
[1] Bell J S 1987 *Speakable and Unspeakable in Quantum Mechanics* (Cambridge: Cambridge Univ. Press)
[2] Khrennikov A 2009 *Contextual Approach to Quantum Formalism* (Berlin-Heidelberg-New York: Springer)
[3] Giustina M, Mech Al, Ramelow S, Wittmann B, Kofler J, Beyer J, Lita A, Calkins B, Gerrits Th, Woo Nam S, Ursin R and Zeilinger A 2013 *Nature* **497** 227
[4] Christensen B G et al. 2013 *Phys. Rev. Lett.* **111** 1304
[5] Aspect A, Dalibard J and Roger G 1982 *Phys. Rev. Lett.* **49** 1804
[6] Aspect A 1983 Three experimental tests of Bell inequalities by the measurement of polarization correlations between photons (Orsay: Orsay Univ. Press)
[7] Weihs G, Jennewein T, Simon C, Weinfurter G, and Zeilinger A 1998 *Phys. Rev. Lett.* **81** 5039
[8] Clauser J F and Horne M A 1974 *Phys. Rev.* D **10** 526
[9] Clauser J F and Shimony A 1978 *Rep. Prog. Phys.* **41** 1881
[10] Selleri F and Zeilinger A 1988 *Found. Phys.* **18** 1141
[11] Fine A 1989 *Found. Phys.* **19** 453
[12] Larsson J-A 1998 *Phys. Rev.* A **57** 3304
[13] Adenier G and Khrennikov A 2007 *J. Phys. B: Atomic, Molecular Opt. Phys.* **40** 131
[14] Kupczynski M 1987 *Phys. Lett.* A **121** 205
[15] Khrennikov A 2012 *J. Phys. A: Math. Theor.* **45** 215301
[16] Eberhard Ph H 1993 *Phys. Rev.* A 477
[17] Khrennikov A 1999 *Interpretation of Probability* (Utrecht: VSP)
[18] Kupczynski M 2002 On the Completeness of Quantum Mechanics, arXiv:quant-ph/028061
[19] Khrennikov A (ed) 2001 *Foundations of Probability and Physics. Quantum Probability and White Noise Analysis* vol 13 (Singapore: WSP)
[20] Khrennikov A (ed) 2002 *Quantum Theory: Reconsideration of Foundations.* Ser. Math. Modelling vol 2 (Växjö: Växjö Univ. Press)
[21] Khrennikov A (ed) 2003 *Quantum Theory: Reconsideration of Foundations-2.* Ser. Math. Modelling vol 10 (Växjö: Växjö Univ. Press)
[22] Khrennikov A (ed) 2005 *Foundations of Probability and Physics-3.* vol 750
[23] Jaeger G, Khrennikov A, Schlosshauer M, Weihs G (eds) 2011 *Advances in Quantum Theory* vol 1327 (Melville, NY: AIP)
[24] Khrennikov A (ed) 2010 *Quantum Theory: Reconsideration of Foundations-5* vol 1232 (Melville, NY: AIP)
[25] Accardi L 2005 Some loopholes to save quantum nonlocality. *Foundations of Probability and Physics-3* vol 750 (Melville, NY: AIP) pp 1-20
[26] Roychoudhuri Ch 2007 Shall we climb on the shoulders of the giants to extend the reality horizon of physics? *Quantum Theory: Reconsideration of Foundations-4* vol 962 (Melville, NY: AIP) 195-205
[27] Adenier G. 2008 *Am. J. Phys.* **76** 147
[28] Hess K, Michielsen K and De Raedt H 2009 *EPL* **87** 60007
[29] Kupczynski M 2005 *J. Russian Laser Research* **26** 514
[30] Kupczynski M Seventy Years of the EPR Paradox. *Albert Einstein Century Conference* vol 861 (Melville, NY: AIP) pp 516-523
[31] Larsson J-Å 2000 *Quantum Paradoxes, Probability Theory, and Change of Ensemble* (Linköping: Linköping Univ. Press)
[32] Adenier G 2008 *Local Realist Approach and Numerical Simulation of Nonclassical Experiments in Quantum Mechanics* (Vxjö: Vxjö Univ. Press)
[33] Khrennikov A 2008 *Entropy* **10** 19
[34] De Muynck W M 2002 *Foundations of Quantum Mechanics, an Empiricists Approach* (Dordrecht: Kluwer)
[35] De Raedt H, De Raedt K, Michielsen K, Keimpema K and Miyashita S 2007 *Phys. Soc. Japan* **76** 104005
[36] Kolmogoroff A N 1933 *Grundbegriffe der Wahrscheinlichkeitsrechnung* (Berlin: Springer Verlag); English translation: Kolmogorov A N 1956 *Foundations of Theory of Probability* (New York: Chelsea Publishing Company)
[37] Bohr N 1987 *The Philosophical Writings of Niels Bohr,* 3 vols (Woodbridge, Conn.: Ox Bow Press)
[38] Avis D, Fischer P, Hilbert A and Khrennikov A 2009 Single, Complete, Probability Spaces Consistent With

EPR-Bohm-Bell Experimental Data, *Foundations of Probability and Physics-5* vol 750 (Melville, NY: AIP) pp 294-301.
[39] Feynman R and Hibbs A 1965 *Quantum Mechanics and Path Integrals* (New York: McGraw-Hill)
[40] Khrennikov A 2002 On foundations of quantum theory. Quantum Theory: Reconsideration of Foundations. Ser. Math. Modelling vol 2 (Växjö: Växjö Univ. Press) pp 163-196
[41] 't Hooft, G 2011 *Herald of Russian Academy of Science* **81** 907
[42] Swinburne R (ed.) 2011 *Free Will and Modern Science* (Oxford: Oxford Univ. Press)
[43] Khrennikov A 1997 *Non-Archimedean Analysis: Quantum Paradoxes, Dynamical Systems and Biological Models* (Dordreht: Kluwer)

Journal of Physics: Conference Series **504** (2014) 012020 doi:10.1088/1742-6596/504/1/012020

Newton force from wave function collapse: speculation and test

Lajos Diósi
Wigner Research Center for Physics, Budapest 114, P.O.Box 49, H-1525 Hungary
E-mail: diosi.lajos@wigner.mta.hu

Abstract. The Diosi-Penrose model of quantum-classical boundary postulates gravity-related spontaneous wave function collapse of massive degrees of freedom. The decoherence effects of the collapses are in principle detectable if not masked by the overwhelming environmental decoherence. But the DP (or any other, like GRW, CSL) spontaneous collapses are not detectable themselves, they are merely the redundant formalism of spontaneous decoherence. To let DP collapses become testable physics, recently we extended the DP model and proposed that DP collapses are responsible for the emergence of the Newton gravitational force between massive objects. We identified the collapse rate, possibly of the order of 1/ms, with the rate of emergence of the Newton force. A simple heuristic emergence (delay) time was added to the Newton law of gravity. This non-relativistic delay is in peaceful coexistence with Einstein's relativistic theory of gravitation, at least no experimental evidence has so far surfaced against it. We derive new predictions of such a 'lazy' Newton law that will enable decisive laboratory tests with available technologies. The simple equation of 'lazy' Newton law deserves theoretical and experimental studies in itself, independently of the underlying quantum foundational considerations.

1. Quantum mechanics of massive degrees of freedom

Apart from the troublesome relativistic quantum gravity, we have no reason to doubt quantum mechanics' validity in the macro-world. If quantum mechanics is universally valid, massive d.o.f. of a bulk system must be quantized together with the light and atomic d.o.f. For massive d.o.f., we should consider hydrodynamic modes in general. Unfortunately, the existence of large quantum uncertainties of massive d.o.f. in the so-called Schrödinger Cat states seems controversial. In a cautious modification of quantum mechanics [1, 2, 3, 4, 5, 6, 7], parametrized by the Newton constant G, spontaneous wave function collapses of massive d.o.f. will forbid the existence of Schrödinger Cats. A lasting problem has been accompanying this (and other) spontaneous collapse models. Spontaneous collapse is not testable, being the redundant formalism of spontaneous decoherence. What extension of the spontaneous collapse theory makes it testable? — we'll see below.

1.1. Schrödinger Cats

We don't discuss the quantum mechanics of hydrodynamical d.o.f. in general, we consider the c.o.m. motion of a single spherical macroscopic object. In free space, the wave function Ψ(c.o.m.) satisfies the free Schrödinger equation. Accordingly, a well-localized initial Ψ_0(c.o.m.) is unlimitedly expanding. Nothing prevents it to develop a macroscopic wave packet, called

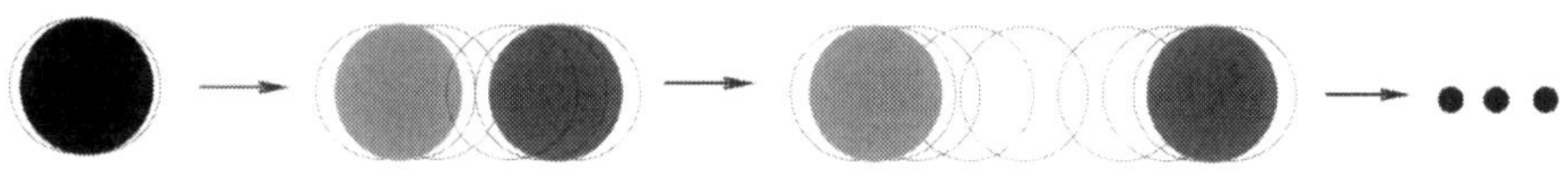

Figure 1. Unlimited macroscopic quantum expansion of the c.o.m. of an isolated massive object.

the Schrödinger Cat (Fig. 1). The interactions with light d.o.f., e.g. with the surrounding thermal gas, will no way block the unlimited expansion of the c.o.m. quantum uncertainty. On the contrary, the composite wave function $\Psi(\text{c.o.m.}, \text{gas})$ will approximately follow the free Schrödinger equation in the c.o.m. [8] while the c.o.m. is getting entangled with the gas (Fig. 2).

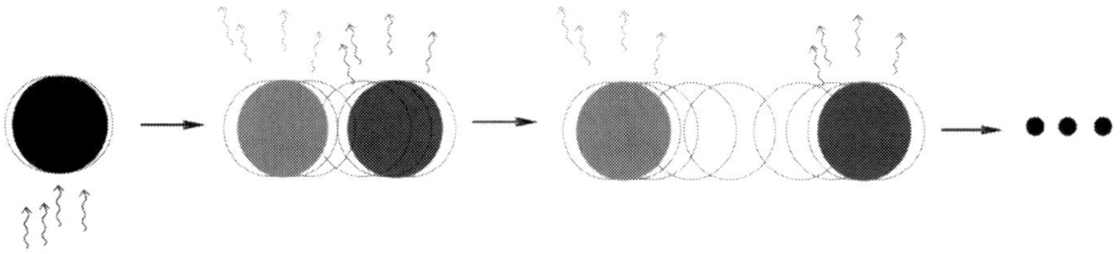

Figure 2. Unlimited macroscopic quantum expansion of the c.o.m. of a massive object entangled with light environment.

Schrödinger Cats are problematic. Some consider the problems are merely metaphysical, not influencing the perfect consistency of quantum mechanics as a physical theory. For others, including the author, the problem becomes physical because the von Neumann collapse of Schrödinger Cats provoke macroscopic violation of all basic conservation laws [9].

Schrödinger Cats can be suppressed by spontaneous collapse models.

1.2. G-related DP spontaneous collapse

Many advocates and critics of different spontaneous collapse models including GRW [10] and CSL [11] as well as DP, would overinterpret their subject constructing complicated notions. We rather propose a shortest definition in standard terms saving the reader's time and patience.

In every spontaneous collapse model, the spontaneous collapses are caused by standard von Neumann measurements with three specific features.

- The devices are weakly measuring certain universal observables.
- The devices are present everywhere and every time.
- The devices are hidden, they decohere all accessible systems yet the measurement outcomes remain inaccessible.

Specifically, in DP theory the hidden devices are universally measuring the local mass density operator of our physical system, the strength parameter of the measurement is scaled by the Newton constant G. Accordingly, the local mass density is universally enduring random collapses (disentanglement) by the hidden devices that forbid macroscopic quantum uncertainties of massive d.o.f. and, in particular, prevent Schrödinger Cats of being created.

Under natural circumstances, unitary evolution by the Schrödinger equation and the random DP spontaneous collapses compete with each other. The unitary evolution makes the c.o.m. uncertainty grow whereas the collapses make it shrink. The two can reach balance at a certain

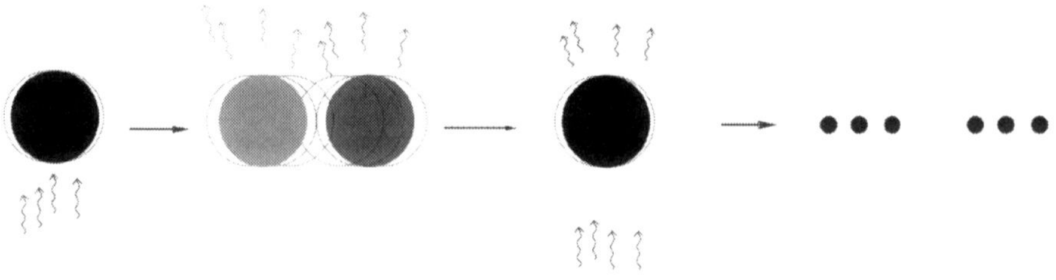

Figure 3. Spontaneous collapse (disentanglement) of the massive d.o.f., leading to random expansion-collapse cycles of characteristic time τ_G.

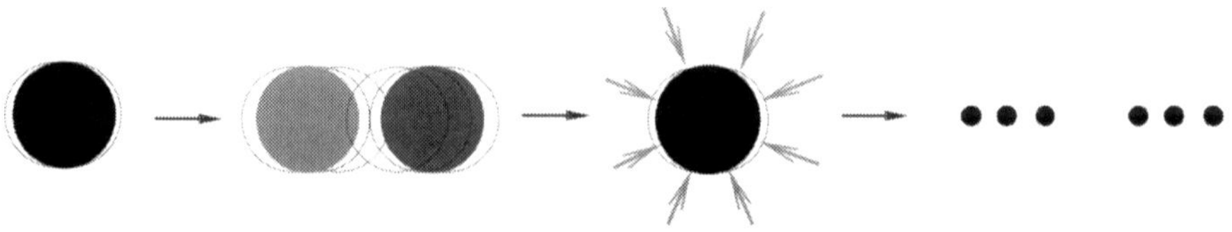

Figure 4. Random spontaneous collapse of massive d.o.f. triggers the Newton gravitational field at a characteristic emergence time τ_G.

equilibrium collapse rate $1/\tau_G$ (where τ_G stands for the equilibrium collapse time scale). The following estimation can be derived:

$$\frac{1}{\tau_G} \sim \sqrt{G\rho^{\text{nucl}}} \sim \frac{1}{\text{millisecond}} \tag{1}$$

where ρ^{nucl} is the nuclear mass density. Details of the estimation, and the particular reasons of the appearance of the nuclear mass density in a literally non-relativistic model, are given in Refs. [9, 12]. We can conclude that in DP theory the massive (hydrodynamic) d.o.f. are, under natural conditions, enduring spontaneous collapses at time scales $\tau_G \sim 1\text{ms}$ (Fig. 3).

1.3. Decoherence vs collapse

According to spontaneous collapse theories, the spontaneous decoherence effects are detectable in principle. Experiments are under way [13, 14, 15, 16, 17] to suppress environmental decoherence that always masks spontaneous decoherence unless we get our massive d.o.f. extremely isolated. Most theoreticians, therefore some experimentalists as well, don't emphasize (or even fail to recognize) that the spontaneous collapse theories are redundant realizations of the corresponding spontaneous decoherence theories. Spontaneous collapse (disentanglement) is never detectable, only the resulting spontaneous decoherence is, see [3]. This becomes obvious from our previous definition of spontaneous collapse theories. Since the hypothetic universal measuring devices are hidden, the measurement outcomes are hidden, one can never detect the collapse of the wave function. What one could detect is the damping of certain well-defined interference terms, i.e.: the phenomenon of spontaneous decoherence.

To let collapse become testable physics, one has to associate it with something physical! Recently we have extended the DP theory, supposed that the Newton force is generated by DP collapses. How should the Newton field emerge from DP collapses? Diverse and highly heuristic arguments have appeared in earlier works [18, 19]. But a detailed model is missing. In fact, we talk about an assumption beyond the DP theory, in minimal formulation.

Suppose the G-related DP collapses induce Newton gravity. Then, independently of the detailed mechanism, the emergence rate/time of Newton gravity is related (proportional) to the wave function collapse rate/time of the sources. Actually, the emergence time scale of Newton gravity must be of the order of the collapse time scale $\tau_G \sim 1\text{ms}$ (Fig. 4). We outline how this hypothesis would lead to a particular modification of the standard Newton theory, offering robust effects compared to an environmentally masked spontaneous decoherence.

2. Lazy Newton force with a delay

Unlike quantum foundational considerations in the previous Section, our resulting task is purely classical, i.e., non-quantum. We have to modify the purely non-relativistic Newton's law to account for a purely non-relativistic delay time τ_G which we think to be of the order of a millisecond. A minimum naive extension of Newton's classical equation would be

$$\phi(r,t) = \int_0^\infty \frac{-GM}{|r - x_{t-\tau}|} \mathrm{e}^{-\tau/\tau_G} \frac{d\tau}{\tau_G} \tag{2}$$

where $\phi(r,t)$ is the Newton potential at location r and time t, of a source of mass M moving along the trajectory x_t. The field at time t depends on the position of the source then and before, including a period of the order of τ_G. For zero delay $\tau_G = 0$, our Eq. (2) restores Newton's.

Unfortunately, for $\tau_G > 0$ the predicted modification would violate the equivalence of inertial frames and the equivalence of accelerated frames with gravity. Consider a source at rest at $x_t \equiv 0$, yielding $\phi(r,t) \equiv -GM/r$. Let's boost our inertial frame by a constant velocity v so that $x_t = -vt$ in the new inertial frame where (2) yields

$$\phi(r,0) = \int_0^\infty \frac{-GM}{|r - v\tau|} \mathrm{e}^{-\tau/\tau_G} \frac{d\tau}{\tau_G}\,. \tag{3}$$

The gravitational field $\phi(r,0)$ has changed from one inertial frame to the other. This shows that our naive equation (2) is not consistent in itself. Without illustrating how the naive equation (2) violates also the equivalence of accelerated frames and gravity, we show the resolution below.

In Ref. [20], we proposed that the naive equation (2) is valid and should be applied in the co-moving-free-falling reference frame. In other words, before we calculate the Newton potential of the moving source at time t, we install an instantaneous reference frame where $\dot{x}_t = 0$ and $M\ddot{x}_t$ is equal to the non-gravitational force if there is any. Having calculated $\phi(r,t)$ in the co-moving-free-falling frame, we can transform the result back to any other frame.

Hence we guarantee that the proposed modification of the Newton law is the same in all inertial frames. One can also show that the equivalence between gravity and an accelerated reference frame is respected. We should emphasize a major consequence of our chosen frame for Eq. (2): Newton law is completely restored in absence of non-gravitational forces. Accordingly, the effect of the delay time τ_G can only be expected with a massive source which is under the influence of a non-gravitational force. The point is not that the source is moving or isn't, the point is the presence of a non-gravitational force. We show three different experimental situations.

2.1. Testable predictions of gravity's laziness I.

Interestingly, there is a universal effect in Earth gravitational acceleration $g = 9.81\text{cm/s}^2$. Let us start from free falling objects first and apply Eq. (2) in the co-moving-free-falling frame as we have to. The result is that the free-falling object creates the standard instantaneous Newton potential. Now we consider an object at rest in the laboratory frame. Again, we use the co-moving-free-falling frame where the object is accelerated by the non-gravitational force through the support or the thread. We can calculate the effect of the delay in (2) in the lowest order

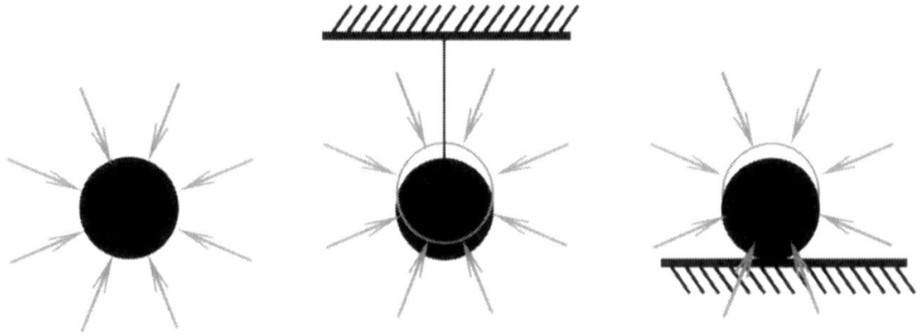

Figure 5. Free-falling object (left) in Earth gravity creates standard Newton field. Static sources, suspended (middle) or supported (right) create standard Newton field shifted slightly upward by a universal height $\delta_G = g\tau_G^2$.

of the delay time τ_G, leading to a surprising result. All static (supported or suspended) objects create Newton forces as if these objects were higher than their geometric position (Fig. 5), by a universal height

$$\delta_G = g\tau_G^2 \,. \tag{4}$$

For $\tau_G \sim 1$ms this universal height is of the order of 10^{-3}cm (or nearly 10^{-2}cm), a surprisingly big value. In a recent Cavendish experiment the positioning uncertainty of the source and test masses is ca. $0.5 \times \mu$m [21]. This would enable detecting a 10^{-3}cm vertical (there lateral) shift of the source's field.

2.2. Testable predictions of gravity's laziness II.

Alternative to static laboratory objects in Earth gravitational field, also ones accelerated by non-gravitational forces can test the delay τ_G. As a typical example, consider a source revolving at small angular frequency $\Omega \ll 1/\tau_G$ on a rope of length R. For simplicity, we disregard the Earth gravity and calculate the effect of mechanical acceleration $R\Omega^2$ only. Applying Eq. (2) in the instantaneous co-moving frame we find that the source creates Newton field as if it were closer than R to the center by δ_G. A lowest order expansion yields a similar result to (4), with $R\Omega^2$ in place of g:

$$\delta_G = R\Omega^2\tau_G^2 \,, \tag{5}$$

valid if $\Omega \ll 1/\tau_G$. Suppose we are interested in the Newton field in the center of the source's orbit. Since the source generates the field from a shorter distance $R-\delta_G$, we get stronger Newton field by the factor $1+\Omega^2\tau_G^2$ [22], a small effect for slow revolution and small acceleration (Fig. 6). Experiments with revolving sources, like [23], might make decisive tests or put an upper limit on τ_G at least.

2.3. Testable predictions of gravity's laziness III.

Direct test of the delay τ_G is possible if we impose extreme non-gravitational acceleration on our source. Let us prepare a massive laboratory source at static location $x_t \equiv 0$ for $t < 0$. It develops the standard static Newton field $-GM/r$ for $t < 0$, apart from the tiny universal shift δ_G (4) discussed previously. Here, again, we disregard this small effect, rather we look for a robust explicit delay of the Newton field. To this end, at $t = 0$ we suddenly displace the object to a new static position $x_t \equiv a$ for $t > 0$. Due to the static sources, and because we ignore the

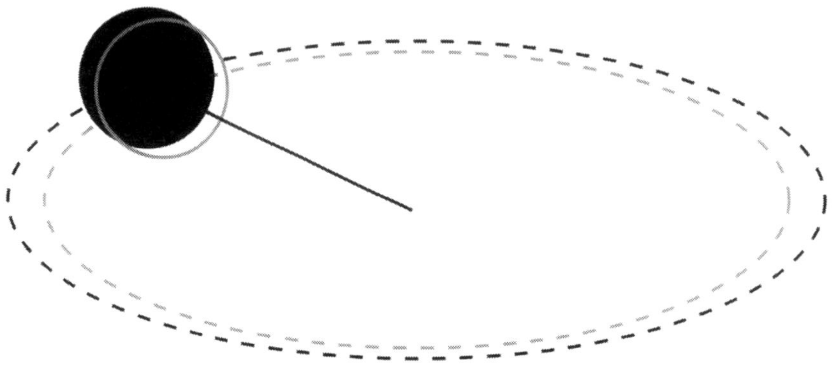

Figure 6. Revolving sources at angular frequency $\Omega \ll 1/\tau_G$ on a circle of radius R create standard Newton field as if the radius were smaller by a factor $1-\Omega^2\tau_G^2$.

effect of the Earth gravity, we can apply Eq. (2) in the laboratory system, yielding

$$\phi(r,t) = \mathrm{e}^{-t/\tau_G}\frac{-GM}{r} + \left(1-\mathrm{e}^{-t/\tau_G}\right)\frac{-GM}{|r-a|} \tag{6}$$

for $t>0$. From the old to the new Newton fields around respectively the old and new locations 0 and a, the change is gradually taking place with a characteristic delay time τ_G. This is exactly what we expect of the modified Newton law (2): the lazy Newton field follows the sudden displacement of the source with the delay τ_G (Fig. 7). In a real well-controlled test, only

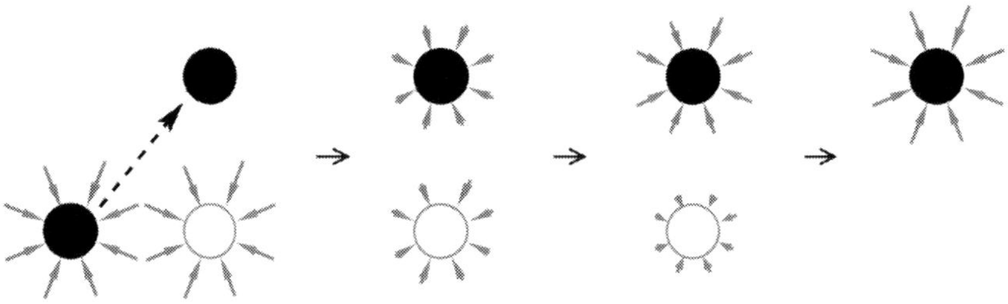

Figure 7. Sudden displacement of the source followed by the delayed redistribution of the Newton field from the old to the new configuration.

small macroscopic objects can be displaced or removed faster than a millisecond. Short-distance gravimetry at fine time-resolution is requesting but not impossible.

3. Summary, outlook

In the first part, we outlined the DP theory of spontaneous collapse of massive d.o.f., yielding an equilibrium time scale $\tau_G \sim 1$ms of the collapse (disentanglement). We emphasized that in spontaneous collapse theories, like DP and all the others, the spontaneous decoherence is the testable effect, the spontaneous collapse (disentanglement) is not. We invoked the recent extension of DP theory [20] that makes collapse testable by associating it with the emergence of the Newton potential. The second part discussed the heuristic modification of Newton law to incorporate the 'lazy' emergence of the Newton potential just with the delay time of the order of τ_G. We showed three different laboratory tests of the delay. The first test is based on the indirect effect of the delay, that static Newton fields of static sources in Earth gravitational field will get shifted upward by a universal height. A second test may exploit the slight modification of the standard Newton field of a source in periodic orbital motion. A direct test of the delay is possible with the sudden (i.e.: faster than τ_G) displacement of a small macroscopic source where the old/new configurations should die/emerge exponentially with the characteristic time τ_G. The realization of these tests seems possible with recent technologies.

We have emphasized that the concept of a lazy emergence of the Newton field is extremely speculative, the model is heuristic and minimal. As a non-relativistic theory, it is in peaceful coexistence with the general relativistic theory of gravitation. On one hand experimental evidences have not yet surfaced against it, and on the other hand it has various detectable predictions.

The modified Newton law—given by Eq. (2) in co-moving-free-falling frame—is a naive classical non-relativistic model, it can be considered both by theorists and experimentalists independently of the quantum mechanical DP theory, so that the value of the delay parameter τ_G could be tested in ranges different from the quantum theoretically predicted milliseconds.

Acknowledgments

This work was supported by the Hungarian Scientific Research Fund under Grant No. 75129 and the EU COST Action MP1006 'Fundamental Problems in Quantum Physics'. The author is indebted to the organizers of EmQM13 for the generous invitation.

References

[1] Diósi L 1986 *A quantum-stochastic gravitation model and the reduction of the wavefunction* (Thesis, in Hungarian)
http://www.rmki.kfki.hu/~diosi/thesis1986.pdf
[2] Diósi L 1987 *Phys. Lett.* **120 A** 377
[3] Diósi L 1989 *Phys. Rev.* **A 40** 1165
[4] Penrose R 1994 *Shadows of the mind* (Oxford: Oxford University Press)
[5] Penrose R 1996 *Gen. Rel. Grav.* **28** 581
[6] Penrose R 1998 *Phil. Trans. R. Soc. Lond.* **A 356** 1927
[7] Penrose R 2004 *The Road to Reality* (London: Jonathan Cape Publishers)
[8] In preparation.
[9] Diósi L 2014 *Found. Phys.* (in press)
[10] Ghirardi G C, Rimini A, Weber T 1986 *Phys. Rev.* **D 34** 470
[11] Ghirardi G C, Pearle P, Rimini A 1990 *Phys. Rev.* **A 42** 78
[12] Diósi L 2013 *J. Phys. Conf. Ser.* **442** 012001
[13] Marshall W, Simon C, Penrose R, Bouwmeester D 2003 *Phys. Rev. Lett.* **91** 130401
[14] Vanner M R, Pikovski I, Cole G D, Kim M S, Brukner Č, Hammerer K, Milburn G J, Aspelmeyer M 2011 *PNAS* **108** 16182
[15] Romero-Isart O 2011 *Phys. Rev.* **A 84** 052121
[16] Li T, Kheifets S, Raizen M G 2011 *Nat. Phys.* **7** 527
[17] Pepper B, Ghobadi R, Jeffrey E, Simon C, Bouwmeester D 2012 *Phys. Rev. Lett.* **109** 023601
[18] Diósi L, Papp T N 2009 *Phys. Lett.* **A 373** 3244
[19] Diósi L 2009 *J. Phys. Conf. Ser.* **174** 012002
[20] Diósi L 2013 *Phys. Lett.* **A 377** 1782
[21] Quinn T, Parks H, Speake C, Davis R 2013 *Phys. Rev. Lett.* **111** 101102
[22] Erratum: factor 1/2 was redundant in the Taylor expansion (4) in [20], also led to a redundant 1/2 within the enhancement factor there.
[23] Gundlach J H, Merkowitz S M 2000 *Phys. Rev. Lett.* **85** 2869

Gravitational decoherence, alternative quantum theories and semiclassical gravity

B L Hu

Maryland Center for Fundamental Physics and Joint Quantum Institute,
University of Maryland, College Park, Maryland 20742-4111 U.S.A.

E-mail: blhu@umd.edu

Abstract. In this report we discuss three aspects: 1) Semiclassical gravity theory (SCG): 4 levels of theories describing the interaction of quantum matter with classical gravity. 2) Alternative Quantum Theories: Discerning those which are derivable from general relativity (GR) plus quantum field theory (QFT) from those which are not 3) Gravitational Decoherence: derivation of a master equation and examination of the assumptions which led to the claims of observational possibilities. We list three sets of corresponding problems worthy of pursuit: a) Newton-Schrödinger Equations in relation to SCG; b) Master equation of gravity-induced effects serving as discriminator of 2); and c) Role of gravity in macroscopic quantum phenomena.

1. Introduction

There is general agreement that general relativity (GR) is an excellent theory describing the large scale structures of spacetime and quantum field theory (QFT) a highly successful theory for matter down to the verifiable subnuclear levels. Yet, it is equally well-accepted that intrinsic contradictions between general relativity and quantum theories exist. There are many serious efforts to reconcile or unify them in the search of *a theory for the microscopic structures of spacetime*, which is what **quantum gravity** (QG) entails – and carries no other meaning, specifically not quantizing general relativity (see, e.g., [1])– but it is fair to say to date no one school can show definitive and complete success in this goal.

1.1. Semiclassical gravity

A modest yet no less productive attempt is to place these two theories together: Q $\oplus$ G, not Q $\otimes$ G, which we don't yet quite understand how to do – GR being a classical theory for the macroscopic realm while QFT a quantum theory for the microscopic world, and see what discrepancies this union may reveal, such as in their mathematical structures, and what new physical insights we may gain. This was the goal set in **quantum field theory in curved spacetime** (QFTCST) [2, 3, 4] which began in the late 60's with cosmological particle creation studies [5] and epitomized in Hawking's 1974 [6] discovery of black hole radiance. Focused efforts in seeking ways to regularize or renormalize the stress energy tensor of quantum fields made it possible to tackle the so-called 'backreaction problem' [7, 8] in finding how quantum matter fields affect the dynamics of spacetime.

Solving the backreaction problem is at the core of **semiclassical gravity** theory (SCG) [9] developed in the 80's based on the *semiclassical Einstein equation* (SCE). Discovery in the 90s of a lawful place for the fluctuations of quantum fields promoted this to the *Einstein-Langevin equation* [10] which enables one to solve for the induced metric fluctuations (Wheeler's poetic 'spacetime foam'). This ushered in a new theory known as **stochastic gravity** [11, 12]. Both theories have since been developed extensively and applied to strong field situations such as structure formation in the early universe and black hole fluctuation and backreaction issues.

As a summary remark, the validity of semiclassical gravity in the form first proposed by Moller and Rosenfeld [13] in the 60's is often raised by authors of Newton-Schrödinger equation, citing the arguments by Page and Geilker [14], Eppley and Hannah [15][1]. Leaving aside the question of whether gravity should be quantized, which had seen much broader and deeper discussions since then, the internal consistency of relativistic semiclassical gravity by itself had since been investigated further and there are better responses to the challenges posed in the early 80s (read e.g. papers by Kibble & Randjbar-Daemi and Duff in [16]). We refer to two substantive papers, one by Flanagan and Wald [17] on semiclassical gravity, the other by Hu, Roura and Verdaguer [18], which also considered the role of the induced metric fluctuations in the criteria.

1.2. *Alternative quantum theories (AQT)*

General relativists following this vein have probed the interplay between gravity and quantum largely from the angle of how quantum matter affects spacetime (Q $\rightarrow$ G). Asking the question in the other direction (G$\rightarrow$ Q), namely, how gravity could have an effect on quantum phenomena, has been going on for just as long (e.g., [19]) mainly by quantum foundation theorists. The foremost issue is why macroscopic objects are found sharply localized in space (their wave functions "collapsed" on definite locales) while those of microscopic objects extend over space. This contradiction is captured in the celebrated Cat of Schrödinger [2]. One can very coarsely place these theories in three groups: The Girahdi-Remini-Weber (GRW)- Pearle models [20] of continuous spontaneous localization (CSL), the Diósi-Penrose theories [21, 22] invoking gravitational decoherence, and the recent trace dynamics theory of Adler [23] which attempts to provide a sub-stratum theory from which quantum mechanics emerges. A nice description of these theories can be found in a recent review by Bassi et al [24].

1.3. *Gravitational decoherence*

One important process where the interplay of gravity with quantum manifest is gravitational decoherence – the mechanism where the quantum coherence of a particle is diminished due to its interaction with an environment, in this case provided by the gravitational field. (The differences between quantum, intrinsic and gravitational decoherence are explained in the Introduction of [28]. See also [29, 30]). Gravitational decoherence is invoked in an important class of AQTs, that by the name of Diósi-Penrose theories. We want to find out the special features of gravitational decoherence, such as the decoherence rate and the associated basis and how gravitational decoherence differs from decoherence by a non-gravitational environment. To perform quantitative analysis of this effect one needs a master equation which has not been

[1] Since many of these theories which combine classical gravity with quantum mechanics are now also referred to as semiclassical gravity by many practitioners, to distinguish from them (which we see below are important) we will call theories based on quantum field theory (second quantized, permitting particle creation) in curved spacetime, which is a well-established form of GR+QFT, Relativistic Semi-Classical Gravity (RSCG).

[2] Note that 'cat-states' have been found for atoms whereas entangled 'dead' and 'alive' state for real cats, which are a little bigger than atoms, have not. The difference between micro and macro objects is crucial insofar as their quantum behaviors are concerned. A missing basic ingredient is nonequilibrium statistical mechanics which helps us interpolate between micro/few body effects and macro/many-body phenomena. Here lies the importance of macroscopic quantum phenomena (MQP), a subject hitherto overlooked by theorists but I feel is essential in understanding why cat-states can never be found for real cats.

derived from first principles until only recently (see papers by Blencowe and by Ananstopoulos & Hu below). Earlier equations have been reasoned out rather than derived from known microscopic physics. The 'reasoning out' process admits inputs based on phenomenological arguments according to the proponents' particular wishes. Whether gravity can be an effective source of decoherence is a reasonable motivation to work out a master equation for such analysis. More of this in Sec. 4.

1.4. Experimental possibilities

While the early universe and black holes are the natural arena where strong field quantum processes play out, which necessitate QFTCST and RSCG, the weak field and nonrelativistic limits are certainly more within the reach of what laboratory experiments can measure. Rapidly improving precision levels of observational possibilities in molecule interferometry, optomechanics and mirco-trap experiments [31] are pushing this closer to reality. While tests of AQTs are understandably high on the agenda of the proponents of such theories, the more modest yet equally if not more important task is to "put SCG to test" (note SCG here is used in the weak field nonrelativistic context) as exemplified by Chen's group's recent derivation of a NS equation for many particles and estimating the predictions of their NS equation as different from that of the standard Schrödinger equation [32] (see below).

Designs of experimental setups and tests are already underway which involve theorists working in QFTCST. Witness the increased activities in precision measurement of Casimir effect, dynamical Casimir effect, for understanding the vacuum energy modified by boundaries, and the parametric amplification thereof, which is the mechanism underlying cosmological particle creation. Jets and bursts of atoms from the controlled collapse of a BEC ("Bosenova") can be used to understand particle creation in inflationary universe [33], an example of laboratory cosmology. Accelerating atoms for testing the Unruh effect and finding analogs of Hawking effect in BECs and moving mirrors are being pursued. These activities belong to the realm of a new field called analog gravity [34]. Theoretical results obtained in the 70s are now being improved on and applied to designs for possible experimental verifications. In our opinion, tests related to RSCG will come next. Now is a good time for researchers in the 80s (on RSCG) and 90s (on semiclassical stochastic gravity) to join force with experimental activities so their expertise can enrich our understanding of the interplay between quantum matter and classical gravity.

1.5. Theoretical preparatory work for observing gravitational decoherence and testing AQTs

AQTs have been in existence for 3 decades, but the tests of such theories are put in practice only recently thanks to the increasing precision required of the measurements of these gravity-induced effects (see, e.g., [31]). Indications of the timeliness for these investigations are seen also in concentrated recent activities, e.g., 6 papers of substance have appeared in the last 6 months. We list five of them here with a short description because we will refer to all of them in this report. Noteworthy also is Adler's recent work incorporating gravity into his trace dynamics theory [35]. Increased effort in theoretical investigations of NS equations and gravitational decoherence are needed to better prepare the ground for measurement possibilities.

1) Dyson's 2012 essay, [36] "Is a graviton detectable?" will be useful for our discussions of gravitational decoherence, e.g., whether a thermal bath of gravitons is for real.

2) Review by Bassi et al [24] has a lucid summary of alternative quantum theories and a clear description of the experimental results in progress (See Table 1, also [37].)

3) Chen's group [32], while assessing experimental observability, attempts to mark out the differences between NS and ordinary QM predictions and give a theoretical derivation of a NS equation for N particles. (**Version C.**) We will show in Sec. 3 the differences between their

equation and that obtained from taking the nonrelativistic limit of SCE equation. We find that the single or many particle NS equations are not from GR + QFT.

4) Anastopoulos and Hu [28] derived a master equation for gravitational decoherence based solely on GR+QFT. We call this theory **Version A**. The procedure and results of this paper are summarized in Section 4.

5) Blencowe [38] derived a master equation using the influence functional method and made claims to the effect that gravitational decoherence is strong enough to soon be within observational range. (We call his master equation **Version B**.) We have reservations in his claim which invokes two assumptions: a thermal bath of gravitons and the increase of decoherence strength with mass. We will comment on these two assumptions in Section 7.

In all, our present work attempts to address the core issues in the interplay between gravity and quantum, places the Newton-Schrödinger equation in the context of relativistic semiclassical gravity and explores the theoretical base for observational possibilities of gravitational decoherence. Our hope is that the theoretical structure developed based on known physics from general relativity and quantum field theory and the results obtained here can provide a definitive standard for all alternative quantum theories to compare in their reasonings and predictions. The new results reported here on the master equation for gravitational decoherence and the NS equation in relation to SCG are based on two recent papers [28, 39]

2. Quantum matter interacting with classical gravity: 4 levels of inquiry

Newton-Schrödinger (NS) equations describe the motion of nonrelativistic quantum particle(s) in a weak gravitational field potential. There are many forms, justified with different rationales, originated from different motivations. Some are in conflict with general relativity as they are not the nonrelativistic limit of relativistic semiclassical gravity (RSCG). From their formal appearance one may view the NS equation as the nonrelativistic limit of the Einstein-Klein-Gordon equation, but NS in structure is closer to the Hartree-Fock equations. Yet, despite the similarity in structure between the NS eqn and Hartree-Fock equations, there are differences between gravito- and electro-statics) in the self energy for a many body system. These issues need be clarified so that one knows what his/her results of calculation pertain to in relation to other theories and in comparison with experiments.

2.1. Level 0: Newton-Schrödinger / Schrödinger-Poisson equations: non-relativistic quantum particle in a weak gravitational field

This is the arena where most of the activities in finding / showing the overt / hidden effects of gravity on quantum mechanics takes place and proposals of alternative quantum theories reside. Because it is for non-relativistic particles in a weak gravitational field, this is also the domain where laboratory experiments are carried out. Most researchers working in the 70s on quantum field theory in curved spacetime (QFTCST) and in the 80s working on (relativistic) semiclassical gravity (RSCG) (see below) have largely ignored this level of activities going on at the same time, albeit sparingly, as much as researchers working on quantum foundational issues largely were unaware of developments in RSCG in the last four decades. The former group's attention was focused on quantum effects in strong gravitational fields as in the early universe and in black holes while the latter group was focusing on proposing alternative quantum theories (AQTs) and their experimental verification possibilities. The latter group's definition of SCG is largely within the realm of Level 0.

So what are the problems of interest at this Ground Level? One can start with an almost textbook-like example of calculating the dispersion of a Gaussian wave packet with initial spread a_0 of a massive (m) particle according to the Schrödinger equation $i\hbar\partial\psi/\partial t = -\frac{\hbar^2}{2m}\nabla^2\psi - mV\psi$ when it moves in a gravitational potential V sourced by its own wavefunction $\nabla^2 V = 4\pi Gm|\psi|^2$.

This set of Newton-Schrödinger equation is imbued with a tension illustrated in this simple example between the natural quantum dispersion of a wavepacket against the gravitational collapse due to its own mass. The rather low critical mass obtained by Salzman and Carlip [40, 41] in 2006 was contested by Giulini and Grossardt in [42] in 2010. By introducing a length scale ℓ the SN equation can be written in terms of a dimensionless coupling constant, $K = 2(\ell/\ell_P)(m/m_P)^3$ where ℓ_P, m_P are the Planck length and mass respectively. Giulini and Grossardt in [42] showed that inhibition of the dispersion becomes significant when the dimensionless coupling constant K becomes of order unity. Their conclusion (quoting from [24])".. leads to an important inference: The models of Karolyhazy, Diósi, and Penrose all agree that if the width of the quantum state associated with an object of mass m becomes greater than of the order $\hbar^2/(Gm^3)$, the quantum-to-classical transition sets in. For the experimentally interesting $a = 0.5\mu m$ this gives m of about 10^9 amu." This estimate puts the actual measurement of this effect beyond today's experimental capability but the intellectual challenge and excitement towards realization of this goal are certainly growing. More examples to expound the interplay between effects of quantum dispersion and gravitational pull in wave packets and more complex systems can be explored at the next level (Level 1) with the help of quantum field theory in curved spacetime techniques.

2.2. Level 1: Relativistic matter fields in strong gravitational field. Einstein-Klein-Gordon equation

We now enter the relativistic realm, both for the quantum field and for gravity: Schrödinger equation is upgraded to Klein-Gordon equation for scalar particles or Dirac equation for spinors. The effect of curvature enters in the wave equation for a scalar field through the Laplace-Beltrami operator $\nabla^2\Phi - m^2\Phi = 0$. This theory has been applied to treat self-gravitating particles [43] or boson stars [44]. Looking for a solution of the metric tensor sourced by the relativistic field requires that the Einstein equation and the KG equation be solved simultaneously in a self-consistent manner, namely, $G_{\mu\nu} = 8\pi G T_{\mu\nu}(\Phi)$ where $G_{\mu\nu}$ is the Einstein tensor and $T_{\mu\nu}(\Phi)$ is the stress energy tensor of the scalar field. Note at this level the field can be viewed either as first quantized or second quantized. As a first quantized field Guzman et al have shown that the E-KG equation reduces in the non-relativistic, weak field limit to the NS equation. Spherically symmetric solutions of the NS equation have been found by e.g., Moroz et al [45].

At the second quantized level truly quantum field theoretical effects like particle creation from the vacuum begin to show up. At the 'test field' level in which a quantum field propagates in a fixed given curved spacetime, this is the realm of quantum field theory in curved spacetime (QFTCST) [2, 4]. How this field affects the background spacetime is described by the semiclassical Einstein equation. One now enters the realm of relativistic semiclassical gravity.

2.3. Level 2: Relativistic semiclassical gravity (RSCG): semiclassical Einstein equation (SCE), including backreaction of quantum matter field

Semiclassical gravity (SCG) has been used for a wide range of theories where gravity is treated classically and the matter field quantum mechanically, including the nonrelativistic NS equation. But the treatment of matter varies from one particle to many-particle systems to quantum fields, and even for quantum, there is a difference between first quantized and second quantized. To avoid confusion we add the word relativistic to SCG to refer to fully relativistic Einstein's theory for gravity valid under strong field conditions, and fully relativistic quantum fields at the second quantized level for matter. One known example of such a theory is that based on the semiclassical Einstein equation: $G_{\mu\nu} = 8\pi G < T_{\mu\nu}(\Phi) >$ where $T_{\mu\nu}(\Phi)$ is the stress energy tensor of the matter field, here represented by a scalar field Φ, and $<>$ denotes taking the expectation value with respect to a certain quantum state.

Note in general because of the sum over all modes there is ultraviolet divergence in this expression. Much of the effort in the field of QFTCST in the mid-70s focused on finding ways to regularize or renormalize these divergences. By 1978 when the results obtained by different regularization approaches more or less converged serious studies of RSCG began, under the theme of "backreaction problems" which went on to the 80's – e.g., the backreaction of vacuum energy of quantum fields (such as the Casimir effect) and particle creation (from the vacuum) on the dynamics of the spacetime. This requires a self consistent solution of both the semiclassical Einstein equation governing the spacetime dynamics and the quantum matter field equation.

From physical considerations, the backreaction of quantum fields brings forth dissipation in the dynamics of spacetime through the SCE Eq. How to reckon the appearance of non-unitary terms in an otherwise unitary evolution dictated by Einstein's equation was the first conceptual challenge. Understanding this issue from the open quantum system viewpoint was helpful in discovering the next level of structure, in the Einstein-Langevin equation.

2.4. Level 3: Stochastic SC gravity (SSCG): Einstein-Langevin equation, including fluctuations in quantum field and metric

Including fluctuations of quantum field as a source driving the semiclassical Einstein equation faces another challenge. Why and how should a noise term appear in the SCE equation? These two issues were resolved by borrowing concepts in nonequilibrium statistical mechanics, namely, the existence of fluctuation-dissipation relations and the use of the (Feynman-Vernon) influence functional formalism to provide an analytic basis for the description of quantum noise. This was how semiclassical stochastic gravity theory came into being [12]. Further proof by Verdaguer et al [46] that the noise can be written in a covariant form and satisfies the divergence-free condition ensures its rightful place in the Einstein-Langevin equation [10].

3. Newton-Schrödinger equation and semiclassical gravity

The Newton-Schrödinger (NS) equations play a prominent role in alternative quantum theories (AQT)[24], emergent quantum mechanics [23], macroscopic quantum mechanics [27], gravitational decoherence [28, 38](as in the Diósi-Penrose models [21, 22]) and semiclassical gravity [12]. The class of theories built upon these equations have drawn increasing attention because experimentalists often use it as the conceptual framework and technical platform for understanding the interaction of quantum matter with classical gravity and to compare their prospective laboratory results. It is thus timely and necessary to explore the assumptions entering into the construction of these equations and the soundness of the theories built upon them, especially in their relations to general relativity (GR) and quantum field theory (QFT), the two well-tested theories governing the dynamics of classical spacetimes and quantum matter.

Since NS are often simplistically conjured as the weak field (WF) limit of GR and the nonrelativistic (NR) limit of QFT, their viability is usually conveniently assumed by proxy, courtesy their well-accepted progenitor theories. We are not convinced of this. In a recent paper [39] Anastopoulos and I show that NSEs do not follow from general relativity (GR) and quantum field theory (QFT), and there are no 'many-particle' NSEs, such as derived recently in [32]. We come to this conclusion from two considerations: 1) Working out a model (see [28]) with matter described by a scalar field interacting with weak gravity, with a procedure the same as in deriving the NR limit of quantum electrodynamics (QED). 2) Taking the NR limit of the semiclassical Einstein equation (SCE), the central equation of relativistic semiclassical gravity (RSCG) (see last section for the four levels of SCG), a fully covariant theory based on GR+QFT with self-consistent backreaction of quantum matter on the spacetime dynamics [12]. The key points are summarized in [47].

Before we explain the differences between theories based on NSE and those obtained from GR+QFT it may be useful to first highlight their differences in physical predictions:

3.1. Problems with the Newton-Schrödinger equations (NSE)

We mention three aspects here.

A. In NSE the *gravitational self-energy* defines non-linear terms in Schrödinger's equation. In Diósi's theory [21], the gravitational self-energy defines a stochastic term in the master equation. With GR+QFT gravitational self-energy only contributes to mass renormalization, at least in the weak field (WF) limit.The Newtonian interaction term at the field level induces a divergent self-energy contribution to the single-particle Hamiltonian. It does not induce nonlinear terms to the Schrödinger equation for any number of particles.

B. The *one-particle NS equation* appears as the Hartree approximation for N particle states as $N \to \infty$. Consider the ansatz $|\Psi\rangle = |\chi\rangle \otimes |\chi\rangle \ldots \otimes |\chi\rangle$ for a N-particle system. At the limit $N \to \infty$ the generation of particle correlations in time is suppressed and one gets an equation which reduces to the NS equation for χ [51, 35][3]. . However, in the Hartree approximation, $\chi(\mathbf{r})$ is *not* the wave-function $\psi(\mathbf{r})$ of a single particle, but a *collective variable* that describes a system of N particles under a mean field approximation.

C. A severe problem of the NSE when applied to a single-particle wave function is its *probabilistic interpretation.* Consider two statistical ensembles of particles one of which is described by the wave-function $\psi_1(\mathbf{r})$ and the other by the wave function $\psi_2(\mathbf{r})$. The ensemble obtained from mixing these ensembles with equal weight is described in standard quantum theory by the density matrix $\rho(\mathbf{r}, \mathbf{r}') = \frac{1}{2}[\psi_1(\mathbf{r})\psi_1^*(\mathbf{r}') + \psi_2(\mathbf{r})\psi_2^*(\mathbf{r}')]$. The usual Schrödinger evolution guarantees that the probabilistic interpretation of the density matrix remains consistent under time evolution $\rho_t(\mathbf{r}, \mathbf{r}') = \frac{1}{2}[\psi_1(\mathbf{r}, t)\psi_1^*(\mathbf{r}', t) + \frac{1}{2}[\psi_2(\mathbf{r}, t)\psi_2^*(\mathbf{r}', t)]$. This property does not apply for non-linear evolutions of the wave-functions. The problem of nonlinearity in quantum mechanics is an old issue which many AQTs are aware of, so we will just mention it here without further pursuit.

In what follows we will show that the only meaningful description of quantum matter interacting with classical gravity is if the matter degrees of freedom are described in terms of quantum fields, *not* in terms of single-particle wave functions in quantum mechanics.

3.2. NS equation not from GR + QFT

The NS equation governing the wave function of a single particle $\psi(\mathbf{r}, t)$ is of the form

$$i\frac{\partial}{\partial t}\psi = -\frac{\hbar^2}{2m}\nabla^2\psi + m^2 V_N[\psi]\psi, \tag{1}$$

where $V_N(\mathbf{r})$ is the (normalized) gravitational (Newtonian) potential given by

$$V_N(\mathbf{r}, t) = -\int d\mathbf{r}' \frac{|\psi(\mathbf{r}', t)|^2}{|\mathbf{r} - \mathbf{r}'|}. \tag{2}$$

It satisfies the Poisson equation

$$\nabla^2 V_N = 4\pi G\mu, \tag{3}$$

where $\mu = m|\psi(\mathbf{r}, t)|^2$ is the mass density, the nonrelativistic (slow motion) limit of energy density $\varepsilon = T_{00}$ (see below).

The Newton-Schrödinger equation predicts spatial localization of the wave-function, and decoherence only as a consequence of spatial localization. This "collapse of the wave function"

[3] Note it is long known [49] that RSCG is the theory obtained as the large N limit of N component quantum fields living in a curved spacetime. Roura and Verdaguer [50] further showed that the next to leading order large N expansion produces stochastic semiclassical gravity.

in space for macroscopic objects is a big 'selling-point' of NS equations in many AQTs. Its desirable attributes aside, the logical foundation of the Newton-Schrödinger equation seems shaky to us. The naive identification of Newton as weak field limit of GR and Schrödinger equation as the nonrelativistic limit of QFT is likely behind the justification of NS equations. Here, one should exercise caution, as illustrated below: E.g., on the GR side, not to identify gravitational potential as dynamical variables, and on the QFT side, not to mistake a field as a collection of particles described by single particle wave functions.

3.3. Non-relativistic weak field limit of SCE equation

The central equation of relativistic semiclassical gravity (RSCG) is the semiclassical Einstein equation (SCE), and when quantum field fluctuations are included, the Einstein-Langevin equation, the centerpiece of stochastic semiclassical gravity [12]. We examine the nonrelativistic limit of SCE and show that it is qualitatively different from the 'many-particle' NS equation derived in [32].

The SCE Equation is in the form [4] $G_{\mu\nu} = 8\pi G\langle\Psi|\hat{T}_{\mu\nu}|\Psi\rangle$, where $\langle\hat{T}_{\mu\nu}\rangle$ is the expectation value of the stress energy density operator $\hat{T}_{\mu\nu}$ with respect to a given (Heisenberg-picture) quantum state $|\Psi\rangle$ of the field.

In the weak field limit the spacetime metric has the form $ds^2 = (1-2V)dt^2 - d\mathbf{r}^2$, and the non-relativistic limit of the semi-classical Einstein equation takes the form

$$\nabla^2 V = 4\pi G\langle\hat{\varepsilon}\rangle, \tag{4}$$

where $\hat{\varepsilon} = \hat{T}_{00}$ is the energy density operator. This can be solved to yield

$$V(\mathbf{r}) = -G\int d\mathbf{r}'\frac{\langle\hat{\varepsilon}(\mathbf{r}')\rangle}{|\mathbf{r}-\mathbf{r}'|}. \tag{5}$$

The expectation value of the stress energy tensor in general has ultraviolet divergences and need be regularized. The procedures have been established since the mid-70's (see, e.g., [2]).

Two key differences between the NR limit of SCE and NSE are: i) the energy density $\hat{\varepsilon}(\mathbf{r})$ is an operator, not a c-number. The Newtonian potential is not a dynamical object in GR, but subject to constraint conditions. ii) the state $|\Psi\rangle$ of a field is a N-particle wave function. Quantum matter is coupled to classical gravity as a mean-field theory, well defined only when N is sufficiently large.

The (misplaced) procedure leading one from SCE to a NS equation is the treatment of $m|\psi(\mathbf{r},t)|^2$ as a mass density for a single particle, while in fact it is a quantum observable that corresponds to an operator $\hat{\epsilon}(\mathbf{r}) = \hat{\psi}^\dagger(\mathbf{r})\hat{\psi}(\mathbf{r})$ in the QFT Hilbert space when the matter degrees of freedom are treated as quantum fields $\hat{\psi}(\mathbf{r})$ and $\hat{\psi}^\dagger(\mathbf{r})$, as they need be. Not treating these quantities as operators bears the consequences A and B.

3.4. Analog to the nonrelativistic limit of QED

To cross check these observations we have carried out an independent calculation for matter described by a scalar field interacting with weak gravity, following the same procedures laid out in [28], namely, solve the constraint, canonically quantize the system, then take the nonrelativistic limit. This procedure is same as in obtaining the non-relativistic limit of QED. We obtain the Schrödinger equation

$$i\partial|\psi\rangle/\partial t = \hat{H}|\psi\rangle, \tag{6}$$

[4] We prefer calling this the semiclassical Einstein equation over the Moller-Rosenfeld equation because, after all, it is Einstein's equation with a quantum matter source.

with

$$\hat{H} = -\frac{\hbar^2}{2m}\int d\mathbf{r}\hat{\psi}^\dagger(\mathbf{r})\nabla^2\hat{\psi}(\mathbf{r}) - G\int\int d\mathbf{r}d\mathbf{r}'\frac{(\hat{\psi}^\dagger\hat{\psi})(\mathbf{r})(\hat{\psi}^\dagger\hat{\psi})(\mathbf{r}')}{|\mathbf{r}-\mathbf{r}'|}. \tag{7}$$

The electromagnetic analog of this equation with the Coulomb potential replacing the gravitational potential here is widely used in condensed matter physics (see [39] for details).

The matrix elements of the operator (7) on the single-particle states $|\chi\rangle$ define the single-particle Hamiltonian:

$$\langle\chi_2|\hat{H}|\chi_1\rangle = -\frac{\hbar^2}{2m}\int d\mathbf{r}\chi_2^*(\mathbf{r})\nabla^2\chi_1(\mathbf{r}) - G\int d\mathbf{r}d\mathbf{r}'\frac{\chi_2^*(\mathbf{r}')\chi_1(\mathbf{r})\delta(\mathbf{r}-\mathbf{r}')}{|\mathbf{r}-\mathbf{r}'|}. \tag{8}$$

It is clear that Eq. (7) is very different from the NS equation (1) when considering a single particle state. For single-particle states the gravitational interaction leads only to a mass-renormalization term (similar to mass renormalization in QED). This is point A we made above. Using the Hartree approximation to Eq. (4) leads to the same result as the NR WF limit of SCE, not NSE. This is Point B we made earlier. Details of this calculation are in [39].

Our analysis via two routes based on GR+QFT shows that NSEs are not derivable from them. Coupling of classical gravity with quantum matter can only be via mean fields. There are no N-particle NSEs. Theories based on Newton-Schrödinger equations assume unknown physics.

4. Gravitational decoherence

4.1. Master equations from GR + QFT: Our analysis and main results

The procedures we took in [28] are as follows:

First step: Start with the classical action of a massive scalar field interacting with gravity described by the Einstein-Hilbert action. Linearize the Einstein-Hilbert action around the Minkowski spacetime. Look at the weak-field regime. We do this for two reasons: a) we want results which can be tested in laboratory experiments at today's low energy (in contrast to strong field conditions, as found in the early universe or late time black holes). b) In the derivation of the master equation for consideration of gravitational decoherence the tracing-out of the gravitational field is not technically feasible, except for linearized gravitational perturbations.

The second step is to perform a 3+1 decomposition of the action and construct the associated Hamiltonian. Identify the constraints of the system and solve them at the classical level, expressing the Hamiltonian in term of the true physical degrees of freedom of the theory, namely, the transverse-traceless perturbations for gravity and the scalar field. The third step is to canonically quantize the scalar field and the gravitational perturbations together, to ensure the consistency between these two sectors from the beginning.

The fourth step (after Eq. (21) of [28]) is to trace over the gravitational field acting as its environment to obtain a master equation for the reduced density matrix of the quantum matter field, including the backreaction of the gravitational degrees of freedom. The system under consideration is formally similar to a quantum Brownian motion (QBM) model [54, 55].

The master equation for the reduced density matrix $\hat{\rho}_1$ of one non-relativistic quantum particle in 3D interacting with weak perturbative gravity, valid to first order in $\kappa = 8\pi G$, is given by

$$\frac{\partial\hat{\rho}_1}{\partial t} = -\frac{i}{2m_R}[\hat{\mathbf{p}}^2, \hat{\rho}_1] - \frac{\kappa\Theta}{18m_R^2}(\delta^{ij}\delta^{kl} + \delta^{ik}\delta^{jl})[\hat{p}_i\hat{p}_j, [\hat{p}_k\hat{p}_l, \hat{\rho}_1]] \tag{9}$$

where p_i are the momentum components of the particle, m_R is renormalized mass and Θ has meaning explained below. This master equation enables gravitational decoherence studies and other related tasks. The fifth and final step is to project to a one particle state, then take the non-relativistic limit. We then use this nonrelativistic master equation for the analysis of gravitational decoherence in a single quantum particle.

Main Results

(i) A special feature of decoherence by the gravitational field (in the non-relativistic limit) is the decoherence in the energy (momentum squared) basis, but not (directly) to decoherence in the position basis. This is a direct challenge to theories (such as that proposed by Diósi and concurred by Penrose), which assume a potential energy term so that decoherence occurs in the position basis. Our analysis shows that this class of theories violates the principles of general relativity.

(ii) Many approaches to gravitational or fundamental decoherence proceed by modeling temporal or spatial fluctuations in terms of stochastic processes. However, such fluctuations correspond to time or space reparameterizations, which are pure gauge variables, with no dynamical content, according to classical GR. The assignment of dynamical content to such reparameterizations implicitly presupposes an underlying theory that violates the fundamental symmetry of classical GR.

(iii) The decoherence rate depends not only on the matter-gravity coupling, but also on the intrinsic properties of the environment, such as its spectral density which reflects to some extent the characteristics of its sub-constituents composition. Measurement of the gravitational decoherence rate, if this effect due to gravity can be cleanly separated from other sources, may provide valuable information about the statistical properties of the sub-constituents, or what we called the "textures", of spacetime.

5. Constraining alternative quantum theories (AQT)

Diósi's theory – Version D [21]

Diósi proposed a master equation of the form

$$\frac{\partial \hat{\rho}}{\partial t} = -i[\hat{H}, \hat{\rho}] - \frac{1}{4}\kappa G \int d\mathbf{r}_1 d\mathbf{r}_2 [\hat{\mu}(\mathbf{r}_1), [\hat{\mu}(\mathbf{r}_2), \hat{\rho}]] \frac{1}{|\mathbf{r}_1 - \mathbf{r}_2|}, \tag{10}$$

where $\hat{\mu}(\mathbf{r})$ is the mass density operator for the system and κ a constant of order unity. Diósi's master equation predicts decoherence of superpositions of macroscopically distinct states X and Y with a decoherence time $\tau_{dec} = 2\hbar/[2U_D(X,Y) - U_D(X,X) - U_D(Y,Y)]$, where

$$U_D(X,Y) = -G \int d\mathbf{r}_1 d\mathbf{r}_2 \frac{f(\mathbf{r}_1; X) f(\mathbf{r}_2; Y)}{|\mathbf{r}_1 - \mathbf{r}_2|}, \tag{11}$$

with X and Y parameterizing the distributions μ. Typically one thinks of X and Y as centers of mass, whence the theory predicts decoherence in position.

One consequence of our investigation is the observation that Diósi's master equation cannot be derived from the framework of GR+QFT. It comes from the following considerations: General Relativity implies that the Newtonian interaction follows from the theory's Hamiltonian constraint. The solution of the constraint leads to a modification of the Hamiltonian through the addition of a Newtonian interaction term (in the non-relativistic limit): $H = H_0 - G\int d\mathbf{r}_1 d\mathbf{r}_2 \frac{f(\mathbf{r}_1) f(\mathbf{r}_2)}{|\mathbf{r}_1 - \mathbf{r}_2|}$. Hence, the consistent quantization of the theory should place the Newtonian interaction term as a part of the quantum *Hamiltonian*, not as part of the non-unitary dynamics. There is no reason to structure the postulated non-unitary terms as a Newtonian interaction term, as is the case in Diósi's master equation. This is forced upon as an assumption which contradicts general relativity.

More generally, we feel that general relativity suggests a very different class of fundamental decoherence models with different reduction basis and type of noise. Working this out explicitly can make the comparison with the alternative models more quantifiable.

Other alternative theories of quantum mechanics 'aided' (or 'interceded') by gravitational effects – at low energy (in contrast to the Planck scale) include the so called 'continuous spontaneous localization' (CSL) models of Girahdi-Remini-Weber (GRW)- Pearle [20] (see also work of Bassi et al [24]). The state reduction in these schemes is often facilitated by considering stochastic processes on the quantum system's Hilbert space and stochastic Schrödinger equations are often suggested as an alternative to quantum mechanics. We will not address them here because the source of noise is phenomenologically motivated. We focus in the above on the Diósi-Penrose theories because it highlights the conflicts between gravity with quantum in a more transparent way - even if finally proven wrong, either way.

6. Role of gravity in macroscopic quantum phenomena

Historically a primary motivation for introducing the continuous collapse models (CLS) is trying to make sense of the readily collapsed wave function of macroscopic objects while preserving the wave function in the microscopic realm. There are two main features of this class of models. They are (from [24] p. 482): nonlinearity (which we also see in the SCE eqn), 2) stochasticity (which we see in stochastic gravity where the noise originates from quantum matter fields). There are also two requirements: 3) no superluminal signaling – this is forbidden from the start in RSCG since the basic principles of quantum field theory and relativity are observed. 4) an amplification mechanism – an important issue which we feel has not been explored enough. This is one aim of Chen's program on macroscopic quantum mechanics (MQM) [27] and our work on macroscopic quantum phenomena [56, 57, 58, 59]. We now turn to this issue.

The main motivation on the theoretical side of the recent work by Chen's group is to derive a NS equation for many particles. They looked into the interaction between particles, the separation of scales in the dynamics of the center of mass variable from other variables. A similar concern was raised in the paper by Chou Hu and Yu [60] where they set out to find the conditions where the "Center of Mass Axiom" is observed and a master equation for the N particles can be derived. Their key finding is, for interaction potentials dependent only on the separation between any two oscillators, the master equation for N oscillators has the same form as the HPZ master equation [55] for a single oscillator. Studying N oscillator systems will enable one to see how their interaction affects the outcome.

Technically the master equation for gravitational decoherence we derived recently [28] applies to configurations with any number of particles. This is because we first derived a master equation for quantum matter fields before projecting it to the single-particle subspace. One can easily project it to any particle number state to obtain a master equation for N particles.

7. Observing gravitational decoherence – contributing factors

A major factor in the surge of attention paid to gravitational decoherence is because several well respected experimental groups showed interest in the measurement of such effects [31]. This is corroborated by some theorists' claims that their predicted values are close to current observable experimental precision levels. It is thus important to examine carefully the assumptions made in theories which assert a significant effect in gravity's power to decohere a quantum particle. For example in the recent paper of Blencowe [38] (which we referred to as Version B) two assumptions were made, as follows:

7.1. Is thermal graviton bath a tenable assumption?

Gravitational decoherence depends strongly on assumptions about the nature of gravitational perturbations. The usual assumption that Minkowski spacetime is the ground state of quantum gravity would imply that gravitational perturbations are very weak and cannot lead to decoherence. This is the result we obtained in [28] where the source of gravitational decoherence is due to weak perturbations off the Minkowski spacetime. However, if general relativity is a hydrodynamic theory and gravity is in the nature of thermodynamics, Minkowski spacetime could presumably be identified with a macrostate (i.e., a coarse-grained state of the micro-structures). In this case, the perturbations are expected to be much stronger and they may act efficiently as agents of decoherence. (This information is contained in the Θ parameter in [28], the former case has $\Theta = 0$, the latter case some large value.)

The underlying issue is whether gravitons are thermalized, and if so what is the graviton bath temperature? The source of gravitons can either be from weak gravitational perturbations in the experiment's environment or as remnants from the early universe. This is not a new issue. For gravitons as quantized perturbations off Minkowski space, which provide the lowest common denominator for a gravitational source in the consideration of decoherence, one can take the graviton scattering processes (see e.g., Papini's review [61]) and calculate their cross sections. Because of the extremely weak nature of their interactions, it will be very small. For gravitons of cosmological origin, Blencowe took the value of 1 degree K citing Kolb and Turner's book [5]. Dyson's lecture [36] also addresses these points and is a good source of reference and comparison.

A somewhat equivalent way to look at this issue is the difference between (in the classical view) a superposition of gravitational waves and (in the quantum view) a mixed state of such superpositions. This essential point need be explicated mathematically.

In a more probing and elaborate investigation of this issue some earlier calculations may be helpful, e.g., by Calzetta and Hu [62] and others on the conditions of thermalization in a $\lambda\phi^4$ theory. (This had been cross-examined by particle physicists before they did the same calculation for non-Abelian theories in heavy ion collision and quark gluon plasma processes.) We can replace the λ by the graviton interaction constant obtained above to get an estimate of whether it makes sense to assign a temperature to gravitons. It may turn out that a 'graviton bath' is quite remote from reality in today's environment.

7.2. Does simple scaling up of quantum attributes apply to macroscopic objects?

In addition to the assumption of a thermal bath for gravitons the other main reason why Blencowe obtained a large number (compared to ours) for the gravitationally induced decoherence rate is because he uses a simple scaling from a quantum particle to a massive object. For an initial superposition of ground and excited states of a single atom the decoherence rate from his formula is $\sim 10^{-45}/sec$. This small rate (meaning it takes a very long time) is commensurate with our claim that for a weak gravitational perturbations background (at zero temperature) there is essentially no gravitational decoherence effect. However, Blencowe continues, "For a matter system comprising an Avogadro's number of atoms $\sim$ 1 gram in a quantum superposition where all of the atoms are either in their ground state or all in their excited state," he got a decoherence rate of $\sim 10^2$/sec. "For a system with mass $\sim$ 1 kg in such a superposition state, the gravitationally induced decoherence rate projects is $\sim 10^8$/sec. We believe Blencowe made an implicit assumption in how a macroscopic system's quantum behavior is directly related to the quantum features of its microscopic constituents. This is a largely unexplored topic under the general subject of macroscopic quantum phenomena (MQP).

[5] We are not sure how that value came about. The likely path of argument is to draw an analogy with neutrinos (e.g., from Weinberg's 1971 book). But neutrinos also interact very weakly. Thus this 1K value needs closer scrutiny.

This issue needs to be addressed before one can assuringly take the results for micro quantum objects and scale it up to macro domains. Interaction strength and quantum coherence amongst the sub-constituents are expected to play a role.

Acknowledgments

The principal organizers, Gerhard Grössing and Jan Walleczek, are to be thanked for making this meeting on foundational issues of physics possible, even lavishly so. The main parts of this talk are based on two papers written with Charis Anastopoulos.

References

[1] B. L. Hu, Emergent /Quantum Gravity: Macro/Micro Structures of Spacetime J. Phys. Conf. Ser. 174 (2009) 012015 [arXiv:0903.0878]

[2] N. D. Birrell and P. C.W. Davies, *Quantum Fields in Curved Space* (Cambridge University Press, Cambridge, U.K., 1982).

[3] R. M. Wald, *Quantum Field Theory in Curved Spacetime and Black Hole Thermodynamics* (University of Chicago Press, 1994)

[4] L. Parker and D. J. Toms, *Quantum Field Theory in Curved Spacetime: Quantized Fields and Gravity* (Cambridge University Press, 2009)

[5] L. Parker, Phys. Rev. 183, 1057 (1969).

[6] S. W. Hawking, Nature **248**, 30 (1974); Commun. Math. Phys. **43**, 199 (1975).

[7] Ya. Zel'dovich and A. Starobinsky, Zh. Eksp. Teor. Fiz **61**, 2161 (1971) [Sov. Phys.- JETP **34**, 1159 (1971)] L. Grishchuk, Ann. N. Y. Acad. Sci. 302, 439 (1976). B. L. Hu and L. Parker, Phys. Lett. 63A, 217 (1977). B. L. Hu and L. Parker, Phys. Rev. **D17**, 933 (1978). F. V. Fischetti, J. B. Hartle and B. L. Hu, Phys. Rev. **D20**, 1757 (1979). J. B. Hartle and B. L. Hu, Phys. Rev. **D20**, 1772 (1979). **21**, 2756 (1980). J. B. Hartle, Phys. Rev. D23, 2121 (1981). P. A. Anderson, Phys. Rev. D28, 271 (1983); D29, 615 (1984).

[8] J. W. York, Jr. "Black hole in thermal equilibrium with a scalar field: The back-reaction" *Phys. Rev.* D **31**, 775 (1985). P. R. Anderson, W. A. Hiscock and D. A. Samuel, Phys. Rev. D **51**, 4337 (1995). W. A. Hiscock, S. L. Larson and P. A. Anderson, "Semiclassical effects in black hole interiors" gr-qc/9701004 Phys. Rev. D **56**, 3571 (1997). S. Bose, L. Parker and Y. Peleg, Phys. Rev. Lett. 76, 861 (1996) Phys. Rev. D53, 7089 (1996) and erarlier references therein. F. C. Lombardo, F. D. Mazzitelli, Phys. Rev. D58 (1998) 024009; F. C. Lombardo, F. D. Mazzitelli and J. G. Russo, Phys. Rev. D59 (1999) 064007.

[9] Semiclassical gravity entails the backreaction problem studied in [7] and [8].

[10] E. Calzetta and B.-L. Hu, *Phys. Rev.* D **49**, 6636 (1994). B. L. Hu and A. Matacz, Phys. Rev. D **51**, 1577 (1995); B. L. Hu and S. Sinha, Phys. Rev. D **51**, 1587 (1995); A. Campos and E. Verdaguer, Phys. Rev. D **53**, 1927 (1996); F. C. Lombardo and F. D. Mazzitelli, Phys. Rev. D **55**, 3889 (1997). R. Martin and E. Verdaguer, Phys. Rev. D **60**, 084008 (1999); D **61**, 084008 (1999)

[11] B. L. Hu and E. Verdaguer,Class. Quant. Grav. 20 (2003) R1-R42 [gr-qc/0211090]

[12] B. L. Hu and E. Verdaguer, Living Review Relativity, 11 (2008) 3 [arXiv:0802.0658]

[13] C. Moller, Les Theories Relativistes de la Gravitation Colloques Internationaux CNRX 91 edited by A Lichnerowicz and M.-A. Tonnelat (Paris: CNRS) (1962). L. Rosenfeld, Nucl. Phys. 40, 353 (1963).

[14] D. N. Page and C. D. Geilker, "Indirect Evidence for Quantum Gravity", Phys. Rev. Lett., 47, 979-982 (1981).

[15] K. Eppley and E. Hannah, "On the Logical Necessity of Quantizing the Gravitational Field", Found. Phys. 7, 51-68 (1977).

[16] C. J. Isham, Roger Penrose, Dennis William Sciama, eds Quantum Gravity 2 (Clarendon Press, 1981)

[17] E. E. Flanagan and R. M. Wald, Phys. Rev. D 54, 6233 (1996)

[18] B. L. Hu, Albert Roura and E. Verdaguer, "Induced Metric Fluctuations and Validity of Semiclassical Gravity", Physical Review D 70 (2004) 044002

[19] F. Karolyhazy, Nuovo Cimento A 42 (1966) 390.

[20] G. C. Ghirardi, A. Rimini and T. Weber, Phys. Rev. D34, 470 (1986). P. Pearle, Phys. Rev. D13, 857 (1976); 29, 235 (1984); 33, 2240 (1986). G. C. Ghirardi, P. Pearle and A. Rimini, Phys. Rev. A42, 78 (1990); G. C. Ghirardi and A. BassiRMP, Phys. Rep. 379, 257 (2003); R. Tumulka, Rev. Math. Phys. 21, 155 (2009).

[21] L. Diósi, Phys. Lett. A105, 199 (1984); Phys. Lett. 120, 377 (1987); Phys. Rev. A40, 1165 (1989).

[22] R. Penrose, in *Quantum Concepts in Space and Time*, R. Penrose and C. J. Isham editors, (Oxford, 1986, Clarendon Press); Gen. Rel. Grav. 28, 581 (1996); Phil. Trans. Roy. Soc. (London) A, 356, 1927 (1998).

[23] S. L. Adler, *Quantum Theory as an Emergent Phenomenon* (Cambridge University Press, 2004).

[24] Angelo Bassi, K. Lochan, S. Satin, T. P. Singh and H. Ulbricht, "Models of wave-function collapse, underlying theories, and experimental tests", Rev. Mod. Phys. 85, 471- 527 (2013)
[25] W. H. Zurek, Phys. Rev. D24, 1516 (1981); D26, 1862 (1982); in *Frontiers of Nonequilibrium Statistical Physics*, ed. G. T. Moore and M. O. Scully (Plenum, N. Y., 1986); Physics Today 44, 36 (1991); W. H. Zurek, Prog. Theor. Phys. 89, 281 (1993); E. Joos and H. D. Zeh, Z. Phys. B59, 223 (1985); A. O. Caldeira and A. J. Leggett, Phys. Rev. A31, 1059 (1985); W. G. Unruh and W. H. Zurek, Phys. Rev. D40, 1071 (1989); B. L. Hu, J. P. Paz and Y. Zhang, Phys. Rev. D 45, 2843 (1992); J. P. Paz, S. Habib and W. H. Zurek, Phys. Rev. D 47, 488 (1993).
[26] A. J. Leggett, J. Phys. C: Solid State Phys. 14 R415 (2002); Phys. Scr. 102, 69 (2002).
[27] Yanbei Chen, "Macroscopic quantum mechanics: theory and experimental concepts of optomechanics" J. Phys. B: At. Mol. Opt. Phys. 46 (2013) 104001
[28] C. Anastopoulos and B. L. Hu, "A Master Equation for Gravitational Decoherence: Probing the Textures of Spacetime" Class. Quant. Grav. 30, 165007 (2013)
[29] C. Anastopoulos and B. L. Hu, "Decoherence in Quantum Gravity: Issues and Critiques" J. Phys. Conf. Ser. 67, 012012 (2007)[gr-qc/0703137]
[30] C. Anastopoulos and B. L. Hu, "Intrinsic and Fundamental Decoherence: Issues and Problems" Class. Quant. Grav. (special issue) 25, 154003 (2008) [arXiv:0803.3447
[31] M. Arndt, O. Nairz, J. Vos-Andreae, C. Keller, G. van der Zouw and A. Zeilinger,' *Nature* **401**, 680 (1999). A. D. Armour, M. P. Blencowe and K. C. Schwab, Phys. Rev. lett. **88**, 148301 (2002). W. Marshall, C. Simon, R. Penrose and D. Bouwmeester, Phys. Rev. Lett. **91**, 130401 (2003). M. P. Blencowe, "Nanomechanical Resonators" Phys. Rep. 395, 159-222 (2004). O. Romero-Isart, A. C. Pflanzer, F. Blaser, R. Kaltenbaek, N. Kiesel, M. Aspelmeyer and J. I. Cirac, Phys. Rev. Lett. 107, 020405 (2011). J. D. Teufel, T. Donner, D. Li, J.W. Harlow, M. S. Allman, K. Cicak, A. J. Sirois, J. D. Whittaker, K.W. Lehnert and R.W. Simmonds, Nature (London) 475, 359 (2011). J. Chan, T. P. Mayer Alegre, A. H. Safavi-Naeini, J. T. Hill, A. Krause, S. Groblacher, M. Aspelmeyer and O. Painter, Nature (London) 478, 89 (2011). M. Aspelmeyer, P. Meystre and K. Schwab, Phys. Today 65, 29 (2012).
[32] Huan Yang, Haixing Miao, Da-Shin Lee, Bassam Helou and Yanbei Chen, "Macroscopic Quantum Mechanics in a Classical Spacetime" Phys. Rev. Lett. 110, 170401 (2013)
[33] E. Calzetta and B. L. Hu, Phys. Rev. A68, 043625 (2003)
[34] C. Barcelo, S. Liberati and M. Visser, Analogue gravity, Living Rev. Relativity 14, 3 (2011). G. E. Volovik, The Universe in a Helium Droplet, (Oxford University Press, Oxford, U.K., 2003).
[35] S. L. Adler, Class. Quan. Grav. 30, 195015 (2013)
[36] Freeman Dyson, "Is a Graviton Detectable?" Poincare Prize Lecture presented at the International Congress of Mathematical Physics, Aalborg, Denmark, August 6, 2012.
[37] O. Romero-Isart, Phys. Rev. A 84, 052121 (2011).
[38] M. P. Blencowe, Phys. Rev. Lett. 111, 021302 (2013)
[39] C. Anastopoulos and B. L. Hu, "Newton-Schrödinger Equation and Semiclassical Gravity"
[40] Carlip, S., 2008, Classical Quantum Gravity 25, 154010.
[41] Salzman, P. and S. Carlip, 2006, arXiv:gr-qc/0606120.
[42] Giulini, D. and A. Grossardt, 2011, Classical Quantum Gravity 28, 195026.
[43] Ruffini, R. and S. Bonazzola, 1969, Phys. Rev. 187, 1767.
[44] F. S. Guzman and L. A. Urẽna-Lopez, Phys. Rev. D 68, 024023 (2003); 69, 124033 (2004).
[45] I. M. Moroz, R. Penrose and P. Tod, Classical Quantum Gravity 15, 2733 (1999). R. Harrison, I. Moroz and K. P. Tod, Nonlinearity 16, 101 (2002).
[46] R. Martin and E. Verdaguer, Phys. Rev. D **60**, 084008 (1999); Phys. Rev. D **61**, 124024 (2000).
[47] C. Anastopoulos and B. L. Hu, Comment on [32]
[48] S. L. Adler, J. Phys. A 40, 2935 (2007)
[49] J. B. Hartle and G. T. Horowitz, Phys. Rev. D **24**, 257 (1981).
[50] A. Roura and E. Verdaguer, unpublished.
[51] R. Alicki and J. Messer, J. Stat. Phys. 32, 299 (1983).
[52] H. Spohn, Rev.Mod.Phys. 53, 569 (1980);
C. Bardos, F. Golse, N. Mauser, Methods Appl.Anal. 2, 275 (2000)
[53] C. Bardos, L. Erds, F. Golse, N. Mauser and H-T Yau, C.R. Acad. Sci. Paris 334, 515 (2002).
[54] See, e.g., A. O. Caldeira and A. J. Leggett, Physica A 121, 587 (1983)
[55] B. L. Hu, J. P. Paz and Y. Zhang, Phys. Rev. D **45**, 2843 (1992); **D 47**, 1576 (1993).
[56] C. H. Chou, B. L. Hu and Y. Subasi, "Macroscopic Quantum Phenomena from the Large N Perspective" in the 5th DICE meeting September 2010, Castello Pasquini/ Castiglioncello (Tuscany), Italy, J. Phys. Conf. Ser. 306 012002 (2011) [arXiv:1106.0556]
[57] C. H. Chou, B. L. Hu and Y. Subasi, " Macroscopic Quantum Phenomena from the Correlation, Coupling

and Criticality Perspectives" in the 2010 IARD Conference, May 2010 Hualien, Taiwan, J. Phys. Conf. Ser. 330, 012003 (2011) [arXiv:1107.3008]

[58] B. L. Hu and Y. Subasi, "Pathways toward understanding Macroscopic Quantum Phenomena" DICE 2012, J. Phys. Conf. Series 442, 012010 (2013) [arXiv:1304.7839]

[59] C. H. Chou, Y. Subasi and B. L. Hu, "Macroscopic Quantum Phenomena from the Coupling Pattern and Entanglement Structure Perspective" submitted to Physica A [arXiv:1308.4225]

[60] C. H. Chou, B. L. Hu and Ting Yu,"Quantum Brownian Motion of a Macroscopic Object in a General Environment", Physica A387 (2008) 432- 444 [arXiv:0708.0882]

[61] G. Papini and S.R. Valluri, Physics Reports 33C, 51 (1977)

[62] E. Calzetta, B. L. Hu, "Thermalization of an Interacting Quantum Field in the CTP-2PI Next-to-leading-order Large N Scheme" (unpublished) [hep-ph/0205271]

Journal of Physics: Conference Series **504** (2014) 012022 doi:10.1088/1742-6596/504/1/012022

Collapse models and perceptual processes

GianCarlo Ghirardi
Emeritus, University of Trieste and The Abdus Salam ICTP, Trieste, Italy
E-mail: ghirardi@ictp.it

Raffaele Romano
Department of Mathematics, Iowa State University, Ames, IA 50011 (USA)
E-mail: rromano@iastate.edu

Abstract. Theories including a collapse mechanism have been presented various years ago. They are based on a modification of standard quantum mechanics in which nonlinear and stochastic terms are added to the evolution equation. Their principal merits derive from the fact that they are mathematically precise schemes accounting, on the basis of a unique universal dynamical principle, both for the quantum behavior of microscopic systems as well as for the reduction associated to measurement processes and for the classical behavior of macroscopic objects. Since such theories qualify themselves not as new interpretations but as modifications of the standard theory they can be, in principle, tested against quantum mechanics. Recently, various investigations identifying possible crucial test have been discussed. In spite of the extreme difficulty to perform such tests it seems that recent technological developments allow at least to put precise limits on the parameters characterizing the modifications of the evolution equation. Here we will simply mention some of the recent investigations in this direction, while we will mainly concentrate our attention to the way in which collapse theories account for definite perceptual process. The differences between the case of reductions induced by perceptions and those related to measurement procedures by means of standard macroscopic devices will be discussed. On this basis, we suggest a precise experimental test of collapse theories involving conscious observers. We make plausible, by discussing in detail a toy model, that the modified dynamics can give rise to quite small but systematic errors in the visual perceptual process.

1. A concise review of the measurement problem

As is well known the paradigmatic case in which an embarrassing situation emerges within a quantum scenario is the macro-objectification or measurement problem. It can be summarized by resorting to the idealized scheme proposed by von Neumann [1].

We consider a microscopic system s_{micro}, we assume that it has been prepared in the eigenstate $|\omega_i\rangle$ of an appropriate observable Ω, and we suppose to be interested in determining the value of this observable by a measurement process. We then consider a macroscopic apparatus S_{macro} prepared in its "ready" state $|S_0\rangle$ and we switch on a system-apparatus interaction inducing, in the time span in which the measurement takes place, the process summarized by the second of the following equations:

$$\Omega|\omega_i\rangle = \omega_i|\omega_i\rangle \tag{1}$$

$$|\omega_i\rangle \otimes |S_0\rangle \quad \rightarrow \quad |\omega_i\rangle \otimes |S_i\rangle, \tag{2}$$

where we have denoted as $|S_i\rangle$ the orthogonal final states of the apparatus which are assumed to correspond to mutually exclusive perceptions of a conscious observer looking, e.g., to the "pointer position" of the apparatus.

Then, if we trigger the apparatus with a state which is a superposition of different eigenvectors of Ω, the linear nature of quantum mechanics implies the following evolution for the system-apparatus system:

$$[\sum_i c_i|\omega_i\rangle] \otimes |S_0\rangle \rightarrow \sum_i c_i|\omega_i\rangle \otimes |S_i\rangle. \tag{3}$$

Equation (3) shows clearly that the system and the apparatus are now entangled, and, in particular, that the apparatus cannot be claimed to possess a precise macro property associated to our definite perceptions when we look at it.[1]

We consider it appropriate to close this section by the sharp statement by J. Bell:

> *Nobody knows what quantum mechanics says exactly about any situation, for nobody knows where the boundary really is between wavy quantum systems and the world of particular events.*

2. Some proposed solutions which do not alter the quantum predictions

We list and briefly comment here some of the most popular proposals to overcome the just mentioned problem, proposals which, however, fully agree with quantum predictions for what concerns the outcomes of measurement processes.

- *Incompleteness.* The statevector is not everything. This way out of the macro-objectification problem considers the quantum specification of the state of a physical system, given by the statevector, as not complete: to have an exhaustive description one has to add to the statevector further variables, whose knowledge would allow a more detailed specification of the objective situation of individual physical systems and would consent more accurate or even fully precise predictions of the outcomes of measurement processes. The paradigmatic example of this position is the de Broglie-Bohm theory, which, in recent years, has seen an impressive set of mathematical investigations aimed to clarify the most delicate points of the theory[2] The ensuing formal scheme is completely satisfactory, it represents a deterministic completion of the theory based on the addition to the state vector of the so called "hidden variables", which are identified with the positions of all particles of the system, and which are, in principle, unaccessible. The theory is logically consistent and, in our opinion, it represents an extremely interesting theoretical scheme which overcomes the foundational problems of quantum mechanics.
- *Limiting observability or invoking deoherence.* This perspective has many facets, as one can see by taking into account the positions of Jauch [4], Daneri *et al* [5], Joos and Zeh [6] and Zurek [7], among others. The philosophy which characterizes it is extremely simple: either in principle (Jauch, Daneri) or for practical reasons (Joos & Zeh, Zurek) one cannot find a way to put into evidence that superpositions of macroscopically and perceptibly different situations actually occur. There are two fundamental objections which can be

[1] It has to be stressed that the unacceptable situation we have just mentioned does not derive from the fact that we have resorted to the (extremely) idealized description of von Neumann of the measurement process. Actually, as it has been recently proved [2], the very requirement that one can ascertain a micro property with a reasonable degree of reliability together with the assumption of the universal linear character of the evolution, implies that superpositions of macroscopically and perceptively distinguishable states unavoidably occur.

[2] For an exhaustive and lucid discussion of the de Broglie-Bohm theory we refer the reader to the book [3].

raised against this position: one derives from remarking that it never happened, in the history of science, that a theory leads to a nonsensical (in the sense of not matching our perceptions) situation - the superposition of perceptibly different states - but, at the same time, that resorting to an approximation one gets an acceptable picture of natural phenomena. Secondly, as remarked, among many others, by S. Adler [8], *the decoherence approach has not solved the measurement problem*, this statement following from the well known fact that the correspondence *physical ensembles* $\rightarrow$ *statistical operators* is infinitely many to one in quantum mechanics, so that, as recognized even by the strongest supporters of the decoherence approach [6]:

> *no unitary treatment of the time dependence can explain why only one of these dynamically independent components is experienced.*

According to these authors the emergence of our definite perceptions *might depend on our local way of perceiving*, a totally unacceptable position, in our opinion, due to the central role which it attributes to the conscious observer.

- *Enriching Reality.* We have decided to use this term to identify both the Many Worlds [9, 10] as well as the Many Minds [11] interpretations. In the first case, it is assumed that when one reaches the macro level and superpositions of macroscopically and perceptibly different states occur, the universe multifurcates, so that there are many universes in which all different macro situations become actual, one of them for each different universe. The second attitude corresponds to assuming the same not for the actual occurrence of different situations, but for the occurrence of all potentially possible perceptions: our brains are built in such a way that there are different "sheets" registering different perceptions, the different sheets of different observers having to be synchronized among themselves in such a way that no intersubjective disagreement might occur.

At this point we consider it interesting to present the completely different position of the adherents to the collapse view of natural processes.

3. Collapse theories

The central idea of this approach consists in contemplating that the linear and deterministic evolution of the standard theory has not a universal validity: the basic Schrödinger's equation must be modified by the addition of nonlinear and stochastic terms, which account for the definite features of the reduction process.

As it is obvious, and as it has been stressed by many scientists, macro-objects are characterized by the fact that they correspond to perceptually different locations of (some) of their macroscopic parts (typically their "pointer"), so that it is quite natural to tackle "the preferred basis problems" by assuming that the modified dynamics strives to make precise the positions of physical objects. Given this, we can be fully precise about the features of collapse models. For simplicity we will make reference to the original proposal of ref.[12].

- *States.* A Hilbert space $\mathcal{H}$ is associated to any physical system and the state of the system at time t is represented by a normalized vector $|\psi_t\rangle$ of $\mathcal{H}$.
- *Dynamics.* The evolution of the system is governed by Schrödinger's equation. In addition, at random times, with a Poissonian distribution with mean frequency λ (which is assumed to be proportional to the mass of the particle under consideration), each particle of any system is subjected to a spontaneous localization process of the form:

$$|\psi_t\rangle \rightarrow \frac{L_n(\mathbf{x})|\psi_\mathbf{t}\rangle}{||L_n(\mathbf{x})|\psi_t\rangle||}; \quad L_n(\mathbf{x}) = (\frac{\alpha}{\pi})^{3/4} \exp[-\frac{\alpha}{2}(\hat{\mathbf{x}}_n - \mathbf{x})^2]. \tag{4}$$

In this equation $\hat{\mathbf{x}}_n$ is the position operator of the $n-th$ particle of the system.

- *Collapse probability.* One assumes that the probability density that the collapse for the $n-th$ particle occurs at the space point $\mathbf{x}$ is given by:

$$p_n(\mathbf{x}) = ||L_n(\mathbf{x})|\psi_t\rangle||^2. \tag{5}$$

- *Ontology.* Let $\psi_t(\mathbf{x}_1, ...\mathbf{x}_N)$ be the wave function in configuration space. Then:

$$m(\mathbf{x},t) \equiv \sum_{n=1}^{N} m_n \int d^3\mathbf{x}_1...d^3\mathbf{x}_N \delta^{(3)}(\mathbf{x}_n - \mathbf{x})|\psi_t(\mathbf{x}_1, ...\mathbf{x}_N)|^2, \tag{6}$$

  is assumed to describe the *density of mass* distribution of the system of the N particles under consideration in three-dimensional space as a function of time.
- *The trigger mechanism.* The nicest feature of the model consists in the fact that the localizations increase with the number of particles. In particular, it can be rigorously proved that any localization of any particle implies a localization of the centre of mass of the whole system. Accordingly, in the case of an almost rigid body, its position suffers a localization with a frequency amplified, with respect to the one of the individual constituents, by the number of the constituents of the body.
- *Choosing the parameters of the theory.* The original choice [12] of the values of the localization accuracy and of the mean frequency of the localization *for a nucleon* has been:

$$\lambda \simeq 10^{-16} sec^{-1}, \quad \alpha \simeq 10^{10} cm^{-2}. \tag{7}$$

  Note that with these choices a microscopic system suffers a localization about every 10^7years, while a macroscopic system ($N \simeq 10^{23}$) one about every 10^{-7} sec. This is why the standard theory is left practically unchanged at the micro-level, while superpositions of macroscopically distinct states are suppressed in extremely short times.

Some important remarks:

- The physics depends essentially only on the product $\alpha\lambda$ with the only proviso that the localization accuracy must be much larger than the atomic dimensions in order that the modified dynamics leaves practically unaffected the internal motion.
- Changing the above product of some orders of magnitude contradicts well established facts or it requires some important modification like the introduction of an appropriate cut-off.
- The theory qualifies itself as a rival theory with respect to quantum mechanics and can be subjected to crucial tests with respect to this theory. Moreover, if one takes it seriously, one can get from it indications concerning where to look for an hypothetical breaking of the superposition principle.
- Recently a lot of attention has been paid to the possibility of devising crucial experimental tests of the theory against quantum mechanics. These tests cover a wide range of experimental situations. We will mention those making reference to the fact that the collapse processes tend to destroy quantum interference (the problem has been discussed in ref. [13, 14, 15]). Other interesting effects follow from the fact that the collapse models imply spontaneous photon emission (see ref. [16, 17, 18]), as well as from the fact that energy is not conserved (see ref. [12, 23]). For a general review of the problem of testing collapse models we refer the reader to [19].

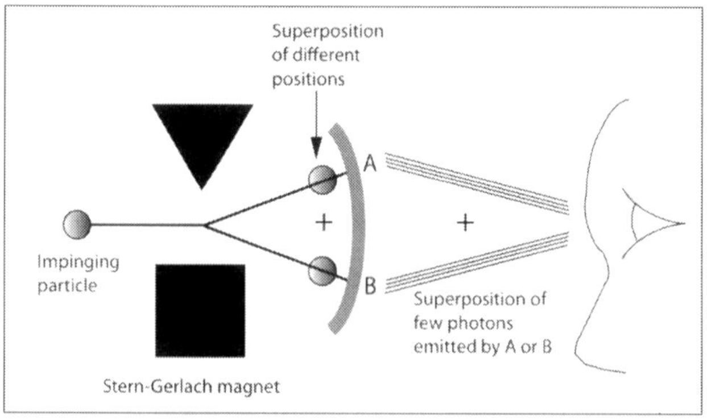

Figure 1. The Albert and Vaidman proposal

4. The problem of definite perceptions within collapse models

In 1989, Albert and Vaidman [20] have raised an important issue concerning the implications of collapse models with respect to the perceptual process itself. The challenge can be easily formulated: one has a neutral atom of spin 1/2 in the spin state corresponding to $\sigma_x = +1$, which traverses a region in which an inhomogeneous magnetic field (along the z-direction, orthogonal to x and to the propagation direction y) generated by a Stern-Gerlach device is present. As we all know, quantum mechanics predicts that the state of the atom, after the region of the apparatus has been traversed, is the equal weights superposition of its being *deflected upwards* and *deflected downwards*. On the considered paths one puts a fluorescent screen with the following properties: when the screen is hit by the impinging atom, a relatively small number of its atoms (of the order of ten) are excited, the excited states having an extremely short life time, so, immediately after the hitting, we have a superposition of two states $|10\ photons\ from\ A\rangle$ and $|10\ photons\ from\ B\rangle$, a notation to denote that the photons originate, in the first case, from a point A, and in the second case, from a point B of the screen. The experimental set up is depicted in Fig.[1].

The argument then goes as follows:

- In the time interval before the photons reach the eye, the collapse dynamics, due to the small number of particles involved and due to the fact that collapses do not involve photons, is not able to induce a collapse, so that we actually have the linear superposition $(1/\sqrt{2})[|10\ photons\ from\ A > +|10\ photons\ from\ B >]$. We assume that one can test, by an interference experiment, that the superposition is actually there.
- Now we call into play a perceiving observer who looks at the screen to check whether he sees a spot from the point A or from the point B. Since the threshold for visual perceptions is of the order of 5-7 photons, one is led to conclude that the observer will have a definite perception concerning the point in which there is a luminous spot.
- The conclusion follows: in the considered situation it is the conscious act of perception of the observer which determines the actual outcome, i.e. it induces the "reduction" of the "superposition" of two different and incompatible perceptions to only one of them.

This point of view has been subsequently stressed by Albert himself [21] in his interesting book *Quantum Mechanics and Experience*:

Suppose we want to stick with the GRW theory anyway. What would that entail? ... Well, we would have to deny that the measurement described above is over even once a recording exists. We would have to insist (and certainly this is an ineluctable fact, when you come right down to it) that no measurement is absolutely over, no measurement absolutely requires an outcome, until there is a "sentient observer" who is actually aware of that outcome.

The inappropriateness of this statement should be clear to anybody. There is no doubt that, within collapse theories, any (non human) detector aimed to check whether there is a spot at A or there is a spot at B (e.g., the firing of one of two Geiger counters located along the paths of the emitted photons, or even an apparatus such that its triggering by the superposition would put it - in the quantum description - in the superposition of having its pointer pointing to "From A" or "From B"), would imply a reduction of the state. No conscious observer is necessary in collapse models. However, the argument of ref. [20] undoubtedly represents an extremely serious and important challenge to collapse theories, a challenge we accepted to face.

Shortly after the appearance of Ref. [20], Aicardi, Borsellino, Ghirardi and Grassi analyzed [22] the perceptual process adopting precisely the point of view of collapse theories [3] .

The perceptual process involves the following steps:

- Transmission of the stimulus from the rods of the retina to the lateral geniculate body and from this to the higher visual cortex, the transmission of the nervous signals taking place along the axons connecting the various regions.
- The transmission mechanism implies the passage of Na and K ions from the internal to the external region of the axon through the Ranvier nodes. The axon is coated by the myelin sheet, which is precisely $10^{-5}cm$ thick (the characteristic reduction distance of the collapse models).
- If one takes seriously this line and assumes a very prudent attitude about the number of particles involved, it turns out that the process involves, *just within perceptual times*, the displacement of a sufficient number of particles in order that the collapse mechanism become effective: one of the two nervous signals is suppressed and a definite perception emerges.

5. Have the parameters of collapse theories been appropriately chosen?

As we have discussed in the previous section, if one makes the choices (7) for the parameters of the model under considerations, one can show that macroscopic measurement processes have outcomes and conscious observers have definite perceptions, even when the perceptual process is triggered by a genuine quantum superposition and no apparatus enters into play. However, as we have just mentioned, in ref. [22], in order to account for conscious perceptions, the whole process of nervous transmission in the brain has been taken into account. This fact does not seem fully satisfactory to some authors, and has been challenged, e.g., by S. Adler [23]. He perfectly agrees on the fact that collapse theories, when few particles are involved, turn out to make predictions in full agreement with those of standard quantum mechanics, as well as on the fact that one must require that when one reaches the macroscopic level, linear superpositions must be suppressed. However, there is a mesoscopic region (from the point of view of the number of involved particles) about which there might be some disagreement. Typically, as already indicated, S. Adler is not fully satisfied with the fact that the whole process of nervous transmission is necessary in order that the collapse takes place, and suggests that reduction must already take place when the rods of the eye register the signal.

Another fact should be mentioned. In the formulation we have adopted, collapse theories are not able to consistently deal with systems containing identical constituents, because collapses do not respect the (anti)symmetry requirements for such constituents. This requires a modification of the formalism and a quite nice reformulation of the original theory which solves this problem has been worked out by Pearle [24] and by Pearle *et al.* [25]. However, taking into account the

[3] Borsellino has been the first and most prominent representative of neurophysicists in Italy.

Journal of Physics: Conference Series **504** (2014) 012022 doi:10.1088/1742-6596/504/1/012022

identity of the elementary constituents, the efficiency of the trigger mechanism is increased[4]. Without entering into the details of the mathematical analysis we summarize the general situation: the localization accuracy $1/\sqrt{\alpha}$ cannot be decreased without inducing unacceptable effects, so, the game we can play is entirely related to possible changes of the mean frequency λ of the localizations. The ensuing situation is the following:

- The choice for λ made in the original collapse model [12] leads to the classical behavior of system with more than 10^{13} particles, while there is a continuous transition to a complete quantum behaviour for numbers appreciably smaller than this. In particular when only $10^4 - 10^5$ particles are involved, the behaviour is certainly still quantum.
- On the other hand, if one pretends, with Adler, the collapse to occur at the level of rods, in which case the displacement of a number of particles of the just indicated order is involved, one must appreciably increase the value of λ. As discussed by Adler a choice 10^8 times greater than the original one is perfectly able to induce the collapse in the time in which the rearrangement of the constituents in the rod takes place. Accordingly, Adler suggests to choose for λ the value $\lambda \simeq 10^{-8\pm2} sec^{-1}$. This request, however, implies a too fast increase of the energy of the universe due to collapses and, as a consequence, it requires some modifications which alter the original simplicity of the theory.

At any rate, the present analysis is useful to point out an essential fact which distinguishes the case of collapses due, typically, to superpositions of different positions of a macroscopic body from those taking place when our apparatus is triggered by a micro superposition whose states induce directly different perceptions. We discuss this point in the next section.

6. A comparison of the characteristic times of the hamiltonian evolution and of the reduction processes

Just to give an idea of the situation, let us take first of all into account a free macroscopic system, like a pointer of $1gr$, initially in a state with a spread of the c.o.m. position of the order of 10^{-5} cm. If we want to get an idea of the changes induced in the statevector by the hamiltonian evolution, a good estimate is given by the time it takes to double its spread, or (considering a physically more meaningful spread) to reach a spread of 10^{-1} cm, which we assume to be smaller than the separation of the pointer positions related to the different outcomes in the measurement. This computation leads to the conclusion that, during the time in which the hamiltonian evolution changes appreciably the state of the object, it suffers, according to the original choice of the parameters, from 10^{25} to 10^{29} localizations. In the case considered by Adler, these numbers have to be amplified of a factor 10^8.

On the contrary, for what concerns the perceptual process, when the perceptual apparatus is triggered by a superposition of microscopically different states (photons from A or from B), both in the original as well as in Adler's case, the reduction takes place just in the relevant times, i.e. the perceptual ones, during which the statevector changes from the superposition to one of its terms. And here a new feature emerges. Important differences in the dynamics occur when the two processes, the hamiltonian evolution and the changes induced by the localizations, become competitive, with respect to the case in which the localizations are enormously more frequent that the time required by hamiltonian changes. We will make clear this point by discussing, in the next section, a specific toy model (which has nothing to do with perceptions) in which we have a competition of the hamiltonian evolution and the reduction process. The study of this case is interesting because it allows to make plausible that collapse models might have consequences on the unfolding of the perceptual process.

[4] We refer the reader to Ref. [2] for a detailed discussion of this problem.

Journal of Physics: Conference Series **504** (2014) 012022 doi:10.1088/1742-6596/504/1/012022

Actually, the arguments we will present have already been outlined in a paper [26] by one of us. However, the treatment of ref.[26] was rather qualitative. The real novelty consists in the fact that here we will present rigorous results supporting the proposal.

7. A toy model making plausible that collapse models can be tested in perceptual experiments

In this section, motivated by the considerations of the previous one, we will discuss in detail a toy model, characterized by an hamiltonian evolution and by a collapse process which can be considered as competing. The aim is to check whether under such conditions some interesting effects might arise. The toy model we will consider has nothing to do with actual processes triggering different perceptions, and is a trivial two dimensional example, the two possible states in which the collapse strives to drive the statevector mimicking, in a sense, the two possible perceptions in an experiment of the kind presented in ref. [20].

Here is the model. We have a spin 1/2 particle evolving according to Schrödinger's equation with the hamiltonian:

$$H = \hbar\omega \begin{pmatrix} 0 & 1 \\ 1 & 0 \end{pmatrix}, \tag{8}$$

and subjected, with mean frequency λ, to reduction processes on the eigenstates of σ_z. We assume that the initial state of the system is:

$$\psi(0) = \begin{pmatrix} a \\ ib \end{pmatrix}, \tag{9}$$

where a and b are real numbers that we choose to take the values[5]

$$a = \sqrt{0.48}, \quad b = \sqrt{0.52}. \tag{10}$$

Let us write the statistical operator at time t in the following form:

$$\rho(t) = \begin{pmatrix} \rho_1(t) & \rho_3(t) \\ \rho_3^*(t) & (1-\rho_1(t)) \end{pmatrix}, \tag{11}$$

and let us call:

$$P_+ = \begin{pmatrix} 1 & 0 \\ 0 & 0 \end{pmatrix}, \; P_- = \begin{pmatrix} 0 & 1 \\ 0 & 0 \end{pmatrix}, \tag{12}$$

the projection operators on the eigenstates of σ_z belonging to the eigenvalues $+1$ and -1, respectively.

The quantum dynamical semigroup equation for the statistical operator is then:

$$\frac{d\rho(t)}{dt} = -\frac{i}{\hbar}[H, \rho(t)] + \lambda P_+\rho(t)P_+ + \lambda P_-\rho(t)P_- - \lambda\rho(t). \tag{13}$$

Note that the time which is necessary in order that the quantum evolution induces appreciable changes of the statevector (e.g. the time in which the spin makes a complete rotation around the z-axis) is $\tau = \pi/2\omega$, while the characteristic time between two reduction processes is $1/\lambda$. Since we want the reduction to occur within times of the order of the perceptual times, we will choose $\lambda = 10^2 sec^{-1}$. The parameter determining the ratio of the inverses of the times of the unitary evolution and of the reductions will be denoted as $\epsilon = \omega/\lambda$.

[5] We have made $a \neq b$ because the effects we are going to analyze depend on the initial state and the choice we have made turns out to be particularly appropriate for the emergence of the discrepancies we are interested in.

With reference to the expression (11) of the statistical operator we call the attention of the reader on the fact that $\rho_1(t)$ gives the probability that, at time t, the spin is in the state corresponding to $\sigma_z = +1$, while the vanishing of $\rho_3(t)$ tells us that reduction has actually taken place. In fact, in the present model it is obvious that reductions take place at the individual level, so that, here, it is legitimate to study the collapse process looking only at the statistical operator.

We analyze now in detail the evolution of the statistical operator for 2 quite different values of ϵ. First of all, to mimic the typical situation of realistic collapse models, we make the reductions much more frequent than the time in which the hamiltonian changes appreciably the statevector by choosing $\epsilon = 10^{-6}$, so that one million reductions occur during the time in which a rotation of 2π of the spin is caused by the hamiltonian. Alternatively we will consider the case in which $\epsilon = 10^{-2}$, corresponding to a much smaller number of reductions ($\simeq 100$) during the same time.

7.1. A remark concerning the asymptotic behavior

In realistic collapse theories and in the case of a macroscopic body in a superposition of macroscopically different states the time evolution goes as follows:

- In times of the order of $\lambda_{macro}^{-1} \simeq 10^{-7} sec$ the reduction process takes actually place leading to one of the two superposed states with the correct quantum probabilities.
- Subsequently, the statevector remains, for an incredibly long time (many times the age of the universe), the one to which the reduction process has led.
- However, asymptotically, i.e. for $t \to +\infty$, the statevector changes and it does not tend to a precise limit. This is a mathematical fact which has no physical relevance.

The situation is rather different in the case we are discussing, i.e. when one has an equation of the quantum dynamical semigroup type which admits a steady solution. In such a case a completely general theorem asserts that any solution must tend to it when $t \to \infty$. Our model manifestly exhibits the steady solution $\rho(t) = (1/2)I$. This implies that the unfolding of the evolution process takes place in this way:

- The reduction time is characterized essentially, as before, by λ^{-1}.
- Subsequently, reduction takes place to one of the superposed states with a probability depending on the initial state and it lasts for a certain period.
- However, at a certain time a new regime occurs, since, as we have stressed, the statistical operator must take the form $\rho(t) = (1/2)I$. It is important to have clear that what really matters is the second step, i.e. the detailed features of the reduction process and the time for which it lasts. In fact, here we are dealing with a toy example, mimicking only a very small part of the real process. Actually, if reduction takes place and lasts even for a short (but long with respect to the perceptual times) period, the very reduction process, i.e. the actual emergence of a definite perception, triggers other changes involving an extremely large number of particles. For instance other parts of the brain are surely involved besides the axons which have transmitted the signal, or, just to make a trivial example, we might ask the person whose eyes have been stimulated by the superposition to write on a piece of paper either A or B according to region from which he or she has seen the luminous spot. This fact has to be taken into account because it has an important consequence: after the perception emerges other macroscopic changes are implied by its specific character (having perceived a spot from A or from B) and such changes are then frozen by the collapse mechanism which is active for all particles of the universe. So, also in the present case, after the collapse corresponding to the emergence of a definite perception, the reduction lasts practically forever.

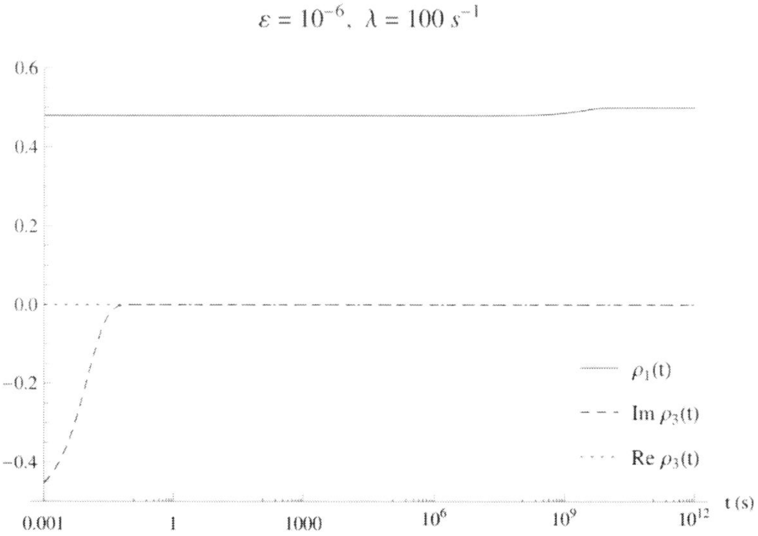

Figure 2. The behavior of the statistical operator in the case $\epsilon = 10^{-6}$.

8. The actual unfolding of the process of Section 7

We now present the results of the detailed analysis we have performed of the time behavior of the statistical operator for the two considered choices of ϵ. We summarize our results in a series of figures in which we have plotted $\rho_1(t)$ and $\rho_3(t)$ of Eq.(11) under the dynamics (13). We start, in Fig.2, with the case in which ϵ is very small (10^{-6}), like in actual collapse processes involving macroscopic systems. The figure shows clearly (the dashed and the dotted lines) that in a time of few hundredth of a second reduction takes place and leads to the eigenstate $\sigma_z = +1$ with the correct quantum probability $a^2 = 0.48$ implied by the form (9) of the initial state. The reduced state remains then unaltered for a quite long time ($\simeq 10^9 sec$). Subsequently the asymptotic behaviour, i.e. the one corresponding to a probability of 50% for the reduction to one of the eigenstates $\sigma_z = \pm 1$, emerges.

We pass now to discuss the most interesting case, i.e. the one in which ϵ is appreciably increased ($\epsilon = 10^{-2}$). Figure 3 (pay attention to the change of scale of the abscissa) looks similar to the previous one but there are some differences. They are not evident from Fig. 3 since use has been made of an inappropriate vertical scale. To put into evidence these features, their relevance and to discuss their physical implications one has to amplify the vertical scale. This has been done in Fig.4 which shows various relevant facts. First of all, after the reduction has taken place (in a time of the order of 10^{-2} sec, as shown in Fig.3), the function $\rho_1(t)$ exhibits a plateau, lasting for about 6-7 seconds, before the asymptotic regime emerges. The noticeable fact is that in this time interval (which, as discussed before is sufficient for inducing other macroscopic changes which are then frozen by the theory) the probability of the outcome $\sigma_z = +1$ amounts to more than 49%, i.e. it differs from the one of 48% required by quantum mechanics. It seems that the fact that the dynamical set up of the perceptions and the reductions due to the collapses become competitive can lead to systematic perceptual discrepancies from the quantum predictions.

9. A summary of the situation

Our analysis gives some clear indications about the perceptual process. If one assumes that in the case in which one triggers the visual apparatus with states corresponding to definite and

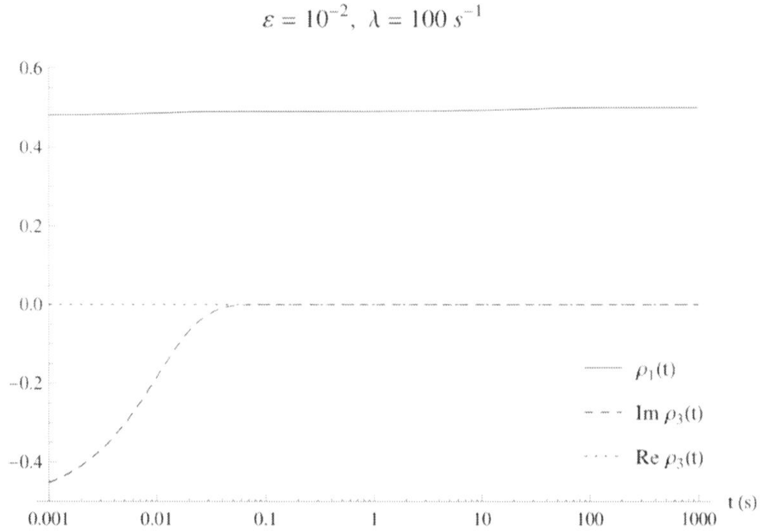

Figure 3. The behavior of the statistical operator in the case $\epsilon = 10^{-2}$.

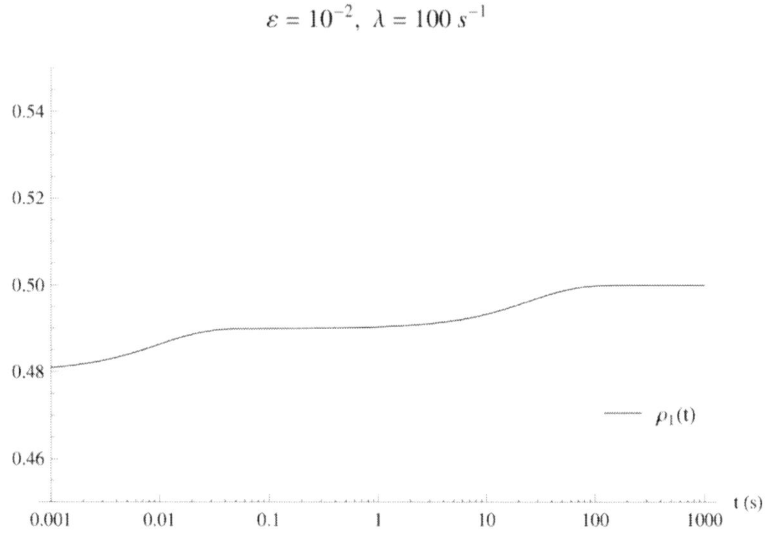

Figure 4. The vertically amplified behavior of the function $\rho_1(t)$ in the case $\epsilon = 10^{-2}$.

different perceptions, randomly distributed with probabilities 48% and 52% between "a spot at A" and "a spot at B" , no perceptual errors occur, the person subjected to the test would give answers reproducing the considered distributions (actually one could also check that in each individual process the perception matches the stimulus). On the other hand, in the case in which one triggers the perceptual apparatus with a state which is a superposition of the just mentioned states which should yield, according to standard quantum theory, once more to a 48%-52% distribution, some systematic errors seems to emerge, implying a violation of the quantum predictions. In a sense, one might pictorially describe the situation by claiming that in the second case, the brain (or the rods) has (have) to make the additional work of "reducing the

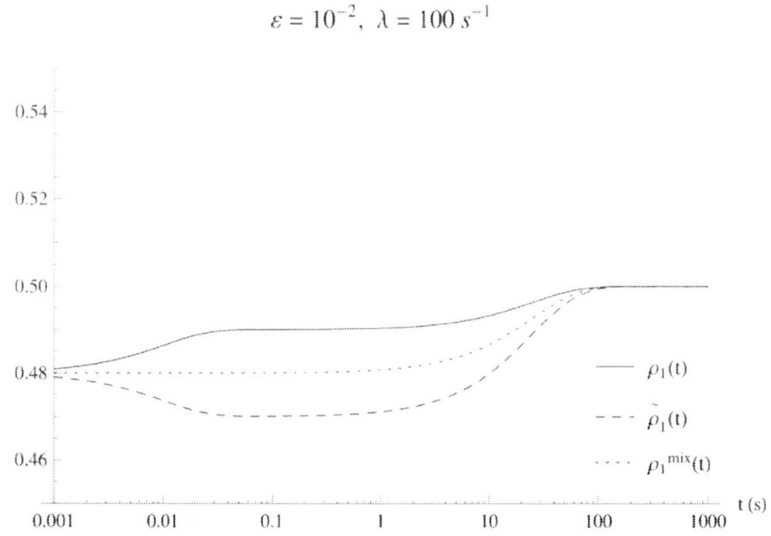

Figure 5. The proof that no faster than light effect can occur

state" and this might affect systematically the occurrence of the outcomes. Our analysis gives some (extremely) vague support to this idea and suggests that among the devised experiments to test collapse models versus the standard theory, also experiments involving perceptual process might and should be performed.

10. Rejecting a possible criticism

There is an aspect of the dynamical reduction mechanism which deserves further investigations. We have been led to the idea that when directly stimulated by a superposition, the perceptual apparatus might yield distribution of outcomes (very slightly) disagreeing with those implied by quantum mechanics. However, if the discrepancies can be actually put into evidence, we must check that the newly suggested peculiarity of the collapse models does not lead to consequences allowing faster than light effects. In fact, suppose the same effect, i.e. the enhancement of perceptions favoring "a spot from A", when triggering the perceptual apparatus with the state we have considered, $(a|10\ photons\ from\ A\rangle + ib|10\ photons\ from\ B\rangle)$, occurs also in the case when one triggers the apparatus with the state $(a|10\ photons\ from\ A\rangle - ib|10\ photons\ from\ B\rangle)$, while no perceptual mistake occurs when triggering the apparatus with a state corresponding to a definite perception (i.e. one of the two of the superposition). In such a case one might entangle the spin states of our particle with two orthogonal states of a further physical system far apart from the region in which the photons beams are, in the following way:

$$|\Psi> = \frac{1}{\sqrt{2}}\{|\chi_1\rangle \otimes [a|A\rangle + ib|B\rangle] + |\chi_2\rangle \otimes [a|A\rangle - ib|B\rangle]\}, \tag{14}$$

where we have used the notation $|A\rangle$ for the state $|10\ photons\ from\ A\rangle$ and $|B\rangle$ for the state $|10\ photons\ from\ B\rangle$, respectively. In this case, the guy, let us call him Bob, who has access to the system whose states are $|\chi_1\rangle$ and $|\chi_2\rangle$ can choose whether to measure an observable which has such states as eigenstates, leading then to reduction to one of the states $[a|A\rangle \pm ib|B\rangle]$. Alternatively, Bob might choose to measure an observable having as eigenstates $[|\chi_1\rangle \pm |\chi_2\rangle]$ inducing then a reduction to either the state $|A\rangle$ or to the state $|B\rangle$ and precisely with the probabilities a^2 and b^2. Then, the far away observer (Alice) would have his/her perceptual apparatus triggered either by the states corresponding, with the desired probabilities, to definite perceptions or by their superposition. However, if in the second case (reduction to any one of the states $[a|A\rangle \pm ib|B\rangle]$) the perceptual process of Alice would be affected in the same way, Alice would perceive more outcomes corresponding to $|A\rangle$ with respect to those implied by quantum mechanics and she might become aware (in the long run) of the kind of measurements that Bob has decided to perform. Obviously, to avoid faster than light effects, we must guarantee that this cannot happen. The request implies that the perceptual errors associated to triggering Alice visual apparatus with the state in which the sign of b is changed should compensate those for which no change has been performed. We know that things must go in this way because there is a general proof [27] that collapse theories do not imply superluminal effects. We have considered it useful to check this fact also with reference to our toy model. In Fig.5 we have plotted the first term of the statistical operator for three different initial conditions:

(i) $\rho_1(t)$: the initial state is $[\sqrt{0.48}|10\, photons\ from\ A\rangle + i\sqrt{0.52}|10\, photons\ from\ B\rangle]$,

(ii) $\tilde{\rho}_1(t)$: the initial state is $[\sqrt{0.48}|10\, photons\ from\ A\rangle - i\sqrt{0.52}|10\, photons\ from\ B\rangle]$,

(iii) $\rho_1^{mix}(t)$: the initial state is the statistical mixture, with weights 0.48 and 0.52 of the states corresponding to definite perceptions associated to $|10\ photons\ from\ A\rangle$ and $|10\, photons\ from\ B\rangle$.

As it is evident, the function for the third case turns out to be precisely the average of the first two cases. No superluminal signal can occur.

11. Conclusions

We have made plausible that, besides the various possible tests of collapse models versus quantum mechanics, also resorting to specific processes involving the direct triggering of the perceptual apparatus by a quantum superposition of microscopic states might help in discriminating between these two descriptions of natural processes.

References

[1] von Neumann J 1932 *Mathematische Grundlagen der Quantenmechanik*, (Berlin: Spinger), English translation 1955 *Mathematical Foundations of Quantum Mechanics* (Princeton: Princeton University Press)
[2] Bassi A and Ghirardi G C 2000 *Phys. Lett.* A **275**, 373
[3] Dürr D and Teufel S 2009 *Bohmian Mechanics* (Berlin Heidelberg: Springer-Verlag)
[4] Jauch J M 1964 *Helv. Phys. Acta* **37**, 293
[5] Daneri A Loinger A and Prosperi G M 1962 *Nucl. Phys.* **33**, 297
[6] Joos E and Zeh H D 1985 *Z. Phys.* B **59**, 223
[7] Zurek W H 1991 *Phys. Today* **44**, 36
[8] Adler S 2003 *Stud. Hist. Phil. Mod. Phys.* **34**, 135
[9] Everett III H 1957 *Rev. Mod. Phys.* **29**, 454
[10] DeWitt B S 1971 *Foundations of Quantum Mechanics*, B. d'Espagnat(ed), (New York: Academic Press)
[11] Albert D Z and Loewer B 1990 *Proceedings of the 1990 Biennial Meeting of the Philosophy of Science Association* v.1, East Leasing
[12] Ghirardi G C Rimini A and Weber T 1986 *Phys, Rev.***34** 470
[13] Arndt M, Nairz O, Vos-Andrease J, Keller C, van der Zouw G and Zeilinger A 1999 *Nature* **401**, 680
[14] Hackermüller M, Hornberger K, Brezker B, Zeilinger A and Arndt M 2004 *Nature* **427**, 711
[15] Gerlich S, Hackermüller L, Hornberger K, Stibor A, Ulbricht H, Gring M, Goldfarb F, Savas T, Müri M, Mayor H and Arndt M 2007 *Nature Phys.*, **3**, 711
[16] Fu Q 1997 *Phys. Rev.*, A **55**, 1806
[17] Adler S, Bassi A and Donadi S 2013 J: Phys. A: Math. Theor. **46** 245304
[18] Adler S and Ramazanoglu F 2007 *J. Phys.*, A **40**, 13395
[19] Adler S and Bassi A 2009 *Science* **325** 275
[20] Albert D and Vaidman L 1989 *Phys. Lett.* **139** 1
[21] Albert D. 2009 *Quantum Theory and Experience* Harvard: Harvard University Press)
[22] Aicardi F, Borsellino A, Ghirardi G C and Grassi R 1991 *Found. Phys. Lett.* **4**, 109

[23] Adler S 2007 *J. Phys. A: Math. Theor.* **40**, 2935
[24] Pearle P 1989 *Phys. Rev.* A **39**, 2277
[25] Ghirardi G C, Pearle P and Rimini A 1990 *Phys. Rev.* A **42**, 78
[26] Ghirardi G C 1999 *Phys. Lett.* A **262**, 1
[27] Bell J S 1987 *Schrödinger: Centenary Celebration of a Polymath*, Kilmister (ed), (Cambridge: Cambridge University Press)

Collapse models: from theoretical foundations to experimental verifications

Angelo Bassi
Department of Physics, University of Trieste, Strada Costiera 11, 34151 Trieste, Italy
Istituto Nazionale di Fisica Nucleare, Trieste Section, Via Valerio 2, 34127 Trieste, Italy
E-mail: bassi@ts.infn.it

Hendrik Ulbricht
School of Physics and Astronomy, University of Southampton, Southampton SO17 1BJ, United Kingdom
E-mail: h.ulbricht@soton.ac.uk

Abstract.
The basic strategy underlying models of spontaneous wave function collapse (collapse models) is to modify the Schrödinger equation by including nonlinear stochastic terms, which tend to localize wave functions in space in a dynamical manner. These terms have negligible effects on microscopic systems—therefore their quantum behaviour is practically preserved. On the other end, since the strength of these new terms scales with the mass of the system, they become dominant at the macroscopic level, making sure that wave functions of macro-objects are always well-localized in space. We will review these basic features. By changing the dynamics of quantum systems, collapse models make predictions, which are different from standard quantum mechanical predictions. Although they are difficult to detect, we discuss the most relevant scenarios, where such deviations can be observed.

1. Introduction

There is a well-known and simple reason why quantum theory, in its standard formulation, is not satisfactory as a theory of nature. When applied to macroscopic systems, it predicts the existence of superpositions of different macroscopic states—Schrödinger's cat being dead + alive. However such superpositions do not match with the world as we perceive it (Fig. 1). Hence the problem: The link between theory and physical phenomena (at least at the macroscopic level) is not clear [1].

Over the decades, several ways to solve the problem have been presented. The textbook solution introduces the collapse of the wave function, which takes places at the end of each measurement process—essentially, when a microscopic systems interacts with a macroscopic object—and prevents the Schrödinger equation from turning microscopic superpositions into macroscopic ones. It is an easy practical way to dismiss the problem, but it cannot be taken too seriously, since it introduces an arbitrary and not better defined division between micro-macro, small-large. A fundamental theory of nature cannot be founded on such a vague and ill-defined basis.

Some advocate decoherence to solve the problem [2, 3, 4]. Large objects like measuring devices, or cats, unavoidably interact with the surrounding environment, which cannot be fully controlled. Not having a full control on part of the universe the object interacts with, forbids in general the possibility of controlling and measuring quantum coherence for that object, hence no possibility of seeing Schrödinger's cat dead + alive. Again, this solution works out very well from the practical point of view, but not from the fundamental one. If the Schrödinger equation rules the universe, than the universe lives in a huge, complicated, highly nonclassical superposition, nothing that looks like the world as we know it [5, 6].

The situation improves slightly if combined with a many-worlds interpretation [7, 8, 9]. The huge, complicated, highly nonclassical superposition the universe lives in, actually refers not just to our universe, but to several "parallel universes", to use an expression taken from science fiction. Our universe corresponds to one term of the superposition. The emergence of probabilities, i.e. the statistics of experiments as we experience them, though, is still problematic.

In recent years, the success of quantum information technologies pushed part of the community to considering the possibility that information itself is the foundation of physics [10, 11, 12]. It is a very fascinating idea, nevertheless it is hard to dismiss the question of J. Bell [1]: "Information about what?", calling for a 'reality', which information is about. And if this is a case, than the fundamental theory should be about that reality, first of all.

There are two other alternatives, which provide a simple and rather straightforward explanation of quantum phenomena: Bohmian mechanics [13, 14, 15] and collapse models [16, 17, 18, 19, 20, 21]. The first theory tells that the world is made of (point-like) particles and gives the equation for the motion of the particles. Everything follows from that. The second theory tells that the world is made of wave-packets, and gives the equation for the motion of these wave-packets. Everything follows from that. At least at the non-relativistic level. Here we show how this is accomplished, in the case of collapse models.

2. How to modify the Schrödinger equation

According to the Ghirardi-Rimini-Weber (GRW) and Continuous Spontaneous Localization (CSL) models [17, 18], perhaps the two most popular collapse models, the Schrödinger equation is modified by adding nonlinear and stochastic terms, which induce the collapse of the wave function in space. The modification is such that these models are compatible with all known experimental data about quantum phenomena, and at the same time they make sure that macroscopic systems are never in funny macrocopic superpositions—Schrödinger's cat is always either dead or alive.

In order to better understand the role played by collapse models as an alternative to standard quantum theory, let us take a broader perspective. Let us suppose for a moment that these models were not already on the market, and let us ask the question: How can we modify the Schrödinger equation in a consistent way, without violating basic principles of physics? This issue was considered repeatedly in the literature, and was given a precise formulation by S. Adler [22]: The only possible modifications to the Schrödinger equation, which preserve the norm of the state vector and do not allow for superluminal signaling, are only those corresponding to collapse models. More recently, also S. Weinberg [23] raised the same question and has given the renewed interest in the subject.

The (more or less) most general form of the collapse equation is:

$$d\psi_t = \left[-iH dt + \sum_{k=1}^{n} (L_k - \ell_{k,t})\, dW_{k,t} - \frac{1}{2} \sum_{k=1}^{n} \left(L_k^\dagger L_k - 2\ell_{k,t} L_k + |\ell_{k,t}|^2 \right) dt \right] \psi_t, \qquad (1)$$

$$\ell_{k,t} \equiv \frac{1}{2} \langle \psi_t, (L_k^\dagger + L_k)\psi_t \rangle. \qquad (2)$$

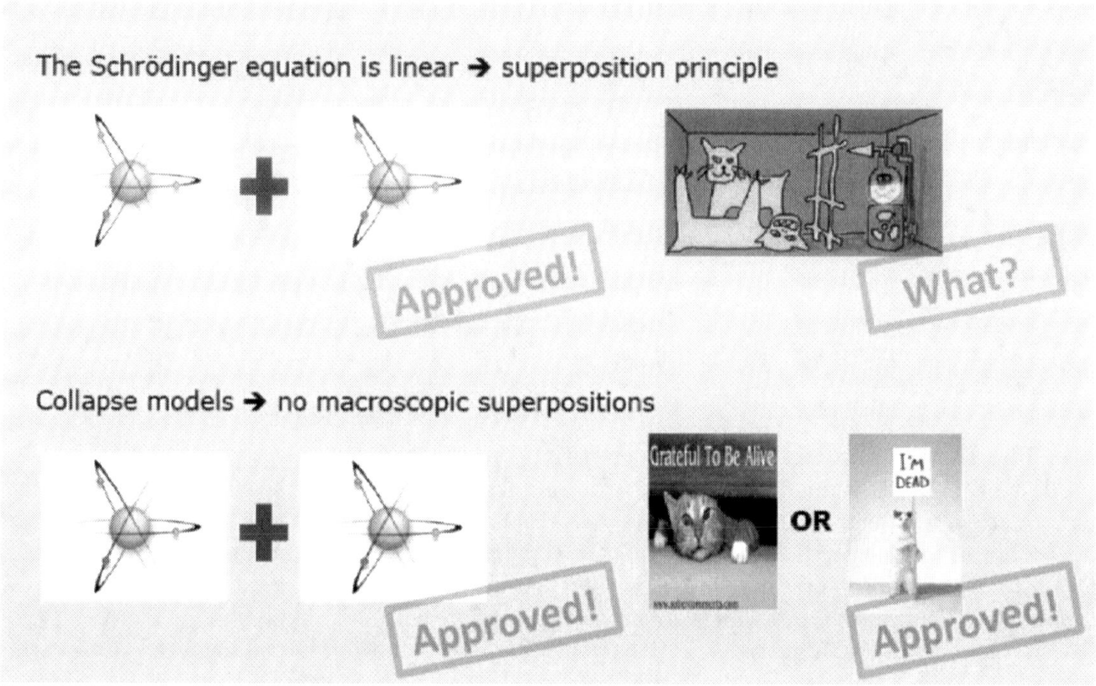

Figure 1. According to standard quantum theory, the superposition principle holds at all scales. Therefore Schrödinger's cat can be in a dead + alive state, which does not correspond to any classical state of it. According to collapse models, the superposition is (almost) right only at the microscopic scale, while it is more and more violated when moving toward larger and larger systems, to the point that macroscopic systems like Schrödinger's cat is always in a classical state, either dead or alive.

The first term on the right-hand-side represents the usual quantum evolution (unitary part). The other terms are responsible for the collapse of the wave function: the wave function is driven towards one of the common eigenstates of the operators L_k (assuming that they exist, as it is the case for all collapse models considered in the literature), with a probability corresponding to the Born rule [18, 27]. Note that these terms are both nonlinear (the quantum average $\langle\psi|...|\psi\rangle$ enters the equation) and stochastic ($W_{k,t}$ are standard independent Wiener processes), as it should be in order to have a collapse process.

3. The CSL model

In the above equation, the only relevant degree of freedom is given by the form of the collapse operators L_k. As stated here above, the nonlinear terms tend to collapse the wave function towards one of the common eigenstates of L_k, therefore choosing these operators amounts to choosing the preferred basis for the collapse.

If we wish to collapse the wave function in the energy basis, we choose $L_k = H$. If we wish to collapse the wave function in momentum, than we pick the momentum operator. But since the main duty of collapse models is to make sure that macroscopic objects are always localized in space, the most direct—and in many ways most natural—choice is to pick the position operator, or some function of it. This is what happens in the GRW model, the CSL model, and all other most relevant collapse models.

Here we report the equation for the CSL model:

$$d|\psi_t\rangle = \left[-\frac{i}{\hbar} H dt + \sqrt{\lambda} \int d^3x (N(\mathbf{x}) - \langle N(\mathbf{x}) \rangle_t) dW_t(\mathbf{x}) - \frac{\lambda}{2} \int d^3x (N(\mathbf{x}) - \langle N(\mathbf{x}) \rangle_t)^2 dt \right] |\psi_t\rangle, \tag{3}$$

where H is the standard quantum Hamiltonian, in the QFT language, $N(\mathbf{x}) = a^\dagger(\mathbf{x})a(\mathbf{x})$ is the particle density operator, λ the coupling constant which sets the strength of the collapse mechanism, and $W_t(\mathbf{x})$ a family of independent standard Wiener processes with zero average and correlation:

$$\mathbb{E}[W_t(\mathbf{x})W_s(\mathbf{y})] = \delta(t-s)e^{-(\alpha/4)(\mathbf{x}-\mathbf{y})^2}. \tag{4}$$

The choice of the particle *density* as the collapse operators corresponds precisely to the desire of having wave functions localized in *space*, i.e. it is a good alternative to the position operator in the formalism of QFT.

4. The collapse mechanism—Choice of the parameters

The CSL mode is defined in terms of two parameters: the collapse strength λ and the spatial correlation function of the noise $r_C = 1/\sqrt{\alpha}$. The first parameter sets the strength of the collapse mechanism for one particle, and through the *amplification mechanism* which we will now explain, also for a macroscopic system. The second parameter tells which kind of superpositions are effectively suppressed, and which not. The situation is pictorially explained in Fig. 2.

Let us consider a single particle (a nucleon) in a superposition of two localized states, whose relative distance in space is d. If d is smaller than r_C, then the noise—so to say—does not see such a superposition as a 'macroscopic' superposition (the two states are too close to each other), and does not modify it in a significant way. On the other hand, if d is much larger than r_C, then the noises sees it as a macroscopic superposition and tends to localized it to one of the two states. The rate is given by λ. Therefore r_C sets the threshold between microscopic and macroscopic superpositions.

Let us now consider a composite system, an object, again in a superposition of two different states, distant relative to each other more than r_C. The collapse mechanism becomes difficult to unfold analytically, due to the complexity of the dynamics for the system. Nevertheless one can work out an easy approximative rule to estimate the effect [28]. The center of mass of the system collapses with a rate

$$\Gamma = \lambda n^2 N, \tag{5}$$

where λ is the collapse rate of a single constituent (a nucleon), n is the number of constituents contained in a volume of radius r_C (therefore it is a measure of the density of the system) and N counts how many such volumes can be accommodated in the system (therefore being a measure of the total volume of the system).

The important message is that the collapse rate of a composite system scales with its size: the bigger the system, the faster the collapse. This is precisely the reason why collapse models can accommodate, within a single dynamical equation, both the quantum properties of microscopic systems (few constituents) and the classical properties of macroscopic objects (many constituents). In doing this, collapse models explain the quantum-to-classical transition, how the probabilistic and wavy nature of atoms and molecules gives rise to the world of classical physics as we experience it, when they glue together to form larger and larger objects.

As a marginal note, one can see that the collapse rate scales partly linearly and partly quadratically with the number of constituents. The quadratic increase is due to the identity of particles: if the noise tries to collapse a particle, it has to collapse all other particles identical to it, in order to preserve the symmetry of the wave function. This quadratic increase is not present in the original GRW model, which deals only with distinguishable particles.

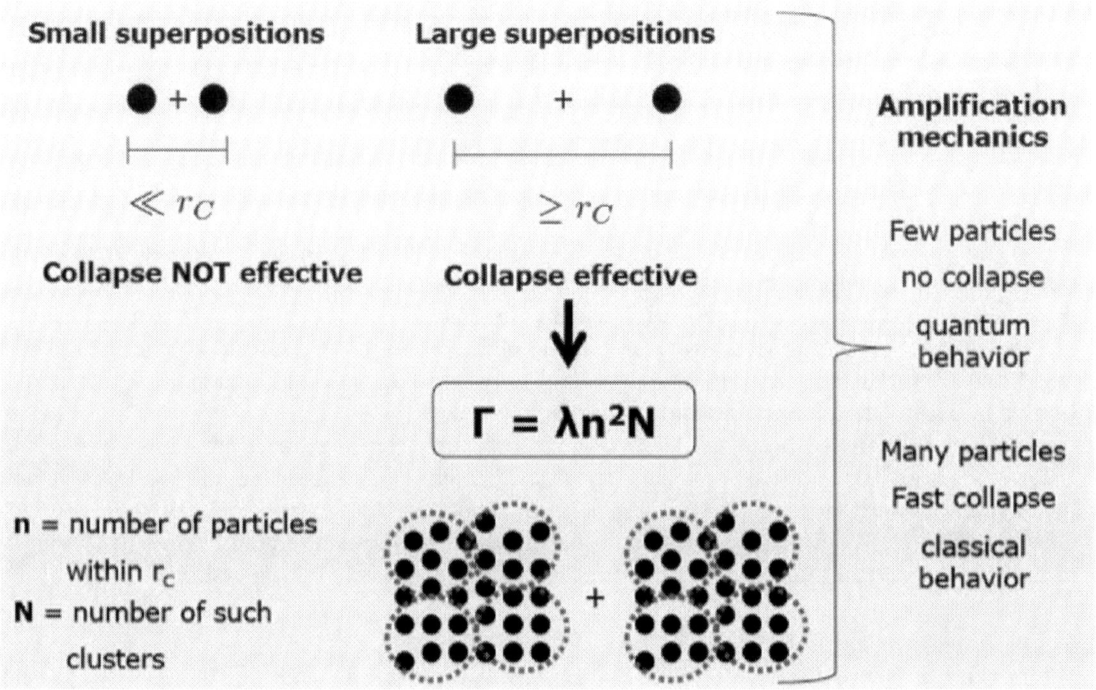

Figure 2. The *amplification mechanism* is the crucial property of collapse models: the collapse rate for the center of mass of a composite object is a function of the size and number of constituents of the system. The larger the system, the stronger the collapse rate. This is how collapse models can accommodate, within one single dynamical equation, both the quantum properties of microscopic systems and the classical properties of macroscopic objects.

Given this, the relevant question is: which values can one reasonably assign to λ and r_C? The question is not easy to answer from the theoretical point of view, as there are only very rough guidelines one can follow. As we explained here above, r_C sets the division between micro and macro distances. A reasonable value to take, first suggested by GRW [17], is:

$$r_C \simeq 10^{-7}\text{m}, \tag{6}$$

which is much bigger than typical inter-atomic distances in matter ($\sim 10^{-10}$m), but much smaller than human-size distances (~ 0.1mm). Of course, this value could fluctuate a few orders of magnitude, but this is the best one can say at the moment.

There is much less of a consensus regarding the value λ should have. In particular, two very different values were proposed. The first one was suggested by GRW:

$$\lambda_{\text{GRW}} = 10^{-16}\text{s}^{-1}, \tag{7}$$

which sets the quantum-to-classical (i.e. no-collapse vs. collapse) transition at the border between the mesoscopic and macroscopic world. This means that particles, atoms, small and

Journal of Physics: Conference Series **504** (2014) 012023 doi:10.1088/1742-6596/504/1/012023

large molecules and even those objects which are invisible to the naked eye, but otherwise are rather complex, all behave in a fully quantum mechanical way. Only when we reach the level of objects of our daily-life experience (dust grains, tables, chairs) the collapse becomes effective and protects wave functions from being in macroscopic superpositions. Therefore, proposed by GRW is more or less the weakest, most conservative value for λ one can take, which guarantees that the macroscopic world behaves in a classical way.

In more recent years, Adler suggested to take a much stronger value [28]:

$$\lambda_{\text{Adler}} = 10^{-8}\text{s}^{-1}. \tag{8}$$

The motivation is that the collapse process should be effective in all those situations which we would refer to as a measurement process, like the process of latent image formation in photography. A simple estimate of the number of atoms involved in this process shows that, in order for the collapse to be effective within the reaction time which leads to the formation of the latent image, one needs to take λ some eight orders of magnitude stronger than what suggested by GRW. In this way, the threshold between quantum and classical is moved to the border between the microscopic and mesoscopic world.

This huge difference between the two proposals might be puzzling, but merely reflects our ignorance about the validity of the superposition principle in a vast domain. We know it holds for elementary particles, up to large molecules. We know that macroscopic systems are never observed in macroscopic superpositions. What happens in between, no one knows. It is a matter of intense theoretical and most of all experimental debate, as we will now discuss.

5. Remarks on experimental tests

While there is a broad range of possible experimental tests for collapse models, we are here focusing on only some of those: matter-wave interferometry experiments where a macroscopic quantum superposition state has to be generated, recently proposed high-resolution spectroscopy tests in the frequency domain and tests based on a tiny violation of energy conservation: the Pearle disc proposal. We shortly summarize the state of the art of realised experiments and how close they are to actually test some of the proposed collapse models. We also emphasise the timeline for the experimental realisation of those tests, which have not yet been realised. A brief summary of the most prominent experimental tests can be found in ref. [35].

5.1. Quantitative measures to classify experiments

Clearly a set of parameters has to be specified to decide if a specific experiment is able to test the superposition principle in the macroscopic domain and therefore collapse models. Such parameters are the mass of the particle, the spatial size of the single-particle superposition state and the time duration of its existence. A recently introduced measure, called macroscopicity μ which includes all three parameters is essential to objectively compare experiments [36].

For the experiment it means the superposition state has to be associated to a large mass, large spatial separation between the positions where the particle is delocalised and a long (life)time for the existence of these spatial superposition states. It becomes clear that not every matter-wave experiment would be sensitive to test macroscopicity as defined above. For instance atomic matter-waves, while having an impressive momentum spread in modern experiments [40], the mass is too light for a test. Also the formation of loosely bound ensembles of many identical atoms (Bose Einstein condensate) cannot be of help. Contrariwise, the very massive mechanical cantilevers, which are used for quantum opto-mechanic experiments [38] and have been shown that they can be prepared in the vibrational ground state, possess typically a too small spatial separation to be really macroscopic [see table in reference [36]]. But it would not be wise to have a last word about this today. Experimental development is vital and due to ever ongoing inventions, which could put back such systems on the map. For instance a way forward for mechanical cantilevers would be to transfer the quantum superposition state on two identical cantilevers as proposed in ref. [41].

5.2. Experiments using matter waves
The interest in testing the limits of validity of quantum mechanics has increased in the last decades. The test of the break of the linearity of the Schrödinger equation, the loss of unitarity are all different names for the same thing, namely to test if quantum mechanics continues to be the correct theory even for macroscopic systems and if the quantum to classical transition can be explained by a universal mass-scaling mechanism. Clearly, while many competing theories and theoretical concepts try to explain this transition two of the main contenders are decoherence theory and the here discussed collapse models. There is an increasing number of experimental groups around the world, rushing towards creating quantum superpositions, which are as macroscopic as possible. In a nutshell, a very massive single particle superposition, with a large spatial and temporal extension has to be generated to test the quantum superposition principle. Here we want to look into some conceptual ideas to experimentally test quantum linearity.

5.3. Test as of today
So far the answer is affirmative, but the systems employed are still rather small. The macroscopicity record is held by the Arndt group in Vienna by molecule interferometry [39] of organic molecules of mass above 10,000 amu (atomic mass units) [42, 43]. The experiment is a so called Talbot-Lau interferometer, where the diffraction grating is a structure made of a retroreflected laser beam, based on the Kapitza-Dirac effect [44, 45]. The natural extension of those particle beam experiments is proposed here for heavy particle and cluster beams. The OTIMA interferometer has been already realised [46, 47]. Now intense heavy-particle gas-phase beams have to be realised [48]. Macroscopicity and how these experiments can test collapse models are worked out in ref. [37].

5.4. New proposals for tests using matter-wave interferometry
These experiments have motivated others to come up with new proposals for macroscopic matter-wave experiments. Others propose the use of levitated particles (optical or magnetic) to generate a spatial quantum superposition state of an even larger (more massive) particle [49, 50, 51]. Ideas range from performing fast tests while the particle stays always in the trap [52, 53] to drop particles after initial state preparation [50, 54]. Recent progress in the manipulations and preparation of the centre of mass motion allows for the cooling to about 10 mK, which is sufficient for one of the proposals to generate a superposition state [54]. However a number of experimental challenges (mainly related to build a nanoparticle source in vacuum) remain before the experiments can be realised.

In the first case the stability of the trap has to be enormous and limits the actual lifetime of the superposition state to typically less than 10 μs. In some schemes internal (spin) states are prepared by a sequence of magnetic pulses in a resonance spectroscopic Ramsey technique and a closed interferometer has to be achieved within the time constraints given by dephasing and decoherence of the spin states. This internal state interferometry is coupled to the centre of mass motion of the magnetic particle and therefore represents a matter-wave interferometer.

The free fall / drop experiments are limited by the acceleration of the particle in Earth's gravitational field. For a particle of increasing mass the de Broglie wavelength of the accelerated particle by g gets so small that matter-wave experiments are almost impossible. The limit on mass is at the one million amu level. Such experiments will be able to test the CSL value with Adler's value for λ [37, 21], but other models like the original GRW can not be approached and need a larger mass for the superposition. Therefore, MAQRO has been proposed to explore possibilities for quantum superposition experiments with nanoparticles in space [55]. Related conceptual and practical aspects have been worked out [56].

A further alternative is the transfer of superposition from matter-wave beam to a pair of mechanical cantilevers. A new concept for generating an almost macroscopic quantum superposition state of a large mass has been proposed by us [22]. The basic idea is that an atomic matter-wave is imprinted on a pair of identical mechanical cantilevers. State of the art quantum optics calculations show that the superposition state can survive this transfer, which if done in a real experiment would boost test in a new regime (10^{13} amu). The status of this project requires now a careful design of the experiment and major funding to set it up. This experiment, if realized, can cover all proposed collapse models [37, 21].

6. Experiments in the frequency domain

The general strategy for the aforementioned matter-wave interferometry tests is to create superpositions in space, and test them with interferometers. Basically all groups follow this strategy. Recently we suggested an alternative route: to test quantum mechanics not in the time domain (see whether macroscopic superpositions exist and persist in time), but in the frequency domain [57, 58]. The idea is that deviations from quantum theory modify the spectral properties of atoms and molecules, and therefore can be checked with sophisticated spectroscopic techniques. The analysis shows that high-resolution experiments in the not too far future should be possible. For more details, see [58]. Currently we are investigating this general effect applied specifically on opto-mechanical systems. The goal is to understand if parameter settings in today's opto-mechanics experiments can test collapse models.

7. Other experiment

A-normal (collapse induced) Brownian motion: Another interesting idea, which is worth revisiting is an effect first described by Philipp Pearle [59]. A micrometer-sized disc is (frictionless) mounted inside an ultra-high vacuum chamber at very low temperatures to avoid heating and collisions of the disc. Then, collapse models predict, that due to stochastic noise, the disc should start to rotate and a quantitative effect is predicted and compared to classical Brownian motion caused by particle collision and the quantum version from e.g. laser shot noise. The parameter regime is practically accessible but various details of the concrete experimental setup have to be explored in a feasibility study.

8. Outlook

Collapse models are a viable alternative to quantum mechanics, avoiding the interpretational problems of the standard formulation of the theory. They are a well founded theory, since they consist of just a dynamical equation for the wave function, and everything else follows from that. This is what theories should be like. In particular, it can be proven [29] that the whole quantum formalism of measurements (observables as self-adjoint operators, the role of eigenvalues and eigenvectors, the Born rule ...) can be deduced, not postulated, from the dynamical equation of collapse models, when applied to those specific physical situations, which we call measurements.

On the other hand, it is somehow clear that collapse models are phenomenological, and call for an underlying theory out of which the quantum world—with the collapse—emerges, perhaps in a similar why to how thermodynamics emerges from classical statistical mechanics. The difficult question is to say something about this possible underlying theory. The only framework we are aware of, is the one provided by Adler with trace dynamics [22]. The work is still in progress, but already represents a promising way of understanding the quantum world as emerging from a deeper level of reality.

Another interesting question is whether the collapse is related to gravity. This idea has been on the market for a long time, and perhaps the strongest input has been given by Penrose [30, 31]. There are good reasons to explore this possibility: gravity is a nonlinear theory, and nonlinearity is precisely what one needs to kill superpositions. Moreover, gravity becomes stronger, the larger the systems, just like the collapse process. Models have been developed [32, 34, 34], but a breakthrough is still needed, in order to make it convincing that the collapse can be really linked to gravitational effects.

Also experimentally we are experiencing an exciting boost in possibilities and ideas on how to test collapse models. We predict that in the next three years some of the CSL models, such as the one proposed by Adler, will have been experimentally tested by matter-wave experiments on Earth while other experiments will be pushed forward to test even more macroscopic quantum states. If a state collapse is observed, specific scaling of loss of quantum features of the superposition state (such as visibility of the quantum interference contrast) with experimental controllable parameters (such as internal and external temperature of the particle in superposition or the number of other particles in its environment) can be used to distinguish the collapse mechanism.

Acknowledgments

AB thanks Gerhard Grössing for organizing the conference EmQM13 on Emergent Quantum Mechanic, and the Fetzer Franklin fund for its financial support. He acknowledges financial support from the EU project NANOQUESTFIT and INFN. He also thanks the COST Action MP1006 "Fundamental Problems in Quantum Physics". HU thanks the UK funding agency EPSRC (EP/J014664/1). Both authors thank the Foundational Questions Institute (FQXi) and the John F Templeton Foundation for financial support.

References

[1] J.S. Bell, *Speakable and Unspeakable in Quantum Mechanics*, Cambridge University Press (1988).
[2] E. Joos, H. D. Zeh, C. Kiefer, D. Giulini, J. Kupsch and I.O. Stamatescu, *Decoherence and the Appearance of a Classical World in Quantum Theory*, Springer-Verlag (2003).
[3] H.-P. Breuer and F. Petruccione, *The Theory of Open Quantum Systems*, Oxford University Press (2002)
[4] M.A. Schlosshauer *Decoherence and the Quantum-to-Classical Transition*, Springer (2008).
[5] A. Bassi and G. C. Ghirardi, Phys. Lett. **A 275**, 373 (2000).
[6] S.L. Adler, Stud. Hist. Philos. Mod. Phys. **34**, 135 (2003).
[7] H. Everett, *Theory of the Universal Wave function*, Thesis, Princeton University, (1956, 1973).
[8] H. Everett, Rev. Mod. Phys. **29**, 462 (1957).
[9] B.S. DeWitt, Phys. Today, **23**, 30 (1970).
[10] G. M. D'Ariano, in: "Philosophy of Quantum Information and Entanglement", edited by A. Bokulich and G. Jaeger, Cambridge University Press, Cambridge (2010).
[11] G. Chiribella, G. M. D'Ariano, P. Perinotti, Phys. Rev. A **84** 012311 (2011).
[12] Časlav Brukner, Physics **4**, 55 (2011).
[13] P.H. Holland, *The Quantum Theory of Motion: An Account of the de Broglie-Bohm Causal Interpretation of Quantum Mechanics*, Cambridge University Press (1995).
[14] D. Dürr and S. Teufel, *Bohmian Mechanics: The Physics and Mathematics of Quantum Theory*, Springer (2009)
[15] D. Dürr, S. Goldstein and N. Zanghì, *Quantum Physics Without Quantum Philosophy*, Springer (2013).
[16] P. Pearle, Phys. Rev. D **13**, 857 (1976).
[17] G.C. Ghirardi, A. Rimini and T. Weber, Phys. Rev. D **34**, 470 (1986).
[18] G.C. Ghirardi, P. Pearle and A. Rimini, Phys. Rev. A **42**, 78 (1990).
[19] A. Bassi and G.C. Ghirardi, Phys. Rept. **379**, 257 (2003).
[20] S.L. Adler and A. Bassi, Science **325**, 275 (2009).
[21] A. Bassi, K. Lochan, S. Satin, T.P. Singh and H. Ulbricht, Rev. Mod. Phys. **85**, 471 (2013).
[22] S.L. Adler, *Quantum Theory as an Emergent Phenomenon: The Statistical Mechanics of Matrix Models as the Precursor of Quantum Field Theory*, Cambridge University Press (2009).
[23] S. Weinberg, Phys. Rev. A **85**, 062116 (2012).

[24] N. Gisin, Hel. Phys. Acta **62**, 363 (1989).
[25] G. Lindblad, Commun. Path. Phys. **48**, 119 (1976).
[26] V. Gorini, A. Kossakowski and E.C.G. Sudarshan, J. Math. Phys. **17**, 821 (1976).
[27] S. L. Adler, D. C. Brody, T. A. Brun and L. P. Hughston, J. Phys. A: Math. Gen. **34**, 8795 (2001).
[28] S.L. Adler, Journ. Phys. A **40**, 2935 (2007).
[29] A. Bassi, G.C. Ghirardi and D.G.M. Salvetti, Journ. Phys. A **40**, 13755 (2007).
[30] R. Penrose, Gen. Rel. Grav. **28**, 581 (1996).
[31] L. Diosi, Phys. Lett. A **120**, 377 (1987).
[32] L. Diosi, Braz. J. Phys. **35**, 260 (2005).
[33] L. Diosi, Phys. Lett. A **105**, 199 (1984).
[34] D. Giulini and A. Grossardt, Class. Quant. Grav. **28**, 195026 (2011).
[35] A. Bassi and S. Adler, Science 325, 275 (2009).
[36] S. Nimmrichter and K Hornberger, Physical Review Letters 110, 160403 (2013).
[37] S. Nimmrichter, K. Hornberger, P. Haslinger and M. Arndt, Phys. Rev. A 83, 043621 (2011).
[38] M. Aspelmeyer, T. J. Kippenberg and F. Marquardt, arXive:1303.0733 (2013).
[39] K. Hornberger, S. Gerlich, P. Haslinger, S. Nimmrichter and M. Arndt, Rev. Mod. Phys. 84, 157 (2012).
[40] S.-W. Chiow, T. Kovachy, H.-C. Chien and M. A. Kasevich Phys. Rev. Lett. 107, 130403 (2011).
[41] A. Xuereb, H. Ulbricht and M. Paternostro, Scien. Rep. 3, 3378 (2013).
[42] S. Gerlich, S. Eibenberger, M. Tomandl, S. Nimmrichter, K. Hornberger, P. J. Fagan, J. Txen, M. Mayor and M. Arndt, Nat. Comm. 2, 263(2011).
[43] S. Eibenberger, S. Gerlich, M. Arndt, M. Mayor and J. Txen, Phys.Chem. Chem. Phys. 15, 14696 (2013).
[44] S. Gerlich, L. Hackermller, K. Hornberger, A. Stibor, H. Ulbricht, F. Goldfarb, T. Savas, M. Mri, M. Mayor and M. Arndt, Nat. Phys. 3, 711 (2007).
[45] K. Hornberger, S. Gerlich, H. Ulbricht, L. Hackermller, S. Nimmrichter, I. V. Goldt, O. Boltalina and M. Arndt, New J. Phys. 11 043032 (2009).
[46] S. Nimmrichter, P. Haslinger, K. Hornberger and M. Arndt, New J. Phys. 13, 075002 (2011).
[47] P. Haslinger, N. Drre, P. Geyer, J. Rodewald, S. Nimmrichter and M. Arndt, Nat. Phys. 9, 144–148 (2013).
[48] T. Juffmann, H. Ulbricht and M. Arndt, Rep. Prog. Phys. 76 086402 (2013).
[49] D. Home and S. Bose, Phys. Lett. A 217, 209 (1996).
[50] O. Romero-Isart, A. C. Pflanzer, F. Blaser, R. Kaltenbaek, N. Kiesel, M. Aspelmeyer and J. I. Cirac, Phys. Rev. Lett. 107, 020405 (2011).
[51] O. Romero-Isart, L. Clemente, C. Navau, A. Sanchez and J. I. Cirac, Phys. Rev. Lett. 109, 147205 (2012).
[52] M. Scala, M. S. Kim, G. W. Morley, P. F. Barker and S. Bose, Phys. Rev. Lett. 111, 180403 (2013).
[53] Z.-Q. Yin, T. Li, X. Zhang, L. M. Duan, arXiv:1305.1701 (2013).
[54] J. Bateman, S. Nimmrichter, K. Hornberger and H. Ulbricht, arXive:1312.0500 (2013).
[55] R. Kaltenbaek, G. Hechenblaikner, N. Kiesel, O. Romero-Isart, K. C. Schwab, U. Johann, M. Aspelmeyer, Exp. Astr. 34 pp 123 (2012).
[56] G. Hechenblaikner, F. Hufgard, J. Burkhardt, N. Kiesel, U. Johann, M. Aspelmeyer and R. Kaltenbaek, arXiv:1309.3234 (2013).
[57] M. Bahrami, S. Donadi, L. Ferialdi, A. Bassi, C. Curceanu, A. Di Domenico, B. C. Hiesmayr, Scie. Rep. 3, 1952 (2013).
[58] M. Bahrami, A. Bassi and H. Ulbricht, arXive:1309.5889 (2013).
[59] B. Collett and P. Pearle, Found. Phys. 33, 1495 (2003).

Heisenberg's uncertainty relation: Violation and reformulation

Masanao Ozawa
Graduate School of Information Science, Nagoya University, Chikusa-ku, Nagoya, 464-8601, Japan

E-mail: ozawa@is.nagoya-u.ac.jp

Abstract. The uncertainty relation formulated by Heisenberg in 1927 describes a trade-off between the error of a measurement of one observable and the disturbance caused on another complementary observable so that their product should be no less than a limit set by Planck's constant. In 1980, Braginsky, Vorontsov, and Thorne claimed that this relation leads to a sensitivity limit for gravitational wave detectors. However, in 1988 a model of position measurement was constructed that breaks both this limit and Heisenberg's relation. Here, we discuss the problems as to how we reformulate Heisenberg's relation to be universally valid and how we experimentally quantify the error and the disturbance to refute the old relation and to confirm the new relation.

1. Heisenberg's error-disturbance relation (EDR)

The discovery of quantum mechanics introduced non-commutativity in algebraic calculus of observables; the *canonical commutation relation* (CCR)

$$[Q, P] = i\hbar \tag{1}$$

is required to hold between a coordinate Q of a particle and its momentum P, where the commutator $[Q, P]$ is defined by $[Q, P] = QP - PQ$. In 1927, Heisenberg proposed an operational meaning of the non-commutativity: "the more precisely the position is determined, the less precisely the momentum is known, and conversely" [1, p. 64].

By the famous γ ray microscope thought experiment he derived the relation

$$\varepsilon(Q)\eta(P) \geq \frac{\hbar}{2}, \tag{2}$$

where $\varepsilon(Q)$ is the "mean error" of a position measurement and $\eta(P)$ is the thereby caused "discontinuous change" in the momentum P:

> Let $\varepsilon(Q)$ be the precision with which the value Q is known ($\varepsilon(Q)$ is, say, the mean error of Q), therefore here the wavelength of the light. Let $\eta(P)$ be the precision with which the value P is determinable; that is, here, the discontinuous change of P in the Compton effect [1, p. 64].

Here, "mean error" is naturally understood to be "root-mean-square (rms) error" as introduced by Gauss [2], and "discontinuous change" is often called "mean disturbance." Heisenberg claimed that Eq. (2) is a "straightforward mathematical consequence" of Eq. (1) [1, p. 65] and gave its mathematical justification [1, p. 69].

2. Heisenberg's derivation

In his mathematical justification of Eq. (2), Heisenberg firstly derived the relation

$$\sigma(Q)\sigma(P) = \frac{\hbar}{2} \tag{3}$$

for the standard deviations $\sigma(Q)$ and $\sigma(P)$ of the position Q and the momentum P in the state described by a Gaussian wave function [1, p. 69], which Kennard [3] subsequently generalized as the relation

$$\sigma(Q)\sigma(P) \geq \frac{\hbar}{2} \tag{4}$$

for arbitrary wave functions. Here, the standard deviation is defined for any observable A by $\sigma(A)^2 = \langle A^2 \rangle - \langle A \rangle^2$, where $\langle \cdots \rangle$ stands for the mean value in a given state. Note that in Ref. [1, p. 69] Heisenberg actually derived the relation

$$\tilde{\sigma}(Q)\tilde{\sigma}(P) = \hbar \tag{5}$$

for $\tilde{\sigma}(Q) = \sqrt{2}\sigma(Q)$ and $\tilde{\sigma}(P) = \sqrt{2}\sigma(P)$ in Gaussian wave functions, and Kennard [3] actually derived the relation

$$\tilde{\sigma}(Q)\tilde{\sigma}(P) \geq \hbar \tag{6}$$

that generalizes Heisenberg's relation to arbitrary wave functions.

Heisenberg secondly applied Eq. (4) to the state just after the measurement assuming:

(H1) *Any measurement with rms error $\varepsilon(A)$ of an observable A leaves the state satisfying $\sigma(A) \leq \varepsilon(A)$.*

(H2) *If an observable A can be measured with the rms error $\varepsilon(A)$ and the rms disturbance $\eta(B)$ of another observable B, then A and B can be jointly measured with the rms errors $\varepsilon(A)$ and $\varepsilon(B) = \eta(B)$, respectively.*

Then, it can be easily seen that Eq. (2) can be derived from Eq. (4) under assumptions (H1) and (H2). In fact, if Q can be measured with $\varepsilon(Q) = \alpha$ and $\eta(P) = \beta$, then by (H2) Q and P can be measured jointly with $\varepsilon(Q) = \alpha$ and $\varepsilon(P) = \beta$, so that by (H1) the state after the joint measurement satisfies $\sigma(Q) \leq \alpha$ and $\sigma(P) \leq \beta$, and hence Eq. (4) concludes Eq. (2).

3. Heisenberg's unsupported assumption

Assumption (H2) is considered to hold in general; see Ref. [4] for a detailed discussion. However, assumption (H1) is not, whereas Heisenberg's contemporaries, including von Neumann, supported assumption (H1):

> We are then to show that if Q, P are two canonically conjugate quantities, and a system is in a state in which the value of Q can be given with the accuracy $\varepsilon[= \sigma(Q)]$ (i.e., by a Q measurement with an error range $\varepsilon[= \varepsilon(Q)]$), then P can be known with no greater accuracy than $\eta[= \sigma(P)] = \hbar/(2\varepsilon)$. Or: a measurement of Q with the accuracy $\varepsilon[= \varepsilon(Q)]$ must bring about an indeterminancy $\eta[= \eta(P)] = \hbar/(2\varepsilon)$ in the value of P [5, pp. 238–239]. (Terms in [...] are supplemented by the present author.)

In those days the repeatability hypothesis was considered a natural requirement for all the precise measurements of an observable A and (H1) is considered as a natural generalization to arbitrary approximate measurements of A. Here, the *repeatability hypothesis* is formulated as follows.

(RH) *If an observable A is measured twice in succession in a system, then we get the same value each time* [5, pp. 335].

Under (RH), any precise measurement of A with $\varepsilon(A) = 0$ changes the state to be an eigenstate of the measured observable A, which satisfies $\sigma(A) = 0$. However, in the light of modern theory of quantum measurement, (RH) has been abandoned as proposed by Davies and Lewis [6]:

> One of the crucial notions is that of repeatability which we show is implicitly assumed in most of the axiomatic treatments of quantum mechanics, but whose abandonment leads to a much more flexible approach to measurement theory [6, p.239].

In fact, in Ref. [7] we have mathematically characterized all the physically possible quantum measurements, shown that (RH) is no longer universally valid, and even more that no precise measurements of continuous observables satisfy (RH). Thus, (H1) does not hold even in the case where $\varepsilon(A) = 0$.

Therefore, in the light of the modern theory of quantum measurement, assumption (H1) cannot be accepted, so that Eq. (2) cannot be considered as an immediate consequence of Eq. (4), although their meanings have often been confused even in standard text books [5, 8, 9, 10]. As above, the original justification of Eq. (2) was limited, but its universal validity has not been refuted in theory until 1980's.

4. Von Neumann's model of position measurement

Until 1980's the only solvable model of position measurement had been given by von Neumann [5]. In what follows, we discuss the von Neumann model and show that this long-standing standard model satisfies the Heisenberg error-disturbance relation (EDR) (2). Thus, the model analysis of position measurement did not lead to refuting the Heisenberg EDR but rather enforced the belief that good position measurements satisfy the Heisenberg EDR (2).

Consider a one-dimensional mass, called an *object*, with position Q and momentum P, described by a Hilbert space $\mathcal{H}$. The measurement of Q is carried out by a coupling between the object $\mathbf{S}$ and a probe $\mathbf{P}$ from time $t = 0$ to $t = \Delta t$. The probe $\mathbf{P}$ is another one-dimensional mass with position $\overline{Q}$ and momentum $\overline{P}$, described by a Hilbert space $\mathcal{K}$. The outcome of the measurement is obtained by measuring the probe position $\overline{Q}$, called the *meter observable*, at time $t = \Delta t$. The total Hamiltonian for the object and the probe is taken to be

$$H_{\mathbf{S}+\mathbf{P}} = H_{\mathbf{S}} + H_{\mathbf{P}} + KH, \tag{7}$$

where $H_{\mathbf{S}}$ and $H_{\mathbf{P}}$ are the free Hamiltonians of $\mathbf{S}$ and $\mathbf{P}$, respectively, H represents the measuring interaction, and K is the coupling constant. We assume that the coupling is so strong ($K \gg 1$) that $H_{\mathbf{S}}$ and $H_{\mathbf{P}}$ can be neglected. We choose Δt as $K\Delta t = 1$. In the von Neumann model the measuring interaction is given by

$$H = Q \otimes \overline{P}. \tag{8}$$

Then, the unitary operator of the time evolution of $\mathbf{S} + \mathbf{P}$ from $t = 0$ to $t = \tau \le \Delta t$ is given by

$$U(\tau) = \exp\left(\frac{-iK\tau}{\hbar} Q \otimes \overline{P}\right). \tag{9}$$

Suppose that the object $\mathbf{S}$ and the probe $\mathbf{P}$ are in the state $|\psi\rangle$ and $|\xi\rangle$, respectively, just before the measurement; we assume that the wave functions $\psi(x) = \langle x|\psi\rangle$ and $\xi(y) = \langle y|\xi\rangle$ are Schwartz rapidly decreasing functions [11], where $|x\rangle$ and $|y\rangle$ are the position bases of $\mathbf{S}$ and $\mathbf{P}$, respectively.

Then, the state of the composite system $\mathbf{S} + \mathbf{P}$ just after the measurement is $U(\Delta t)|\psi, \xi\rangle$. By solving the Schrödinger equation, we have

$$\langle x, y|U(\Delta t)|\psi, \xi\rangle = \langle x|\psi\rangle\langle y - x|\xi\rangle. \tag{10}$$

If the observer observes the meter observable $\overline{Q}(\Delta t)$ just after the measuring interaction, the probability distribution of the outcome is given by

$$\Pr\{a < \mathbf{x} \le b\} = \int_a^b dy \int_{-\infty}^{+\infty} |\langle x|\psi\rangle|^2 \, |\langle y - x|\xi\rangle|^2 \, dx. \tag{11}$$

This shows that if the probe initial wave function $\xi(y)$ approaches the Dirac delta function $\delta(x)$, the output probability distribution approaches the correct Born probability distribution for the observable Q at time $t = 0$.

In the Heisenberg picture, we denote

$$\begin{aligned} Q(\tau) &= U(\tau)^\dagger (Q \otimes I) U(\tau), \quad P(\tau) = U(\tau)^\dagger (P \otimes I) U(\tau), \\ \overline{Q}(\tau) &= U(\tau)^\dagger (I \otimes \overline{Q}) U(\tau), \quad \overline{P}(\tau) = U(\tau)^\dagger (I \otimes \overline{P}) U(\tau). \end{aligned}$$

Solving the Heisenberg equations of motion, we have

$$Q(\Delta t) = Q(0), \tag{12}$$
$$\overline{Q}(\Delta t) = Q(0) + \overline{Q}(0), \tag{13}$$
$$P(\Delta t) = P(0) - \overline{P}(0), \tag{14}$$
$$\overline{P}(\Delta t) = \overline{P}(0). \tag{15}$$

5. Root-mean-square error and disturbance

In order to define "root-mean-square error" of this measurement, we recall classical definitions. Suppose that the true value is given by $X = x$ and its measured value is given by $Y = y$. For each pair of values $(X, Y) = (x, y)$, the error is defined as $y - x$. To define the "mean error" with respect to the joint probability distribution $\mu^{X,Y}(dx, dy)$ of X and Y, Gauss [2] introduced the *root-mean-square error* $\varepsilon_G(X, Y)$ of Y for X as

$$\varepsilon_G(X, Y) = \left(\iint_{\mathbf{R}^2} (y - x)^2 \mu^{X,Y}(dx, dy) \right)^{1/2}, \tag{16}$$

which Gauss [2] called the "mean error" or the "mean error to be feared," and has long been accepted as a standard definition for the "mean error."

In the von Neumann model, the value of the observable $Q(0)$ is measured by the value of the meter observable $\overline{Q}(\Delta t)$. Since $Q(0)$ and $\overline{Q}(\Delta t)$ commute, as seen from Eq. (13), we have the joint probability distribution $\mu^{Q(0),\overline{Q}(\Delta t)}(dx, dy)$ of the values of $Q(0)$ and $\overline{Q}(\Delta t)$ as

$$\mu^{Q(0),\overline{Q}(\Delta t)}(dx, dy) = \langle E^{Q(0)}(dx) E^{\overline{Q}(\Delta t)}(dy) \rangle, \tag{17}$$

where E^A stands for the spectral measure of an observable A [12], and $\langle \cdots \rangle$ stands for the mean value in the state $|\psi, \xi\rangle$. Then, from Eq. (16) the *root-mean-square error* $\varepsilon(Q)$ of $\overline{Q}(\Delta t)$ for $Q(0)$ in $|\psi\rangle$ is given by

$$\begin{aligned} \varepsilon(Q) &= \varepsilon_G(Q(0), \overline{Q}(\Delta t)) \\ &= \left(\iint_{\mathbf{R}^2} (y - x)^2 \mu^{Q(0),\overline{Q}(\Delta t)}(dx, dy) \right)^{1/2} \\ &= \langle (\overline{Q}(\Delta t) - Q(0))^2 \rangle^{1/2} \\ &= \langle \overline{Q}(0)^2 \rangle^{1/2}. \end{aligned} \tag{18}$$

Since $P(0)$ and $P(\Delta t)$ commute, as seen from Eq. (14), we have the joint probability distribution $\mu^{P(0),P(\Delta t)}(dx, dy)$ of the values of $P(0)$ and $P(\Delta t)$ as

$$\mu^{P(0),P(\Delta t)}(dx, dy) = \langle E^{P(0)}(dx) E^{P(\Delta t)}(dy) \rangle. \tag{19}$$

The *root-mean-square disturbance* $\eta(P)$ of P from $t = 0$ to $t = \Delta t$ is defined as the root-mean-square error of $P(\Delta t)$ for $P(0)$ given by

$$\begin{aligned} \eta(P) &= \varepsilon_G(P(0), P(\Delta t)) \\ &= \left(\iint_{\mathbf{R}^2} (y - x)^2 \mu^{P(0),P(\Delta t)}(dx, dy) \right)^{1/2} \\ &= \langle (P(\Delta t) - P(0))^2 \rangle^{1/2} \\ &= \langle \overline{P}(0)^2 \rangle^{1/2}. \end{aligned} \tag{20}$$

Then, by the Kennard inequality (4) we have

$$\begin{aligned} \varepsilon(Q)\eta(P) &= \langle \overline{Q}(0)^2 \rangle^{1/2} \langle \overline{P}(0)^2 \rangle^{1/2} \\ &\geq \sigma(\overline{Q}(0))\sigma(\overline{P}(0)) \geq \frac{\hbar}{2}. \end{aligned} \tag{21}$$

Thus, the von Neumann model satisfies the Heisenberg EDR (2).

Since only the von Neumann model is available as a mathematically solvable mode of position measurement until 1980's, model analysis of position measurement did not lead to refuting the Heisenberg EDR but rather enforced the belief that good position measurements satisfy the Heisenberg EDR (2); see for example Refs. [13, 14].

This belief was also enforced by Arthurs and Kelly [15] suggesting that all the joint unbiased measurement of position and momentum satisfy the Heisenberg error tradeoff relation. A *joint position-momentum measurement* can be modeled by a triple $(M_Q, M_P, |\xi\rangle)$ consisting of commuting meter observables M_Q and M_P in the composite system $\mathbf{S} + \mathbf{P}$ described by $\mathcal{H} \otimes \mathcal{K}$ with the initial state $|\xi\rangle \in \mathcal{K}$ of the probe $\mathbf{P}$. Then, the joint measurement $(M_Q, M_P, |\xi\rangle)$ is called *unbiased* if the mean values of meter observables M_Q and M_P coincides with the mean values of Q and P, respectively, i.e., $\langle M_Q \rangle = \langle Q \rangle$ and $\langle M_P \rangle = \langle P \rangle$, in any input states of $\mathbf{S}$. Then, Arthurs and Kelly [15] showed that the relation

$$\sigma(M_Q)\sigma(M_P) \geq \hbar \tag{22}$$

holds for any unbiased joint position-momentum measurement $(M_Q, M_P, |\xi\rangle)$ in any input state of $\mathbf{S}$. It is explained that the lower bound is twice as much as the lower bound for $\sigma(Q)\sigma(P)$ because of inevitable errors included in the value of M_Q and M_P. In Ref. [16], their result was reformulated so that the Heisenberg error tradeoff relation

$$\varepsilon(Q)\varepsilon(P) \geq \frac{\hbar}{2} \tag{23}$$

holds for any unbiased joint position-momentum measurement $(M_Q, M_P, |\xi\rangle)$ in any input state of $\mathbf{S}$, where $\varepsilon(Q)$ and $\varepsilon(P)$ are the rms errors of joint measurement of position Q and momentum P defined through the joint probability distribution of Q and M_Q and that of P and M_P, which always exist in unbiased case; see for another approach Ref. [17]. Thus, if we consider unbiased joint measurements as good joint measurements, we can say that every good joint measurement satisfies the Heisenberg error tradeoff relation (23).

However, this does not imply that every good position measurement satisfies the Heisenberg EDR (2). Given a position measurement with the meter observable M in the probe, if the momentum is measured just after the position measurement, we have a joint position-momentum measurement $(M_Q, M_P, |\xi\rangle)$ with $M_Q = M(\Delta t)$ and $M_P = P(\Delta t)$. In this case, the error tradeoff relation (23) is equivalent with the EDR (2) with $\varepsilon(P) = \eta(P)$, and we can consider unbiased position measurements as good position measurements. Nevertheless, we cannot conclude that every good position measurement satisfies the Heisenberg EDR (2), since an unbiased position measurement followed by a precise momentum measurement not necessarily satisfies $\langle P(\Delta t)\rangle = \langle P(0)\rangle$ or $\langle M_P\rangle = \langle P\rangle$.

6. Measurement violating the Heisenberg EDR

In 1980, Braginsky, Vorontsov, and Thorne [18] claimed that the Heisenberg EDR (2) leads to a sensitivity limit, called the *standard quantum limit* (SQL), for gravitational wave detectors of interferometer type, which make use of free-mass position monitoring. Subsequently, Yuen [19] questioned the validity of the SQL, and then Caves [14] defended the SQL by giving a new proof of the SQL without a direct appeal to Eq. (2). Eventually, the conflict was reconciled in Refs. [20, 21] by pointing out that Caves [14] used (unfounded) assumption (H1) in his derivation of the SQL, and a solvable model of an error-free position measurement was constructed that breaks the SQL (see also Ref. [22]); later this model was shown to break the Heisenberg EDR (2) [23].

In what follows, we introduce this model by modifying the measuring interaction of the von Neumann model. In this new model, the object, the probe, and the probe observables are the same systems and the same observable as the von Neumann model. The measuring interaction is taken to be [20]

$$H = \frac{\pi}{3\sqrt{3}}(2Q \otimes \overline{P} - 2P \otimes \overline{Q} + QP \otimes I - I \otimes \overline{QP}). \tag{24}$$

The coupling constant K and the time duration Δt are chosen as before so that $K \gg 1$ and $K\Delta t = 1$. Then, the unitary operator $U(\tau)$ for $0 \le \tau \le \Delta t$ is given by

$$U(\tau) = \exp\left[\frac{-i\pi K\tau}{3\sqrt{3}\hbar}(2Q \otimes \overline{P} - 2P \otimes \overline{Q} + QP \otimes I - I \otimes \overline{QP})\right]. \tag{25}$$

Solving the Heisenberg equations of motion for $t < t + \tau < t + \Delta t$, we obtain

$$Q(\tau) = \frac{2}{\sqrt{3}}Q(0)\sin\frac{(1+K\tau)\pi}{3} + \frac{-2}{\sqrt{3}}\overline{Q}(0)\sin\frac{K\tau\pi}{3}, \tag{26}$$

$$\overline{Q}(\tau) = \frac{2}{\sqrt{3}}Q(0)\sin\frac{K\tau\pi}{3} + \frac{-2}{\sqrt{3}}\overline{Q}(0)\sin\frac{(1-K\tau)\pi}{3}, \tag{27}$$

$$P(\tau) = \frac{-2}{\sqrt{3}}P(0)\sin\frac{(1-K\tau)\pi}{3} + \frac{-2}{\sqrt{3}}\overline{P}(0)\sin\frac{K\tau\pi}{3}, \tag{28}$$

$$\overline{P}(\tau) = \frac{2}{\sqrt{3}}P(0)\sin\frac{K\tau\pi}{3} + \frac{2}{\sqrt{3}}\overline{P}(0)\sin\frac{(1+K\tau)\pi}{3}. \tag{29}$$

For $\tau = \Delta t = 1/K$, we have

$$Q(\Delta t) = Q(0) - \overline{Q}(0), \tag{30}$$

$$\overline{Q}(\Delta t) = Q(0), \tag{31}$$

$$P(\Delta t) = -\overline{P}(0), \tag{32}$$

$$\overline{P}(\Delta t) = P(0) + \overline{P}(0). \tag{33}$$

As in the von Neumann model, the value of the observable $Q(0)$ is measured by the value of the meter observable $\overline{Q}(\Delta t)$. Since $Q(0)$ and $\overline{Q}(\Delta t)$ commute, as seen from Eq. (31), we have the joint probability distribution $\mu^{Q(0),\overline{Q}(\Delta t)}(dx, dy)$ of the values of $Q(0)$ and $\overline{Q}(\Delta t)$ by Eq. (17). Then, from Eq. (16) the rms error $\varepsilon(Q)$ of $\overline{Q}(\Delta t)$ for $Q(0)$ in $|\psi\rangle$ is given by

$$\begin{aligned}\varepsilon(Q) &= \left(\iint_{\mathbf{R}^2}(y-x)^2\mu^{Q(0),\overline{Q}(\Delta t)}(dx, dy)\right)^{1/2}\\ &= \langle(\overline{Q}(\Delta t) - Q(0))^2\rangle^{1/2}\\ &= 0. \end{aligned}\tag{34}$$

Since $P(0)$ and $P(\Delta t)$ commute, as seen from Eq. (32), we have the joint probability distribution $\mu^{P(0),P(\Delta t)}(dx, dy)$ of the values of $P(0)$ and $P(\Delta t)$ by Eq. (19). The rms disturbance $\eta(P)$ of P from $t = 0$ to $t = \Delta t$ is given by

$$\begin{aligned}\eta(P) &= \left(\iint_{\mathbf{R}^2}(y-x)^2\mu^{P(0),P(\Delta t)}(dx, dy)\right)^{1/2}\\ &= \langle(P(\Delta t) - P(0))^2\rangle^{1/2}\\ &= \langle(\overline{P}(0) + P(0))^2\rangle^{1/2} < \infty. \end{aligned}\tag{35}$$

Consequently, we have

$$\varepsilon(Q)\eta(P) = 0. \tag{36}$$

Therefore, our model obviously violates the Heisenberg EDR (2).

Taking advantage of the above model, the argument was refuted that the uncertainty principle generally leads to the SQL claimed in Ref. [18] for monitoring free-mass position [19, 20].

If $\langle P(0)^2\rangle \to 0$ and $\langle \overline{P}(0)^2\rangle \to 0$ (i.e., $|\psi\rangle$ and $|\xi\rangle$ tend to the momentum eigenstate with zero momentum) then we have even $\eta(P(t)) \to 0$ with $\varepsilon(Q) = 0$. Thus, we can measure position precisely without effectively disturbing momentum in a near momentum eigenstate; see Ref. [24] for detailed discussion on the quantum state reduction caused by the above model.

As shown above, the Heisenberg EDR (2) is taken to be a breakable limit [25], but then the problem remains: what is the unbreakable constraint between error and disturbance, which Heisenberg originally intended?

7. Universally valid EDR

In 2003, the present author [26, 4, 27] showed the relation

$$\varepsilon(A)\eta(B) + |\langle[n(A), B]\rangle + \langle[A, d(B)]\rangle| \geq \frac{1}{2}|\langle[A, B]\rangle|, \tag{37}$$

which is universally valid for any observables A, B, any system state, and any measuring apparatus, where $n(A)$ and $d(B)$ are system observables representing the first moments of the error and the disturbance for A and B, respectively. ¿From Eq. (37), it is concluded that if the error and the disturbance are statistically independent from system state, then the Heisenberg EDR

$$\varepsilon(A)\eta(B) \geq \frac{1}{2}|\langle[A, B]\rangle| \tag{38}$$

holds, extending the previous results [28, 29, 16, 17]. The additional correlation term in Eq. (37) allows the error-disturbance product $\epsilon(A)\eta(B)$ to violate the Heisenberg EDR (38). In general, the relation

$$\epsilon(A)\eta(B) + \epsilon(A)\sigma(B) + \sigma(A)\eta(B) \geq \frac{1}{2}|\langle[A, B]\rangle| \tag{39}$$

holds for any observables A, B, any system state, and any measuring apparatus [26, 4, 27, 30, 31, 32].

The new relation (39) leads to the following new constraints for error-free measurements and non-disturbing measurements: if $\varepsilon(A) = 0$ then

$$\sigma(A)\eta(B) \geq \frac{1}{2}\left|\langle[A, B]\rangle\right|, \tag{40}$$

and if $\eta(B) = 0$ then

$$\epsilon(A)\sigma(B) \geq \frac{1}{2}\left|\langle[A, B]\rangle\right|. \tag{41}$$

Note that if $\langle[A, B]\rangle \neq 0$, Heisenberg EDR (38) leads to divergences in both cases. The new error bound Eq. (41) was used to derive conservation-law-induced limits for measurements [27, 33] (see also [34, 35]) quantitatively generalizing the Wigner-Araki-Yanase theorem [36, 37, 38, 39] and was used to derive an accuracy limit for quantum computing induced by conservation laws [27] (see also [40, 41, 42, 43, 44, 45, 46]).

8. Operator formalism for error and disturbance

To derive the above relations, consider a measuring process $\mathbf{M} = (\mathcal{K}, |\xi\rangle, U, M)$ determined by the probe system $\mathbf{P}$ described by a Hilbert space $\mathcal{K}$, the initial probe state $|\xi\rangle$, the unitary evolution U of the composite system $\mathbf{S} + \mathbf{P}$ during the measuring interaction, and the meter observable M of the probe $\mathbf{P}$ to be directly observed [4]. We assume that the measuring interaction turns on at time $t = 0$ and turns off at time $t = \Delta t$. In the Heisenberg picture, we write

$$A_1(0) = A_1 \otimes I, \quad A_2(0) = I \otimes A_2, \quad A_{12}(\Delta t) = U^\dagger A_{12}(0)U,$$

for an observable A_1 of $\mathbf{S}$, an observable A_2 of $\mathbf{P}$, and an observable $A_{12}(0)$ of $\mathbf{S} + \mathbf{P}$.

The *error observable* $N(A)$ representing the difference between the measured observable $A(0)$ and the meter observable $M(\Delta t)$ to be read and the *disturbance observable* $D(A)$ representing the change in B caused by the measuring interaction are defined by

$$N(A) = M(\Delta t) - A(0), \tag{42}$$

$$D(B) = B(\Delta t) - B(0). \tag{43}$$

The *mean error operator* $n(A)$ and the *mean disturbance operator* $d(B)$ in Eq. (37) are defined by

$$n(A) = \langle\xi|N(A)|\xi\rangle, \tag{44}$$

$$d(B) = \langle\xi|D(B)|\xi\rangle. \tag{45}$$

The *(root-mean-square) error* $\varepsilon(A, \rho)$ and the *(root-mean-square) disturbance* $\eta(B, \rho)$ for observables A, B and state (density operator) ρ on a Hilbert space $\mathcal{H}$ were defined by

$$\varepsilon(A, \rho)^2 = \mathrm{Tr}[N(A)^2\rho \otimes |\xi\rangle\langle\xi|], \tag{46}$$

$$\eta(B, \rho)^2 = \mathrm{Tr}[D(B)^2\rho \otimes |\xi\rangle\langle\xi|]. \tag{47}$$

The definition of $\varepsilon(A) = \varepsilon(A, \rho)$ is uniquely derived from the classical notion of root-mean-square error if $M(\Delta t)$ and $A(0)$ commute [47], as in the models discussed in the previous sections. Otherwise, it is considered as a natural quantization of the notion of classical root-mean-square error. It is also pointed out that $\varepsilon(A)$ coincides with the root-mean-square error of a quantum estimator for an orthogonal pure state estimation problem with the uniform prior distribution

[48]. The definition of $\eta(B) = \eta(B, \rho)$ is derived analogously, although there are recent debates on alternative approaches [49, 50, 51, 47].

In particular, Busch, Heinonen, and Lahti [49] pointed out that there is a case in which $\varepsilon(A) = 0$ holds but A cannot be considered to be measured precisely. In response to this, we characterized the case where A is measured precisely as follows [52, 53].

We say that the measuring process $\mathbf{M}$ *precisely measures* an observable A in a state ρ if observables $A(0)$ and $M(\Delta t)$ commute in the state $\rho\otimes|\xi\rangle\langle\xi|$ and the joint probability distribution $\mu^{A(0),M(\Delta t)}$ of $A(0)$ and $M(\Delta t)$ concentrates on the diagonal, i.e.,

$$\mu^{A(0),M(\Delta t)}(\{(x, y) \in \mathbf{R}^2 \mid x \neq y\} = 0. \tag{48}$$

The *weak joint distribution* $\mu_W^{A(0),M(\Delta t)}$ of $A(0)$ and $M(\Delta t)$ in a state ρ is defined by

$$\mu_W^{A(0),M(\Delta t)}(dx, dy) = \langle E^{A(0)}(dx)E^{M(\Delta t)}(dy)\rangle. \tag{49}$$

The joint probability distribution $\mu^{A(0),M(\Delta t)}$ exists only when $A(0)$ and $M(\Delta t)$ commute in the state $\rho \otimes |\xi\rangle\langle\xi|$, while the weak joint distribution $\mu_W^{A(0),M(\Delta t)}$ always exists. The *cyclic subspace* $\mathcal{C}(A, \rho)$ generated by A and ρ is defined as the closed subspace of $\mathcal{H}$ generated by $\{E^A(\Delta)|\phi\rangle \mid \Delta \in \mathcal{B}(\mathbf{R}), |\phi\rangle \in \rho\mathcal{H}\}$, where $\mathcal{B}(\mathbf{R})$ is the Borel σ-field of the real line $\mathbf{R}$. A *generating subset* of $\mathcal{C}(A, \rho)$ is a set $\mathcal{S}$ of vector states $|\phi\rangle \in \mathcal{H}$ such that $\mathcal{S}^{\perp\perp} = \mathcal{C}(A, |\phi\rangle)$, where $\perp$ stands for the orthogonal complement. Then, the following theorem holds [52, 53].

Theorem 1 *Let* $\mathbf{M} = (\mathcal{K}, |\xi\rangle, U, M)$ *be a measuring process for the system* $\mathbf{S}$ *described by a Hilbert space* $\mathcal{H}$. *Let* A *be an observable of* $\mathbf{S}$ *and* ρ *be a state of* $\mathbf{S}$. *Then, the following conditions are equivalent.*

(i) The measuring process $\mathbf{M}$ *precisely measures observable* A *in state* ρ.

(ii) The weak joint distribution $\mu_W^{A(0),M(\Delta t)}$ *in state* ρ *concentrates on the diagonal, i.e.,*

$$\langle E^{A(0)}(\Delta)E^{M(\Delta t)}(\Gamma)\rangle = 0$$

if $\Delta \cap \Gamma = \emptyset$.

(iii) $\varepsilon(A, |\phi\rangle) = 0$ *for all* $|\phi\rangle \in \mathcal{C}(A, \rho)$.

(iv) There exists a generating subset $\mathcal{S}$ *of* $\mathcal{C}(A, \rho)$ *such that* $\varepsilon(A, |\phi\rangle) = 0$ *for all* $|\phi\rangle \in S$.

We say that the measuring process $\mathbf{M}$ *does not disturb* an observable B in a state ρ if observables $B(0)$ and $B(\Delta t)$ commute in the state $\rho\otimes|\xi\rangle\langle\xi|$ and the joint probability distribution $\mu^{B(0),B(\Delta t)}$ of $B(0)$ and $B(\Delta t)$ concentrates on the diagonal. The non-disturbing measuring processes defined above can be characterized analogously.

¿From the above theorem, we can conclude that $\varepsilon(A)$ is negatively biased in the sense that positive error $\varepsilon(A) > 0$ always implies that the A cannot be measured precisely. Thus, a non-zero lower bound for $\varepsilon(A)$ indicates a limitation for precise measurements. For $\eta(B) > 0$ we have an analogous conclusion. Moreover, the above characterizations of precise and non-disturbing measurements lead to the following definitions of the *locally uniform root-mean-squre error* $\overline{\varepsilon}(A, \rho)$ and the *locally uniform root-mean-squre disturbance* $\overline{\eta}(B, \rho)$ [54]:

$$\overline{\varepsilon}(A, \rho) = \sup_{|\phi\rangle\in\mathcal{S}} \varepsilon(A, |\phi\rangle), \tag{50}$$

$$\overline{\eta}(B, \rho) = \sup_{|\phi\rangle\in\mathcal{S}} \eta(B, |\phi\rangle), \tag{51}$$

where $\mathcal{S} = \mathcal{C}(A,\rho)$ or $\mathcal{S}$ is a generating subset of $\mathcal{C}(A,\rho)$. Then, it is shown that $\overline{\varepsilon}(A,\rho) = 0$ if and only if the measurement precisely measures A in ρ, and that $\overline{\eta}(B,\rho) = 0$ if and only if the measurement does not disturb B in ρ. For those quantities, Heisenberg's EDR

$$\overline{\varepsilon}(Q,\rho)\overline{\eta}(P,\rho) \geq \frac{\hbar}{2} \qquad (52)$$

is still violated by a linear position measurement [54], and the relation

$$\overline{\varepsilon}(A)\overline{\eta}(B) + \overline{\varepsilon}(A)\sigma(B) + \sigma(A)\overline{\eta}(B) \geq \frac{1}{2}|\langle[A,B]\rangle| \qquad (53)$$

holds universally [54], where $\overline{\varepsilon}(A) = \overline{\varepsilon}(A,\rho)$ and $\overline{\eta}(B) = \overline{\eta}(B,\rho)$.

9. Experimental tests

There has been a controversy [55, 56] on the question as to whether the rms error and rms disturbance are experimentally accessible without knowing the details of measuring process $(\mathcal{K}, |\xi\rangle, U, M)$ and the state ρ. To clear this question two methods have been proposed so far: the "three-state method" proposed by the present author [31] and the "weak-measurement method" proposed by Lund-Wiseman [57] based on the relation between the rms error/distrubance and the weak joint distribution [20, 16, 52]. The three-state method was demonstrated for qubit systems: projective measurement of a neutron-spin qubit [58, 59] and generalized measurement of a photon-polarization qubit [60]. The weak-measurement method was demonstrated for generalized measurement of a photon-polarization qubit by Rozema *et al* [61], Baek *et al* [62], and Ringbauer *et al.* [63]. All the above experiments observed that the Heisenberg EDR (38) does not hold, while the universally valid EDR (39) and a new stronger universally valid EDR recently proposed by Branciard [64] hold. Very recently, we have proposed the third method, the "two-point quantum correlator method," for measuring rms error and disturbance with an experimental proposal for quit measurements in Ref. [65].

Acknowledgments

This work was supported by MIC SCOPE, No. 121806010, and the John Templeton Foundation, ID #35771.

References

[1] Heisenberg W 1983 *Quantum Theory and Measurement* ed Wheeler J A and Zurek W H (Princeton, NJ: Princeton University Press) pp 62–84 [Translation into English by J. A. Wheeler and W. H .Zurek, 1981. Originally published under the title, "Über den anschaulichen Inhalt der quantemtentheoretischen Kinematik und Mechanik," Zeitschrift für Physik, 43, 172-98 (1927).]
[2] Gauss C F 1995 *Theoria Combinationis Observationum Erroribus Miinimis Obnoxiae, Pars Prior, Pars Posterior, Supplementum (Societati Regiae Scientiarum Exhibita, Feb. 15, 1821) [Theory of the Combination of Observations Least Subject to Errors: Part One, Part Two, Supplement, translated by G. W. Stewart]* (Philadelphia, USA: Society for Industrial and Applied Mathematics)
[3] Kennard E H 1927 *Z. Phys.* **44** 326–352
[4] Ozawa M 2003 *Phys. Rev. A* **67** 042105
[5] von Neumann J 1955 *Mathematical Foundations of Quantum Mechanics* (Princeton, NJ: Princeton University Press) [English translation of *Mathematische Grundlagen der Quantenmechanik* (Springer, Berlin, 1932)]
[6] Davies E B and Lewis J T 1970 *Commun. Math. Phys.* **17** 239–260
[7] Ozawa M 1984 *J. Math. Phys.* **25** 79–87
[8] Bohm D 1951 *Quantum Theory* (New York: Prentice-Hall)
[9] Messiah A 1959 *Mécanique Quantique* vol I (Paris: Dunod) [*Quantum Mechanics*, Vol. I (North-Holland, Amsterdam, 1959)]

[10] Schiff L I 1968 *Quantum Mechanics* (New York: MacGraw-Hill)
[11] Reed M and Simon B 1980 *Methods of Modern Mathematical Physics, I: Functional Analysis (Reivsed and Enlarged Edition)* (New York: Academic)
[12] Halmos P R 1951 *Introduction to Hilbert Space and the Theory of Spectral Multiplicity* (New York: Chelsea)
[13] Caves C M, Thorne K S, Drever R W P, Sandberg V D and Zimmermann M 1980 *Rev. Mod. Phys.* **52** 341–392
[14] Caves C M 1985 *Phys. Rev. Lett.* **54** 2465–2468
[15] Arthurs E and Kelly, Jr J L 1965 *Bell. Syst. Tech. J.* **44** 725–729
[16] Ozawa M 1991 *Quantum Aspects of Optical Communications* ed Bendjaballah C, Hirota O and Reynaud S (Berlin: Springer) pp 3–17
[17] Ishikawa S 1991 *Rep. Math. Phys.* **29** 257–273
[18] Braginsky V B, Vorontsov Y I and Thorne K S 1980 *Science* **209** 547–557
[19] Yuen H P 1983 *Phys. Rev. Lett.* **51** 719–722 [see also *ibid.* p. 1603]
[20] Ozawa M 1988 *Phys. Rev. Lett.* **60** 385–388
[21] Ozawa M 1989 *Squeezed and Nonclassical Light* ed Tombesi P and Pike E R (New York: Plenum) pp 263–286
[22] Maddox J 1988 *Nature* **331** 559
[23] Ozawa M 2002 *Phys. Lett. A* **299** 1–7
[24] Ozawa M 2001 *Phys. Lett. A* **282** 336–342
[25] Giovannetti V, Lloyd S and Maccone L 2004 *Science* **306** 1330–1336
[26] Ozawa M 2003 *Phys. Lett. A* **318** 21–29
[27] Ozawa M 2003 *Int. J. Quant. Inf.* **1** 569–588
[28] Arthurs E and Goodman M S 1988 *Phys. Rev. Lett.* **60** 2447–2449
[29] Raymer M G 1994 *Am. J. Phys.* **62** 986–993
[30] Ozawa M 2004 *Phys. Lett. A* **320** 367–374
[31] Ozawa M 2004 *Ann. Phys. (N.Y.)* **311** 350–416
[32] Ozawa M 2005 *J. Opt. B: Quantum Semiclass. Opt.* **7** S672–S681
[33] Ozawa M 2004 *AIP Conf. Proc.* **734** 95–98
[34] Ozawa M 2002 *Phys. Rev. Lett.* **88** 050402
[35] Busch P and Loveridge L 2011 *Phys. Rev. Lett.* **106** 110406
[36] Wigner E P 1952 *Z. Phys.* **133** 101–108
[37] Araki H and Yanase M M 1960 *Phys. Rev.* **120** 622–626
[38] Yanase M M 1961 *Phys. Rev.* **123** 666–668
[39] Ozawa M 1991 *Phys. Rev. Lett.* **67** 1956–1959
[40] Ozawa M 2002 *Phys. Rev. Lett.* **89** 057902
[41] Gea-Banacloche J 2002 *Phys. Rev. Lett.* **89** 022308
[42] Ozawa M 2003 *Proceedings of the Sixth International Conference on Quantum Communication, Measurement and Computing* ed Shappiro J H and Hirota O (Princeton: Rinton Press) pp 175–180
[43] Gea-Banacloche J and Ozawa M 2005 *J. Opt. B: Quantum Semiclass. Opt.* **7** S326–S332
[44] Gea-Banacloche J and Ozawa M 2006 *Phys. Rev. A* **74** 060301(R)
[45] Karasawa T and Ozawa M 2007 *Phys. Rev. A* **75** 032324
[46] Karasawa T, Gea-Banacloche J and Ozawa M 2009 *J. Phys. A: Math. Theor.* **42** 225303
[47] Ozawa M 2013 Disproving Heisenberg's error-disturbance relation arXiv:1308.3540v1 [quant-ph]
[48] Buscemi F, Hall M J, Ozawa M and Wilde M M 2014 *Phys. Rev. Lett.* **112** 050401
[49] Busch P, Heinonen T and Lahti P 2004 *Phys. Lett. A* **320** 261–270
[50] Busch P, Lahti P and Werner R F 2013 *Phys. Rev. Lett.* **111** 160405
[51] Rozema L A, Mahler D H, Hayat A and Steinberg A M 2013 A note on different definitions of momentum disturbance arXiv:1307.3604v1 [quant-ph]
[52] Ozawa M 2005 *Phys. Lett. A* **335** 11–19
[53] Ozawa M 2006 *Ann. Phys. (N.Y.)* **321** 744–769
[54] Ozawa M 2006 *Proc. SPIE* **6244** 62440Q
[55] Werner R F 2004 *Quantum Inf. Comput.* **4** 546–562
[56] Koshino K and Shimizu A 2005 *Phys. Rep.* **412** 191–275
[57] Lund A P and Wiseman H M 2010 *New J. Phys.* **12** 093011
[58] Erhart J, Sponar S, Sulyok G, Badurek G, Ozawa M and Hasegawa Y 2012 *Nature Phys.* **8** 185–189
[59] Sulyok G, Sponar S, Erhart J, Badurek G, Ozawa M and Hasegawa Y 2013 *Phys. Rev. A* **88** 022110
[60] Baek S Y, Kaneda F, Ozawa M and Edamatsu K 2013 *Sci. Rep.* **3** 2221
[61] Rozema L A, Darabi A, Mahler D H, Hayat A, Soudagar Y and Steinberg A M 2012 *Phys. Rev. Lett.* **109** 100404
[62] Kaneda F, Baek S Y, Ozawa M and Edamatsu K 2014 *Phys. Rev. Lett.* **112** 020402

[63] Ringbauer M, Biggerstaff D N, Broome M A, Fedrizzi A, Branciard C and White A G 2014 *Phys. Rev. Lett.* **112** 020401
[64] Branciard C 2013 *Proc. Natl. Acad. Sci. USA* **110** 6742–6747
[65] Buscemi F, Dall'Arno M, Ozawa M and Vedral V 2013 Direct observation of any two-point quantum correlation function arXiv:1312.4240 [quant-ph]

Journal of Physics: Conference Series **504** (2014) 012025 doi:10.1088/1742-6596/504/1/012025

Investigations of fundamental phenomena in quantum mechanics with neutrons

Yuji Hasegawa
Atominstitut, Technische Universität Wien, Stadionallee 2, A-1020 Wien, Austria
E-mail: Hasegawa@ati.ac.at

Abstract. Neutron interferometer and polarimeter are used for the experimental investigations of quantum mechanical phenomena. Interferometry exhibits clear evidence of quantum-contextuality and polarimetry demonstrates conflicts of a contextual model of quantum mechanics á la Leggett. In these experiments, entanglements are achieved between degrees of freedom in a single-particle: spin, path and energy degrees of freedom are manipulated coherently and entangled. Both experiments manifest the fact that quantum contextuality is valid for phenomena with matter waves with high precision. In addition, another experiment is described which deals with error-disturbance uncertainty relation: we have experimentally tested error-disturbance uncertainty relations, one is derived by Heisenberg and the other by Ozawa. Experimental results confirm the fact that the Heisenberg's uncertainty relation is often violated and that the new relation by Ozawa is always larger than the limit. At last, as an example of a counterfactual phenomenon of quantum mechanics, observation of so-called quantum Cheshire Cat is carried out by using neutron interferometer. Experimental results suggest that pre- and post-selected neutrons travel through one of the arms of the interferometer while their magnetic moment is located in the other arm.

1. Introduction

Quantum mechanics (QM) is one of the most successful theories in physics and its predictions are verified with high precision in a wide range of the field by experiments. From the beginning, however, QM is providing an extraordinary and strange view of nature, which is different from that in classical physics. For instance, a particle such as an electron, a neutron and a positron can behave as non-locally located wave: wave-particle duality postulates that all particles exhibit both wave and particle properties. The deBroglie relation, $\lambda = \hbar/mv$ connects the wavelength λ with the mass m and the velocity v of the particle. Furthermore, the uncertainty principle forms the basis of indeterminacy in QM. It prohibits and describes the limitations of simultaneous measurements of certain pairs of observables [1]: there are physical properties, which can be accessed only with limitation. Moreover, superposition of macroscopic systems like Schrödinger's cat [2] as well as the non-local correlation suggested by Einstein, Podolsky, and Rosen (EPR) [3] are not intuitively understandable and seem to be paradoxical at first sight. Both are proposed as thought experiments to derive (seemingly) contradictory conclusions. Fundamental investigations of QM have been revealing its peculiarities, some are regarded as a resource of new technology in future.

From the very beginning, neutron interferometer experiments are established as one of the most powerful tools for investigations of quantum mechanical phenomena on a very fundamental basis [4, 5]. Over the last decades the neutron interferometry has provided excellent opportunities for many different types of interferometer experiments with neutrons, ranging from fundamental quantum investigations to application measurements, such as precise measurements of coherent neutron scattering lengths. The former exploits the neutron interferometry as a matter-wave interference experiment and the latter is frequently required for other neutron scattering spectroscopy. Consequences of the nonrelativistic Schrödinger equation for matter-waves can be studied, for instance, with electrons, atoms, ions, and molecules. Features of neutron interferometry, such as macroscopic-scale experiments, high detector-efficiency, low decoherence-rate, and high-efficiency manipulation rate, make it a unique strategy for quantum mechanical investigations. Recently neutron polarimeter experiments have turned out to serve as another tool to verify the basic concepts of quantum mechanics. There, basis used in the experiment are spanned not by two paths $|\mathrm{I}\rangle$ and $|\mathrm{II}\rangle$, but by spin eigenstate $|\uparrow\rangle$ and $|\downarrow\rangle$. With this device, for instance, the noncommutation of the Pauli spin operator [6] and a number of geometric phase measurements [7] are carried out. The implicit polarization interference scheme allows us to perform textbook like demonstrations of QM with high efficiency and stability.

In this paper, recent experiments with the neutron interferometer and the polarimeter are presented. The interferometry studies properties of quantum system triply-entangled between degrees of freedom [8] and the polarimetry investigates a contextual model á la Leggett [9]. In addition, successive measurements of neutron's spin [10, 11] exhibits the violation of the naive error-disturbance relation by Heisenberg and confirms the validity of a new universally valid error-disturbance relation by Ozawa [12, 13]. Furthermore, a recent experiment to observe a quantum Cheshire Cat [14, 15] in neutron interferometer setup is described: neutron and its magnetic moment are observed to be disembodied and spatially separated [63].

2. Entanglement between degrees of freedom of single neutrons

2.1. Entanglement studied with neutron interferometer

It was Einstein, Podolsky, and Rosen (EPR) [3] and afterwards Bell [17] who shed light on the non-local properties between subsystems in quantum mechanics. Bell inequalities are constraints imposed by local hidden-variable theories (LHVTs) on the results of spacelike separated experiments on distant systems. The conflict between LHVTs and QM is even more apparent in tri- or multi-partite quantum systems which was analysed by Greenberger, Horne and Zeilinger (GHZ) [18, 19]. There, the contradiction leads to nonstatistical predictions in contrast to common Bell-inequalities. Mermin [20] showed that this conflict can be converted into a larger violation of a Bell-like inequality between three or more separated systems. Experimental tests of these inequalities were reported with photons [21, 22] as well as with ions [23]. Kochen and Specker [24] were the first to analyse the concept of contextuality in QM and find striking phenomena predicted by quantum theory. Note that LHVTs form a subset of a larger class of hidden-variable theories known as noncontextual hidden-variable theories (NCHVTs): in NCHVTs the result of a measurement of an observable is assumed to be predetermined and not affected by a (previous or simultaneous) suitable measurement of any other compatible or co-measurable observable.

In single particle systems, different degrees of freedom (DOFs) can be entangled. In this scenario, the conflict arises not between QM and LHVTs but a violation confirms the impossibility of NCHVTs [25]. The violation of Bell-like inequality is reported by using entanglement between spin and path DOF of single neutrons [26, 5], as well as single photons [27], both confirming quantum contextuality. In addition, tests of Kochen-Specker theorem are carried out [28, 29], which again favor quantum contextuality. Later, quantum tomographic ananlysis of the Bell-like state is accomplished [30], approving the quality of the generated Bell-like state of DOFs. For further investigations of entanglement of DOFs, it is essential to development methods to manipulate other DOFs: the use of interaction between neutron's

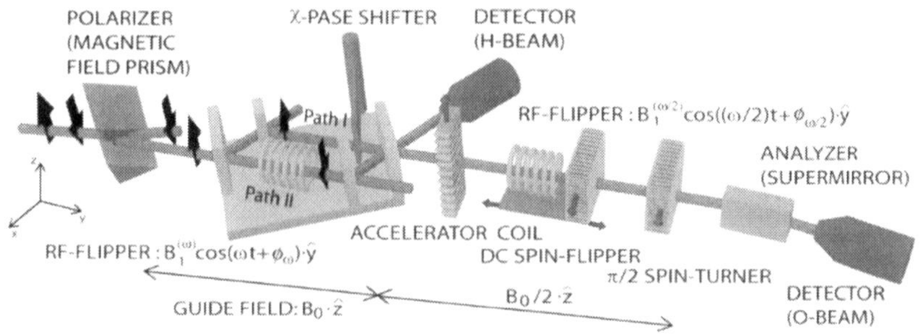

Figure 1. Setup for the study of tripartite entangled GHZ-like state in neutron interferometry.

spin and a time-dependent magnetic field enables a coherent energy-manipulation scheme. The experiment confirms high quality of the method [31], which is expected to allow advanced studies of multipartite entanglements in single particles.

2.1.1. Tripartite entangled Greenebrger-Horne-Zeilinger (GHZ) state

Here, we describe the experimental realization of tripartite entanglement for single neutrons [8] where one external degree of freedom (path states in the interferometer) is entangled with two internal degrees of freedom (spin and energy) leading to a violation of a Mermin-like inequality [20]. It was Greenberger, Horne and Zeilinger who first proposed the Greenebrger-Horne-Zeilinger (GHZ) state for four spin-1/2 particles [18, 19]. Later Mermin [32] presented a version with three spin-1/2 particles; in Ref. [33] he pointed out the use of the GHZ-states to reveal the relation between Kochen-Specker theorem [24] and Bell theorem [17].

Imagine a perfect crystal neutron interferometer experiment [4, 5], where the up-polarized incident neutron beam passes through the beam-splitter. Here, the state of neutron's path is transformed into a superposition of two path-states, $\frac{1}{\sqrt{2}}(|\mathrm{I}\rangle + |\mathrm{II}\rangle)$. In the interferometer, a RF spin-flipper is inserted in the path II, which flips neutron's spin by a time-dependent interaction: this induces energy transitions from the initial energy state $|E_0\rangle$ to states $|E_0-\hbar\omega\rangle$ by photon exchange [34]. Consequently, one can generate neutrons in a triply entangled GHZ-like state, given by

$$|\Psi_N^{GHZ}\rangle = \frac{1}{\sqrt{2}}\Big(|\uparrow\rangle\otimes|\mathrm{I}\rangle\otimes|E_0\rangle + |\downarrow\rangle\otimes|\mathrm{II}\rangle\otimes|E_0-\hbar\omega\rangle\Big). \tag{1}$$

The state of neutrons is characterized by three (the spin, path and energy) DOFs: all of them are described simply by two-level quantum systems. Measurement represented by operators σ_x and σ_y in each DOF are accomplished simply by phase manipulations between the basis states in each DOF. In particular,

(i) The spin-phase α is adjusted by a magnetic field oriented along +z direction ('accelerator' coil) together with a DC-flipper in $\pi/2$-flipping mode.

(ii) The phase manipulation of the path is accomplished with an auxiliary phase shifter χ made of a parallel-sided Si plate and the last plate of the interferometer.

(iii) The so-called zero-field precession phase γ [35] is employed for the phase manipulation of the energy. The second RF-flipper together with a DC-flipper are used. (An experimentally convenient method to manipulate individually the Larmor phase α and the zero-field phase γ was found and reported in [36].)

Mermin derived an inequality suitable for experimental tests to distinguish between predictions of QM and LHVTs [20]. In a similar way, assuming a tripartite system and taking

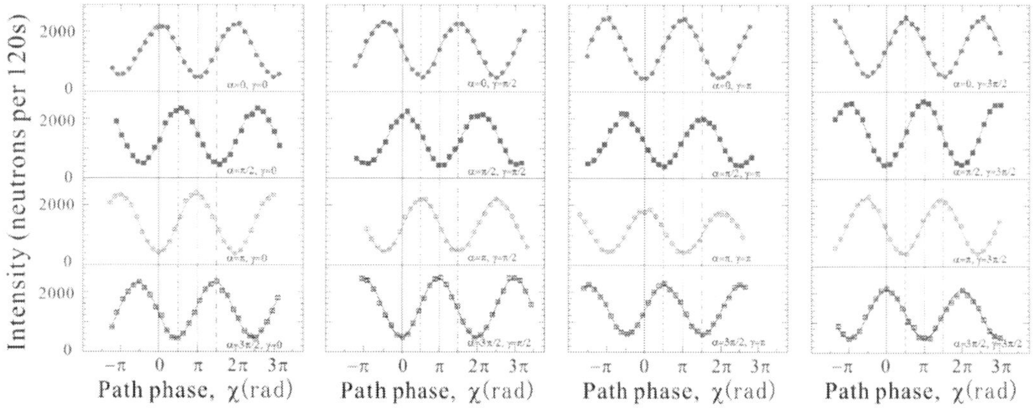

Figure 2. Typical intensity modulations by scanning the phase shifter at various angles for spin (α) and energy (γ) degrees of freedom.

$\sigma_x^s\sigma_x^p\sigma_x^e)$	$\sigma_x^s\sigma_y^p\sigma_y^e)$	$\sigma_y^s\sigma_x^p\sigma_y^e)$	$\sigma_y^s\sigma_y^p\sigma_x^e)$
+0.659(2)	-0.632(2)	-0.603(2)	-0.664(2)

Table 1. Finally obtained expectation values

the assumption in the conditionally independent form due to NCHVTs instead of LHVTs, the border for a sum of expectation values of certain product observables is obtained. The sum of expectation values of product observables, called M, is defined as

$$M \equiv E[\sigma_x^s\sigma_x^p\sigma_x^e] - E[\sigma_x^s\sigma_y^p\sigma_y^e] - E[\sigma_y^s\sigma_x^p\sigma_y^e] - E[\sigma_y^s\sigma_y^p\sigma_x^e] \quad (2)$$

where $E[\ldots]$ represents expectation values, and σ_j^s, σ_j^p, and σ_j^e represent Pauli operators for the two-level systems in the spin, path, and energy DOF, respectively. NCHVTs demands $|M| \leq 2$, while quantum theory predicts an upper bound of 4: any measured value of M that is larger than 2 decides in favor of quantum contextuality.

The experiment was carried out at the neutron-interferometer beam line S18 at the high flux reactor at the Institute Laue Langevin (ILL) [8]. A schematic view of the experimental setup is shown in Fig.1. To determine the expectation values in M in Eq.(2), we performed 16 independent path phase χ scans by tuning the spin phase α, and the energy phase γ each at 0, $\pi/2$, π and $3\pi/2$. Typical sinusoidal intensity modulations are depicted in Fig.2. The finally obtained expectation values are shown in Tab.1. From these values, the final M-value was calculated as $M = 2.558 \pm 0.004$, which clearly exhibits a violation of the Mermin-like inequality, $|M| \leq 2$, and confirms the invalidity of the assumption of non-contextuality. The deviation from the ideal value of 4 is solely due to the reduced contrast of the interferograms. Recently a further study of tripartite entanglement of neutron's DOF appeared [37]: experimental evidence of the generation of distinct types of genuine multipartite entanglement is confirmed. By using appropriate nonlinear witness, even finer properties of the type of the genuine multipartite entanglement are revealed. In the analysis, the extraordinarily high fidelity of the generated entangled states is verified.

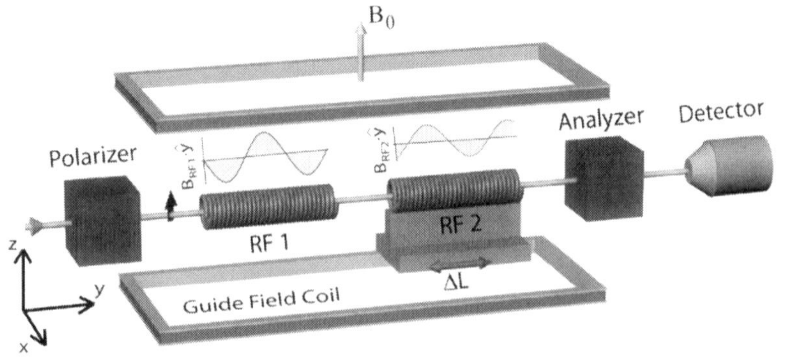

Figure 3. Setup of the neutron polarimetric test of contextual realistic model á la Leggett.

2.2. Entanglement studied with neutron polarimeter

For investigations on the foundations of quantum mechanics, the neutron interferometer is established as an ideal tool of studies with matter waves. Although such an interferometer is suitable for highly intuitive proof-of-principle demonstrations of quantum phenomena [4, 5], the device is extremely delicate: interferogram can easily disappear due to vibrations and thermal disturbance. An alternative approach to perform quantum optical experiments with neutrons is the use of neutron polarimeter, where the interference between spin eigenstates is observed. The advantange of this approach is (i) instrument is very robust: pretty insensitive to the environmental disturbances, (ii) highly efficient ($> 99\%$) manipulations are possible: final contrast reaching at about 99%. Neutron polarimetry has been used to demonstrate fundamental quantum-mechanical properties. For studies of entanglement, A Bell-test with bipartite entanglement [38] as well as GHZ-state with tripartite enantanglement [39] are reported.

2.2.1. Leggett's model tested with matter waves

Here, we describe a falsification of a contextual realistic model á la Leggett using neutrons [9]. In 2003, Leggett proposed a non-local realistic model [40], and an extension of Leggett's model appeared later [41]. The experiments with photons show conflicts with Leggett's model [42, 43, 44, 45], but all experimental tests were performed with photons. Thus, it is important to test a model á la Leggett with neutrons, in particular with matter waves. It is worth noting here that this experiment demands extremely high contrast of interferograms, i.e., more than 97.4%; such a high contrast is only accessible with polarimeter in neutron optics.

In our polarimetric test, we followed the criteria used in the first experimental study by Gröblacher et al. [42]. In particular, the contextual theory to be tested here is based on the following assumptions. (i) All the values of measurements are predetermined. (ii) States are a statistical mixture of subensembles. (iii) The expectation values taken for each subensemble obey cosine dependence.. Assumptions (i) and (ii) are common to experimental tests of ordinary non-contextual theories and assumption (iii) is a peculiarity of this model. Here, the result of the final measurement of B (A) depends on the setting of the previous measurement of A (B): a realistic contextual model is tested in our experiment. Denoting the measurement settings for observables A and B by $\vec{a}_1$, $\vec{a}_2$ and $\vec{b}_1$, $\vec{b}_2$, respectively, the Leggett-like inequality is given by

$$S_{\text{Legg}} \equiv \left|E_1(\vec{a}_1;\phi) + E_1(\vec{a}_1;0)\right| + \left|E_2(\vec{a}_2;\phi) + E_2(\vec{a}_2;0)\right| \leq 4 - \frac{4}{\pi}\left|\sin\frac{\phi}{2}\right|, \tag{3}$$

where $E_j(\vec{a}_j;\phi)$ represent expectation values of joint (correlation) measurements at settings $\vec{a}_j$

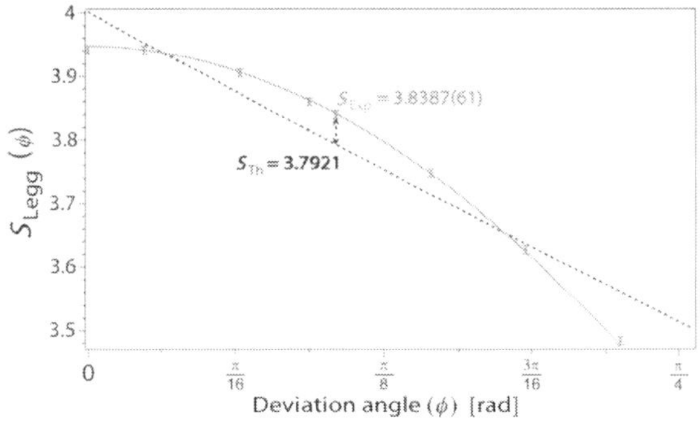

Figure 4. S-values as a function of the deviation angle ϕ to test Leggett's alternative model.

and $\vec{b}_j$ with relative angle ϕ. We assume settings $\vec{a}_1$, $\vec{a}_2$, and $\vec{b}_1$ to lie in a single (equatorial) plane and only $\vec{b}_2$ lies in a plane perpendicular to it: expectation values E_1 and E_2 are given by correlations in planes perpendicular to each other. Maximum violation is expected at $\phi_{\max} \sim 0.1\pi$, resulting in a bound of the Leggett-like inequality $S_{\text{Legg}} = 3.797$ and a quantum value of $S_{\text{QM}} = 3.899$.

The experiment has been carried out at the research reactor facility TRIGA Mark II of the Vienna University of Technology. The experimental setup is depicted in Fig.3. Passing through a bent Co-Ti super mirror array, the beam is highly polarized. The same technique is employed to analyze the polarization. Two identical radio-frequency (RF) spin rotators are employed. Both are put in a homogeneous and static magnetic guide field. In the present experiment, a maximally entangled Bell-like state

$$|\Psi_{\text{N}}^{\text{Bell}}\rangle \propto \frac{1}{\sqrt{2}}\Big(|\uparrow\rangle|E_0\rangle - |\downarrow\rangle|E_0 - \hbar\omega\rangle\Big)\,, \tag{4}$$

with spin basis states, $|\uparrow\rangle$ and $|\downarrow\rangle$, as well as energy basis states, $|E_0\rangle$ and $|E_0 - \hbar\omega\rangle$, is generated and affected by successive energy and spin measurements. Initial Larmor-precession scan exhibit sinusoidal intensity modulations with more than 99% contrast. By tuning the rotation angle of RF1 to $\pi/2$, a maximally entangled Bell-like state $|\Psi_N^{\text{Bell}}\rangle$ is generated. Denoting unit vectors representing measurement directions as $\vec{x}[\theta, \varphi]$ with polar angle θ and azimuthal angle φ, the analyzer and RF2 enable to set $\vec{a}_j[\pi/2, \gamma_j]$ for energy and $\vec{b}_k[\alpha_k, \beta_k]$ for spin.

For a measurement of the Leggett-like inequality, Eq. (3), we require correlation measurements between settings outside the single plane. The maximum discrepancy between Leggett's model and quantum mechanics is expected at the angle $\phi_{\max} \sim 0.1\pi$ with directions $\vec{a}_1[\pi/2, 0]$, $\vec{a}_2[\pi/2, \pi/2]$, $\vec{b}_1[\pi/2, -\phi_{\max}]$ and $\vec{b}_2[\pi/2 - \phi_{\max}, \pi/2]$. The final S_{Legg}-value of the Leggett-like inequality is determined as $S_{\text{Legg}} = 3.8387(61)$ at $\phi = 0.104\pi$, which is clearly larger than the boundary 3.7921: the violation is more than 7.6 standard deviations. In order to see the tendency of the violations, the parameter ϕ is tuned to 8 different values between 0 and 0.226π. Figure 4 shows a plot of the experimentally determined S_{Legg} together with the limit of the contextual model as well as the quantum mechanical prediction, calculated for a contrast of 99%. The experimental values follow the quantum mechanical prediction, and this clearly confirms the violation of Leggett's model.

3. Error-disturbance uncertainty relation

The uncertainty principle refers to intrinsic indeterminacy of quantum mechanics and ranks among the most famous statements of modern physics [46]. It was Heisenberg who first formulated the uncertainty relation as a limitation of accuracies of position and momentum measurements [1]. Later on, the uncertainty relation was reformulated in terms of standard deviations[47, 48], which denotes only the statistical quantity and neglects neither the disturbance due to interactions in a quantum measurement nor measurement error. It was known that the validity of Heisenberg's original relation is justified only under limited circumstances and Ozawa proposed a new universally valid error-disturbance uncertainty relation [12, 13]. Here, we describe a successive spin measurement of neutrons that allows determining the error of a spin-component measurement and the disturbance caused on another spin-component measurement [10]. The results confirm that both error and disturbance completely obey the Ozawa's relation but often violate the Heisenberg's relation.

3.1. From Heisenberg to universally valid uncertainty relation

In 1927, Heisenberg proposed the uncertainty relation for the error $\epsilon(Q)$ of an electron's position measurement and the disturbance $\eta(P)$ of the momentum measurement in a form $\epsilon(Q)\eta(P) \sim \frac{\hbar}{2}$ where $\hbar$ is Planck's constant divided by 2π [1] (here, we use $\frac{\hbar}{2}$ for consistency with modern treatments). The reciprocal relation $\sigma(Q)\sigma(P) \geq \frac{\hbar}{2}$ for standard deviations $\sigma(Q)$, $\sigma(P)$ of position and momentum was proved by Kennnard [47], which was generalized to arbitrary pairs of observables A, B by Robertson[48] as $\sigma(A)\sigma(B) \geq \frac{1}{2}|\langle\psi|[A,B]|\psi\rangle|$, in any states. Here, $[A,B]$ represents the commutator $[A,B] = AB - BA$ and the standard deviation is defined as $\sigma(A)^2 = \langle\psi|A^2|\psi\rangle - \langle\psi|A|\psi\rangle^2$. Robertson's relation with standard deviations has a mathematical basis. Nevertheless, the proof of the reciprocal relation for the error, a generalized form of Heisenberg's error-disturbance uncertainty relation

$$\epsilon(A)\eta(B) \geq \frac{1}{2}|\langle\psi|[A,B]|\psi\rangle|, \tag{5}$$

is not straightforward. Recently, rigorous and general theoretical treatments of quantum measurements have revealed the failure of Heisenberg's relation Eq.(5), and derived a new universally valid uncertainty relation[12, 13] given by

$$\epsilon(A)\eta(B) + \epsilon(A)\sigma(B) + \sigma(A)\eta(B) \geq \frac{1}{2}|\langle\psi|[A,B]|\psi\rangle|. \tag{6}$$

Here, the error $\epsilon(A)$ is defined as the root-mean-square (rms) of the difference between the output operator O_A actually measured and the observable A to be measured, whereas the disturbance $\eta(B)$ is defined as the rms of the change in the observable B during the measurement. Note that the additional second and third terms imply a new accuracy limitation, which does not necessarily follow the trade off relation of error and disturbance.

3.2. Uncertainty relation in successive spin measurements

Here, the validities of two forms of error-disturbance relations, Eqs.(5) and (6) are experimentally tested with neutron's successive spin-measurements. The experimental setup is depicted in Fig.5. Observables A and B are set as σ_x and σ_{ϕ_B} (an observable lying on the equator with the azimuthal angle ϕ_B of the Bloch sphere). The initial state $|\Psi\rangle$ is set to be +z spin state, $|+z\rangle$. In order to observe dependence of the error $\epsilon(A)$ and the disturbance $\eta(B)$ on the output observable, $O_A = \sigma_x \cos\phi_{OA} + \sigma_y \sin\phi_{OA}$ (instead of exactly measuring $A = \sigma_x$), the apparatus M1 is designed to actually carry out measurements of adjustable observables. For determination of the error $\epsilon(A)$ and the disturbance $\eta(B)$, the method proposed in Ref. [13] is used.

The experiment was carried out at the research reactor facility TRIGA Mark II of the TU-Vienna. The monochromatic neutron beam is polarized crossing a super-mirror polarizer and two other super-mirrors are used as analyzers. The guide field together with four DC spin

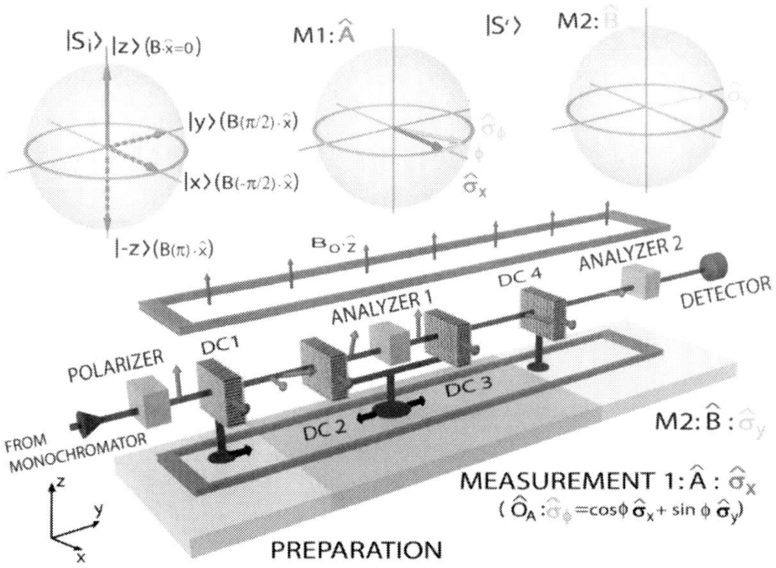

Figure 5. Setup of the neutron optical test of error-disturbance uncertainty relation

rotator coils, induces Larmor precession to allow state preparation and projective measurements of O_A in M1 and B in M2. To test the error-disturbance uncertainty relation in Eqs.(5) and (6), the standard deviations $\sigma(A)$, $\sigma(B)$, the error $\epsilon(A)$ and the disturbance $\eta(B)$ are determined from the experimentally obtained data:the measurements of the standard deviations $\sigma(A)$ and $\sigma(B)$ are carried out by M1 and M2 separately, whereas error $\epsilon(A)$ and disturbance $\eta(B)$ are determined by successive projective measurements utilizing M1 and M2.

Results of the measurements are shown in two cases: (i) $B = \sigma_y$ and (ii) $B = \sigma_x \cos(5\pi/6) + \sigma_y \sin(5\pi/6)$. The azimuthal angle of ϕ_{OA} of the output observalbe O_A is varied from 0 to 2π. From the obtained values of error $\epsilon(A)$, disturbance $\eta(B)$, standard deviations $\sigma(A)$ and $\sigma(B)$, the Heisenberg error-disturbance product $\epsilon(A)\eta(B)$ and the universally valid expression $\epsilon(A)\eta(B) + \epsilon(A)\sigma(B) + \sigma(A)\eta(B)$ are plotted as a function of the detuned azimuthal angle ϕ_{OA} in Fig.6. The figures illustrate the fact that the universally valid expression is always larger than the limit whereas the Heisenberg product is often below the limit. In particular in the range $\phi_{OA} = [0, \pi/2]$ of Fig.6(a), a trade-off relation between the error and the disturbance is observed and the Heisenberg product is always below the limit, which is reported in [10, 11] in more detail. In Fig.6(b), the situation is observed where the universal expression actually touches the limit, which corresponds to the case where the equal sign of the inequality Eq. (6) really occurs.

After the publication of the first experimental confirmation of the new error-disturbance uncertainty relation by us [10], several papers have followed: for instance, experimental demonstration with photonic system [49, 50] as well as theoretical improvement of the inequality [51] appeared. Note that, among them, a theoretical letter was published, which claims that, by defining error and disturbance in different manners, one can consider that the Heisenberg'e relation is still valid [52]. New definition used in the letter is "state independent, each giving the worst-case estimates across all states": this allows overestimate of the measurement error and disturbance. This claim is immediately criticized by two papers of some authors of the experimental papers [53, 54]: physical analysis in the former figures out the new definition as "disturbance power" and mathematical consideration in the latter reveals breakdowns of the

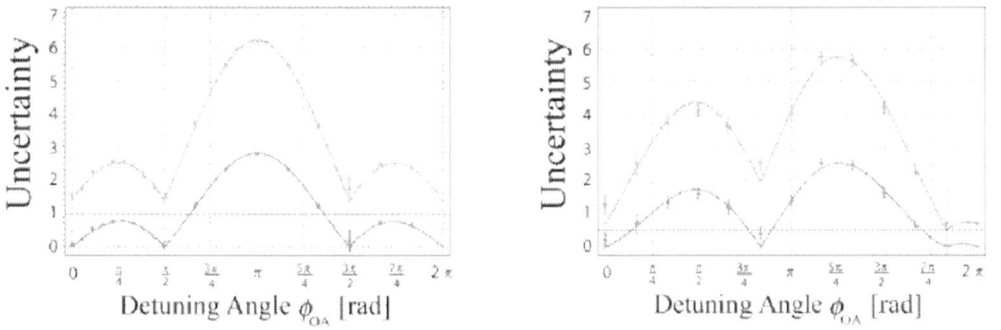

Figure 6. Experimentally determined values of the universally valid uncertainty relation: (i) $\epsilon(A)\eta(B)+\epsilon(A)\sigma(B)+\sigma(A)\eta(B)$ (orange) and (ii) $\epsilon(A)\eta(B)$ (red) are plotted as a function of the detuning angle ϕ for the cases (a)$B=\sigma_y$ (left) and (b) $B=\sigma_x\cos(5\pi/6)+\sigma_y\sin(5\pi/6)$ (right).

new definition. It should be emphasized here that, in practical circumstances, one can easily assume (and actually one has pretty often) a measurement apparatus, which is *optimized* for measurements in certain region of the result and/or of certain state: in particular, the device actually measures *only* certain regions (or states) of the sample. In such cases, it is rather natural to consider that the error and disturbance of this kind of device has no influence of measurements not optimized for the device or what actually will not be done, i.e. they can be *state-dependent*. From these considerations, we believe that the worst-case estimates is unsuitable measure of error and disturbance in practice: Ozawa's treatment and definition of error and disturbance reflect the sprit of the original argument of uncertainty relation by Heisenberg.

4. Quantum Cheshire cat: paradoxical phenomenon in quantum mechanics

Quantum mechanics is still capable to exhibit counterfactual phenomena. For instance, Hardy paradox describes a contradiction between classical picture and the outcome of quantum mechanics [55]: joint weak measurements of trajectories of a photon pair on a post-selected state in a pair of Mach-Zehnder interferometers reveal a negative value for a joint probability of locations [56, 57]. Here, a weak measurement is a technique proposed by Aharonov, Albert and Vaidman (AAV) [58], where a weak value defined as $\hat{A}_w = \langle\Psi_{fin}|\,\hat{A}\,|\Psi_{ini}\rangle/\langle\Psi_{fin}|\Psi_{ini}\rangle$ can be obtained with certain pre- and post-selected systems, represented by $|\Psi_{ini}\rangle$ and $|\Psi_{fin}\rangle$ with minimal disturbance on the measured system. Note that weak values lie over the range of eigenvalues [59, 60] and may even be complex [61]. The weak value was also used as an amplifier to discover new physical effects that could not be otherwise detected[62]. Recently, another counterfactual paradox, called quantum Cheshire cat, attracted attention: in pre- and post-selected circumstances, a cat, i.e. a particle, is in one place and its grin, e.g. a spin is in another [14, 15].

4.1. Quantum Cheshire cat in a neutron interferometer experiment

In our neutron interferometer experiment, neutron plays a role of the cat and its spin does the grin [63]. The concept of quantum Cheshire cat as well as the experimental setup is depicted in Fig.7. The pre- and post-selected states are set as

$$\begin{cases} |\Psi_i\rangle = \frac{1}{\sqrt{2}}\{|+x\rangle\,|I\rangle + |-x\rangle\,|II\rangle\} \\ |\Psi_f\rangle = \frac{1}{\sqrt{2}}\,|-x\rangle\,\{|I\rangle + |II\rangle\}. \end{cases} \quad (7)$$

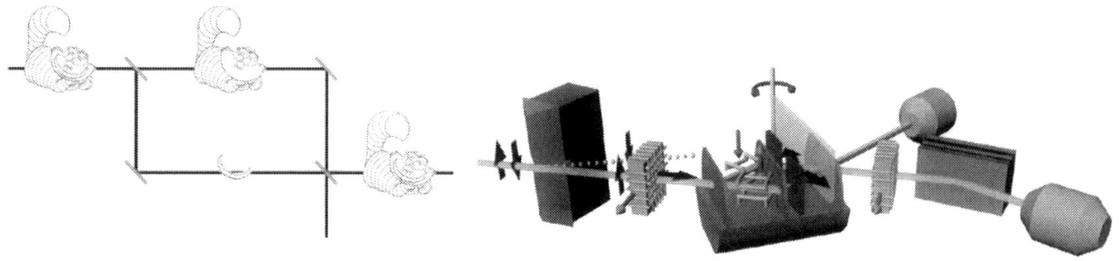

Figure 7. Quantum Cheshire cat in concept (on the left) and the experimental setup with neutron interferometer (on the right). In the Mach-Zehnder interferometer, a cat is in the upper beam path while the grin is in the lower beam path. In the neutron interferometric version, an incident beam is polarized and falls on the interferometer. The pre-selected state $|\Psi_i\rangle$ is generated in the interferometer and the post selection on the state $|\Psi_f\rangle$ is carried out on the beam leaving the interferometer.

To characterize the neutron's population in the interferometer and the location of its spin, weak values of the observables, $\hat{\Pi}_j \equiv |j\rangle\langle j|$ and $\langle\hat{\sigma}_z^s\hat{\Pi}_j\rangle$ with $j = I$ and II are determined. Theory predicts the values, $\langle\hat{\Pi}_I\rangle_w = 0$, $\langle\hat{\Pi}_{II}\rangle_w = 1$, $\langle\hat{\sigma}_z^s\hat{\Pi}_I\rangle_w = 1$, and $\langle\hat{\sigma}_z^s\hat{\Pi}_{II}\rangle_w = 0$.

In the experiment, the incident neutron beam is polarized by using magnetic prisms and the initial state $|\Psi_i\rangle$ is generated: a pair of water-cooled spin-rotators are employed in the interferometer [64]. After the relative phase χ between the two beams is adjusted by the phase shifter and the beams are recombined at the last plate of the interferometer, the O-beam (interfering beam leaving the interferometer in forward direction) is affected by spin analysis: the post-selection is carried out in combination of the phase shifter and the spin analysis system. Weak measurements of the neutron's population and the spin's location are performed by the use of absorbers and additional spin rotation in one of the beams in the interferometer.

Here, we explain the experimental results qualitatively. First, in the measurements of neutron's population, an absorber is inserted in one of the beam paths in the interferometer. Typical results are shown in Fig. 8: the absorbers in the beam path I (lower path) does not affect the final intensity of the O-beam with a spin-analysis, while intensity of the path II (upper path) decreases according to the strength of the absorber. This suggests that neutrons are traveling through the interferometer, following the beam path I. Next, in the measurement of the location of the neutron's spin, a fairly weak magnetic field is applied in one of the beam path in the interferometer. Typical results are shown in Fig. 9: the magnetic field in the beam path II (upper path) does not affect the final intensity of the O-beam with a spin-analysis, while sinusoidal intensity modulation appears by applying the magnetic field in the path I (lower path). This suggests that neutron's spin, in turn, is traveling through the interferometer, following the beam path II. These results are consistent with the theoretical prediction [15]: neutrons, which are affected by appropriate pre- and post-selection, seem to travel in one of the paths, while their spin is disembodied and located in the other path.

5. Concluding remarks

Experimental tests of quantum mechanics and demonstrations of counterfactual phenomena in quantum mechanics are performed with neutron interferometer and polarimeter: efficient manipulations of the neutron's quantum system enable measurements with high precision. Quantum contextuality is manifested in the studies of bi-partite and tri-partite entanglements between degrees of freedom. In particular, entanglements between different degrees of freedom

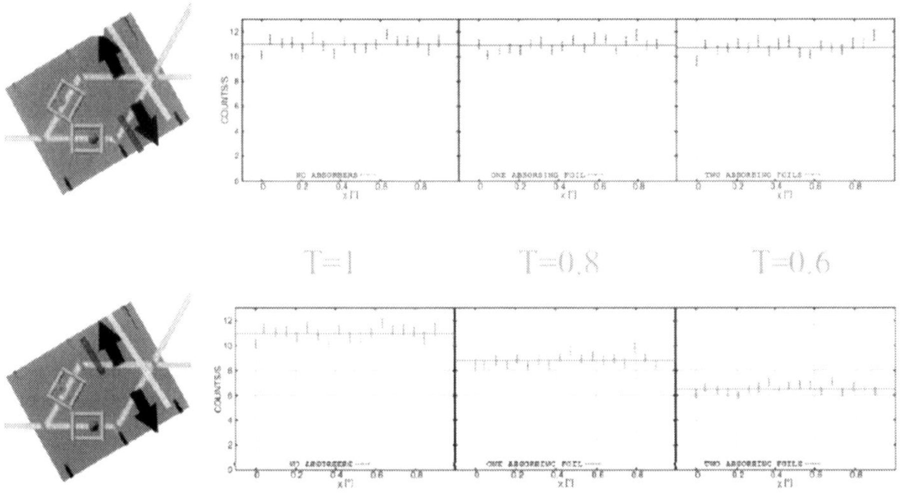

Figure 8. Weak measurements of the neutron's population of the path in the interferometer. Absorbers with transmissivity of 1, 0.8 and 0.6 are inserted in one of the beam path in the interferometer. In the upper panel: absorbers in the path I (lower path) affect the intensity. In the lower panel, the intensity of the path II (upper path) decreases by inserting stronger absorbers. Now, one finds neutrons in the path I.

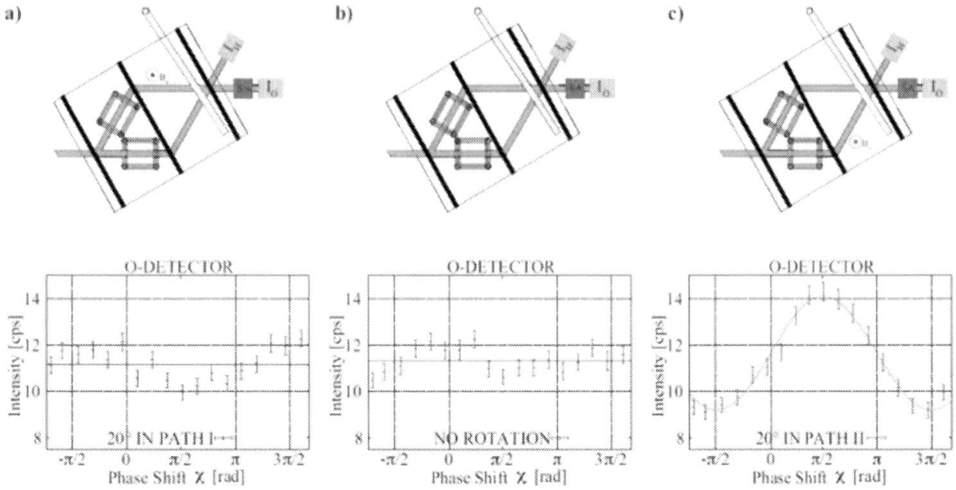

Figure 9. Weak measurements of the neutron's population of the path in the interferometer. Absorbers with transmissivity of 1, 0.8 and 0.6 are inserted in one of the beam path in the interferometer. In the upper panel: absorbers in the path I (lower path) affect the intensity. In the lower panel, the intensity of the path II (upper path) decreases by inserting stronger absorbers. Now, one finds neutrons in the path I.

in single particles are realized and various forms of quantum contextuality have been tested: our investigations with matter waves falsified various hidden-variable theories. Studies of quantum entanglement with matter waves can be extended to generate and utilize a W-like state, a general mixture and even more. The results agreed well with predictions of standard quantum mechanics, although some interpretations are still under discussion. The test of the error-disturbance uncertainty relation, presented here, is actually the first experimental test of this kind: the demonstration is the first experimental evidence for the invalidity of the old (by Heisenberg) and validity of the new (by Ozawa) uncertainty relation. It should be emphasized here that this experiment opens up a new era of uncertainty relation where, after more than 80 years of the Heisenberg's original publication, the topic has revived and come again under hot discussions both from the theoretical and experimental point of view. Our result clarifies a long-standing problem of describing the relation between measurement accuracy and disturbance, and sheds light on fundamental limitations of quantum measurements. One cannot too much emphasize the importance of amending such a fundamental concept not only from a purely academic but also a practical point of view. The studies of counterfactual phenomena of quantum mechanics are presented, where quantum Cheshire cat is generated and observed by using a neutron interferometer setup: neutron and its spin are disembodied. Note that weak values, which are defined and obtained by pre- and post-selection together with special estimation strategy of the intermediate state, allow to create the quantum Cheshire cat. Further investigations of fundamental questions, both conceptual and applied ones, concerning the quantum Cheshire cat, will be possible in the near future.

Acknowledgments
We thank all our colleagues who were involved in carrying our the experiments presented here; in particular, we appreciate G. Badurek, H. Bartosik, B. Demirel, T. Denkmayr, K. Durstberger-Rennhofer, J. Erhart, H. Geppert, A. Hosoya, J. Klepp, H. Lemmel, A. Matzkin, M. Ozawa, H. Rauch, C. Schmitzer, S. Sponar, G. Sulyok and J.Tollaksen. This work was partially supported by the Austrian FWF (Fonds zur Föderung der Wissenschaftlichen Forschung) through grant numbers P25795-N20.

References
[1] W. Heisenberg, Z. Phys. **43**, 172 (1927).
[2] E. Schrödinger, Naturwissenschaften, **23** 807 (1935).
[3] A. Einstein, B. Podolsky and N. Rosen, Phys. Rev. **47**, 777 (1935).
[4] H. Rauch and S.A. Werner, Neutron interferometry, Clarendon Press, Oxford (2000).
[5] Y. Hasegawa and H. Rauch, New J. Phys **13**, 115010 (2011).
[6] Y. Hasegawa and G. Badurek, Phys. Rev. A **59**, 4614 (1999).
[7] For instance, J. Klepp, S. Sponar, S. Filipp, M. Lettner, G. Badurek and Y. Hasegawa, Phys. Rev. Lett. **101**, 150404 (2008).
[8] Y. Hasegawa, R. Loidl, G. Badurek, K. Durstberger-Rennhofer, S. Sponar and H. Rauch, Phys. Rev. A **81**, 032121 (2010).
[9] Y. Hasegawa, C. Schmitzer, H. Bartosik, J. Klepp, S. Sponar, K. Durstberger-Rennhofer and G. Badurek, New J. Phys. **14**, 023039 (2012).
[10] J. Erhart, S. Sponar, G. Sulyok, G. Badurek, M. Ozawa and Y. Hasegawa, Nature Phys. **8**, 185 (2012).
[11] G. Sulyok, S. Sponar, J. Erhart, G. Badurek, M. Ozawa and Y. Hasegawa, Phys. Rev. A **88**, 022110 (2013).
[12] M. Ozawa, Phys. Rev. A **67**, 042105 (2003); ibid., Phys. Lett. A **318**, 21 (2003).
[13] M. Ozawa, Ann. Phys. (N.Y.) **311**, 350 (2004).
[14] Y. Aharonov and D. Rohrlich, Quantum Paradoxes: Quantum Theory for the Perplexed, Wiley-VCH (2005).
[15] Y. Aharonov, S. Popescu, D. Rohrlich, P. Skrzypczyk, New J. Phys. **15**, 113018 (2013).
[16] T. Denkmayr, H. Geppert, S. Sponar, H. Lemmel, A. Matzkin, J. Tollaksen and Y. Hasegawa (in preparation).
[17] J. S. Bell, Physics **1**, 195 (1964).
[18] D.M. Greenberger, M.A. Horne, A. Zeilinger, in M. Kafatos (Ed.), Bell's Theorem, Quantum Theory and Conceptions of the Universe, Kluwer Academic, Dordrecht (1989), p. 69-72.

[19] D. M. Greenberger, M. A. Horne, A. Shimony and A. Zeilinger, Am. J. Phys. **58**, 1131 (1990).
[20] N. D. Mermin, Phys. Rev. Lett. **65**, 1838 (1990).
[21] J. W. Pan, D. Bouwmeester, M. Daniell, H. Weinfurter and A. Zeilinger, Nature **403**, 515 (2000).
[22] Z. Zhao, Y. A. Chen, A. N. Zhang, T. Yang, H. J. Briegel and J. W. Pan, Nature **430**, 54 (2004).
[23] C. A. Sackett, D. Kielpinski, B. E. King, C. Langer, V. Meyer, C. J. Myatt, M. Rowe, W. A. Turchette, W. M. Itano, D. J. Wineland and C. Monroe, Nature **404**, 256 (2000).
[24] S. Kochen and E. P. Specker, J. Math. Mech. **17**, 59 (1967).
[25] N. D. Mermin, Rev. Mod. Phys. **65**, 803 (1990).
[26] Y. Hasegawa, R. Loidl, G. Badurek, M. Baron and H. Rauch, Nature (London) **425**, 45 (2003).
[27] E. Amselem, M. Radmark, M. Bourennane and A. Cabello, Phys. Rev. Lett. **103**, 160405 (2009).
[28] Y.Hasegawa, R. Loidl, G. Badurek, M. Baron and H. Rauch, Phys. Rev. Lett. **97**, 230401 (2006).
[29] H. Bartosik, J. Klepp, C. Schmitzer, S. Sponar, A. Cabello, H. Rauch and Y. Hasegawa, Phys. Rev. Lett. **103**, 040403 (2009).
[30] Y. Hasegawa, R. Loidl, G. Badurek, S. Filipp, J. Klepp and H. Rauch, Phys. Rev. A **76**, 05210 (2007).
[31] S. Sponar, J. Klepp, R. Loidl, S. Filipp, G. Badurek, Y. Hasegawa and H. Rauch, Phys. Rev. A **78**, 061604(R) (2008).
[32] N. D. Mermin, Am. J. Phys. **58**, 731 (1990).
[33] N. D. Mermin, Phys. Rev. Lett. **65**, 3373 (1990).
[34] J. Summhammer, Phys. Rev. A **47**, 556 (1993).
[35] R. Golub, R. Gähler and T. Keller, Am. J. Phys. **62**, 779 (1994).
[36] S. Sponar, J. Klepp, R. Loidl, S. Filipp, G. Badurek, Y. Hasegawa and H. Rauch, Phys. Lett. A **372**, 3153 (2008).
[37] D. Erdösi, M. Huber, B.C Hiesmayr and Y. Hasegawa, New J. Phys. **14**, 023039 (2013).
[38] S. Sponar, J. Klepp, C. Zeiner, G. Badurek and Y. Hasegawa, Phys. Lett. A **374** 431 (2010).
[39] S. Sponar, J. Klepp, K. Durstberger-Rennhofer, C. Schmitzer, H. Bartosik, H. Geppert, M. Both, G. Badurek and Y. Hasegawa, New J. Phys. **15**, 023033 (2013).
[40] A.J. Leggett, Found. Phys. **33**, 1469 (2003).
[41] R. Colbeck and R. Renner, Phys. Rev. Lett. **101**, 050403 (2008).
[42] S. Gröblacher, T. Paterek, R. Kaltenbaek, Č. Brukner, M. Żukowski, M. Aspelmeyer and A. Zeilinger, Nature **446**, 871 (2007); ibid, Nature **449**, 252 (2007) (corrigendum).
[43] T. Paterek, A. Fedrizzi, S. Gr¨oblacher, T. Jennewein, M. Żukowski, M. Aspelmeyer and A. Zeilinger, Phys. Rev. Lett. **99**, 210406 (2007).
[44] C. Branciard, A. Ling, N. Gisin, C. Kurtsiefer, A. Lamas-Linares and V. Scarani, Phys. Rev. Lett. **99**, 210407 (2007).
[45] C. Branciard, N. Brunner, N. Gisin, C. Kurtsiefer, A. Lamas-Linares and V. Scarani, Nature Phys. **4**, 681 (2008).
[46] J.A. Wheeler and W.H. Zurek, (eds) Quantum Theory and Measurement (Princeton Univ. Press, 1983).
[47] E.H. Kennard, Z. Phys. **44**, 326 (1927).
[48] H.P. Robertson, *Sitzungsberichte der Preussischen Akademie der Wissenschaften.* **14**, 296 (1930); *ibid.* Rev. Mod. Phys. **42**, 358 (1970).
[49] L.A. Rozema, A. Darabi, D.H. Mahler, A. Hayat, Y.Soudagar and A.M. Steinberg, Phys. Rev. Lett. **109**, 100404 (2012).
[50] S.Y. Baek, F. Kaneda, M. Ozawa and K. Edamatsu, Sci. Rep. **3**, 02221 (2013).
[51] C. Branciard, Proc. Natl. Acad. Sci. USA **110**, 6742 (2013).
[52] P. Busch, P. Lahti and F. Werner, Phys. Rev. Lett. **111**, 160405 (2013).
[53] L.A. Rozema, D.H. Mahler, A. Hayat and A.M. Steinberg, arXiv:1307.3604 (2013).
[54] M. Ozawa, arXiv:1308.3540 (2013).
[55] L. Hardy, Phys. Rev. Lett. **68**, 2981 (1992).
[56] J.S. Lundeen and A.M. Steinberg, Phys. Rev. Lett. **102**, 020404 (2009).
[57] K. Yokota, T. Yamamoto, M. Koashi and N. Imoto, New J. Phys. **11**, 033011 (2009).
[58] Y. Aharonov, D.Z. Albert and L.H. Vaidman, Phys. Rev. Lett. **60**, 1351 (1988).
[59] I.M. Duck, P.M. Stevenson and E.C.G.Sudarshan, Phys. Rev. D **40**, 2112 (1989).
[60] N.W.M. Ritchie, J.G. Story and R.G. Hulet, Phys. Rev. Lett. **66**, 1107 (1991).
[61] R. Jozsa, Phys. Rev. A **76**, 044103 (2007).
[62] O. Hosten and P. Kwiat, Science, **319**, 787 (2008).
[63] T. Denkmayr, H. Geppert, S. Sponar, H. Lemmel, A. Matzkin, J. Tollaksen and Y. Hasegawa, arXiv:1312.3775 (2013).
[64] H. Geppert, T. Demkmayr, S. Sponar, H. Lemmel and Y. Hasegawa (in preparation).

Event-based simulation of neutron experiments: interference, entanglement and uncertainty relations

Kristel Michielsen[1,2] **and Hans De Raedt**[3]

[1] Institute for Advanced Simulation, Jülich Supercomputing Centre, Forschungszentrum Jülich, D-52425 Jülich, Germany
[2] RWTH Aachen University, D-52056 Aachen, Germany
[3] Department of Applied Physics, Zernike Institute for Advanced Materials, University of Groningen, Nijenborgh 4, NL-9747 AG Groningen, The Netherlands

E-mail: k.michielsen@fz-juelich.de, h.a.de.raedt@rug.nl

Abstract. We discuss a discrete-event simulation approach, which has been shown to give a unified cause-and-effect description of many quantum optics and single-neutron interferometry experiments. The event-based simulation algorithm does not require the knowledge of the solution of a wave equation of the whole system, yet reproduces the corresponding statistical distributions by generating detection events one-by-one. It is showm that single-particle interference and entanglement, two important quantum phenomena, emerge via information exchange between individual particles and devices such as beam splitters, polarizers and detectors. We demonstrate this by reproducing the results of several single-neutron interferometry experiments, including one that demonstrates interference and one that demonstrates the violation of a Bell-type inequality. We also present event-based simulation results of a single neutron experiment designed to test the validity of Ozawa's universally valid error-disturbance relation, an uncertainty relation derived using the theory of general quantum measurements.

1. Introduction

The mathematical framework of quantum theory allows for the calculation of numbers which can be compared with experimental data as long as these numbers refer to statistical averages of measured quantities. However, as soon as an experiment records individual detector clicks which contribute to the statistical average of a quantity then a fundamental problem, related to the quantum measurement problem [1,2], appears. An example of such experiments are interference experiments in which the interference pattern is built up by successive discrete detection events. Another example are Bell-test experiments in which the correlations between two degrees of freedom are computed as averages of pairs of detection events which are seen to take values that are reminiscent of those of the singlet state in the quantum theoretical description of the thought experiment.

An intriguing question to be answered is why individual entities which do not interact with each other can exhibit the collective behavior that gives rise to the observed interference pattern and why two degrees of freedom can show correlations corresponding to those of the singlet state. Since quantum theory postulates that it is fundamentally impossible to go beyond the description in terms of probability distributions, an answer in terms of a cause-and-effect

Journal of Physics: Conference Series **504** (2014) 012026 doi:10.1088/1742-6596/504/1/012026

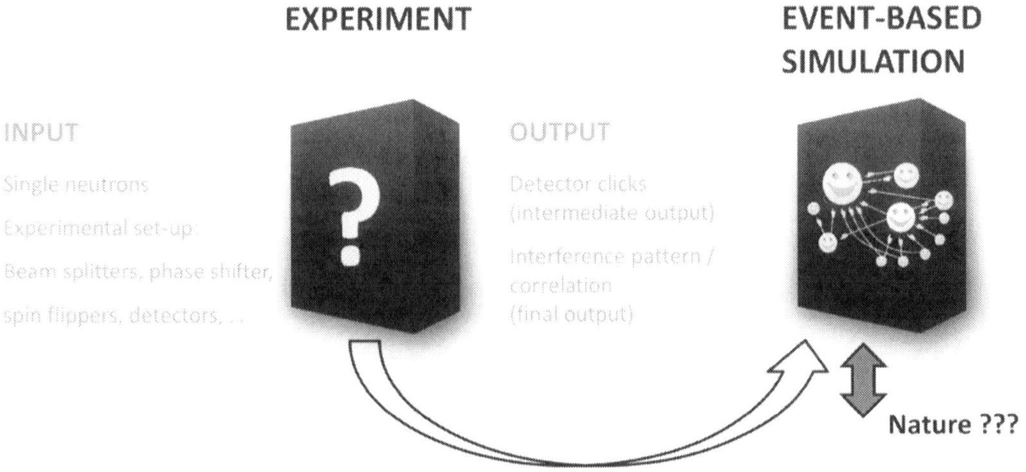

Figure 1. Schematic of the working principle of the event-based simulation approach. The first step consists of an analysis of the experiment, keeping in mind that the devil is in the detail. Information about the input, such as characteristics of the particle source and all other components in the experimental setup, and about the output such as the detector clicks (intermediate output) and the interference pattern or correlation (final output) including the data analysis procedure, is collected. It is assumed that no information is available about how the input is transformed into the output. In a second step the "black box" that connects input and output in the experiment is replaced by a set of simple rules that transform the same input into the same output. The question whether the rules describe what is going in Nature cannot be answered since information necessary to answer this question is not available.

description of the observed phenomena cannot be given within the framework of quantum theory. We provide an answer by constructing an event-based simulation model that reproduces the statistical distributions of quantum (and Maxwell's) theory without solving a wave equation but by modeling physical phenomena as a chronological sequence of events whereby events can be actions of an experimenter, particle emissions by a source, signal generations by a detector, interactions of a particle with a material and so on [3–5]. The underlying assumption of this simulation approach is that current scientific knowledge derives from the discrete events which are observed in laboratory experiments and from relations between those events. Hence, the event-based simulation approach concerns what we can say about these experiments but not what "really" happens in Nature.

The general idea of the event-based simulation method (see Fig. 1) is that simple rules define discrete-event processes which may lead to the behavior that is observed in experiments. The basic strategy in designing these rules is to carefully examine the experimental procedure and to devise rules such that they produce the same kind of data as those recorded in experiment, while avoiding the trap of simulating thought experiments that are difficult to realize in the laboratory. Evidently, mainly because of insufficient knowledge, the rules are not unique. Hence, the simplest rules one could think of can be used until a new experiment indicates otherwise. The method may be considered to be entirely classical since it only uses concepts of the macroscopic world, but some of the rules are not those of classical Newtonian dynamics. An overview of the method and its applications can be found in Refs. [3–5]. Here we present an application of the method to single-neutron experiments showing interference [6,7], the violation of a Bell-type inequality [8], and the validity of Ozawa's universally valid error-disturbance relation [9,10].

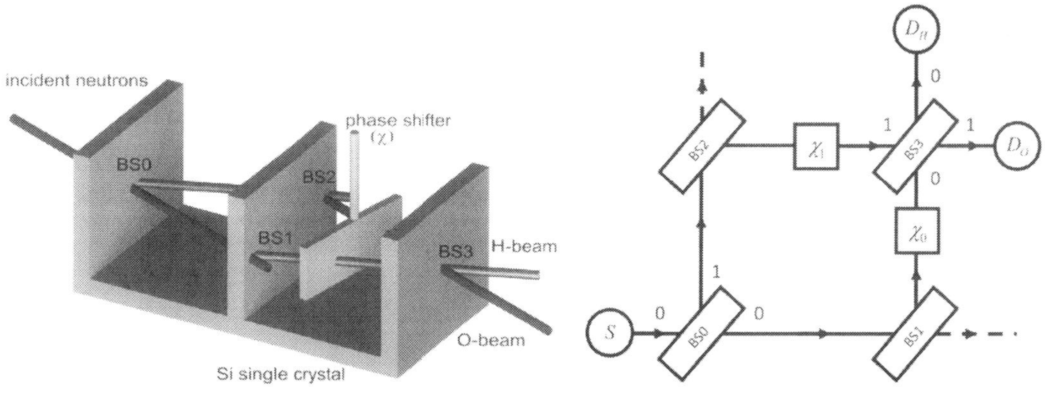

Figure 2. Left: Schematic picture of the silicon-perfect-crystal neutron interferometer [6]. BS0, ..., BS3: beam splitters; phase shifter (χ): aluminum foil; neutrons transmitted by BS1 or BS2 leave the interferometer and do not contribute to the interference signal. Detectors count the number of neutrons in the O- and H-beam. Right: Event-based network of the interferometer shown on the left. S: single neutron source; BS0, ... , BS3: beam splitters; χ_0, χ_1: phase shifters; D_O, D_H: detectors counting all neutrons leaving the interferometer via the O- and H-beam, respectively. Neutrons transmitted by BS1 or BS2 (dashed lines) are not counted.

2. Neutron interference

One of the most fundamental experiments in quantum physics is the single-particle double-slit experiment. Feynman stated that the phenomenon of electron diffraction by a double-slit structure is "impossible, *absolutely* impossible, to explain in any classical way, and has in it the heart of quantum mechanics. In reality it contains the only mystery." [11] While Young's original double-slit experiment helped to establish the wave theory of light [12], variants of the experiment over the years with electrons [13–20], single photons [21–24], neutrons [6, 25, 26], atoms [27, 28], molecules [29–31] and droplets [32] provided illustrations of the necessity of wave-particle duality in quantum theory [2].

In classical optics, both diffraction and interference play a role in the two-slit experiment and are manifestations of the wave character of these phenomena. However, an interference pattern identical in form to that of classical optics can be observed by collecting many detector spots or clicks which are the result of electrons, photons, neutrons, atoms or molecules travelling one-by-one through a double-slit structure. In these experiments the so-called interference pattern is the statistical distribution of the detection events (spots at or clicks of the detector). Hence, in these particle-like experiments, only the correlations between detection events reveal interference. In what follows we use the term interference pattern for the statistical distribution of detection events.

Various experiments have been designed to study interference phenomena. In simple terms, an interferometer is a device containing a beam splitter that splits an incident beam into two or more mutually coherent beams. These beams are then recombined, usually by employing mirrors and/or additional beam splitters, on a detection screen showing interference fringes. A well-known type of interferometer which is frequently used is the Mach-Zehnder interferometer (MZI). In a MZI the incoming beam is split by a beam splitter. The two resulting beams are each reflected by a mirror. One of the beams passes through a phase shifter, then both beams pass through a second beam splitter and impinge on two detectors which are placed behind the two output beams of the second beam splitter.

In neutron optics there exist various realizations of the MZI type of interferometer, but here we only consider a triple Laue diffraction type of interferometer [6, 26, 33]. Figure 2 (left) shows the experimental configuration [6]. The three parallel crystal plates act as beam splitters and are assumed to be identical, which means that they have the same transmission and reflection properties [26]. The three plates have to be parallel to high accuracy [6] and the whole device needs to be protected from vibrations in order to observe interference [7]. The rotatable-plate phase shifter between the second and third plate can for example be an aluminum foil which hardly absorbs neutrons [26]. Minute rotations of the foil about an axis perpendicular to the base plane of the interferometer induce large variations in the phase difference $\chi = \chi_0 - \chi_1$ [26, 34]. The neutron detectors placed in the so-called H-beam or O-beam have a very high, almost 100%, efficiency [26].

A coherent monoenergetic neutron beam is split by the first plate (BS0). Neutrons refracted by beam splitters BS1 and BS2 (second plate), playing the role of the mirrors in the optical MZI, are directed to the third plate (BS3) thereby first passing through the phase shifter, and are detected by one of the two detectors. Neutrons which are not refracted by BS1 and BS2 leave the interferometer and do not contribute to the detection counts. The intensities in the O- and H-beam, obtained by counting individual neutrons for a certain amount of time, exhibit sinusoidal variations as a function of the phase shift χ, a characteristic of interference [26].

The experiment can be interpreted in different ways. In the quantum-corpuscular view a wave packet is associated with each individual neutron. At BS0 the wave packet splits in two components, one directed towards BS1 and one towards BS2. At BS1 and BS2 these two components each split in two. Two of the in total four components leave the interferometer and the other two components are redirected towards each other at BS3 where they recombine. At BS3 the recombined wave packet splits again into two components. Only one of these two components triggers a detector. It is a mystery how four components of a wave packet can conspire to do such things. Assuming that only a neutron, not merely a part of it can trigger the nuclear reaction that causes the detector to "click", on elementary logical grounds, the argument that was just given rules out a wave-packet picture for the individual neutron (invoking the wave function collapse only adds to the mystery).

In the statistical interpretation of quantum mechanics there is no such conflict of interpretation [2, 35]. As long as we consider descriptions of the statistics of the experiment with many neutrons, we may think of one single "probability wave" propagating through the interferometer and as the statistical interpretation of quantum theory is silent about single events, there is no conflict with logic either [26]. However, the statistical interpretation does not offer an explanation of the fact that each measurement gives a definite result.

Therefore, the question to answer is: Can one model single detection events that give rise to an interference pattern if the detection events are the result of non-interacting neutrons arriving one-by-one at the detector? In what follows we demonstrate that it is possible to construct a logically consistent, cause-and-effect description in terms of discrete-event, particle like processes which produce results that agree with those of neutron interferometry experiments (individual detection events **and** an interference pattern after many single detection events have been collected) and the quantum theory thereof (interference pattern only).

2.1. Event-based model

We construct a model for the neutrons and for the various units in the block diagram (see Fig. 2 (right)) representing the setup of the neutron interferometry experiment (see Fig. 2 (left))

- *Neutrons:* A neutron is regarded as a messenger carrying a message. In the following we use a pictorial description, which should not be taken literally, to determine in terms of macroscopic physics the minimal content of the message. From other neutron experiments we deduce that the neutrons have a magnetic moment and that they move

Journal of Physics: Conference Series **504** (2014) 012026 doi:10.1088/1742-6596/504/1/012026

from one point in space to another within a certain time period, the time of flight [26]. Hence, both the magnetic moment and the time of flight have to be encoded in the message. We model the neutron as a tiny classical magnet spinning around the direction $\mathbf{m} = (\cos\delta\sin\theta, \sin\delta\sin\theta, \cos\theta)$, relative to a fixed frame of reference defined by a static magnetic field. Apart from δ and θ, which completely specify the magnetic moment of the neutron, a third degree of freedom is required to specify the time of flight. This is conveniently done by representing the message by the two-dimensional unit vector

$$\mathbf{u} = (e^{i\psi^{(1)}}\cos(\theta/2), e^{i\psi^{(2)}}\sin(\theta/2)), \tag{1}$$

where $\psi^{(i)} = \nu t + \delta_i$, for $i = 1, 2$ and $\delta = \delta_1 - \delta_2 = \psi^{(1)} - \psi^{(2)}$. Here, t specifies the time of flight of the neutron and ν is an angular frequency which is characteristic for a neutron that moves with a fixed velocity v. The direction of the magnetic moment $\mathbf{m}$ follows from Eq. (1) through $\mathbf{m} = \mathbf{u}^T\sigma\mathbf{u}$, where $\sigma = (\sigma^x, \sigma^y, \sigma^z)$ with $\sigma^x, \sigma^y, \sigma^z$ denoting the Pauli matrices (here we use the isomorphism between the algebra of Pauli matrices and rotations in three-dimensional space).

Within the present model, the state of the neutron is fully determined by the angles $\psi^{(1)}$, $\psi^{(2)}$ and θ and by rules (to be specified), by which these angles change as the neutron travels through the network of devices.

The rules are as follows. A messenger with message $\mathbf{u}$ at time t and position $\mathbf{r}$ that travels with velocity v, along the direction $\mathbf{q}$ during a time interval $t' - t$, changes its message according to $\psi^{(i)} \leftarrow \psi^{(i)} + \phi$ for $i = 1, 2$, where $\phi = \nu(t' - t)$. Hence, the new message becomes $\mathbf{w} = e^{i\phi}\mathbf{u}$. In the presence of a magnetic field $\mathbf{B} = (B_x, B_y, B_z)$, the magnetic moment rotates about the direction of $\mathbf{B}$ according to the classical equation of motion $d\mathbf{m}/dt = \mathbf{m} \times \mathbf{B}$. Hence, in a magnetic field the message $\mathbf{u}$ is changed into the message $\mathbf{w} = e^{ig\mu_N T\sigma\cdot\mathbf{B}/2}\mathbf{u}$, where g denotes the neutron g-factor, μ_N the nuclear magneton and T the time during which the neutron experiences the magnetic field.

- *Source:* From the experiment we deduce that it is very unlikely that there is more than one neutron in the interferometer at a time [26]. Therefore, we model the source such that it creates messengers one-by-one. The source waits until the messenger's message has been processed by the detector before creating the next messenger. Hence, there can be no direct communication between the messengers. When the source creates a messenger, its message is initialized. This means that the three angles $\psi^{(1)}$, $\psi^{(2)}$ and θ are specified. The specification depends on the type of source that has to be simulated. A monochromatic beam of incident neutrons is assumed to consist of neutrons that all have the same value of ν [26]. For a fully coherent spin-polarized beam of neutrons, the three angles have different but the same values for all the messengers being created. Hence, three random numbers are used to specify $\psi^{(1)}$, $\psi^{(2)}$ and θ for all messengers.
- *Beam splitters* BS0, ... , BS3: To model the beam splitter for a neutron beam we exploit the similarity between the magnetic moment of the neutron and the polarization of a photon. Therefore we use a model for the beam splitter that is similar to the one used for polarized photons, thereby assuming that neutrons with spin up and spin down have the same reflection and transmission properties, while photons with horizontal and vertical polarization have different reflection and transmission properties [36]. The coordinate system is defined by the static magnetic guiding field.

  In optics, dielectric plate beam splitters are often used to partially transmit and partially reflect an incident light beam. From classical electrodynamics we know that if an electric field is applied to a dielectric material the material becomes polarized [36]. Assuming a linear response, the polarization vector of the material is given by $\mathbf{P}(\omega) = \chi(\omega)\mathbf{E}(\omega)$ for a monochromatic wave with frequency ω. In the time domain, this relation expresses the fact that the material retains some memory about the incident field, $\chi(\omega)$ representing the memory kernel that is characteristic for the material used. We use a similar kind of memory effect in our algorithm to model the beam splitter for neutrons. Note that memory effects are also essential for uncertainty and quantization phenomena observed in a system of droplets bouncing on a liquid bath [37–39].

A beam splitter has two input and two output channels labeled by 0 and 1 (see Fig 2 (right)). Note that in the neutron interferometry experiment, for beamsplitter BS0 only entrance port $k=0$ is used. In the event-based model, the beam splitter has two internal registers $\mathbf{R}_{k,n}=(R_{0,k,n},R_{1,k,n})$ (one for each input channel) with $R_{i,k,n}$ for $i=0,1$ representing a complex number, and an internal vector $\mathbf{v}_n=(v_{0,n},v_{1,n})$ with the additional constraints that $v_{i,n}\geq 0$ for $i=0,1$ and that $v_{0,n}+v_{1,n}=1$. As we only have two input channels, the latter constraint can be used to recover $v_{1,n}$ from the value of $v_{0,n}$. These three two-dimensional vectors $\mathbf{v}_n$, $\mathbf{R}_{0,n}$ and $\mathbf{R}_{1,n}$ are labeled by the message number n because their content is updated every time the beam splitter receives a message. Before the simulation starts we set $\mathbf{v}_0=(v_{0,0},v_{1,0})=(r,1-r)$, where r is a uniform pseudorandom number. In a similar way we use pseudo-random numbers to initialize $\mathbf{R}_{0,0}$ and $\mathbf{R}_{1,0}$.

When the nth messenger carrying the message $\mathbf{u}_{k,n}$ arrives at entrance port $k=0$ or $k=1$ of the beam splitter, the beam splitter first stores the message in the corresponding register $\mathbf{R}_{k,n}$ and updates its internal vector according to the rule

$$\mathbf{v}_n=\gamma\mathbf{v}_{n-1}+(1-\gamma)\mathbf{q}_n, \tag{2}$$

where $0<\gamma<1$ and $\mathbf{q}_n=(1,0)$ ($\mathbf{q}_n=(0,1)$) if the nth event occurred on channel $k=0$ ($k=1$). Note that exactly ten floating point numbers can be stored in the registers $\mathbf{R}_{k,n}$ and the internal vector $\mathbf{v}_n$. The beam splitter stores information about the last message only. The information carried by earlier messages is overwritten by updating the internal registers and internal vector. By construction $v_{i,n}\geq 0$ for $i=0,1$ and $v_{0,n}+v_{1,n}=1$. Hence the update rule Eq. (2) preserves the constraints on the internal vector. Obviously, these constraints are necessary if we want to interpret the $v_{k,n}$ as (an estimate of) the relative frequency for the occurrence of an event of type k.

From Eq. (2), one could say that the internal vector $\mathbf{v}$ (corresponding to the material polarization $\mathbf{P}$ in classical electromagnetism) is the response of the beam splitter to the incoming messages represented by the vectors $\mathbf{q}$ (corresponding to the electric field $\mathbf{E}$). Elsewhere [40] we have shown that the update rule Eq. (2) for the internal vector leads to the Debye model for the interaction between material and electric field. In other words, the beam splitter "learns" so to speak from the information carried by the messengers. The characteristics of the learning process depend on the parameter γ (corresponding to the response function χ). Since this learning mechanism is completely deterministic we call this type of message processor a deterministic learning machine (DLM) with a parameter γ that controls the pace and accuracy of the learning process.

The next step is to use the ten numbers stored in $\mathbf{R}_{k,n}$ and $\mathbf{v}_n$ to calculate four complex numbers

$$\begin{pmatrix} h_{0,n}\\ h_{1,n}\\ h_{2,n}\\ h_{3,n}\end{pmatrix}=\begin{pmatrix}\sqrt{\mathcal{T}} & i\sqrt{\mathcal{R}} & 0 & 0\\ i\sqrt{\mathcal{R}} & \sqrt{\mathcal{T}} & 0 & 0\\ 0 & 0 & \sqrt{\mathcal{T}} & i\sqrt{\mathcal{R}}\\ 0 & 0 & i\sqrt{\mathcal{R}} & \sqrt{\mathcal{T}}\end{pmatrix}$$
$$\times\begin{pmatrix}\sqrt{v_{0,n}} & 0 & 0 & 0\\ 0 & \sqrt{v_{1,n}} & 0 & 0\\ 0 & 0 & \sqrt{v_{0,n}} & 0\\ 0 & 0 & 0 & \sqrt{v_{1,n}}\end{pmatrix}\begin{pmatrix}R_{0,0,n}\\ R_{0,1,n}\\ R_{1,0,n}\\ R_{1,1,n}\end{pmatrix}$$

$$= \begin{pmatrix} \sqrt{v_{0,n}}\sqrt{\mathcal{T}}R_{0,0,n} + i\sqrt{v_{1,n}}\sqrt{\mathcal{R}}R_{0,1,n} \\ i\sqrt{v_{0,n}}\sqrt{\mathcal{R}}R_{0,0,n} + \sqrt{v_{1,n}}\sqrt{\mathcal{T}}R_{0,1,n} \\ \sqrt{v_{0,n}}\sqrt{\mathcal{T}}R_{1,0,n} + i\sqrt{v_{1,n}}\sqrt{\mathcal{R}}R_{1,1,n} \\ i\sqrt{v_{0,n}}\sqrt{\mathcal{R}}R_{1,0,n} + \sqrt{v_{0,n}}\sqrt{\mathcal{T}}R_{1,1,n} \end{pmatrix}, \tag{3}$$

where the reflectivity $\mathcal{R}$ and transmissivity $\mathcal{T} = 1 - \mathcal{R}$ are real numbers which are considered to be parameters to be determined from experiment.
It is clear that the computation of the numbers $h_{0,n}, \ldots, h_{3,n}$ plays the role of the matrix-vector multiplication in the quantum-theoretical description of the beam splitter. Note however that the input and output amplitudes are constructed event-by-event and only under certain conditions ($\gamma \to 1^-$, sufficiently large number of input events N, stationary sequence of input events) they correspond to their quantum theoretical counterparts.
In the final step $h_{0,n}, \ldots, h_{3,n}$ are used to create an output event. To this end a uniform random number $0 < r_n < 1$ is generated. If $|h_{0,n}|^2 + |h_{2,n}|^2 > r_n$, a message

$$\mathbf{w}_{0,n} = (h_{0,n}, h_{2,n})/\sqrt{|h_{0,n}|^2 + |h_{2,n}|^2}, \tag{4}$$

is sent through output channel 1. Otherwise a message

$$\mathbf{w}_{1,n} = (h_{1,n}, h_{3,n})/\sqrt{|h_{1,n}|^2 + |h_{3,n}|^2}, \tag{5}$$

is sent through output channel 0.

- *Phase shifter* χ_0, χ_1: A phase shifter is simulated without DLM. The device has only one input and one output port and transforms the nth input message $\mathbf{u}_n$ into an output message

$$\mathbf{w}_n = e^{i\chi_j}\mathbf{u}_n \quad j = 0, 1, \tag{6}$$

where χ_j denotes the phase.
- *Detector:* Detectors count all incoming neutrons and hence have a detection efficiency of 100%. This is an idealization of real neutron detectors which can have a detection efficiency of 99% and more [7]. After detection the neutron is destroyed.

2.2. Simulation results

If γ, the parameter which controls the learning pace of a DLM-based processor, approaches one the event-by-event simulation reproduces the results of quantum theory [3, 5, 41, 42]. In the neutron interferometry experiment the parameter γ, controlling the learning pace of the beam splitters BS0, ..., BS3, can be used to account for imperfections of the neutron interferometer [5].

In Fig. 3, we present a comparison of simulation data with experimental data using two different experimental data sets. For Fig. 3 (left) we use experimental data of an experiment performed in 2000 by Kroupa *et al.* [7]. For Fig. 3 (right) we use data from a more recent experiment. Both figures show that the event-based simulation model reproduces, quantitatively, the experimental results. We only had to use different values for γ to get a good fit to the experimental curves. For the oldest experiment we used $\gamma = 0.5$ and for the more recent one $\gamma = 0.7$. This could indicate that the more recent experimental setup suffers from less imperfections (note also the difference in neutron counts) so that a better agreement with quantum theory is found. In this respect γ could also be interpreted as a kind of "(de)coherence parameter".

In conclusion, the event-based simulation approach models the experimentally observed one-by-one build-up process of the interference pattern as a chronological sequence of different events including the source emitting neutrons, the neutrons interacting with the beam splitters and the neutrons arriving at the detector thereby generating detection events.

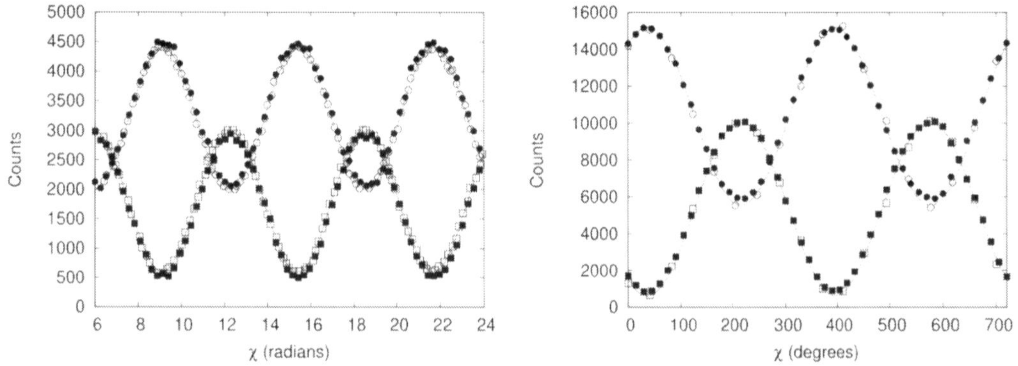

Figure 3. Comparison between the counts of neutrons per second and per square cm in the beams of a neutron interferometry experiment (open symbols) and the number of neutrons per sample leaving the interferometer in an event-by-event simulation (solid symbols). Circles: counts in the H-beam; squares: counts in the O-beam. Lines through the data points are guides to the eye. Left: experimental data extracted from Fig. 2 of Ref. [7]; simulation parameters: $\mathcal{R} = 0.22$, $\gamma = 0.5$, number of incident neutrons per angle χ: $N = 22727$. Right: experimental data provided by H. Lemmel and H. Rauch (private communication); simulation parameters: $\mathcal{R} = 0.22$, $\gamma = 0.7$, number of incident neutrons per angle χ: $N = 72727$. The parameters $\mathcal{R}$ and γ have been adjusted by hand to obtain a good fit.

3. Neutron spin-path entanglement

Now that we have given an event-based description of the fact that individual neutrons which do not interact with each other exhibit the collective behavior that gives rise to an interference pattern, another intriguing question to answer is whether one can model single detection events that give rise to entanglement, another quantum mechanical phenomenon, if the detection events are the result of non-interacting neutrons arriving one-by-one at the detector.

The single-neutron interferometry experiment of Hasegawa *et al.* [8] demonstrates that the correlation between the spatial and spin degrees of freedom of neutrons violates a Bell-CHSH (Clauser-Horne-Shimony-Holt) inequality. In this section we construct an event-based model that reproduces this correlation. We show that the event-based model reproduces the exact results of quantum theory if $\gamma \to 1^-$ and that by changing γ it can also reproduce the numerical values of the correlations, as measured in experiments [8,43]. Note that this Bell-test experiment involves two degrees of freedom of one particle, while the Einstein-Podolsky-Rosen-Bohm (EPRB) thought experiment [44] and EPRB experiments with single photons [45–48] involve two degrees of freedom of two particles. Hence, the Bell-test experiment with single neutrons is not performed according to the CHSH protocol [49] because the two degrees of freedom of one particle are not manipulated and measured independently.

Figure 4 (left) shows a schematic picture of the single-neutron interferometry experiment. Incident neutrons pass through a magnetic-prism polarizer (not shown) which produces two spatially separated beams of neutrons with their magnetic moments aligned parallel (spin up), respectively anti-parallel (spin down) with respect to the magnetic axis of the polarizer which is parallel to the guiding field **B**. The spin-up neutrons impinge on a silicon-perfect-crystal interferometer [26]. On leaving the first beam splitter BS0, neutrons are transmitted or refracted. A mu-metal spin-turner changes the orientation of the magnetic moment of the neutron from parallel to perpendicular to the guiding field **B**. Hence, the magnetic moment of the neutrons

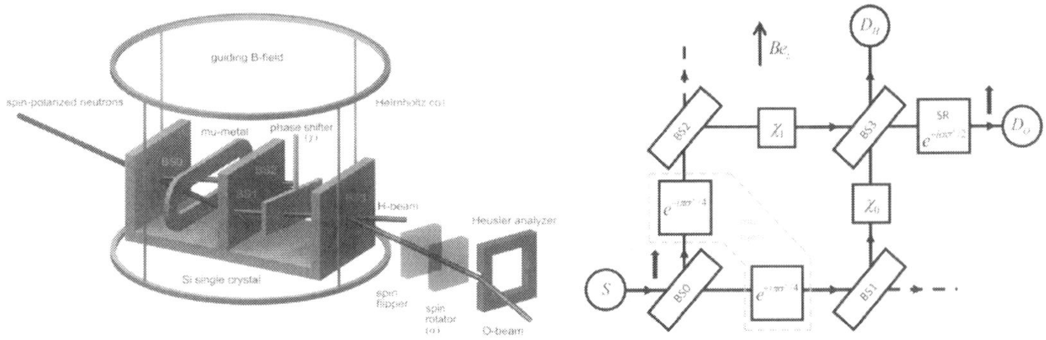

Figure 4. Left: Schematic picture of the single-neutron interferometry experiment to test a Bell inequality violation (see also Fig. 1 in Ref. [8]). BS0, ..., BS3: beam splitters; phase shifter χ: aluminum foil; neutrons transmitted by BS1 or BS2 leave the interferometer and do not contribute to the interference signal. Detectors count the number of neutrons in the O- and H-beam. Right: Event-based network of the experimental setup shown on the left. $B\mathbf{e}_z$: magnetic guiding field along the z-direction, S: single neutron source; BS0, ... , BS3: beam splitters; $e^{+i\pi\sigma^y/4}$, $e^{-i\pi\sigma^y/4}$: spin rotators modeling the action of a mu metal; χ_0, χ_1: phase shifters; SR $e^{i\alpha\sigma^x/2}$: spin rotator; D_O, D_H: detectors counting all neutrons leaving the interferometer via the O- and H-beam, respectively. Neutrons leaving the interferometer via the dashed lines are not counted.

following path H (O) is rotated by $\pi/2$ ($-\pi/2$) about the y axis. Before the two paths join at the entrance plane of beam splitter BS3, a difference between the time of flights along the two paths can be manipulated by a phase shifter χ. The neutrons which experience two refraction events when passing through the interferometer form the O-beam and are analyzed by sending them through a spin rotator and a Heusler spin analyzer. If necessary, to induce an extra spin rotation of π, a spin flipper is placed between the interferometer and the spin rotator. The neutrons that are selected by the Heusler spin analyzer are counted with a neutron detector (not shown) that has a very high efficiency ($\approx 99\%$). Note that neutrons which are not refracted by the second plate leave the interferometer without being detected.

The single-neutron interferometry experiment yields the count rate $N(\alpha,\chi)$ for the spin-rotation angle α and the difference χ of the phase shifts of the two different paths in the interferometer [8]. The correlation $E(\alpha,\chi)$ is defined by [8]

$$E(\alpha,\chi) = \frac{N(\alpha,\chi) + N(\alpha+\pi,\chi+\pi) - N(\alpha+\pi,\chi) - N(\alpha,\chi+\pi)}{N(\alpha,\chi) + N(\alpha+\pi,\chi+\pi) + N(\alpha+\pi,\chi) + N(\alpha,\chi+\pi)}. \tag{7}$$

3.1. Event-based model

A minimal, discrete event simulation model of the single-neutron interferometry experiment requires a specification of the information carried by the neutrons, of the algorithm that simulates the source and the interferometer components (see Fig. 4 (right)), and of the procedure to analyze the data. Various ingredients of the simulation model have already been described in Sect. 2. In the following, we specify the action of the remaining components.

- *Magnetic-prism polarizer:* transforms the unknown magnetic moment of an incoming neutron into a magnetic moment that is either parallel (spin up) or antiparallel (spin down) with respect to the z-axis (which by definition is parallel to the guiding field $\mathbf{B}$). In the experiment, only a neutron with spin up is injected into the interferometer. Therefore, as

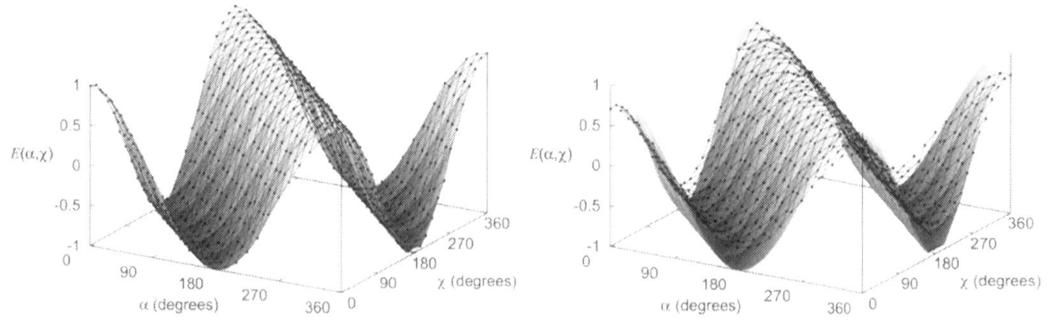

Figure 5. Correlation $E(\alpha, \chi)$ between spin and path degree of freedom as obtained from an event-based simulation of the experiment depicted in Fig. 3. Solid surface: $E(\alpha, \chi) = \cos(\alpha + \chi)$ predicted by quantum theory; dots: simulation data. Model parameters: $\mathcal{R} = 0.2$, $\gamma = 0.99$ (left) and $\gamma = 0.55$ (right). The lines connecting the markers are guides to the eye. For each pair (α, χ), four times 10000 particles were used to determine the four counts $N(\alpha, \chi)$, $N(\alpha+\pi, \chi+\pi)$, $N(\alpha, \chi + \pi)$ and $N(\alpha + \pi, \chi + \pi)$.

a matter of simplification, we assume in the event-based simulation that the source S only creates messengers with spin up. Hence, we assume that $\theta = 0$ in Eq. (1).

- *Mu metal spin turner:* rotates the magnetic moment of a neutron that follows the H-beam (O-beam) by $\pi/2$ ($-\pi/2$) about the y axis. In the event-based simulation, the processor (not a DLM) that accomplishes this takes as input the direction of the magnetic moment, represented by the message $\mathbf{u}$ and performs the rotation $\mathbf{u} \leftarrow e^{i\pm\pi\sigma^y/4}\mathbf{u}$. We emphasize that we use Pauli matrices as a convenient tool to express rotations in three-dimensional space, not because in quantum theory the magnetic moment of the neutron is represented by spin-1/2 operators.
- *Spin-rotator and spin-flipper:* rotates the magnetic moment of a neutron by an angle α about the x axis. We model this component by a processor (not a DLM) that takes as input the message $\mathbf{u}$ and performs the rotation $\mathbf{u} \leftarrow e^{+i\alpha\sigma^x/2}\mathbf{u}$. The spin flipper is a spin rotator with $\alpha = \pi$.
- *Heusler spin analyzer:* In the experiment the spin analyzer selects neutrons with spin up, after which they are counted by a detector. The model of this component is a device that lets the neutron pass if $r \leq (1 + m_z)/2$, where $0 < r < 1$ is a uniform pseudo random number, and destroys the neutron otherwise.

3.2. *Simulation results*

In Fig. 5 (left) we present simulation results for the correlation $E(\alpha, \chi)$, assuming that the experimental conditions are very close to ideal and compare them to the quantum theoretical result [8] $E_{\rm O}(\alpha, \chi) = \cos(\alpha + \chi)$ with $\chi = \chi_0 - \chi_1$, independent of the reflectivity $\mathcal{R}$ of the beam splitters (which have been assumed to be identical). The fact that $E_{\rm O}(\alpha, \chi) = \cos(\alpha + \chi)$ implies that the state of the neutron cannot be written as a product of the state of the spin and the phase. In other words, in quantum language, the spin- and phase-degree-of-freedom are entangled [8, 50]. If the mu-metal would rotate the spin about the x-axis instead of about the y-axis, we would find $E_{\rm O}(\alpha, \chi) = \cos\alpha\cos\chi$, a typical expression for a quantum system in a product state.

As shown by the markers in Fig. 5 (left), disregarding the small statistical fluctuations, there is close-to-perfect agreement between the event-based simulation data for nearly ideal experimental conditions ($\gamma = 0.99$ and $\mathcal{R} = 0.2$) and quantum theory. However, the laboratory experiment suffers from unavoidable imperfections, leading to a reduction and distortion of the interference fringes [8]. In the event-based approach it is trivial to incorporate mechanisms for different sources of imperfections by modifying or adding update rules. However, to reproduce the available data it is sufficient to use the parameter γ to control the deviation from the quantum theoretical result. For instance, for $\gamma = 0.55$, $\mathcal{R} = 0.2$ the simulation results for $E(\alpha, \chi)$ are shown in Fig. 5 (right).

In order to quantify the difference between the simulation results, the experimental results and quantum theory it is customary to form the Bell-CHSH function [49, 51]

$$S = S(\alpha, \chi, \alpha', \chi') = E_{\rm O}(\alpha, \chi) + E_{\rm O}(\alpha, \chi') - E_{\rm O}(\alpha', \chi) + E_{\rm O}(\alpha', \chi'), \tag{8}$$

for some set of experimental settings α, χ, α', and χ'. If the quantum system can be described by a product state, then $|S| \leq 2$. If $\alpha = 0$, $\chi = \pi/4$, $\alpha' = \pi/2$, and $\chi' = \pi/4$, then $S \equiv S_{max} = 2\sqrt{2}$, the maximum value allowed by quantum theory [52].

For $\gamma = 0.55$, $\mathcal{R} = 0.2$ the simulation results yield $S_{max} = 2.05$, in excellent agreement with the value 2.052 ± 0.010 obtained in experiment [8]. For $\gamma = 0.67$, $\mathcal{R} = 0.2$ the simulation yields $S_{max} = 2.30$, in excellent agreement with the value 2.291 ± 0.008 obtained in a similar, more recent experiment [53].

In conclusion, since experiment shows that $|S| > 2$, according to quantum theory it is impossible to interpret the experimental result in terms of a quantum system in the product state [2]. The system must be described by an entangled state. However, the event-based simulation approach which makes use of classical, Einstein-local and causal event-by-event processes can reproduce all the features of this experiment.

4. Ozawa's error-disturbance relation

Very recently, a neutron-optical experiment has been reported [9, 10] of which the results confirm a new error-disturbance uncertainty relation [54], derived using the theory of general quantum measurements. In what follows we demonstrate that the event-based simulation approach can also model this single neutron experiment.

The universally valid error-disturbance uncertainty relation derived by Ozawa reads [54]

$$\epsilon(A)\eta(B) + \epsilon(A)\Delta(B) + \Delta(A)\eta(B) \geq |\langle\psi|[A, B]|\psi\rangle|/2, \tag{9}$$

for every measurement and every state $|\psi\rangle$. For a projective measurement, the error $\epsilon(A)$ on the measurement of the observable A is defined as the root mean squared of the difference between the output operator O_A actually measured and the observable A to be measured and is given by $\epsilon(A) = \|(O_A - A)|\psi\rangle\|$ [9]. The disturbance $\eta(B)$ on the measurement of the observable B caused by the measurement of A is defined as the root mean squared of the change in observable B during the measurement and reads $\eta(B) = \|[O_A, B]|\psi\rangle\|$ [9]. The variance $\Delta^2(X)$ of an observable X is defined as $\Delta^2(X) = \langle X^2\rangle - \langle X\rangle^2$. Uncertainty relation Eq. (9) is a "universally valid" version of the inequality

$$\epsilon(A)\eta(B) \geq |\langle\psi|[A, B]|\psi\rangle|/2, \tag{10}$$

which is the result of an interpretation [9, 10, 54–56] of Heisenberg 's original writings [57], whereby it is assumed, without solid justification, that $\epsilon(A)$ and $\eta(B)$ correspond to the "uncertainties" which Heisenberg had in mind, see also [58, 59]. Unlike Eq. (9), inequality Eq. (10) can be violated since it lacks a mathematical rigorous basis [60].

In the neutron experiment the idea is to first measure the error of the x- and then the disturbance on the y-component of the neutron spin $S = 1/2$ if the state $|\psi\rangle = |\uparrow\rangle$. Hence, the

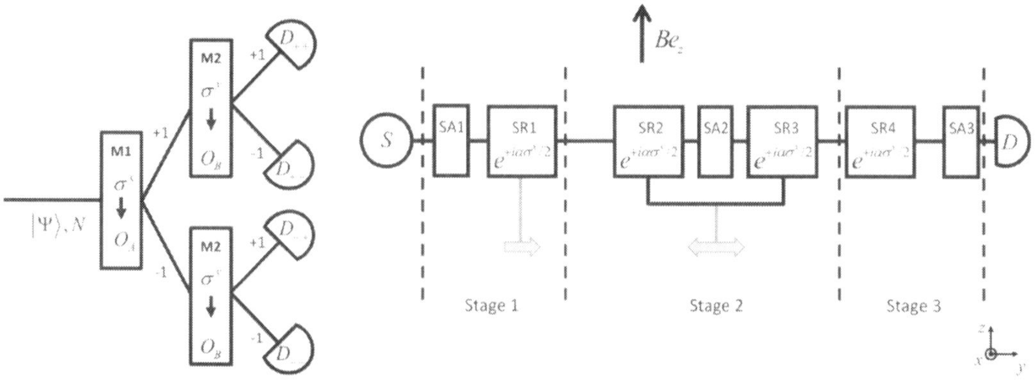

Figure 6. Left: Conceptual experiment to validate Eq. (9). The input consists of neutrons prepared in a state $|\psi\rangle$. Apparatuses M1 and M2 perform Stern-Gerlach-like selections of the neutrons based on the direction of their magnetic moments. A successive measurement by M1 and M2 results in a click by one of the four detectors corresponding to the four possible outcomes of the experiment. Repeated measurement for N neutrons allows for a quantitative determination of the error $\epsilon(A)$ and the disturbance $\eta(B)$ (see text). Right: Block diagram of the laboratory experimental setup, see also Fig. 2 of [9]. S: source; SA1, SA2, SA3: spin analyzers; SR1, ..., SR4: spin rotators, rotating the magnetic moment of the neutron by an angle α about the x axis; D: detector; $B\mathbf{e}_z$: guiding magnetic field along the z-direction. The positions of SR1 and of the pair (SR2,SR3) are variable.

observables A and B in Eq. (9) correspond to $A = \sigma^x$ and $B = \sigma^y$, respectively. In order to control $\epsilon(A)$ and $\eta(B)$ an error model is introduced so that instead of measuring $A = \sigma^x$ one measures $O_A = \sigma^\phi = \sigma^x \cos\phi + \sigma^y \sin\phi$, where the angle ϕ, also called the detuning angle, is controlled by the experimenter. The error-disturbance relation Eq. (9) then reads [9, 10]

$$2\sqrt{2}\cos\phi\sin(\phi/2) + 2\sin(\phi/2) + \sqrt{2}\cos\phi \geq 1. \tag{11}$$

However, the application of Eq.(9) to the neutron experiment is not straightforward. With some clever manipulations [9, 10], it is possible to express the unit operators that appear in Eq.(9) in terms of dynamical variables, the expectations of which can be extracted from the data of single-neutron experiments. If the state of the spin-1/2 system is described by the density matrix $\rho = |z\rangle\langle z|$, we have [9, 10]

$$\begin{aligned} \epsilon^2(A) &= 2 + \langle z|O_A|z\rangle + \langle -z|O_A| - z\rangle - 2\langle x|O_A|x\rangle, \\ \eta^2(B) &= 2 + \langle z|O_B|z\rangle + \langle -z|O_B| - z\rangle - 2\langle y|O_B|y\rangle, \end{aligned} \tag{12}$$

where $O_B = ((1+\sigma^\phi)\sigma^y(1+\sigma^\phi) + (1-\sigma^\phi)\sigma^y(1-\sigma^\phi))/4 = \sin\phi\sigma^\phi = \sin\phi O_A$.

Figure 6 (left) shows the conceptual experiment to determine the error and disturbance in the neutron experiment. The error $\epsilon(A)$ is determined by data from apparatus $M1$ carrying out the projective measurement of O_A, and the disturbance $\eta(B)$ is determined by the data from apparatus $M2$ carrying out the projective measurement of B on the neutron spin state after the measurement by $M1$. A succesive measurement of O_A and B has four possible outcomes. The outcome is registered by one of the four detectors D_{++}, D_{+-}, D_{-+} and D_{--}, where the first (second) subscript refers to the two possible outcomes ± 1 of apparatus $M1$ ($M2$). Repeating the single neutron measurement N times results in the counts N_{++}, N_{+-}, N_{-+} and N_{--} on the respective detectors with $N = N_{++} + N_{+-} + N_{-+} + N_{--}$.

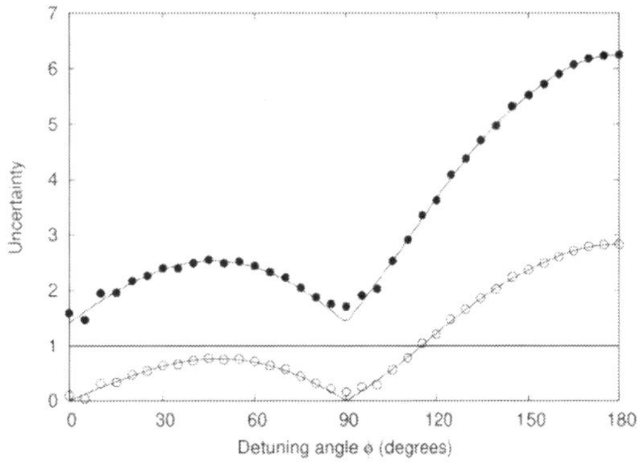

Figure 7. Simulation results for the uncertainties $\epsilon(A)\eta(B) + \epsilon(A)\Delta(B) + \Delta(A)\eta(B)$ (solid circles) and $\epsilon(A)\eta(B)$ (open circles) as a function of the detuning angle ϕ as obtained from the event-by-event simulations of the single-neutron experiment [9, 10]. Simulation parameters: $N = 10000$. Lines through the data points: quantum theoretical prediction as obtained from Eq. (11). Horizontal line: lower bound in Eq. (11).

From these counts the error $\epsilon(A)$ and disturbance $\eta(B)$ can be determined quantitatively [9]. Apparatus $M1$ is performing a Stern-Gerlach type of measurement and is described by the measurement operator $E_\phi(S_1) = (1 + S_1\sigma^\phi)/2$ with $S_1 = \pm 1$ and $O_A = \sum_{S_1=\pm 1} S_1 E_\phi(S_1)$ so that $\langle\psi|O_A|\psi\rangle = (N_{++} + N_{+-} - N_{-+} - N_{--})/N$. The modified measurement operator of apparatus $M2$ for the initial state of $M1$ is $O_B = \sum_{S_2=\pm 1} E_\phi(S_2)BE_\phi(S_2)$. Hence, $\langle\psi|O_B|\psi\rangle = (N_{++} + N_{-+} - N_{+-} - N_{--})/N$. Computation of $\epsilon(A)$ and $\eta(B)$ requires four different experiments with initial states $|\psi\rangle = |z\rangle$, $|\psi\rangle = |-z\rangle$, $|\psi\rangle = |x\rangle$ and $|\psi\rangle = |y\rangle$ (see Eq. (12)).

The block diagram of the laboratory single-neutron experiment is shown in Fig. 6 (right), see also Fig. 2 of [9]. In stage 1, spin analyzer SA1 selects neutrons with spin up, that is with a magnetic moment pointing in the direction of the guiding magnetic field. Spin rotator SR1 rotates the magnetic moment of the neutron by an angle α about the x axis. After SR1 the spin performs a Larmor precession about the $+z$ axis (due to the static magnetic guiding field). By varying the position of SR1 arbitrary initial states (up to an irrelevant phase factor) can be produced at the end of stage 1. The state $|+z\rangle$ is produced by switching off SR1, the state $|-z\rangle$ by setting $\alpha = \pi$, the state $|+y\rangle$ by setting $\alpha = \pi/2$, the state $|+x\rangle$ by setting $\alpha = \pi/2$ and changing the position of SR1 by one-quarter of the Larmor rotation period. In stage 2, corresponding to measurement apparatus M1, the spin of the neutron in the prepared state performs a Larmor precession about the $+z$ axis. By properly placing spin rotator SR2, that is by properly chosing the time that the spin performs Larmor precessions, the spin component to be measured can be projected towards the $+z$ direction. After passing SA2 in the $|+z\rangle$ state, SR3 produces the $|\pm\phi >$ eigenstate of σ^ϕ so that M1 makes the projective measurement of O_A. In a similar way M2 measures B on the eigenstate $|\pm\phi\rangle$ in stage 3.

An important difference between the conceptual and laboratory experiment is that the spin analyzers select neutrons based on the directions of their magnetic moments. Hence, one has to perform different experiments for the four different settings $(S_1 = \pm 1, S_2 = \pm 1)$ of spin analyzers (SA2, SA3), which results in sixteen different experiments in total. Also note that the number of input neutrons N for each of those experiments is different from the number of detector clicks.

4.1. Event-based model and simulation results

A minimal, discrete-event simulation model of the single-neutron experiment depicted in Fig. 6 (right) requires a specification of the information carried by the neutrons, of the algorithm that simulates the devices used in the experimental setup, and of the procedure to analyze the data. All ingredients for simulating this experiment have already been described in the previous two sections. Note that in this case no use of DLMs is required, since only simple spin manipulations (rotations and selections) are involved.

For each pair of settings (S_1, S_2) of the spin analyzers (SA2, SA3) and each position of the pair of spin rotators (SR2, SR3) represened by a rotation of ϕ about the z axis, the source sends N neutrons through the network (see Fig. 6 (right)). The source only sends a new neutron if the previous one has been counted and destroyed by the detector or if the previous one was destroyed by one of the spin analyzers. Counting the detection events allows for the calculation of $\epsilon^2(A)$ and $\eta^2(B)$. The simulation results are presented in Fig. 7. It is clear that the simulation results are in excellent agreement with the quantum theoretical prediction Eq. (11) and with experimental results [9, 10] (not shown).

In conclusion, the event-based simulation approach produces the same results as the single neutron experiment for the error of a spin-component measurement and the disturbance caused on another spin-component and therefore also obeys the error-disturbance relation Eq. (9), derived using the theory of general quantum measurements. In the event-based model no use is made of any concept of quantum theory. The "uncertainty" in the event-based model comes from the spin analyzers which transform time series of magnetic moments into random time series of +1's and −1's. This causes a loss of "classical" certainty. The event-based simulations of the neutron experiment thus demonstrate that the uncertainty relations are not necessarily a signature of quantum physics. This is completely in agreement with the fact that for real data sets, uncertainty relations provide theoretical bounds on the statistical uncertainties in the data. Hence, the relevance of "quantum theoretical" uncertainty relations to laboratory experiments yielding real data needs to be scrutinized.

5. Conclusion

We have presented an event-based simulation method which allows for a mystery-free, particle-only description of interference and entanglement phenomena observed in single-neutron interferometry experiments, and of a single neutron experiment designed to test the validity of Ozawa's error-disturbance relation, a "quantum theoretical" uncertainty relation. The method provides an "explanation" and "understanding" of what is going on in terms of elementary events, logic and arithmetic. Note that a cause-and-effect simulation on a digital computer is a "controlled experiment" on a macroscopic device which is logically equivalent to a mechanical device. Hence, an event-by-event simulation that produces results which are usually thought of being of quantum mechanical origin, but emerge from a time series of discrete events generated by causal adaptive systems, shows that there exists a macroscopic, mechanical model that mimics the underlying physical phenomena.

Acknowledgments

We would like to thank Helmut Rauch for many stimulating discussions.

References

[1] Home D 1997 *Conceptual Foundations of Quantum Physics* (New York: Plenum Press)
[2] Ballentine L E 2003 *Quantum Mechanics: A Modern Development* (Singapore: World Scientific)
[3] Michielsen K, Jin F and De Raedt H 2011 *J. Comp. Theor. Nanosci.* **8** 1052 – 1080
[4] De Raedt H and Michielsen K 2012 *Ann. Phys. (Berlin)* **524** 393 – 410
[5] De Raedt H, Jin F and Michielsen K 2012 *Quantum Matter* **1** 1 – 21
[6] Rauch H, Treimer W and Bonse U 1974 *Phys. Lett. A* **47** 369 – 371
[7] Kroupa G, Bruckner G, Bolik O, Zawisky M, Hainbuchner M, Badurek G, Buchelt R J, Schricker A and Rauch H 2000 *Nucl. Instrum. Methods Phys. Res. A.* **440** 604 – 608
[8] Hasegawa Y, Loidl R, Badurek G, Baron M and Rauch H 2003 *Nature* **425** 45 – 48
[9] Erhart J, Sponar S, Sulyok G, Badurek G, Ozawa M and Hasegawa Y 2012 *Nat. Phys.* **8** 185 – 189
[10] Sulyok G, Sponar S, Erhart J, Badurek G, Ozawa M and Hasegawa Y 2013 *Phys. Rev. A* **88** 022110
[11] Feynman R P, Leighton R B and Sands M 1965 *The Feynman Lectures on Physics, Vol. 3* (Reading MA: Addison-Wesley)
[12] Young T 1802 *Phil. Trans. R. Soc. Lond.* **92** 12
[13] Donati O, Missiroli G P and Pozzi G 1973 *Am. J. Phys.* **41** 639 – 644

[14] Merli P G, Missiroli G F and Pozzi G 1976 *Am. J. Phys.* **44** 306 – 307
[15] Tonomura A, Endo J, Matsuda T, Kawasaki T and Ezawa H 1989 *Am. J. Phys.* **57** 117 – 120
[16] Frabboni S, Gazzadi G C and Pozzi G 2008 *Appl. Phys. Lett.* **93** 073108
[17] Hasselbach F 2010 *Rep. Prog. Phys.* **73** 016101
[18] R Rosa 2012 *Phys. Perspect.* **14** 178 – 195
[19] Frabboni S, Gabrielli A, Gazzadi G C, Giorgi F, Matteucci G, Pozzi G, Cesari N S, Villa M and Zoccoli A 2012 *Ultramicroscopy* **116** 73 – 76
[20] Bach R, Pope D, Liu S and Batelaan H 2013 *New J. Phys.* **15** 033018
[21] Jacques V, Wu E, Toury T, Treussart F, Aspect A, Grangier P and Roch J F 2005 *Eur. Phys. J. D* **35** 561 – 565
[22] Saveliev I G, Sanz M and Garcia N 2002 *J. Opt. B: Quantum Semiclass. Opt.* **4** S477 – S481
[23] Garcia N, Saveliev I G and Sharonov M 2002 *Phil. Trans. R. Soc. Lond. A* **360** 1039 – 1059
[24] Kolenderski P, Scarcella C, Johnsen K, Hamel D, Holloway C, Shalm L, Tisa S, Tosi A, Resch K and Jennewein T Time-resolved double-slit experiment with entangled photons arXiv:1304.4943
[25] Zeilinger A, Gähler R, Shull C G, Treimer W and Mampe W 1988 *Rev. Mod. Phys.* **60** 1067 – 1073
[26] Rauch H and Werner S A 2000 *Neutron Interferometry: Lessons in Experimental Quantum Mechanics* (London: Clarendon)
[27] Keith D W, Ekstrom C R, Turchette Q A and Pritchard D E 1991 *Phys. Rev. Lett.* **66** 2693 – 2696
[28] Carnal O and Mlynek J 1991 *Phys. Rev. Lett.* **66** 2689 – 2692
[29] Arndt M, Nairz O, Vos-Andreae J, Keller C, van der Zouw G and Zeilinger A 1999 *Nature* **401** 680 – 682
[30] Brezger B, Hackermüller L, Uttenthaler S, Petschinka J, Arndt M and Zeilinger A 2002 *Phys. Rev. Lett.* **88** 100404
[31] Juffmann T, Milic A, M Müllneritsch, Asenbaum P, Tsukernik A, Tüxen J, Mayor M, Cheshnovsky O and Arndt M 2012 *Nature Nanotechnology* **7** 297 – 300
[32] Couder Y and Fort E 2006 *Phys. Rev. Lett.* **97** 154101
[33] Hasegawa Y and Rauch H 2011 *New J. Phys.* **13** 115010
[34] Lemmel H and Wagh A G 2010 *Phys. Rev. A* **82** 033626
[35] Allahverdyan A E, Balian R and Nieuwenhuizen T M 2013 *Phys. Rep.* **525** 1 – 166
[36] Born M and Wolf E 1964 *Principles of Optics* (Oxford: Pergamon)
[37] Fort E, Eddi A, Boudaoud A, Moukhtar J and Couder Y 2010 *PNAS* **107** 17515–17520
[38] Eddi A, Sulta E, Moukhtar J, Fort E, Rossi M and Couder Y 2011 *J. Fluid Mechan.* **674** 433–463
[39] Couder Y and Fort E 2012 *J. Phys.: Conf. Ser.* **361** 012001
[40] Jin F, Yuan S, De Raedt H, Michielsen K and Miyashita S 2010 *J. Phys. Soc. Jpn.* **79** 074401
[41] De Raedt K, De Raedt H and Michielsen K 2005 *Comp. Phys. Comm.* **171** 19 – 39
[42] De Raedt H, De Raedt K and Michielsen K 2005 *Europhys. Lett.* **69** 861 – 867
[43] Bartsch C and Gemmer J 2009 *Phys. Rev. Lett* **102** 110403
[44] Bohm D 1951 *Quantum Theory* (New York: Prentice-Hall)
[45] Weihs G, Jennewein T, Simon C, Weinfurther H and Zeilinger A 1998 *Phys. Rev. Lett.* **81** 5039 – 5043
[46] Weihs G 2000 *Ein Experiment zum Test der Bellschen Ungleichung unter Einsteinscher Lokalität* Ph.D. thesis University of Vienna `http://www.uibk.ac.at/exphys/photonik/people/gwdiss.pdf`
[47] Hnilo A, Peuriot A and Santiago G 2002 *Found. Phys. Lett.* **15** 359 – 371
[48] Agüero M B, Hnilo A A, Kovalsksy M G and Larotonda M A 2009 *Eur. Phys. J. D* **55** 705 –709
[49] Clauser J F, Horne M A, Shimony A and Holt R A 1969 *Phys. Rev. Lett.* **23** 880 – 884
[50] Basu S, Bandyopadhyay S, Kar G and Home D 2001 *Phys. Lett. A* **279** 281 – 286
[51] Bell J S 1993 *Speakable and Unspeakable in Quantum Mechanics* (Cambridge: Cambridge University Press)
[52] Cirel'son B S 1980 *Lett. Math. Phys.* **4** 93 – 100
[53] Bartosik H, Klepp J, Schmitzer C, Sponar S, Cabello A, Rauch H and Hasegawa Y 2009 *Phys. Rev. Lett.* **103** 040403
[54] Ozawa M 2003 *Phys. Rev. A* **67** 042105
[55] Fujikawa K 2012 *Phys. Rev. A* **85**(6) 062117
[56] Fujikawa K and Umetsu K 2013 *Prog. Theor. Exp. Phys.* **2013** 013A03
[57] Heisenberg W 1927 *Z. Phys.* **43** 172–198
[58] Busch P, Heinonen T and Lahti P 2007 *Physics Reports* **452** 155 – 176
[59] Busch P, Lahti P and Werner R F 2013 *Phys. Rev. Lett.* **111** 160405
[60] Ballentine L E 1970 *Rev. Mod. Phys.* **42**(4) 358 – 381

Hamiltonian flows, short-time propagators and the quantum Zeno effect

Maurice A de Gosson
Universität Wien, NuHAG, Fakultät für Mathematik, A-1090 Wien

Basil J Hiley
TPRU, Birkbeck, University of London, London, WC1E 7HX

Abstract. In a recent paper we have examined the short-time propagator for the Schrödinger equation of a point source. An accurate expression modulo Δt^2 for the propagator showed that it was independent of the quantum potential implying that the quantum motion is classical for very short times. In this paper we apply these results to the experiment of Itano, Heinzen, Bollinger and Wineland which demonstrates the quantum Zeno effect in beryllium. We show that the transition is inhibited because the applied continuous wave radiation suppresses the quantum potential necessary for the transition to occur. This shows there is no need to appeal to wave function collapse.

1. Introduction

In a recent paper we have shown that there is a one-to-one correspondence between Hamiltonian flows and the quantum evolution group which only becomes apparent when one uses some of the less well known properties of symplectic geometry [15]. Indeed we have shown that the Schrödinger equation can be *mathematically* derived in a *rigorous* manner without recourse to any physical 'quantum mechanical' hypotheses. This result exploits the two-fold covering group of the symplectic group, where the global properties, normally handled by phase changes, can be discussed within the dynamics itself.

When viewed in this way, Bohmian dynamics emerges as a feature of the Hamiltonian flow, where what is normally described by the wave function ψ is now discussed in terms of a Hamiltonian, $H^\psi = H + Q^\psi$. Here H is the usual Hamiltonian operator and Q^ψ is given by

$$Q^\psi = -\frac{\hbar^2}{2m}\frac{1}{\sqrt{\rho}}\frac{\partial^2\sqrt{\rho}}{\partial x^2}$$

This naturally gives rise to the quantum Hamilton-Jacobi equation[1]

$$\frac{\partial S}{\partial t} + \frac{1}{2m}\left(\frac{\partial S}{\partial x}\right)^2 + Q^\psi(x) + V(x) = 0 \quad (1.1)$$

[1] We will only consider a single particle (one dimensional) formalism.

In the standard approach to quantum mechanics, S is interpreted as a phase and $\rho(x)$ is the probability of the particle *being* at the position coordinate x. In our approach the phase is treated as a dynamical variable. The fact that this approach can describe the behaviour of atoms and molecules assumed to be Schrödinger particles is by now well established [5, 22, 13, 10].

In this paper we want to examine the quantum Zeno effect from the perspective of Hamiltonian flows. The reason for considering this effect emerged from a recent paper that was concerned with short-time quantum propagators when analysed in terms of Bohmian dynamics [16]. By studying the short time propagator, we find that any classical potential kicks in immediately i.e. to $O((\Delta t)^2)$. On the other hand the quantum potential does not contribute to the equation of motion to this order. In other words the short time propagator is totally classical in *all* quantum processes. This has a direct consequence for the Zeno effect. In this paper we explain how the Bohm approach explains this phenomenon without appealing to any wave function collapse.

2. The quantum Zeno effect

2.1. The experimental appearance of the Zeno effect

It has been claimed by Misra and Sudarshan [32, 8] that the quantum Zeno effect leads to the conclusion that, and we quote, "an unstable particle observed continuously whether it has decayed or not will *never* be found to decay". These authors confine their discussion specifically to the case of α-decay in a cloud or bubble chamber. Peres [34] has considered a more general question, namely, "If an unstable quantum system is kept under continuous observation, will it decay?"

Rather than considering α-decay, let us consider the specific example that forms the basis of an actual experiment [25]. This experiment involves the suppression of an atomic transition in beryllium when it is subjected to a series of rf pulses. The key question here is whether these pulses can be regarded as 'measurements that collapse the wave function'. The answer is in dispute. Itano *et al* [25], Cook [9] and Porrati and Putterman [35] all claim that the pulses do, in fact, induce a 'collapse' of the wave function. On the other hand Peres [34] and Block and Berman [3] argue that it is the dynamics of the atom plus the field system alone that suppresses the transition. This is the main question we will address in this paper.

The Bohm approach, as discussed in Bohm and Hiley [5], is based on the assumption that the world has a continuous existence with each particle having simultaneously a definite position and a definite momentum. This momentum is not an eigenvalue of the momentum operator for the state under consideration, but the real part of the weak value of the momentum operator at a point in space. This was shown in detail by Leavens [29] and in Hiley [20]. Furthermore these values, and this is a very significant point, can now be determined by experiment as shown in Flack and Hiley [11].

How then do we explain the suppression of a transition in this approach? Bohm and Hiley [5] offered an explanation but that simply relied on a re-interpretation of the standard quantum formalism. In this paper we use a different approach based on the dynamics of quantum Hamiltonian flows, the details of which have already been discussed in Gosson and Hiley [15].

The basic idea depends upon a general mathematical result reviewed by Guillemin and Sternberg [17]. What they show is that the Schrödinger equation appears naturally as a one-parameter sub-group in the metaplectic group, this group being the double cover of the symplectic group. This argument depends solely on the mathematical structure of the symplectic group and its two-fold cover. This fact has wider implications since it implies that both classical and quantum mechanics depend on the same mathematical object, namely the Hamiltonian flow, viewed as an abstract group. A detailed mathematical proof of this structure has been presented in Gosson and Hiley [15].

When we make the group act on the points of phase space, via the symplectic representation, we obtain Hamiltonian mechanics, but if it acts on functions, via the metaplectic representation, we obtain quantum mechanics. In both cases we have an associated theory of motion: in the symplectic representation, the motion is governed by Hamilton's equations. In the metaplectic representation the motion is governed by Bohm's equations. Not only does this structure contain the quantum Hamilton-Jacobi equation (1.1), but it also contains the so-called 'guidance condition'

$$p = \nabla S. \tag{2.1}$$

These equations now emerge from the equation of motion governing the time evolution. In adopting the method of Hamiltonian flows, we have, of course, assumed that every particle actually exists at a point in space-time and, furthermore, is assumed to have simultaneously a well defined momentum.

Before jumping to the conclusion that this approach must necessarily be incorrect because it seems to violate the uncertainty principle, we stress once again that the momentum beable is not an eigenvalue of the momentum operator. Not only is it the real part of the weak value of the momentum operator but this value is also related to the $(0,k)^{th}$ component of the energy-momentum tensor, $T^{\mu\nu}(x,t)$ of the quantum field [21] as was first shown by Takabayasi [36] and re-derived and extended to apply also to the Pauli and Dirac particles by Hiley and Callaghan [21]. Using the Hamiltonian flow approach we find this momentum emerges from the dynamics itself.

It should be emphasised that in the method of Hamiltonian flows, we do not adopt the standard view that the wave function is the only way to specify the state of the quantum system. The wave function merely encodes the information about the dynamics of the system. Hence the appearance of the term, Q^{ψ}, in the quantum Hamilton-Jacobi equation (1.1) is of key significance here since it contains information about the overall experimental environment. Thus although we have a localised object, its behaviour depends on the overall experimental conditions, a point that was strongly emphasised by Bohr [7].

The positive feature of the method of Hamiltonian flows is that, unlike the standard approach, it has no measurement problem. The arguments for this position follow closely the extensive discussions given in Bohm and Hiley [4]. There it is pointed out that there is no need for a collapse of the wave function, the world unfolds in its environment objectively. Thus there is no need for the observer to play any active role in "collapsing the wave function". Of course, if the observer wants to know what has actually happened, he/she must set up a piece of apparatus to display some macroscopic, irreversible effect to reveal which process out of the ensemble of possible processes has actually taken place. This involves making a von Neumann measurement, which merely signifies a change in the wave function, but the world itself evolves regardless of whether we 'look' at the results or not.

In summary then, the main purpose of this paper is to explain the experimentally observed fact [25] that, when we introduce an external process to continuously 'observe' an unstable state, its decay is inhibited.

A preliminary answer to this question has already been sketched out in Bohm and Hiley [6]. Here we use a rigorous mathematical result to confirm that original conjecture. It is our hope that this presentation will help clarify why we feel the method of Hamiltonian flows will provide a clear understanding of this phenomena. Our approach provides support for the conclusion that it is the dynamics of the atom plus the field system that is responsible for the suppression in the quantum transitions thus supporting the conclusions of Peres [34] and of Block and Berman [3].

2.2. The standard approach

Let us first recall the standard argument that shows why a constantly 'observed' or 'watched' transition will not occur. Consider an excited state of some α-active nuclei $|\psi_u(t)\rangle$ and let H

be the Hamilton governing the transition to one of a set of possible decay states $|\phi_{d_i}(t)\rangle$. Then after time t, the state of the nuclei will be given by

$$|\psi(t)\rangle = e^{-iHt}|\psi_u(0)\rangle.$$

The probability that the state remains un-decayed after a time t will be

$$P_u(t) = |\langle\psi_u(t)|e^{-iHt}|\psi_u(0)\rangle|^2 = 1 - \sigma^2t^2 + O(t^4).$$

Here $\sigma^2 = \langle H^2\rangle - \langle H\rangle^2$. Thus for a short time interval, $\Delta t = t/n$, the probability that the state is still excited is simply

$$P_u(t/n) \approx 1 - \sigma^2(t/n)^2 + O((t/n)^4).$$

Suppose now we repeat the observation n times so that we have

$$P_u(t) = [P_u(t/n)]^n = \left[1 - (\sigma t/n)^2\right]^n.$$

Let us now consider the limit as $n \to \infty$ by writing

$$\ln P_u(t) = n\ln\left[1 - (\sigma t/n)^2\right] = n\left[-(\sigma t/n)^2 + O(n^{-4})\right].$$

Then by taking the limit $n \to \infty$, we see that the state does not decay since $P_u(t) \to 1$. This result is well known [1], and is sometimes called the 'watched pot' or even the 'watchdog' effect [27].

Central to this argument is the collapse of the wave function, but as we have already pointed out, there is no collapse in the traditional Bohm approach outlined in [5]. How then is the above result explained in this approach? A preliminary answer has already been presented in Bohm and Hiley [5] where this effect was discussed in the context of the Auger effect. In this paper we will show how these ideas can be used on the actual experimental example first proposed by Cook [9] and later carried out by Itano *et al* [25].

The experiment of Itano *et al* uses three particular atomic states of a beryllium ion placed in a magnetic field. To make our discussion as simple as possible, we will assume level one, $|1\rangle$ is a ground state, level two, $|2\rangle$ is a metastable state with a life time of order 1 sec. The third level, $|3\rangle$ when excited, rapidly relaxes ($\sim 10^{-8}$ sec) only to the ground state emitting a photon in the UV range.

To discuss the principles involved, let us first consider a single Be ion with an electron in $|1\rangle$. The ion is subjected to a optical UV (313nm) source, polarised perpendicular to the applied magnetic field. This causes the electron to jump into level $|3\rangle$. It then quickly relaxes to the ground state emitting a photon, which is then detected. As long as the optical UV source is switched on and the rf source switched off, a continuous burst of optical photons is seen.

Suppose now we switch off the optical source and, instead, switch on a pulse of the rf source lasting time T. This will drive a transition between $|1\rangle$ and $|2\rangle$. If we again start with the atom in state $|1\rangle$, we can choose T so that the probability of the electron being in $|2\rangle$ is $P_2(T) = 1$. If Ω is the Rabi frequency of the transition in the applied magnetic field, the time T will be given by $T = \pi/\Omega$. When the electron 'jumps' to the metastable state $|2\rangle$ and the optical pulse is on, no optical photons will be emitted from the atom.

Suppose now with the rf pulse switched on, we allow n short bursts of the optical source to be applied in the time T. The duration of the pulse is chosen to be much shorter than the time between pulses. With equally spaced pulses the time at which the r-th pulse starts is $\tau = rT/n$. Here $r = 1, 2, \ldots, n$. What we require to find is the probability $P_2(T)$ of finding level $|2\rangle$ occupied after n such pulses. Using the rotating wave approximation, Itano *et al* [25] find

$$P_2(T) = [1 - \cos^n(\pi/n)]/2$$

Thus $P_2(T) \to 0$ as $n \to \infty$. In other words increasing the number of optical pulses stops the electron making a transition into the state $|2\rangle$. Alternatively if the experiment is started with the electron in the state $|2\rangle$, the rf-pulses will prevent the electron relaxing into state $|1\rangle$. In the first case we will detect the emission of a burst of optical photons, while in the second case, no optical photons will be detected. This will allow us to infer in which state the electron has been trapped. Itano *et al* claim that the observation of photons constitutes a "measurement" while the failure to detect any photons constitutes a "null measurement".

It is clear that we have an observation because we can tell in which state the electron is, but does detecting/not detecting the optical photons actively 'collapse' the wave function? This is where a difference of opinion arises. Cook [9], Itano *et al* [25] and Porrati and Putterman [35] all claim that a 'collapse' of the wave function has taken place. On the other hand, Peres [34] and Block and Berman [3] claim it is determined solely by the dynamics of the atom plus the field system.

Thus we are faced with the vexed question concerning the exact meaning of the word 'measurement' in the context of quantum mechanics. In classical physics, there is an unfolding process which can be observed continuously without disturbing it, or if the means of observation does slightly change the unfolding process, a correction can be made in principle. This is not possible in quantum processes described solely in terms of a wave function. As stressed by Bohr, the word 'phenomena' cannot be defined without including "...an account of the whole experimental arrangement" [7] p. 73. In this context, Bohr includes in the experimental conditions a "suitable final amplification device producing, for example, permanent marks on a photographic plate caused by the penetration of electrons into the emulsion". Without the mark or some other irreversible result, we can only talk about the time development of the wave function.

Since the Schrödinger equation is linear and describes a unitary evolution, the final wave function now includes a linear combination of products of wave functions of the system, $\psi_n(x,t)$ and the apparatus, $\phi_n(y,t)$, viz:

$$\Psi(x,y,t) = \sum c_n(t)\psi_n(x,t)\phi_n(y,t). \tag{2.2}$$

The instrument will be found in one of its eigenstates, say, $\phi_j(y,t)$. Thus we say the wave function has 'collapsed'. This then defines what Bohr calls "a closed indivisible phenomena". Whether we can find a dynamical cause for a collapse is, of course, an open question that has received much attention [33] [12] [26], but we will not pursue this discussion in this paper.

When we apply these criteria to the example above, we already see a problem. While the detection of the burst of optical photons provides the necessary irreversible effect, the recording of no photons does not. Nevertheless the linear wave function (2.2) will, in this case, contain the two terms ϕ_{fired} and $\phi_{unfired}$. The collapse argument immediately 'closes the phenomena', but without any appeal to irreversibility.

Clearly the recording or otherwise of the optical photons is not where the core of the Zeno effect resides. It is the rapid succession of externally applied continuous wave optical pulses, while the rf-source remains switched on, that traps the electron either in state $|1\rangle$ or $|2\rangle$. But in what sense is this a 'measurement'? It cannot be argued that the pulses 'cause' the wave function to 'collapse' into one state or the other, because no non-unitary process is involved. It is for this reason that Block and Berman [3] prefer to talk about the results in terms of the dynamics of the atom plus the field system. In fact they trace the reason for the trapping to the rapid decay of the probability amplitude for the transition $1 \leftrightarrow 2$. This follows from a detailed examination of the full density matrix equations of the motion for the three level system and the two resonant continuous wave fields. Interested readers should consult their paper [3] for the details.

On the other hand, this whole process does constitute a kind of 'observation' since it enables one to decide which state the electron is in at any given time. Rather than being a *direct* measurement with wave function collapse, we are essentially making an *inference* about the behaviour of the system. This is much closer to the kind of inference we use in classical physics and, indeed, the kind of inference we make when using the Bohm approach where there is no measurement problem [4]. In this approach one assumes that the system unfolds with the particles following well defined, although unknown, trajectories. Because the trajectories are continuous, it avoids the slippery notion of 'jumps' that arise in quantum measurement which this particular example highlights. An extremely enlightening discussion of the measurement problem relevant to our discussion in this paper can be found in Bell [2]. Indeed Bell felt so strongly about the confusion caused by the word 'measurement' that he called for a ban on the use of the word when discussing quantum phenomena.

3. The Hamiltonian character of Bohmian trajectories

3.1. The Hamiltonian $H^\psi = H + Q^\psi$

The key feature of the Bohm approach through particle trajectories is to realise the whole process is Hamiltonian. The general case is slightly more subtle and we refer to the papers by Holland [23, 24] for a thorough discussion of the interpretation of Bohmian trajectories from the Hamiltonian point of view. Here we will present a simple, more direct way of demonstrating this fact.

Let us start from the basic dynamical equation of the Bohm approach, namely, the quantum Hamilton-Jacobi equation (1.1). We will now show that this equation leads to Hamilton's equations of motion provided we introduce the Hamiltonian $H^\Psi = H + Q^\Psi$ so that (1.1) has the form of a classical Hamilton–Jacobi equation

$$\frac{\partial S}{\partial t} + H^\Psi\left(x, \frac{\partial S}{\partial x}, t\right) = 0. \tag{3.1}$$

Differentiating the latter with respect to $p = \partial S/\partial x$ yields, using the chain rule,

$$\frac{\partial^2 S}{\partial x_0 \partial t} + \frac{\partial H^\Psi}{\partial p}\frac{\partial^2 S}{\partial x_0 \partial x} = 0 \tag{3.2}$$

and differentiating the equation (1.1) with respect to time yields

$$\frac{\partial^2 S}{\partial x_0 \partial t} + \frac{\partial^2 S}{\partial x \partial x_0}\dot{x} = 0. \tag{3.3}$$

Subtracting (3.3) from (3.2) we get

$$\frac{\partial^2 S}{\partial x \partial x_0}\left(\frac{\partial H^\Psi}{\partial p} - \dot{x}\right) = 0$$

which produces the first Hamilton equation

$$\dot{x} = \nabla_p H^\Psi(x, p, t).$$

We have assumed that $\partial^2 S/\partial x \partial x_0 \neq 0$ in this derivation. Let us next show that the second Hamilton equation is also satisfied. For this we differentiate the quantum Hamilton–Jacobi equation (1.1) with respect to x, which yields

$$\frac{\partial^2 S}{\partial x \partial t} + \frac{\partial H^\Psi}{\partial x} + \frac{\partial H^\Psi}{\partial p}\frac{\partial^2 S}{\partial x^2} = 0. \tag{3.4}$$

Introducing the canonical momentum $p(t) = \nabla_x S(x(t), x_0; t, t_0)$ and differentiating with respect to t we get

$$\frac{\partial^2 S}{\partial t \partial x} = -\dot{p}(t) - \frac{\partial^2 S}{\partial x^2}\dot{x} \tag{3.5}$$

and hence the equation (3.4) can be rewritten

$$-\dot{p}(t) - \frac{\partial^2 S}{\partial x^2}\dot{x} + \frac{\partial H^\Psi}{\partial x} + \frac{\partial H^\Psi}{\partial p}\frac{\partial^2 S}{\partial x^2} = 0.$$

Taking into account the relation $\dot{x} = \partial H^\Psi/\partial p$ established above we have

$$-\dot{p}(t) - \frac{\partial H^\Psi}{\partial x} = 0$$

which is precisely the second Hamilton equation. This means that the Bohmian trajectory for a particle initially sharply localised at a point x_0 is Hamiltonian, and is governed by the Hamilton equations:

$$\dot{x} = \frac{\partial H^\Psi}{\partial p}(x, p, t) \quad , \quad \dot{p} = -\frac{\partial H^\Psi}{\partial x}(x, p, t). \tag{3.6}$$

In conclusion we have thus shown that:

Bohm's equation of motion (1.1) is equivalent to Hamilton's equations (3.6).

3.2. Short-time propagators

In order to proceed, we need first to study the short-time solutions of the Schrödinger equation

$$i\hbar\frac{\partial \Psi}{\partial t} = \left[-\frac{\hbar^2}{2m}(\partial^2 S/\partial x^2) + V(x)\right]\Psi \quad , \quad \Psi(x, t_0) = \delta(x - x_0) \tag{3.7}$$

The solution can be written

$$\Psi(x, t) = \int K(x, x_0; t)\Psi_0(x_0)dx_0 \tag{3.8}$$

where the kernel K is the quantum propagator:

$$K(x, x_0; t) = \langle x|\exp(-i\widehat{H}t/\hbar)|x_0\rangle. \tag{3.9}$$

Schrödinger's equation (3.7) is then equivalent to

$$i\hbar\frac{\partial K}{\partial t} = \widehat{H}(x, -i\hbar\partial/\partial x, t)K \ , \ K(x, x_0; 0) = \delta(x - x_0) \tag{3.10}$$

where δ is the Dirac distribution. Physically this equation describes an isotropic source of point-like particle emanating from the point x_0 at initial time $t_0 = 0$. We want to find an asymptotic formula for K for short time intervals Δt. Referring to the usual literature, such approximations are given by expressions of the type

$$K(x, x_0; \Delta t) = \left(\frac{1}{2\pi i\hbar}\right)^{1/2}\sqrt{\rho(x, x_0; \Delta t)}\exp\left(\frac{i}{\hbar}S(x, x_0; \Delta t)\right)$$

where $S(x, x_0; \Delta t)$ is the action along the classical trajectory from x_0 to x in time Δt and

$$\rho(x, x_0; \Delta t) = -\frac{\partial^2 S(x, x_0; \Delta t)}{\partial x \partial x_0}$$

is the corresponding Van Vleck density. We will need the precise short-time behaviour of the action, S. In this regard, Makri and Miller [30, 31] have shown that the asymptotic expression for the generating function is given by

$$S(x, x_0; \Delta t) = \frac{m}{2\Delta t}(x - x_0)^2 - \widetilde{U}(x, x_0)\Delta t + O(\Delta t^2) \tag{3.11}$$

where $\widetilde{U}(x, x_0, 0)$ is the average value of the potential over the straight line joining x_0 at time t_0 to x at time t with constant velocity:

$$\widetilde{U}(x, x_0) = \int_0^1 U(\lambda x + (1 - \lambda)x_0, 0)d\lambda. \tag{3.12}$$

Introducing the following notation,

$$\widetilde{S}(x, x_0; \Delta t) = m\frac{(x - x_0)^2}{2\Delta t} - \widetilde{U}(x, x_0)\Delta t, \tag{3.13}$$

leads us to the Makri and Miller approximation (formula (17c) in [30]) for the short-time propagator:

$$K(x, x_0; \Delta t) = \left(\frac{1}{2\pi i\hbar}\right)^{1/2} \sqrt{\rho(x, x_0; \Delta t)} \exp\left(\frac{i}{\hbar}\widetilde{S}(x, x_0; \Delta t)\right) + O(\Delta t^2) \tag{3.14}$$

where

$$\rho(x, x_0; \Delta t) = -\frac{\partial^2 \widetilde{S}(x, x_0; \Delta t)}{\partial x_j \partial x_0}.$$

It turns out that this formula can be somewhat improved. The Van Vleck density $\rho(x, x_0; \Delta t)$ is explicitly given, taking formula (3.13) into account, by

$$\rho(x, x_0; \Delta t) = -\frac{m}{\Delta t} - \widetilde{U}''_{x,x_0}(x, x_0)\Delta t$$

where

$$\widetilde{U}''_{x,x_0} = -\frac{\partial^2 \widetilde{U}(x, x_0)}{\partial x_j \partial x_k}.$$

Writing

$$\begin{aligned}\left(-\frac{m}{\Delta t} - \widetilde{U}''_{x,x_0}(x, x_0)\Delta t\right) &= -\frac{m}{\Delta t}[1 - m^{-1}\widetilde{U}''_{x,x_0}(x, x_0)\Delta t^2] \\ &= -\frac{m}{\Delta t}[1 + O(\Delta t^2)],\end{aligned}$$

we thus have

$$\rho(x, x_0; \Delta t) = \frac{m}{\Delta t}\left(1 + O(\Delta t^2)\right). \tag{3.15}$$

Writing

$$\widetilde{\rho}(\Delta t) = \frac{m}{\Delta t} \tag{3.16}$$

which is just the Van Vleck density for the free particle Hamiltonian, we have

$$\rho(x, x_0; \Delta t) = \widetilde{\rho}(\Delta t)(1 + O(\Delta t^2)) \tag{3.17}$$

and hence we can rewrite formula (3.14) as

$$K(x, x_0; \Delta t) = \left(\frac{1}{2\pi i\hbar}\right)^{1/2} \sqrt{\widetilde{\rho}(\Delta t)} \exp\left(\frac{i}{\hbar}\widetilde{S}(x, x_0; \Delta t)\right) + O(\Delta t^2). \tag{3.18}$$

We will see below that this formula allows an easy study of the quantum potential for K.

4. Short-time Bohmian trajectories

Let us determine the quantum potential Q corresponding to the propagator $K = K(x, x_0; t)$ using the asymptotic formulas above. Recall that it describes an isotropic source of point-like particles emanating from the point x_0 at initial time $t_0 = 0$. We assume for notational simplicity that the number of degrees of freedom is $n = 1$ (for the general case see [16]). We have, by definition,

$$Q = -\frac{\hbar^2}{2m\sqrt{\rho}}\frac{\partial^2\sqrt{\rho}}{\partial x^2}$$

which we can rewrite

$$Q = -\frac{\hbar}{2m}{}^2\frac{\nabla_x^2\sqrt{\rho}}{\sqrt{\rho}}.$$

We have, using (3.17),

$$\sqrt{\rho} = \sqrt{\widetilde{\rho}(\Delta t)}(1 + O(\Delta t^2))$$

and hence

$$\frac{\partial^2\sqrt{\rho}}{\partial x^2} = O((\Delta t)^2)).$$

From this it follows that the quantum potential associated with the propagator satisfies

$$Q(x, x_0; \Delta t) = O(\Delta t^2). \tag{4.1}$$

The suggests that the quantum trajectory of a sharply located particle should be identical with the classical (Hamiltonian) trajectory for short times. Let us show that this is, indeed, the case. If we want to monitor the motion of such a particle, we have of course to specify its initial momentum which gives its direction of propagation at time $t_0 = 0$; we set

$$p(0) = p_0. \tag{4.2}$$

In view of formula (2.1), the trajectory in position space is obtained by solving the system of differential equations

$$\dot{x} = \frac{\hbar}{m}\operatorname{Im}\left(\frac{1}{K}\frac{\partial K}{\partial x}\right) \quad , \quad x(0) = x_0. \tag{4.3}$$

Replacing the propagator K with its approximation

$$\widetilde{K}(x, x_0; \Delta t) = \left(\frac{1}{2\pi i\hbar}\right)^{1/2}\sqrt{\widetilde{\rho}(\Delta t)}\exp\left(\frac{i}{\hbar}\widetilde{S}(x, x_0; \Delta t)\right)$$

we have, since $K - \widetilde{K} = O(\Delta t^2)$ in view of (3.18),

$$\dot{x} = \frac{\hbar}{m}\operatorname{Im}\left(\frac{1}{\widetilde{K}}\frac{\partial\widetilde{K}}{\partial x}\right) + O(\Delta t^2).$$

A straightforward calculation, using the expression (3.13) for the approximate action $\widetilde{S}(x, x_0; \Delta t)$, leads to the equation

$$\dot{x}(\Delta t) = \frac{x(\Delta t) - x_0}{\Delta t} - \frac{1}{m}\frac{\partial\widetilde{U}}{\partial x}(x(\Delta t), x_0)\Delta t + O(\Delta t^2) \tag{4.4}$$

(*cf.* the proof of Lemma 248 in [14]). This equation is singular at time $t = 0$ hence the initial condition $x(0) = x_0$ is not sufficient for finding a unique solution; this is of course consistent with the fact that (4.4) describes an arbitrary particle emanating from x_0; to single out one quantum trajectory we have to use the additional condition (4.2) giving the direction of the particle at time $t = 0$ (see the discussion in Holland [22], §6.9). We thus have

$$x(\Delta t) = x_0 + \frac{p_0}{m}\Delta t + O(\Delta t^2);$$

in particular $x(\Delta t) = x_0 + O(\Delta t)$ and hence, by continuity,

$$\frac{\partial \widetilde{U}}{\partial x}(x(\Delta t), x_0) = \frac{\partial \widetilde{U}}{\partial x}(x_0, x_0) + O(\Delta t).$$

Let us calculate $(\partial \widetilde{U}/\partial x)(x_0, x_0)$. We have, taking definition (3.12) into account,

$$\frac{\partial \widetilde{U}}{\partial x}(x, x_0) = \int_0^1 \lambda \frac{\partial U}{\partial x}(\lambda x + (1-\lambda)x_0, 0)d\lambda$$

and hence

$$\frac{\partial \widetilde{U}}{\partial x}(x_0, x_0) = \int_0^1 \lambda \frac{\partial U}{\partial x}(x_0, 0)d\lambda = \frac{1}{2}\frac{\partial U}{\partial x}(x_0, 0).$$

We can thus rewrite equation (4.4) as

$$\dot{x}(\Delta t) = \frac{x(\Delta t) - x_0}{\Delta t} - \frac{1}{2m}\frac{\partial U}{\partial x}(x_0, 0)\Delta t + O(\Delta t^2).$$

Let us now differentiate both sides of this equation with respect to Δt:

$$\ddot{x}(t) = \frac{x(t) - x_0}{(\Delta t)^2} + \frac{\dot{x}(t)}{\Delta t} - \frac{1}{2m}\frac{\partial U}{\partial x}(x_0, 0) + O(\Delta t) \tag{4.5}$$

that is, replacing $\dot{x}(\Delta t)$ by the value given by (4.4),

$$\dot{p}(t) = m\ddot{x}(t) = -\frac{\partial U}{\partial x}(x_0, 0) + O(\Delta t). \tag{4.6}$$

Solving this equation we get

$$p(t) = p_0 - \frac{\partial U}{\partial x}(x_0, 0)\Delta t + O(\Delta t^2). \tag{4.7}$$

Summarizing, the solutions of the Hamilton equations are given by

$$x(\Delta t) = x_0 + \frac{p_0}{m}\Delta t + O(\Delta t^2) \tag{4.8}$$

$$p(\Delta t) = p_0 - \frac{\partial U}{\partial x}(x_0, 0)\Delta t + O(\Delta t^2). \tag{4.9}$$

These equations are, up to the error terms $O(\Delta t^2)$ the equations of motion of a classical particle moving under the influence of the potential U; there is no trace of the quantum potential, which is being absorbed by the terms $O(\Delta t^2)$. The motion is thus identical with the classical motion on time scales of order $O(\Delta t^2)$

4.1. The Zeno effect in the Bohm approach

Now let us consider how Zeno's paradox emerges from the Bohm approach as outlined in the previous section. Consider the transition process $|1\rangle \leftrightarrow |2\rangle$ to illustrate the principle involved. This electric dipole transition is driven by an electric field $\boldsymbol{E} = \boldsymbol{E}_1 \cos\omega t$ so that

$$|\psi_{12}(t)\rangle = a(t)|1\rangle + b(t)|2\rangle. \tag{4.10}$$

Since $|a|^2 + |b|^2 = 1$, we need only calculate $a(t)$ using the Schrödinger equation $ida/dt = Ha$. Here $H = -\Omega = \mu|E_1|$ for the rotating-wave model [35].

In order to use the Bohm model on this transition[2], we must rewrite equation (4.10) in the form

$$\Psi(\boldsymbol{x}, t) = a(t)\psi_1(\boldsymbol{x}, t) + b(t)\psi_2(\boldsymbol{x}, t) = R^{\Psi}(\boldsymbol{x}, t)e^{\frac{i}{\hbar}S^{\Psi}(\boldsymbol{x}, t)}$$

so that the momentum of the electron undergoing the transition is

$$p^{\Psi}(\boldsymbol{x}, t) = \nabla S^{\Psi}(\boldsymbol{x}, t). \tag{4.11}$$

while the quantum potential driving the transition is

$$Q^{\Psi}_{12}(\boldsymbol{x}, t) = -\frac{\hbar^2}{2m}\frac{\nabla^2 R^{\Psi}(\boldsymbol{x}, t)}{R^{\Psi}(\boldsymbol{x}, t)}. \tag{4.12}$$

When $a(t) > 0 \neq 1$ equation (4.11) shows there is a non-zero current between $|1\rangle$ and $|2\rangle$, which means that the electron is moving from level 1 to level 2 or vice versa depending on the sign of the current. Equation (4.12) gives the quantum energy driving the transition. If $a(t) = 0$, the electron is in the state $\psi_2(\boldsymbol{x}, t)$. Equation (4.11) then shows its momentum is the momentum it would have in state $|2\rangle$, while equation (4.12) gives the value of its quantum potential in that state. There is no longer any quantum energy driving a transition so the electron stays in that level until a new perturbation again forces a transition. What this means is that if we can prevent the quantum potential Q^{Ψ}_{12} from developing, the electron in state 2 becomes trapped there.

Thus the quantum potential plays an essential role in the 'watched pot' effect. In order to prevent the transition taking place we must introduce a suitable *physical process* which prevents this quantum potential from developing. Thus the intervention is not simply 'watching'. We need a process that actually interacts with the electron. This process must be repudiative and quick enough to prevent the quantum potential developing. This is the role played by the continuous wave optical pulses in the experiment of Itano *et al* [25]. This is an example of the participatory nature of a 'measurement' proposed by Wheeler [37] and by Bohm and Hiley [5]. However notice it is not simply the process of 'watching'.

5. Conclusion

In this paper we have shown how the question raised by Misra and Sudarshan [32] in their discussion of α-decay can be answered using the Bohm-Hiley approach [5]. In order to inhibit α-decay occurring, we must use a process that directly interacts with the unstable nuclei. It is not sufficient to surround the nuclei with a passive detection device like a bubble or cloud chamber. Merely detecting the α-particle outside the nucleus is not sufficient to inhibit the decay.

We have shown that Bohm's dynamics resides in the notion of a Hamiltonian flow. In such an approach, both the classical and quantum behaviours emerge from the same mathematical

[2] We will only present a very simplified discussion of this transition here. A full detailed account of such a transition has been given by Lam and Dewdney [28].

structure, the symplectic structure. It is the appearance of the quantum potential, Q^{Ψ}, in equation (1.1) that differentiates the two types of mechanics. When Q^{Ψ} is negligible compared with the kinetic energy, the equation simply reduces to the classical Hamilton-Jacobi equation and classical mechanics emerges. Hiley and Aziz Mufti [19] have illustrated how this occurs in a specific example based on an 'upside-down' harmonic potential, a potential that was used to illustrate the ideas behind the inflationary cosmological model of Guth and Pi [18].

At first sight it might seem rather ironic that the suppression of the quantum potential leads to classical behaviour and at the same time prevents atomic transitions. However it must be remembered that quantum mechanics was introduced to explain the stability of matter and the Bohm approach shows that the quantum potential plays a key role in stationary states. Indeed if one examines the ground states of atoms and nuclei, we find that the quantum potential plays a key role in determining their energy values. In the ground state of the Schrödinger hydrogen atom, for example, the energy eigenvalue *is all* quantum potential energy.

Thus in order to make a transition between energy levels, a change in the quantum potential energy is essential. Then clearly if we want to prevent such a transition from occurring, we need to stop the quantum potential energy from changing. To achieve this, we need some actual physical process to interact with the system. This then is the explanation of the quantum Zeno effect in the Bohmian approach.

Acknowledgments

One of us (BJH) would like to thank Bob Callaghan and David Robson for valuable discussion. Maurice de Gosson has been supported by a research grant from the Austrian Research Agency FWF (Projektnummer P23902-N13).

References

[1] Ballentine, L. E., The Statistical Interpretation of Quantum Mechanics, Rev. Mod. Phys. **42** (1970) 358-81.
[2] Bell, J. S., Against Measurement, *Physics World* (1990) 33-40. Reproduced in *John Bell on the Foundations of Quantum Mechanics*, ed. Bell, J. Gottfried, K. and Veltman, M, World Scientific, Singapore, (2001).
[3] Block, E. and Berman, P. R., Quantum Zeno effect and quantum Zeno paradox in atomic physics, *Phys. Rev.,* bf A 44 (1991) 1466-72.
[4] Bohm, D. and Hiley, B. J. Measurement Understood Through the Quantum Potential Approach, *Found. Phys.*, **14**, (1984) 255-74.
[5] Bohm, D. and Hiley, B. J., *The Undivided Universe: an Ontological Interpretation of Quantum Theory*, Routledge, London 1993.
[6] Bohm, D. and Hiley, B. J., *ibid* section 6.9.
[7] Bohr, N., *Atomic Physics and Human Nature*, p. 73, Science Editions, New York 1961.
[8] Chiu, C. B., Sudarshan, E. C. G. and Misra, B., Time evolution of unstable quantum states and a resolution of Zeno's paradox, *Phys. Rev.*, **D16** (1977) 520-29.
[9] Cook, R. J., What are Quantum Jumps?, *Phys. Scrip.* bf T21 (1988) 49-51.
[10] Dürr, D., *Bohmian Mechanics and the meaning of the wave function*, 1997.
[11] Flack, R. and Hiley B. J., Weak Measurement and its Experimental Realisation, This volume. 2014.
[12] Ghirardi, G. C., Rimini, A. and Weber, T., Unified dynamics for microscopic and macroscopic systems, *Phys. Rev.*, **D34** (1986) 470
[13] de Gosson, M., *The Principles of Newtonian and Quantum Mechanics: The Need for Planck's Constant*, Imperial College Press, London, 2001.
[14] de Gosson, M.: On the classical and quantum evolution of Lagrangian half-forms in phase space. *Ann. Inst. H. Poincaré*, **70** (1999), 547–573
[15] de Gosson, M. and Hiley, B. J., Imprints of the Quantum World in Classical Mechanics, *Found. Phys.* **41**, (2011), 1415-1436.
[16] de Gosson, M. and Hiley, B. J., Short Time Quantum Propagator and Bohmian Trajectories, Physics Letters A 377 (2013) 3005–3008.
[17] Guillemin, V. and Sternberg, S., *Symplectic Techniques in Physics* Cambridge University Press, Cambridge, (1984).

[18] Guth, A. H. and Pi, S., Quantum mechanics of the scalar field in the new inflationary universe, *Phys. Rev.* **32D** (1985) 1899.
[19] Hiley, B. J. and Aziz Mufti, A. H., The Ontological Interpretation of Quantum Field Theory Applied in a Cosmological Context, in *Fundamental Problems in Quantum Physics*, ed. M. Ferrero and A. van der Merwe, 141-156,. Kluwer, Dordrecht, (1995).
[20] Hiley, B. J., Weak Values:Approach through the Clifford and Moyal Algebras, (2011). quant-ph/1111.6536
[21] Hiley, B. J. and Callaghan, R. E., The Clifford Algebra approach to Quantum Mechanics A: The Schrödinger and Pauli Particles. arXiv:Maths-ph: 1011.4031.
[22] Holland, P.R.:*The quantum theory of motion. An account of the de Broglie-Bohm causal interpretation of quantum mechanics*. Cambridge University Press, Cambridge (1995)
[23] Holland, P.: Hamiltonian theory of wave and particle in quantum mechanics I: Liouville's theorem and the interpretation of the de Broglie-Bohm theory. *Nuovo Cimento* bf B 116 (2001), 1043–1070
[24] Holland, P.: Hamiltonian theory of wave and particle in quantum mechanics II: Hamilton-Jacobi theory and particle back-reaction. *Nuovo Cimento* **B 116**, 1143–1172 (2001)
[25] Itano, W. M., Heinzen, D. J., Bollinger, J. J. and Wineland, D. J., Quantum Zeno Effect, *Phys. Rev.*, **A 41** (1990) 2295-2300.
[26] Joos, E., Continuous measurement: Watchdog Effect and the Golden Rule, *Phys. Rev.*, **D29** (1984) 1626-33.
[27] Kraus, K., Measuring Processes in Quantum Mechanics I. Continuous Observation and the Watchdog Effect, *Found. Phys.*, **11** (1981) 547-76.
[28] Lam, M. M. and Dewdney, C., *Found. Phys.*, **24**, (1994) 3-27 and 29-60.
[29] Leavens, C. R., Weak Measurements from the point of view of Bohmian Mechanics, *Found. Phys.*, **35** (2005) 469-91 doi: 10.1007/s10701-004-1984-8
[30] Makri. N., Miller W.H.: Correct short time propagator for Feynman path integration by power series expansion in Δt, *Chem. Phys. Lett.* **151** (1988), 1-8
[31] Makri. N., Miller W.H.: Exponential power series expansion for the quantum time evolution operator. *J. Chem. Phys.* **90** (1989), 904–911.
[32] Misra, B. and Sudarshan, E. C. G., The Zeno's paradox in quantum theory, *J. Maths. Phys.*, **18** (1977) 756-63.
[33] Pearle, P., Towards explaining why events occur, *Int. J. Theor. Phys.* **18** (1979) 489-518.
[34] Peres, A., Zeno paradox in quantum theory, *Am. J. Phys.* **48** (1980) 931-2.
[35] Porrati, M and Putterman, S., Wave function collapse due to null measurements: the origin of intermittent atomic fluorescence, *Phys. Rev.*, **A36** (1987) 929-32.
[36] Takabayasi, T., Remarks on the Formulation of Quantum Mechanics with Classical Pictures and on Relations between Linear Scalar Fields and Hydrodynamical Fields, *Prog. Theor. Phys.* **9** (1953) 187-222 .
[37] Wheeler, J. A., Beyond the Black Hole, in *Some Strangeness in the Proportion* ed. Wolff, H., pp. 341-375, Addison Wesley, Reading Mass 1980. Also reproduced in Wheeler J. A., *At Home in the Universe*, p.271-94, AIP Press, New York (1991).

Particles, waves and trajectories: 210 years after Young's experiment

Ángel S Sanz
Instituto de Física Fundamental (IFF–CSIC), Serrano 123, 28006 - Madrid, Spain
E-mail: asanz@iff.csic.es

Abstract. Mermin's "shut up and calculate!" somehow summarizes the most widely accepted view on quantum mechanics. This conception has led to a rather constraining way to think and understand the quantum world. Nonetheless, a closer look at the principles and formal body of this theory shows that, beyond longstanding prejudices, there is still room enough for alternative tools. This is the case, for example, of Bohmian mechanics. As it is discussed here, there is nothing contradictory or wrong with this hydrodynamical representation, which enhances the dynamical role of the quantum phase to the detriment (to some extent) of the probability density. The possibility to describe the evolution of quantum systems in terms of trajectories or streamlines is just a direct consequence of the fact that Bohmian mechanics (quantum hydrodynamics) is just a way to recast quantum mechanics in the more general language of the theory of characteristics. Misconceptions concerning Bohmian mechanics typically come from the fact that many times it is taken out of context and considered as an alternative theory to quantum mechanics, which is not the case. On the contrary, an appropriate contextualization shows that Bohmian mechanics constitutes a serious and useful representation of quantum mechanics, at the same level as any other quantum picture, such as Schrödinger's, Heisenberg's, Dirac's, or Feynman's, for instance. To illustrate its versatility, two phenomena will be briefly considered, namely dissipation and light interference.

1. Introduction

In 2013 we have celebrated the 100th anniversary of Bohr's atomic model [1, 2], which led to the development of quantum mechanics in the 1920s. However, we have also celebrated the 210th anniversary of Young's famous two-slit experiment. In 1803 Thomas Young presented to the Royal Society his experimental "proof of the general law of the interference of two portions of light" [3, 4]; more than one hundred years later this experiment has become one of the most influential ones in physics [5], particularly due to its tight connection with quantum mechanics. The fact is vividly expressed at the beginning of the third volume of the *Feynman's Lectures on Physics* [6], where we read:

> In this chapter we shall tackle immediately the basic element of the mysterious behavior in its most strange form. We choose to examine a phenomenon which is impossible, *absolutely* impossible, to explain in any classical way, and which has in it the heart of quantum mechanics. In reality, it contains the *only* mystery. We cannot make the mystery go away by "explaining" how it works. We will just *tell* you how it works. In telling you how it works we will have told you about the basic peculiarities of all quantum mechanics.

The phenomenon referred to here, namely quantum interference, is precisely the same phenomenon observed by Young —although for light instead of massive particles.

Nowadays the enormous success of quantum mechanics is indisputable. Not only this theory explains the most fundamental aspects of the physical world, but it has also given rise to technological applications with a direct impact on our daily life. Now, 210 years after Young's experiment, what do we really know about quantum systems? Unfortunately, not much (if anything at all). That is, although we have a very accurate theory, our understanding of this theory still relies on the ideas prevailing in the late 1920s and 1930s, strongly linked to the experimental capabilities at that moment. At present fine experiments can be performed in the time domain, reproducing Young's experiment particle by particle. Even though the general conception of quantum systems is still anchored in somewhat old-fashioned, self-imposed constraints, which have more to do with positivist prejudices than with limitations of the theoretical framework of quantum mechanics [7, 8]. Mermin's quotation "shut up and calculate!" [9, 10] accurately summarizes this position. Obviously, this has constituted (and still constitutes) an important obstacle to the development and advance of new ways to understand the quantum world, particularly those relying on the concept of trajectory. This is the case, for example, of Bohmian mechanics.

The purpose of this communication is to show that there is nothing contradictory or wrong with Bohmian mechanics. Rather than a matter of taste, the discussion will show that this approach is just another representation (a hydrodynamical one) of quantum mechanics, at the same level as other more standard representations, e.g., Schrödinger's, Heisenberg's, Dirac's, Feynman's, etc. Actually, although it goes beyond the scopes of this work, leaving aside aspects commonly associated with Bohmian mechanics, like the possibility of hidden variables or the ontology of the wave function, it can readily be seen that it is just a direct translation of quantum mechanics into the more general language of the theory of characteristics [11]. Now, why Bohmian mechanics? Because it focuses on the quantum phase, which is not a quantum observable, although it plays a decisive role on quantum system dynamics. Bearing this in mind, this work is organized as follows. The essential elements of this approach are introduced in Sec. 2. In Sec. 3 a brief overview on how this approach and similar ones have been applied to different physical problems is presented. This makes readily apparent the nature of Bohmian mechanics as a theory of characteristics, even if the latter's formalism is not explicitly discussed. In Sec. 4 Bohmian mechanics is applied to two different phenomena, namely dissipation and light interference, in order to show its versatility. To conclude, in Sec. 5 a series of final remarks are summarized.

2. Waves, trajectories and quantum mechanics

In appearance, the formulation of quantum mechanics in the Schrödinger representation keeps a close analogy with classical wave theory. Within this representation, quantum systems are described by a probability amplitude or wave function, Ψ, which evolves according to the partial differential equation

$$i\hbar \frac{\partial \Psi}{\partial t} = -\frac{\hbar^2}{2m} \nabla^2 \Psi + V\Psi. \tag{1}$$

This equation describes the transport of probability instead of energy, as it happens with classical waves. Nonetheless, as Eq. (1) is formulated, this is not evident, since it rather displays the form of a typical diffusion equation (with a complex diffusion coefficient). After Max Born proposed the statistical interpretation, we say that the probability density, $\rho \equiv |\Psi|^2$, gives the probability that the quantum system has (or is in) a particular configuration. Therefore, a representation of ρ in time gives us information on how the configurational probability of the system evolves (is transported) in time, i.e., which are the configurations (e.g., positions) where we have more or less chance to find the system. If we look at a typical outcome obtained from an interference experiment, even in the case of very large and complex molecular systems [12], we find that these particles distribute according to ρ, in agreement with Born's statistical interpretation. However, particles are detected one by one at different places. The individual positions (evolutions) of these particles, however, are no quantum observables.

Motivated by that fact, in 1952 David Bohm proposed [13] a model that could explain such individual arrivals at the same time that could also account for the collective particle behavior, all without appealing to von Neumann's reduction postulate. He regarded the individual systems as "hidden variables". To develop this interpretative model, Bohm considered Schrödinger's equation (1) plus the nonlinear (polar) transformation

$$\Psi(\mathbf{r},t) = \rho^{1/2}(\mathbf{r},t)e^{iS(\mathbf{r},t)/\hbar}. \tag{2}$$

This transformation relates the complex-valued fields (Ψ, Ψ^*) with the real-valued fields (ρ, S), where S describes the local variation of the quantum phase. Substituting (2) into the time-dependent Schrödinger equation and then separating the real and imaginary parts of the resulting equation, one finds

$$\frac{\partial \rho}{\partial t} + \nabla\cdot\left(\rho\frac{\nabla S}{m}\right) = 0, \tag{3}$$

$$\frac{\partial S}{\partial t} + \frac{(\nabla S)^2}{2m} + V - \frac{\hbar^2}{2m}\frac{\nabla^2\rho^{1/2}}{\rho^{1/2}} = 0. \tag{4}$$

The continuity (or conservation) equation (3) rules the ensemble dynamics, i.e., the number of particles described by the probability density has to remain constant; the quantum Hamilton-Jacobi equation (4) describes the time-evolution of the phase field, where the last term on the right-hand side is the so-called quantum potential. These two equations make more apparent the physical meaning of Eq. (1) as a transport equation. In analogy to the classical Hamilton-Jacobi equation, if S is identified with the classical action, then a momentum $\mathbf{p} = \nabla S$ can be postulated (Bohm's momentum). This momentum can be expressed as $\mathbf{p} = m\dot{\mathbf{r}}$, which gives rise to the guidance equation of motion

$$\dot{\mathbf{r}} = \frac{\nabla S}{m}. \tag{5}$$

The local velocity field $\mathbf{v} = \dot{\mathbf{r}}$, which depends directly on the quantum phase, thus governs the individual system dynamics and, in virtue of Eq. (3), gives rise to averages that are in agreement with the results obtained directly from Schrödinger's equation. It is interesting to note at this point that Eq. (5) does not need to be postulated, but it arises automatically within the theory of characteristics [11]. The integration of this equation in whichever evolution parameter (e.g., time) generates the corresponding characteristics perpendicular to S-surfaces of constant phase.

Previous to Bohm's model, de Broglie formulated a similar one [14] where particles were assumed to be singularities guided by the wave. Both models pursued essentially the same idea, namely to explain the ensemble behavior of quantum systems at the same time that their individual motion was also properly described. The meaning of individual system, however, is rather uncomfortable, since its (individual) evolution is not accessible. To understand this idea with a simple example, think of the stream of a river: its precise characterization does not provide any information at all on the individual motion of its molecular constituents. In this regard, it is probably more precise the approach developed by Erwin Madelung in 1926. Shortly after Schrödinger proposed his equation, Madelung provided a clear prescription [15] to reformulate such an equation in hydrodynamic form. Accordingly, quantum systems could be visualized in terms of a series of streamlines, which would follow the flow associated with the system probability density. That is, we know nothing about the precise motion of individual systems

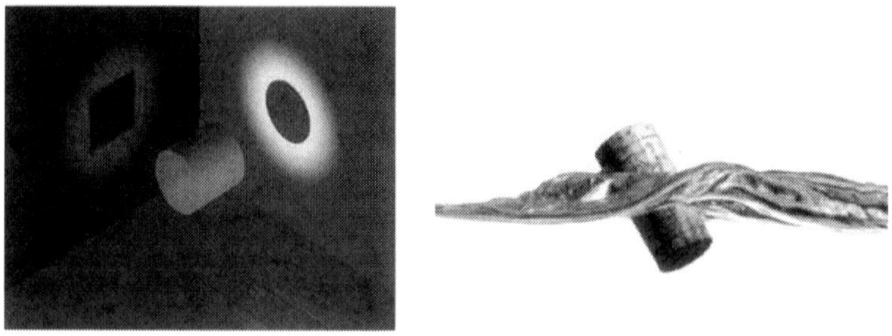

Figure 1. The different representations of quantum mechanics provide us with a different description of quantum systems. Typically quantum systems are thought from a complementary viewpoint (left). Bohmian mechanics, however, gives preference to the configuration space and emphasizes transport-related properties (right).

(they could be evolving in a Brownian-like fashion, for example), but still we can determine how the ensemble probability flows throughout the corresponding configuration space.

Following Madelung's view, Eq. (5) needs not be postulated, but it is a consequence of the fact that Eq. (3) is a (real-valued) transport equation. This allows us to establish a direct analogy with classical fluid dynamics. Hence, making use of the local probability current density,

$$\mathbf{J} = \frac{\hbar}{m} \mathrm{Im}\,(\Psi^* \nabla \Psi) = \mathbf{v}\rho, \tag{6}$$

where $\mathbf{v}$ is a local (hydrodynamic) velocity field, one finds the expression

$$\mathbf{v} = \frac{\mathbf{J}}{\rho} = \frac{\nabla S}{m}, \tag{7}$$

which is equivalent to the above Bohmian velocity. The integration in time (or whichever parameter) of Eq. (7) generates a family of streamlines or paths (for each wave function considered) along which the quantum fluid propagates, just as in the case of a classical fluid. As it can be inferred, the first equality goes beyond Bohmian mechanics and allows to define streamlines in any system characterized by a certain density and a vector that transports it through the corresponding configuration space, regardless whether such a density describes a quantum system or not. Notice that in this case, instead of the set of equations (3) and (4), we have a set of hydrodynamic equations, as shown by Takabayasi [16–18].

Taking into account the fact that Bohmian mechanics is a different theory, but only a reformulation of quantum mechanics, one may wonder why it is worth using. The answer is very simple, it is just a matter of which aspect of the physical system one wishes to stress. This is nicely illustrated in Fig. 1. Typically, we understand quantum systems in a somewhat dichotomic way, e.g., either wave or particle, position or momentum, energy or time, etc, emphasizing the so-called complementarity (see left panel). However, we may also be interested in transport issues, for which Bohmian mechanics constitutes a more convenient picture (right panel), since it describes the probability flow in configuration space without caring about any other complementary aspect. In this regard, we find, for example, that probably one of the most relevant and distinctive properties revealed by this quantum hydrodynamic formulation, is the non-crossing [19, 20], i.e., the fact that quantum fluxes (described in terms of bunches of Bohmian trajectories) cannot cross in configuration space at the same time. This is in compliance with the outcomes obtained from a recent photon-by-photon realization of Young's experiment [21]. As a direct consequence of the non-crossing, one is able to establish well-defined quantum probability

Journal of Physics: Conference Series **504** (2014) 012028 doi:10.1088/1742-6596/504/1/012028

tubes [22, 23], i.e., tubes in configuration space along which the integral of the probability density at a given time remains constant.

3. Hydrodynamic approaches in the literature[1]

As seen above, the hydrodynamic language of Bohmian mechanics enables a visualization of quantum systems in terms of streamlines that follow the flow of the probability density. This simple pictorial representation attracted the attention of the chemical physics community immediately after the first wave-packet propagation schemes were developed and used in the field by the end of the 1960s. Pioneers in this field were McCullough and Wyatt, who studied the quantum dynamics of collinear atom-diatom reactions in hydrodynamic terms [24–26], just in a period when classical trajectories were in fashion. The following quotation, extracted from [25], gives a very clear idea of what this community was looking for:

> Classical mechanics gives an amazingly good description of the probability density and flux patterns during most of the reaction; however, the classical and quantal descriptions begin to diverge near the end of the reaction. Essentially, the classical reaction terminates before the quantal reaction. The dynamic behavior of the reaction is hydrodynamically turbulent, as shown by transient whirlpool formation on the inside of the reaction path.

This ended up with the development of the so-called quantum-trajectory methods [27] at the end of the 1990s, just something that David Bohm himself thought to be the path to follow to describe the physical systems.

In above case, a hydrodynamical viewpoint was adopted, without making an explicit calculation of streamlines or trajectories (hydrodynamical vector fields were used instead). Nevertheless, a few years later, in the middle of the 1970s, Hirschfelder and coworkers materialized Madelung's ideas when trying to recast solutions of Schrödinger's equation in a pictorial way for tunneling and vortical dynamics. As before, the underlying motivation was seeking for a better understanding of the phenomena studied [28]:

> This paper has resulted from an effort to get a better understanding of quantum mechanics by making a thorough study of a very simple problem, [...]. The mathematics is simple, but the analysis is far-reaching.

Again, these studies came from the field of chemical physics. In this sense, it is interesting to note that the first studies of barrier tunneling came from this community [28], thus predating those that later on appeared following Bohm's ideas [29].

In the 1980s we find a convergence to Bohmian mechanics from the field of molecular magnetism [30, 31], where the quantum hydrodynamics was also in fashion. In particular, in [30] we read:

> A representation of the electron flow induced by the external field can be extremely useful to understand molecular magnetism. To this end, maps reporting modulus and trajectory of quantum-mechanical current density revealed a fundamental tool, whose importance could be hardly overestimated.

By means of these representations one can observe very convolved trajectories describing the electron current densities induced by external magnetic fields acting on different types of molecular systems [31]. At an applied level, "these tools provide fundamental help for rationalization of magnetic response properties, such as magnetizability and nuclear magnetic shielding" [31].

[1] The references here provided are rather scarce, but I think that they will be enough for the reader interested in further enquiring about the use of streamlines/trajectories in the literature.

The same motivation can be found even earlier in electromagnetism, where the tradition of explaining phenomena in terms of rays was stronger and therefore there was not an urgent need to appeal to the Madelung-Bohm scheme. Thus, in 1952, the same year that Bohm published his work on hidden variables, Braunbek and Laukien published [32] a work where they studied the diffraction by an edge (a perfectly conducting half-plane) by means of lines of average electromagnetic energy flow, obtained from the analytical solution provided to this problem by Sommerfeld in 1896 [33]. About 20 years later (again in the 1970s, when the computational tools were already more sophisticated), Prosser produced [34] the first trajectories for Young's two slit experiment and provided an explanation in terms of "photon" trajectories [35]. These trajectories preceded those obtained by Dewdney *et al.* for matter waves using Bohm's model [36]. Later on, different authors have treated the problem of electromagnetism in terms of trajectories (an account can be found in [37]), until in 2011 Steinberg and coworkers performed an experiment from which the first trajectories were inferred experimentally [21]. At a more applied level, and in consonance with Madelung's viewpoint, we also find works dealing with streamlines in wave optics [38, 39] or transport through billiards [40], for example.

Of course, one could regard the trajectories for electromagnetic fields (radiation) as similar to those for matter waves (massive particles). In the end, Maxwell's equations can be seen as equivalent to Schrödinger's one [41]. However, we can move apart from these scenarios, and still we find analogous streamline-based descriptions with similar purposes. This is the case with sound waves, for example, some of which started appearing in the middle 1980s [42–46]. In particular, in Ref. [42] we read:

> A method is presented for computing the energy streamlines of a sound source. This enables charts to be plotted showing, as continuous lines, the flow paths of the sound energy from the vibrating surface to the nearfield and beyond. Energy streamlines appear to be a new construct; they have some similarities to the velocity streamlines used in fluid dynamics. Examples of the energy streamlines are given for the point-driven plate in water. [...] These streamlines make it easer for the eye to follow the energy flow from the source into the nearfield and beyond. These paths are complicated, in some cases, but are of considerable interest from several points of view.

This is precisely the role played by Bohmian mechanics in the study of quantum mechanical systems! This role is clearer and clearer as one goes through Ref. [42] in more detail. While searching through the literature, one finds remarkable the fact that in other areas of physics dealing with waves (other than quantum mechanics or, by extension, quantum optics) the visualization of flows is particularly relevant. Physicists have tried to develop methods based on characteristics, like those described above, or just stroboscopic interferometric ones [47], in order to explore and understand the behavior of the fluid qualitative and quantitatively. This is in contrast with the reluctancy found in quantum mechanics to treat systems on equal footing.

According to the above discussion, the interest generated by the calculation of streamlines associated with the transport of some quantity —e.g., probability, electromagnetic energy, pressure energy— has led to a series of approaches that all converge to the very same need: dealing with a tool that allows one to objectively monitor the transport of such a quantity. In this regard, if quantum mechanics is separated from any other wave theory, one may end up concluding that Bohmian trajectories represent real trajectories pursued by real particles (or, in general, degrees of freedom), i.e., they are hidden variables. However, if this theory is properly contextualized, we find that there is nothing that allows us to establish a direct connection between the possible motion of real particles and this type of trajectories [20, 48]. Comparing with other theories, and as it will be seen in the next Section, as much we can say that a Bohmian particle is a particle that obeys a Bohmian dynamics, i.e., according to a (local) average drift momentum, which provides the particle with nonlocal (global) hydrodynamic-like information. Such a particle allows us to infer dynamical properties of the quantum fluid, which are usually "hidden" when studied by means of the wave function formulation. That is, Bohmian particles are the quantum equivalent of classical tracer particles (or just tracers) that can be found in other areas of physics and chemistry.

4. Some applications

4.1. Dissipative Bohmian mechanics

The first context where we are going to apply Bohmian mechanics in the sense described above, namely as a hydrodynamic picture of quantum mechanics, is that of dissipation. Thus, consider the well-known classical dissipative equation

$$m\ddot{x} + m\gamma\dot{x} + \frac{\partial V(x)}{\partial x} = 0, \qquad (8)$$

where m is the system mass and γ is the friction coefficient. According to this equation, the dissipation undergone by the system is proportional to its speed at a given time, $\dot{x}$. In order to determine the quantum analog, i.e., to specify the Schrödinger equation corresponding to Eq. (8), we need to find a suitable Hamiltonian. One way to proceed is by considering [49] the change of variables

$$X = x, \qquad P = me^{\gamma t}\dot{x} = pe^{\gamma t}, \qquad (9)$$

where (x, p) denote the physical variables, with $p = m\dot{x}$, and (X, P) the canonical ones. The latter allow us to define the conservative Hamiltonian

$$\mathcal{H}_{(X,P)} = \dot{X}P - \mathcal{L} = \frac{P^2}{2m}\,e^{-\gamma t} + V(X)e^{\gamma t}, \qquad (10)$$

and satisfy the usual canonical relations

$$\dot{X} = \frac{\partial \mathcal{H}_{(X,P)}}{\partial P}, \qquad \dot{P} = -\frac{\partial \mathcal{H}_{(X,P)}}{\partial X}, \qquad (11)$$

which enable the conservation of the total energy. If the initial energy is E_0 and therefore $\mathcal{H}_{(X,P)} = E_0$, the inverse change to the physical coordinates gives

$$\mathcal{H}_{(x,p)} = \frac{p^2}{2m} + V(x) = E_0 e^{-\gamma t}, \qquad (12)$$

i.e., the energy is lost exponentially at a constant rate (γ), as expected from the physical dissipative system described by Eq. (8). The dissipative model given by Eq. (10) is known as the Caldirola-Kanai model [50, 51] and constitutes one of the former attempts to express (8) in a Hamiltonian form.

In order to find the quantum analog of the Caldirola-Kanai model, we now make use of the standard quantization procedure, and associate the operators $\hat{X}$ and $\hat{P} = -i\hbar\partial/\partial\hat{X}$ with the canonical variables X and P, respectively. Because of their canonicity, these operators satisfy the usual commutation relation $[\hat{X}, \hat{P}] = i\hbar$ —which does not hold for the "physical" operators, $[\hat{x}, \hat{p}] = i\hbar e^{-\gamma t}$. Accordingly, the quantum Caldirola-Kanai Hamiltonian arises by replacing X and P in Eq. (10) by the corresponding operators,

$$\hat{\mathcal{H}}_{(\hat{X},\hat{P})} = -\frac{\hbar^2}{2m}\,e^{-\gamma t}\,\frac{\partial^2}{\partial X^2} + e^{\gamma t}\hat{V}(\hat{X}), \qquad (13)$$

and the corresponding Schrödinger equation, in the physical variable x, reads as

$$i\hbar\,\frac{\partial \Psi}{\partial t} = -\frac{\hbar^2}{2m}\,e^{-\gamma t}\,\frac{\partial^2 \Psi}{\partial x^2} + e^{\gamma t}V(x)\Psi. \qquad (14)$$

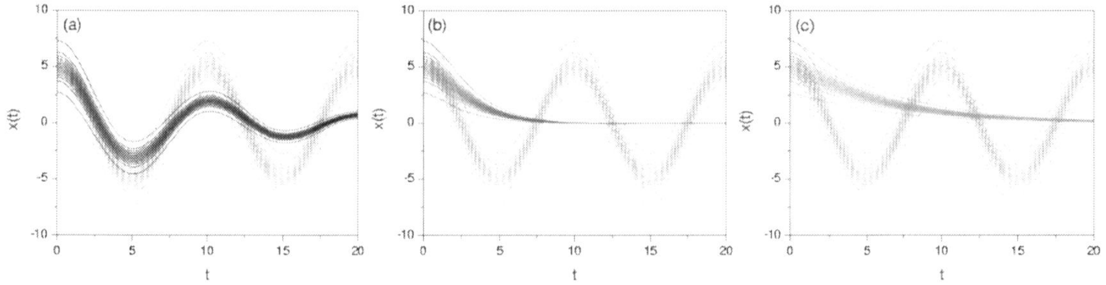

Figure 2. Dissipative Bohmian trajectories for: (a) $\gamma = 0.3\omega_0$, (b) $\gamma = 2\omega_0$, and (c) $\gamma = 4\omega_0$, with $\omega_0 = 2\pi/\tau_0 \approx 0.628$ ($\tau_0 = 10$). To compare with, the frictionless Bohmian trajectories have also been included in each panel (gray dashed lines). The initial positions have been distributed according to the initial Gaussian probability density.

It is easy to show that the associated (dissipative) Bohmian trajectories obey the modified guidance equation

$$\dot{x} = \frac{J}{\rho} = \frac{1}{m}\frac{\partial S}{\partial x}\, e^{-\gamma t}. \tag{15}$$

Note the similarity between this equation and the classical analog, $\dot{x} = pe^{-\gamma t}/m$.

For potential functions $V(x)$ which are polynomials with a degree equal or smaller than two, one can find analytical solutions for initial wave functions with the shape of a Gaussian wave packet [49, 52]. Without entering details, just to illustrate the dissipative dynamics described by Eq. (14), we are going to consider a harmonic oscillator. The dissipative Bohmian trajectories for different values of the friction coefficient are displayed in Fig. 2. As it can be noticed, in the physical coordinates not only there is a clear violation of the usual commutation relation, which manifests in a vanishing dispersion of the wave packet, but the system approaches the bottom part of the potential well, thus going down the zero-point energy. This is a pathological behavior (in the physical variables) typical of the Caldirola-Kanai model, which is based on a continuous dissipation of the system energy due to the lack of a proper quantization. The latter only applies to the canonical variables X and P. A similar behavior can be found, for example, when dealing with beables [53, 54]. The same correspondence can also be found if one considers an initial wave-packet superposition inside the harmonic potential, as shown in Fig. 3.

In order to avoid such an inconvenience, other models have been proposed in the literature as suitable quantum candidates of Eq. (8) [55–59]. In analogy to Brownian-type wave-function models [60], these models use to include a nonlinear term, thus also making the corresponding Schrödinger equation nonlinear. Of course, this nonlinear term is usually accompanied by a stochastic term that accounts for the random fluctuations of the medium that give rise to such a nonlinearity (just as in the Langevin equation for Brownian motion, for example). For example, in Kostin's model [61] a nonlinear term depending on the phase of the wave function is considered, with the Schrödinger equation reading as

$$i\hbar\frac{\partial\Psi}{\partial t} = -\frac{\hbar^2}{2m}\frac{\partial\Psi}{\partial x^2} + V\Psi + V_R\Psi + \gamma\left(S - \int \rho S dx\right)\Psi, \tag{16}$$

with $S = (\hbar/2i)\ln(\Psi/\Psi^*)$ and where V_R is a random potential. Here we have focused on this model in particular because of the link that can be established between this model and Bohmian mechanics precisely through S, as it will be seen below.

In the full dissipative case, where $V_R = 0$, if we substitute the usual polar ansatz into Eq. (16) and then proceed as in standard Bohmian mechanics, we reach the modified quantum Hamilton-

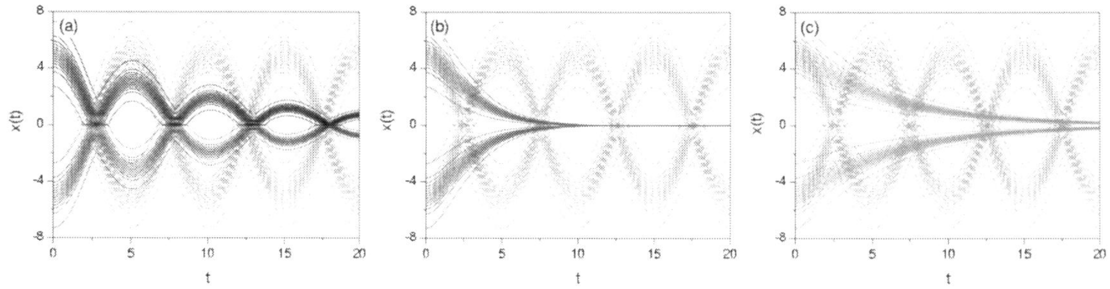

Figure 3. Dissipative Bohmian trajectories for: (a) $\gamma = 0.3\omega_0$, (b) $\gamma = 2\omega_0$, and (c) $\gamma = 4\omega_0$, with $\omega_0 = 2\pi/\tau_0 \approx 0.628$ ($\tau_0 = 10$). To compare with, the frictionless Bohmian trajectories have also been included in each panel (gray dashed lines). The initial positions have been distributed according to the initial Gaussian probability density.

Jacobi equation

$$\frac{\partial S}{\partial t} + \frac{1}{2m}\left(\frac{\partial S}{\partial x}\right)^2 + V + Q + \gamma S = 0, \tag{17}$$

where Q denotes the quantum potential (see Sec. 2). Assuming that the Bohmian trajectories are obtained from the usual equation of motion, $p = m\dot{x} = \partial S/\partial x$, and differentiating Eq. (17) with respect to x, we reach

$$m\ddot{x} + m\gamma\dot{x} + \frac{\partial(V + Q)}{\partial x} = 0, \tag{18}$$

where we have made use of the Lagrangian derivative

$$\frac{d}{dt} = \frac{\partial}{\partial t} + \dot{x}\,\frac{\partial}{\partial x} = \frac{\partial}{\partial t} + \frac{1}{m}\left(\frac{\partial S}{\partial x}\right)\frac{\partial}{\partial x}. \tag{19}$$

Note that Eq. (18) has the same functional form as Eq. (8), except for the fact that it also includes a purely quantum force (given in terms of the space derivative of the quantum potential). That is, Bohmian mechanics shows a direct path to find the quantum analog of Eq. (8), consisting of just including the quantum potential. Kostin's model has been proven very useful to determine energy bound states due to its convergence by dissipation [62].

4.2. "Photon" trajectories

Another field of interest concerning the applications of Bohmian mechanics is electromagnetism, as mentioned above, where a time-independent trajectory approach can be readily developed starting from Maxwell's equations [37, 63]. Within this approach, where the time-dependence is removed by averaging over the oscillating electromagnetic field, the trajectories describe the flow of electromagnetic energy (analogous to the probability accounted for by the Bohmian trajectories). As shown by Prosser [34], this is possible by directly considering Maxwell's equations defined by an electric field $\mathbf{E}(\mathbf{r})$ and a magnetic field $\mathbf{H}(\mathbf{r})$, the two key elements necessary to define the corresponding streamlines are the time-averaged electromagnetic energy density and Poynting vector (electromagnetic current density),

$$U(\mathbf{r}) = \frac{1}{4}\left[\epsilon_0\mathbf{E}(\mathbf{r})\cdot\mathbf{E}^*(\mathbf{r}) + \mu_0\mathbf{H}(\mathbf{r})\cdot\mathbf{H}^*(\mathbf{r})\right], \tag{20}$$

$$\mathbf{S}(\mathbf{r}) = \frac{1}{2}\,\mathrm{Re}\left[\mathbf{E}(\mathbf{r})\times\mathbf{H}^*(\mathbf{r})\right], \tag{21}$$

respectively; $\mathbf{E}(\mathbf{r})$ and $\mathbf{H}(\mathbf{r})$ denote respectively the spatial part of the electric and magnetic field vectors, which satisfy Maxwell's equations and have been assumed to be harmonic, i.e.,

$$\begin{aligned}\tilde{\mathbf{E}}(\mathbf{r}) &= \mathbf{E}(\mathbf{r})e^{-i\omega t},\\ \tilde{\mathbf{H}}(\mathbf{r}) &= \mathbf{H}(\mathbf{r})e^{-i\omega t}.\end{aligned} \tag{22}$$

Since the electromagnetic energy density is transported through space in the form of the Poynting vector, a local velocity field can be defined [64] in analogy to (7), which reads as

$$\mathbf{S}(\mathbf{r}) = U(\mathbf{r})\mathbf{v}. \tag{23}$$

The electromagnetic energy flow lines or "photon" paths are then obtained by integrating the equation

$$\frac{d\mathbf{r}}{ds} = \frac{1}{c}\frac{\mathbf{S}(\mathbf{r})}{U(\mathbf{r})} \tag{24}$$

along the arc-length coordinate s (which can be referred to a proper time $\tau = s/c$, with c being the speed of light). The analogy between light and massive particles becomes more apparent if the spatial parts of the electric and magnetic fields are expressed in terms of a scalar function Ψ that satisfies the Helmholtz equation and the corresponding boundary conditions [37, 63].

In order to show such an analogy, consider a monochromatic electromagnetic wave in vacuum incident onto a plate with two slits. The plate is on the XY plane, at $z = 0$. For simplicity, it is assumed that the fields are independent of the y coordinate. This assumption is well justified if the slits are parallel to the y axis and their width along this axis is much larger than along x. In such a case from Maxwell's equations one obtains two independent sets of equations. One involves the components H_x, H_z, and E_y of the electromagnetic field and is commonly regarded as E-polarized. The other involves the components E_x, E_z, and H_y, and is known as H-polarization. The electric and magnetic fields behind the slits can be expressed as [63]

$$\mathbf{E}(\mathbf{r}) = -\frac{i\beta}{k}\frac{\partial\Psi}{\partial z}\mathbf{e}_x + \frac{i\beta}{k}\frac{\partial\Psi}{\partial x}\mathbf{e}_z + \alpha\Psi\mathbf{e}_y, \tag{25}$$

$$\mathbf{H}(\mathbf{r}) = \frac{i\alpha}{\omega\mu_0}\frac{\partial\Psi}{\partial z}\mathbf{e}_x - \frac{i\alpha}{\omega\mu_0}\frac{\partial\Psi}{\partial x}\mathbf{e}_z + \frac{k\beta e^{i\varphi}}{\omega\mu_0}\Psi\mathbf{e}_y. \tag{26}$$

The scalar field Ψ at any z can be expressed as a Fresnel-Kirchhoff integral [65], which in general has to be numerically integrated once the initial condition is established. In particular, if one considers a grating with two Gaussian slits [66], the initial condition can be assumed to be a coherent superposition of the two diffracted waves,

$$\Psi(x, 0) = \psi_1(x, 0) + \psi_2(x, 0), \tag{27}$$

where

$$\psi_i(x, 0) = \left(\frac{1}{2\pi\sigma_i^2}\right)^{1/4} e^{-(x-x_{0,i})^2/4\sigma_i^2} W(x - x_{0,i}, w_i), \tag{28}$$

with $i = 1, 2$. A window function, $W(x, w)$, has been added to each wave packet in order to provoke a truncation and, therefore, to analyze the eventual effects on the final interference pattern. In this case, this function is such that it is one within the extension covered by the corresponding slit (i.e., between $-w_i$ and w_i), and zero everywhere else.

A series of electromagnetic energy flow lines or averaged "photon" trajectories are displayed in Fig. 4 (upper panel). As in the previous example, the initial positions have also been distributed according to the Gaussian weight. To compare with, the numerical values considered have been

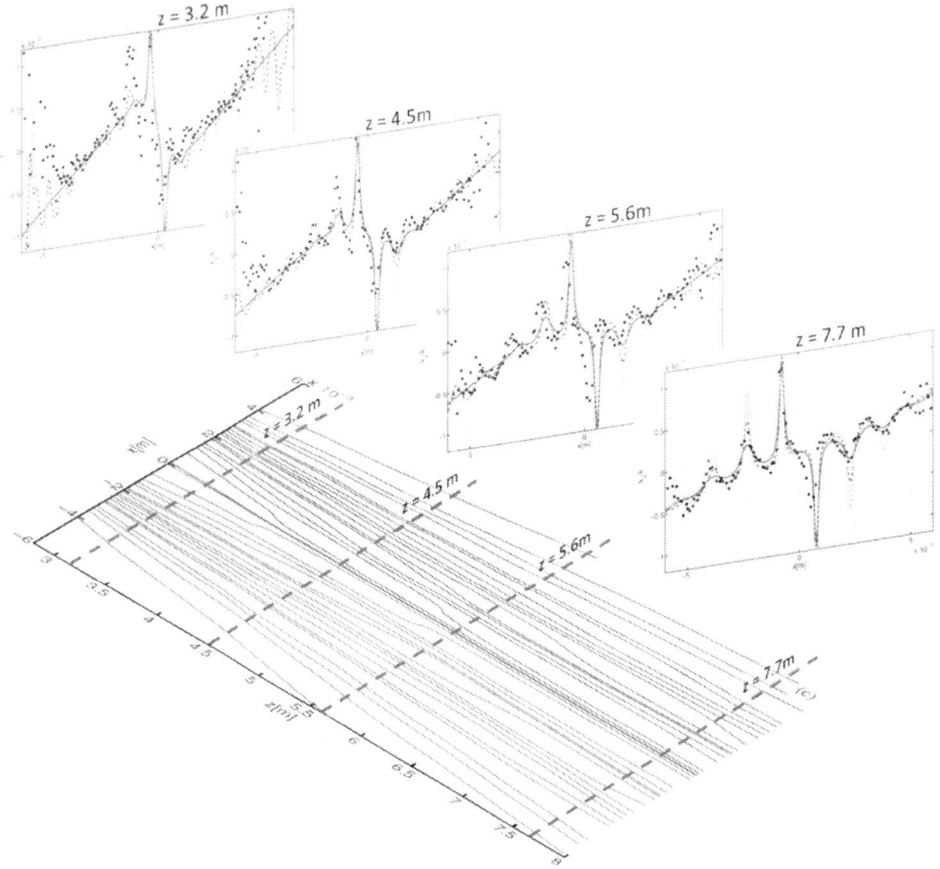

Figure 4. Lower panel: Electromagnetic energy streamlines or average "photon" trajectories behind two Gaussian slits [66]. The numerical parameters used are [65]: $\sigma_1 = \sigma_2 = 0.3$ mm, $x_{0,1} = -x_{0,2} = 2.35$ mm, and $\lambda = 943$ nm. The trajectories have been distributed according to the corresponding initial Gaussian probability densities behind each slit. Upper panels: Transverse momentum, k_x/k, evaluated at the four distances reported in Ref. [21] (see red dashed lines in the lower panel) for full Gaussian slits (red solid lines) and slits truncated at $w_i = 1.9\sigma_i$ (blue dotted lines) and $w_i = 1.5\sigma_i$ (green dashed lines). To compare with, the experimental data (black circles) are also displayed. In these four calculations, asymmetric Gaussians were used, with parameters: $\sigma_1 = 0.307$ mm, $\sigma_2 = 0.301$ mm, $x_{0,1} = 2.335$ mm, $x_{0,2} = -2.355$ mm, and $\lambda = 943$ nm.

chosen as in the experiment reported in Ref. [21]. Although the model is rather simple and makes no explicit use of the concept of "weak measurement", there is a good agreement between the simulation and the trajectories inferred from the experimental data, which shows the suitability of Maxwell's equations to describe this type of experiments. This adequacy is also seen when the transversal momentum obtained from the Poynting vector is compared to the corresponding experimental data, as shown in the upper panels for four different distances from the two slits. Notice that these two quantities are connected through the relation [67]

$$\frac{k_x}{k} = \frac{S_x}{S}, \tag{29}$$

where S_x and S refer to the x-component and modulus of the Poynting vector, respectively. In particular, in the four upper panels displayed in Fig. 4 the Gaussians have been considered to be asymmetric in order to find a better fitting with the experimental data (black circles). It can be seen that as the Gaussians are more severely truncated, the oscillations of k_x/k undergo a remarkable increase. This means a stronger action of the quantum potential between adjacent interference fringes according to the usual Bohmian interpretation [36, 68].

5. Final remarks

In general, the position maintained with respect to Bohmian mechanics is that it constitutes an alternative interpretation to quantum mechanics (or even an alternative theory). From the above discussion, it is difficult to find arguments sustaining such claims —often used in a pejorative sense to refute any work in this area. We have seen that Bohmian mechanics should be rather regarded as an alternative and complementary representation of quantum mechanics, particularly when we note that it is the direct translation of the latter into the language of the theory of characteristics, which has nothing to do with hidden variables or ontological views.

In that sense, it is worth thinking for a while about the different representations that we have in classical and quantum mechanics. For the former, we admit different formulations, each one emphasizing a different physical aspect of the systems described. For example, Newton's formulation relies on the relationship between the motion displayed by objects and the external forces that act on them. On the other hand, we also have Hamilton's formulation, which is based on the concept of energy conservation. This allows us to tackle physical problems in a rather flexible way, choosing the formulation that better fits our needs, and at the same time to understand the same phenomenon from different perspectives.

Similarly in quantum mechanics there are also different formulations or ways to tackle the same problem, which are chosen according to their suitability —either analytical or numerical. For example, Schrödinger's formulation stresses the time-evolution of the system (through its wave function) under the influence of a given Hamiltonian; any property of the system is synthesized from its wave function either at a given time (probability densities) or at two different times (correlation functions). With Heisenberg's formulation, on the contrary, one focuses on the evolution in time of the operators associated with observables and how they act on a given (time-independent) system state. Half-way between both, the Dirac or interaction representation is more convenient to analyze the dynamics of systems interacting with other systems. Feynman's path-integral representation, relying on the concept of classical action and trajectory, is suitable in the treatment of large systems. Other representations, such as the Wigner-Moyal or the Husimi ones, stress the role of the density matrix in phase space. All these well-known examples are equivalent formulations of quantum mechanics, which provide a different description of the same system. In the same way, Bohmian mechanics stresses the role of the quantum phase, which has determining consequences for quantum systems even though it is not an observable. Figure 1 summarizes this descriptive complementarity.

The trajectories that we obtain through Bohmian mechanics help us to visualize and understand the physics underlying quantum systems and phenomena by monitoring the flow of the probability. This appealing feature has been used in different contexts, not necessarily connected to Bohmian mechanics or, in general, to quantum mechanics. Here, for example, the dynamics of a dissipative system has been analyzed, making evident the pathologies of a well-known model and how they can be avoided. In this sense, notice that at a practical level the calculation of Bohmian trajectories is more convenient even if the Schrödinger equation has to be solved in order to compute them. First, the analysis of the system dynamics in terms of trajectories is simpler than in terms of probability densities, specially for two or more dimensions. Second, if instead of probabilities, one considers expectation values, the trajectories are still very appropriate, because the former only provide us with averaged information at every time, while the latter allow us to visualize the dynamics of each particular part of the quantum state.

Furthermore, the same concepts and tools can be extended to other areas of physics, as we have seen in the example of the "photon" trajectories. Although the differential equations are different and the fields propagated have a different physical meaning, the underlying ideas are exactly the same. This is actually the way how the recent experimental realization of Young's two slit experiment has been explained [21]. Among the many different ways that one could devise to join the transversal momentum function at different distances from the two slits, Bohmian mechanics provides a very clear prescription of how to do it, without incurring any kind of approximation or including any external element to quantum mechanics.

In that latter regard, I shall finish here with the following consideration. In order to explain his outcomes from the two-slit experiment, Young used the so-called Huygens' construction. Accordingly, the position of a wavefront at a given time can be obtained by considering a wavefront at a previous time. Each point on this second wavefront is assumed to be a source of secondary waves, whose interference gives rise to the wavefront at the later time. The direction of travel of these wavefronts is what we call a ray. In the case of plane waves, the wavefronts are perpendicular to rays; in the case of interference, the shape of the wavefronts varies from point to point, that making intractable a ray description ...from an analytical viewpoint! Should Young have had a computer, he would have been able to evaluate the rays locally, at each point ...and would have discovered Bohmian mechanics more than 100 years earlier, for Bohmian trajectories constitute the convoluted generalization of the concept of ray. In other words, they are the characteristics that correspond to surfaces describing waves.

Acknowledgments

The author would like to thank the organizers of EmQM13 for their kind invitation, as well as to Gerhard Grössing, Johannes Mesa Pascasio, Herbert Schwabl, and Siegfried Fussy for their hospitality and interesting discussions during his stay in Vienna. Profs. Helmut Rauch and Dieter Schuch are also acknowledged for stimulating discussions during the conference. Financial support from the Ministerio de Economía y Competitividad (Spain) under Project FIS2011-29596-C02-01 and a "Ramón y Cajal" Research Fellowship, and from the COST Action MP1006 (*Fundamental Problems in Quantum Physics*) is acknowledged.

References

[1] New Feature 2013 *Nature* **498** 21
[2] Special Issue 2013 *Nature* **498**
[3] Young T 1804 *Phil. Trans. R. Soc. Lond.* **94** 1–16
[4] Young T 1807 *A Course of Lectures on Natural Philosophy and the Mechanical Arts* (London: Joseph Johnson)
[5] Shamos M H (ed) 1987 *Great Experiments in Physics* (New York: Dover)
[6] Feynman R P, Leighton R B and Sands M 1965 *The Feynman Lectures on Physics* vol 3 (Reading, MA: Addison-Wesley)
[7] Forman P 1971 *Hist. Stud. Phys. Sci.* **3** 1–115
[8] Jammer M 1966 *The Conceptual Development of Quantum Mechanics* (New York: McGraw-Hill)
[9] Mermin N D 1989 *Physics Today* **42** 9–11
[10] Mermin N D 2004 *Physics Today* **57** 10–11
[11] Courant R and Hilbert D 1966 *Methods of Mathematical Physics* vol 2 (New York: John Wiley & Sons) ch 2
[12] Juffmann T, Milic A, Müllneritsch M, Asenbaum P, Tsukernik A, Tüxen J, Mayor M, Cheshnovsky O and Arndt M 2012 *Nat. Nanotech.* **7** 297–300
[13] Bohm D 1952 *Phys. Rev.* **85** 166–179
[14] de Broglie L 1926 *Comptes Rendus* **183** 447–448
[15] Madelung E 1926 *Z. Phys.* **40** 322–326
[16] Takabayasi T 1952 *Prog. Theor. Phys.* **8** 143–182
[17] Takabayasi T 1953 *Prog. Theor. Phys.* **9** 187–222
[18] Takabayasi T 1983 *Prog. Theor. Phys.* **69** 1323–1344

[19] Sanz A S and Miret-Artés S 2008 *J. Phys. A: Math. Theor.* **41** 435303(1–23)
[20] Sanz A S and Miret-Artés S *A Trajectory Description of Quantum Processes. I. Fundamentals* vol 850 (Berlin: Springer)
[21] Kocsis S, Braverman B, Ravets S, Stevens M J, Mirin R P, Shalm L K and Steinberg A M 2011 *Science* **332** 1170–1173
[22] Sanz A S and Miret-Artés S 2011 *J. Phys. A: Math. Theor.* **44** 485301(1–17)
[23] Sanz A S and Miret-Artés S 2013 *Ann. Phys.* **339** 11–21 (*Preprint* 1104.1296)
[24] McCullough E A and Wyatt R E 1969 *J. Chem. Phys.* **51** 1253–1254
[25] McCullough E A and Wyatt R E 1971 *J. Chem. Phys.* **54** 3578–3591
[26] McCullough E A and Wyatt R E 1971 *J. Chem. Phys.* **54** 3592–3600
[27] Wyatt R E 2005 *Quantum Dynamics with Trajectories* (New York: Springer)
[28] Hirschfelder J O, Christoph A C and Palke W E 1974 *J. Chem. Phys.* **61** 5435–5455
[29] Dewdney C and Hiley B J 1982 *Found. Phys.* **12** 27–48
[30] Lazzeretti P, Rossi E and Zanasi R 1984 *Int. J. Quantum Chem.* **25** 929–940
[31] Pelloni S and Lazzeretti P 2011 *Int. J. Quantum Chem.* **111** 356–367
[32] Braunbek W and Laukien G 1952 *Optik* **9** 174–179
[33] Sommerfeld A 1896 *Math. Ann.* **47** 317–374
[34] Prosser R D 1976 *Int. J. Theor. Phys.* **15** 169–180
[35] Prosser R D 1976 *Int. J. Theor. Phys.* **15** 181–193
[36] Philippidis C, Dewdney C and Hiley B J 1979 *Nuovo Cimento* **52B** 15–28
[37] Sanz A S, Davidović M, Božić M and Miret-Artés S 2010 *Ann. Phys.* **325** 763–784
[38] Żakowicz W 2001 *Phys. Rev. E* **64** 066610(1–14)
[39] Sanz A S, Campos-Martínez J and Miret-Artés S 2012 *J. Opt. Am. Soc. A* **29** 695–701
[40] Sadreev A F 2004 *Phys. Rev. E* **70** 016208(1–7)
[41] Scully M O and Zubairy S 1997 *Quantum Optics* (Cambridge: Cambridge University Press)
[42] Waterhouse R V, Yates T W, Feit D and Liu Y N 1985 *J. Acoust. Soc. Am.* **78** 758–762
[43] Waterhouse R V and Feit D 1986 *J. Acoust. Soc. Am.* **80** 681–684
[44] Skelton E A and Waterhouse R V 1986 *J. Acoust. Soc. Am.* **80** 1473–1478
[45] Waterhouse R V, Crighton D G and Ffowcs-Williams J E 1987 *J. Acoust. Soc. Am.* **81** 1323–1326
[46] Waterhouse R V 1987 *J. Acoust. Soc. Am.* **82** 1782–1791
[47] Chevalerias R, Latron Y and Veret C 1957 *J. Opt. Soc. Am.* **47** 703–706
[48] Sanz A S and Miret-Artés S 2012 *Am. J. Phys.* **80** 525–533
[49] Sanz A S and Miret-Artés S *A Trajectory Description of Quantum Processes. II. Applications* vol 831 (Berlin: Springer)
[50] Caldirola P 1941 *Nuovo cim.* **18** 393–400
[51] Kanai E 1948 *Prog. Theor. Phys.* **3** 440–442
[52] Sanz A S, Martínez-Casado R, Peñate-Rodríguez H, Rojas-Lorenzo G and Miret-Artés S 2013 Dissipative Bohmian mechanics: A trajectory analysis of wave-packet dynamics in viscid media *Preprint* 1306.6607
[53] Vink J C 1993 *Phys. Rev. A* **48** 1808–1818
[54] Lorenzen F, de Ponte M A and Moussa M H Y 2009 *Phys. Rev. A* **80** 032101(1–8)
[55] Hasse R W 1975 *J. Math. Phys.* **16** 2005–2011
[56] Schuch D, Chung K M and Hartmann H 1983 *Phys. Rev. A* **24** 1652–1660
[57] Schuch D, Chung K M and Hartmann H 1984 *Phys. Rev. A* **25** 391–410
[58] Schuch D, Chung K M and Hartmann H 1984 *Phys. Rev. A* **25** 3086–3092
[59] Schuch D 1997 *Phys. Rev. A* **55** 935–940
[60] Sanz A S 2013 Effective Markovian description of decoherence in bound systems *Can. J. Chem.* doi:10.1139/cjc-2013-0399
[61] Kostin M D 1972 *J. Chem. Phys.* **57** 3589–3591
[62] Garashchuk S, Dixit V, Gu B and Mazzuca J 2013 *J. Chem. Phys.* **138** 054107(1–7)
[63] Davidović M, Sanz A S, Arsenović D, Božić M and Miret-Artés S 2009 *Phys. Scr.* **T135** 014009(1–5)
[64] Born M and Wolf E 1999 *Principles of Optics. Electromagnetic Theory of Propagation, Interference and Diffraction of Light* 7th ed (Cambridge: Cambridge University Press)
[65] Davidović M, Sanz A S, Božić M, Arsenović D and Dimić D 2013 *Phys. Scr.* **T153** 014015(1–5)
[66] Feynman R P 2006 *Path Integral Approach to Quantum Mechanics* (Sausalito, CA: University Science Books)
[67] Barnett S M and Loudon R 2010 *Phil. Trans. R. Soc. A* **368** 927–939
[68] Sanz A S, Borondo F and Miret-Artés S 2002 *J. Phys.: Condens. Matter* **14** 6109–6145

New insights on emergence from the perspective of weak values and dynamical non-locality

Jeff Tollaksen

Institute for Quantum Studies and Schmid College of Science and Technology, Chapman University, 1 University Drive, Orange, California USA 92866

E-mail: `tollaksen@chapman.edu` and `quantum.chapman.edu`

Abstract. In this article, we will examine new fundamental aspects of "emergence" and "information" using novel approaches to quantum mechanics which originated from the group around Aharonov. The two-state vector formalism provides a complete description of pre- and post-selected quantum systems and has uncovered a host of new quantum phenomena which were previously hidden. The most important feature is that any weak coupling to a pre- and post-selected system is effectively a coupling to a "weak value" which is given by a simple expression depending on the two-state vector. In particular, weak values, are the outcomes of so called "weak measurements" which have recently become a very powerful tool for ultra-sensitive measurements. Using weak values, we will show how to separate a particle from its properties, not unlike the Cheshire cat story: "Well! I've often seen a cat without a grin," thought Alice; "but a grin without a cat! It's the most curious thing I ever saw in all my life!" Next, we address the question whether the physics on different scales "emerges" from quantum mechanics or whether the laws of physics at those scales are fundamental. We show that the classical limit of quantum mechanics is a far more complicated issue; it is in fact dramatically more involved and it requires a complete revision of all our intuitions. The revised intuitions can then serve as a guide to finding novel quantum effects. Next we show that novel experimental aspects of contextuality can be demonstrated with weak measurements and these suggest new restrictions on hidden variable approaches. Next we emphasize that the most important implication of the Aharonov-Bohm effect is the existence of non-local interactions which do not violate causality. Finally, we review some generalizations of quantum mechanics and their implications for "emergence" and "information." First, we review an alternative approach to quantum evolution in which each moment of time is viewed as a new "universe" and time evolution is given by correlations between different moments. Next, we present a new solution to the measurement problem involving future boundary conditions placed on the universe as a whole. Finally, we introduce another fundamental approach to quantum evolution which allows for tremendous richness in the types of allowable Hamiltonians.

1. Introduction

The conference's title "Emergent Quantum Mechanics" (EmQM) was defined as "...a possible *deeper-level theory* that interconnects three fields of knowledge: emergence, the quantum, and information." In this article, we will examine this from a number of novel approaches originating from the group around Aharonov. We will also try to address several ways in which EmQM conference participants interpreted the meaning of EmQM. In particular: 1) **Quantum Mechanics (QM) emerges from a deeper underlying theory:** accepting that the predictions of non-relativistic QM are correct, perhaps QM could emerge as a statistical theory from a deeper, underlying theory (potentially involving hidden-variables); 2) **Generalizations of QM:** assuming that someday non-relativistic QM is disproven (although there is as yet no experimental evidence to support this), what are possible generalizations of QM, 3) **Whole-part relations:** as Gerhard Grössing best put it: "Emergence through the co-evolution of 'microscopic,' local processes and of 'macroscopic' boundary conditions," this could involve novel arrows of causation, the traditional "bottom-up" arrow combined with another "top-down" arrow; and finally 4) **Emergence and the origin of laws:** i.e. whether the physics on different scales "emerges" from QM or whether the laws of physics at those scales are fundamental.

Often these inquiries are approached by seeking new physics beyond the core of standard non-relativistic QM. While we too will examine such an approach, first in §2 we ask what novel perspectives can be gleaned just from standard non-relativistic QM. Then, in §3, we go beyond existing QM and suggest several speculative generalizations of QM and their potential implications for EmQM.

2. Novel aspects of emergence based on standard non-relativistic QM

In this section, we assume that QM, which to-date has been vindicated by all experiments, is correct and need not be recast with hidden variables or classical/stochastic elements. There are still some logical shortcomings such as the apparently non-unitary evolution which occurs during measurements with the ensuing "collapse" of the wave function. Nevertheless, in this section, we are not espousing any of the existing approaches which try to explain such issues. Instead, we find it most beneficial to keep prying and searching for new effects within, and new re-formulations of, the existing quantum theory. Rather than impose new rules, in this section, we utilize the established rules, trying to better understand the underlying logic and their potential relevance to the above mentioned issues. Cognizant of the popular saying "if your only tool is a hammer, then you tend to treat everything as if it were a nail," we have used new measurement paradigms, such as weak measurements, protective measurement, and the measurement of deterministic operators, to investigate these issues.

Entanglement and nonlocality are now considered to be some of the most basic aspects of QM. In particular they are at the core of quantum information science, and, more generally, in all applications of quantum coherence. As such, entanglement has been studied extensively, and it is one of the most active fields of research in theoretical physics. To date, QM indicates non-locality in two fundamental, empirically proven ways: kinematic (e.g. Einstein-Podolsky-Rosen a/k/a EPR) and dynamic (e.g. the Aharonov-Bohm effect a/k/a AB). The AB effect demonstrated that an electromagnetic field inside a confined region can have a measureable impact on the interference pattern of a charged particle which never traveled inside that region. The EPR non-locality is evidenced when the various parts of a quantum system are spatially and/or temporally separated. Bell's theorem shows us that the correlations between these parts cannot be explained by any local-realistic theory. Some have argued that the non-factorizability of a many body state into a product of one body states implies a kind of holism. As Shimony elegantly stated: [42]: "The parts-wholes problem has an ontological aspect, which concerns the properties of the components and the composite system without explicit consideration of how knowledge of them is obtained... are there properties of composite systems which are radically different from those of the components, and which might properly be characterized as 'emergent'?"

We consider novel aspects of emergence using 2 different approaches to standard QM originated by the group around Aharonov. First, in §2.1, we review a time-symmetric re-formulation of quantum mechanics which led to weak measurements performed on pre- and post-selected ensembles. It also led to the "weak value" [29]. Secondly in §2.2 we apply weak values to EmQM. Next in §2.3 we use weak values to introduce new restrictions on hidden variables and then, in §2.4 we discuss a new approach to information, which we call "weak

Journal of Physics: Conference Series **504** (2014) 012029 doi:10.1088/1742-6596/504/1/012029

information," and in §2.6 we introduce a new "internal" quantum frame-of-reference. In the final sub-section in this chapter concerning standard QM, namely §2.5, we consider dynamical non-locality originating from the AB effect as described by modular variables which are particularly useful when considering quantum states composed of several lumps involving interference.

2.1. Review of time-symmetry, weak values and weak measurements

The "time-asymmetry" in QM involves the assumption that measurements only have consequences **after** they are performed, i.e. towards the future. Nevertheless, a positive spin was placed on QM's non-trivial relationship between initial and final conditions by Aharonov, Bergmann and Lebowitz (ABL) [1] who showed that the new information obtained from subsequent measurements were also relevant for the **past** of every quantum-system and not just the future. This inspired ABL to re-formulate QM in terms of *Pre- and Post-Selected ensembles* (PPS). The traditional paradigm for ensembles is to simply prepare systems in a particular state and thereafter subject them to a variety of experiments. These are "pre-selected-only-ensembles." For pre-**and-post-selected**-ensembles, we add one more step, a subsequent measurement or post-selection. By collecting only a subset of the outcomes for this later measurement, we see that the "pre-selected-only-ensemble" can be divided into sub-ensembles according to the results of this subsequent "post-selection-measurement." Because PPS ensembles are the most refined quantum ensemble, they are of fundamental importance and subsequently led to the *two-vector* or *Time-Symmetric re-formulation of Quantum Mechanics* (TSQM) [9,11]. TSQM provides a complete description of a quantum-system at a given moment by using two-wavefunctions, one evolving from the past towards the future (the one utilized in the standard paradigm) and a second one, evolving from the future towards the past.

While TSQM is a new conceptual point-of-view that has predicted novel, verified effects which **seem** impossible according to standard QM, TSQM is in fact a *re-formulation* of QM. Therefore, experiments cannot prove TSQM over QM (or vice-versa). The motivation to pursue such re-formulations, then, depends on their usefulness. TSQM fulfills several criterion in order to be useful and interesting:

- TSQM is consistent with all the predictions made by standard QM (§2.1.1),
- TSQM has revealed new features of QM that were missed before (§2.1.2.1),
- TSQM has led to new mathematics, simplifications in calculations, and stimulated discoveries in other fields (§2.1.3),
- TSQM suggests generalizations of QM (§3).

2.1.1. The main idea behind this re-formulation of QM

TSQM contemplates measurements which occur at the present time t while the state is known both at $t_{\rm in} < t$ (past) and at $t_{\rm fin} > t$ (future). More precisely, we start at $t = t_{\rm in}$ with a measurement of a nondegenerate operator $\hat{O}_{\rm in}$. This yields as one potential outcome the state $|\Psi_{\rm in}\rangle$, i.e. we prepared the "pre-selected" state $|\Psi_{\rm in}\rangle$. At the later time $t_{\rm fin}$, we perform another measurement of a nondegenerate operator $\hat{O}_{\rm fin}$ which yields a number of possible outcomes and we select one of them, the post-selected state $|\Psi_{\rm fin}\rangle$. At an intermediate time $t \in [t_{\rm in}, t_{\rm fin}]$, we measure a non-degenerate (for simplicity) observable $\hat{A}$, with eigenvectors $\{|a_j\rangle\}$. We wish to determine the conditional probability of a_j, given that we have both boundary conditions, $|\Psi_{\rm in}\rangle$ and $\langle\Psi_{\rm fin}|$. [1] To answer this, we use the time displacement operator: $U_{t_{\rm in}\to t} = \exp\{-iH(t - t_{\rm in})\}$ where H

[1] Such an arrangement has long been considered in actual experiments such as a bubble chamber/scattering experiment. The incoming particle, $|\Psi_{\rm in}\rangle$, interacts with a target and then evolves into various outgoing states, $|\Psi_{\rm fin}\rangle_1$, $|\Psi_{\rm fin}\rangle_2$, etc. Typically, photographs are not taken for every target-interaction, but only for certain ones that were triggered by subsequently interacting with detectors. In classical mechanics, there is (in principle) a

is the Hamiltonian for the free system. For simplicity, we assume H is time independent and set $\hbar = 1$. The standard theory of collapse states that the system collapses into an eigenstate $|a_j\rangle$ *after* the measurement at t with an amplitude $\langle a_j|U_{t_{\rm in}\to t}|\Psi_{\rm in}\rangle$. The amplitude for our series of events is $\alpha_j \equiv \langle\Psi_{\rm fin}|U_{t\to t_{\rm fin}}|a_j\rangle\langle a_j|U_{t_{\rm in}\to t}|\Psi_{\rm in}\rangle$ which is illustrated in figure 1.a. This means that the conditional probability to measure a_j given that $|\Psi_{\rm in}\rangle$ is pre-selected and $|\Psi_{\rm fin}\rangle$ will be post-selected is given by the ABL formula [1]:

$$Pr(a_j, t|\Psi_{\rm in}, t_{\rm in}; \Psi_{\rm fin}, t_{\rm fin}) = \frac{|\langle\Psi_{\rm fin}|U_{t\to t_{\rm fin}}|a_j\rangle\langle a_j|U_{t_{\rm in}\to t}|\Psi_{\rm in}\rangle|^2}{\sum_n |\langle\Psi_{\rm fin}|U_{t\to t_{\rm fin}}|a_n\rangle\langle a_n|U_{t_{\rm in}\to t}|\Psi_{\rm in}\rangle|^2} \tag{2.1}$$

As a first step toward understanding the underlying time-symmetry in the ABL formula, we consider the time-reverse of the numerator of Eq. (2.1) and consequently the time reverse of figure 1.a. First we apply $U_{t\to t_{\rm fin}}$ on $\langle\Psi_{\rm fin}|$ instead of on $\langle a_j|$. We note that $\langle\Psi_{\rm fin}|U_{t\to t_{\rm fin}} = \langle U^\dagger_{t\to t_{\rm fin}}\Psi_{\rm fin}|$ by using the well-known QM symmetry $U^\dagger_{t\to t_{\rm fin}} = \{e^{-iH(t_{\rm fin}-t)}\}^\dagger = e^{iH(t_{\rm fin}-t)} = e^{-iH(t-t_{\rm fin})} = U_{t_{\rm fin}\to t}$. We also apply $U_{t_{\rm in}\to t}$ on $\langle a_j|$ instead of on $|\Psi_{\rm in}\rangle$ which yields the time-reverse reformulation of the numerator of Eq. (2.1), $\langle U_{t_{\rm fin}\to t}\Psi_{\rm fin}|a_j\rangle\langle U_{t\to t_{\rm in}}a_j|\Psi_{\rm in}\rangle$ as depicted in Fig. 1.b. Further work is needed to formulate what we mean by the 2-vectors in TSQM. E.g. if we

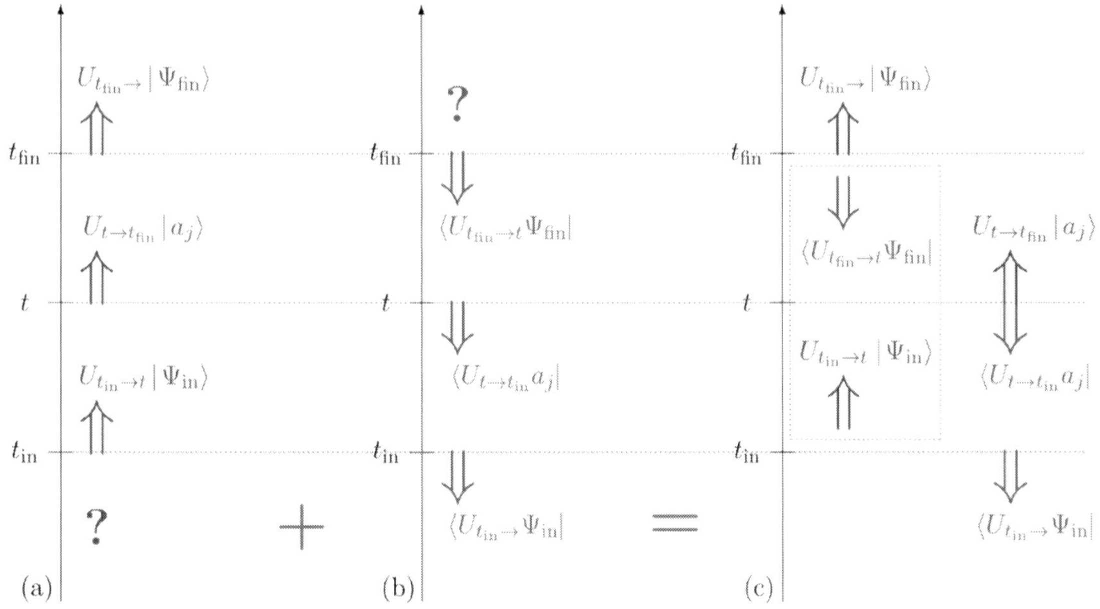

Figure 1. Time-reversal symmetry in probability amplitudes

are interested in the probability for possible outcomes of a_j at time t, we must consider both $U_{t_{\rm in}\to t}|\Psi_{\rm in}\rangle$ and $\langle U_{t_{\rm fin}\to t}\Psi_{\rm fin}|$, since these expressions propagate the pre- and post-selection to the present time t (see the conjunction of both figures 1.a and 1.b giving 1.c; Note: these 2-vectors are not just the time-reverse of each other). This represents the basic idea behind TSQM:

$$Pr(a_j, t|\Psi_{\rm in}, t_{\rm in}; \Psi_{\rm fin}, t_{\rm fin}) = \frac{|\langle U_{t_{\rm fin}\to t}\Psi_{\rm fin}|a_j\rangle\langle a_j|U_{t_{\rm in}\to t}|\Psi_{\rm in}\rangle|^2}{\sum_n |\langle U_{t_{\rm fin}\to t}\Psi_{\rm fin}|a_n\rangle\langle a_n|U_{t_{\rm in}\to t}|\Psi_{\rm in}\rangle|^2} \tag{2.2}$$

one-to-one mapping between incoming states and outgoing states, whereas in QM, it is one-to-many. By selecting a single outcome for the post-selection-measurement, we define the pre- and post-selected-ensemble that has no classical analog.

While this mathematical manipulation clearly proves that the predictions of TSQM is consistent with standard QM, it nevertheless suggests a very different interpretation. For example, the action of $U_{t_{\rm fin}\to t}$ on $\langle\Psi_{\rm fin}|$ (i.e. $\langle U_{t_{\rm fin}\to t}\Psi_{\rm fin}|$) can be interpreted as sending $\langle\Psi_{\rm fin}|$ back in time from $t_{\rm fin}$ to the present, t. In summary, ABL clarified a number of issues in QM. E.g. in this formulation, both the probability and the amplitude are symmetric under the exchange of $|\Psi_{\rm in}\rangle$ and $|\Psi_{\rm fin}\rangle$. Therefore, the possibility of wavefunction collapse in QM does not necessarily imply irreversibility of an arrow of time at the QM level. TSQM suggested a number of new experimentally observable effects, one important example of which are *weak measurements* (§2.1.2.1), which we now begin to motivate by considering strange pre- and post-selection effects.

2.1.2. **Pre- and post-selection, weak measurements and weak values of spin-1/2 systems** One of the simplest, surprising, examples of pre- and post-selection is to pre-select a spin-1/2 system with $|\Psi_{\rm in}\rangle = |\hat{\sigma}_x = +1\rangle = |\uparrow_x\rangle$ at time $t_{\rm in}$. After the pre-selection, spin measurements in the direction perpendicular to x yields complete uncertainty so if we post-select at time $t_{\rm fin}$ in the y-direction, we obtain $|\Psi_{\rm fin}\rangle = |\hat{\sigma}_y = +1\rangle = |\uparrow_y\rangle$ one-half of the time. Since the particle is free, the spin is conserved in time and thus for any $t \in [t_{\rm in}, t_{\rm fin}]$, an ideal-measurement of either $\hat{\sigma}_x$ or $\hat{\sigma}_y$, yields +1 for this pre- and post-selection. This by itself, two non-commuting observables known with certainty, is a most surprising property which no pre-selected-only-ensemble could possess.

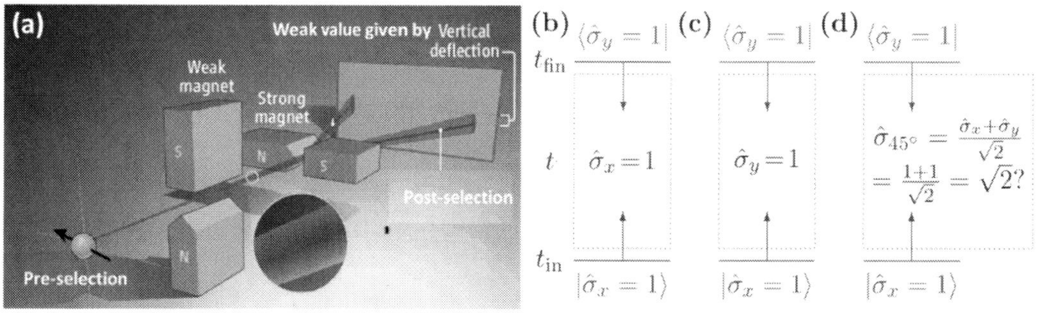

Figure 2. Stern-Gerlach apparatus both prepares (pre-selects) a spin-1/2 particle at time $t_{\rm in}$ to be $|\hat{\sigma}_x = 1\rangle$, and post-selects at $t_{\rm fin}$ to be $\langle\hat{\sigma}_y = 1|$ and weakly measures the spin (from [43]; (b) During the intermediate time $t \in [t_{\rm in}, t_{\rm fin}]$, ABL tells us that an ideal-measurement of either $\hat{\sigma}_x$ or (c) $\hat{\sigma}_y$, yields +1 with certainty, suggesting that such a particle has well-defined values of the two non-commuting spin components (d) It would seem to follow that the spin component $\hat{\sigma}_{45^\circ}$ would have to be $\sqrt{2}$ which is not an allowed eigenvalue.

We now ask a slightly more complicated question about the spin in a direction $\xi = 45^\circ$ relative to the $x - y$ axis. This yields:

$$\hat{\sigma}_\xi = \hat{\sigma}_x \cos 45^\circ + \hat{\sigma}_y \sin 45^\circ = \frac{\hat{\sigma}_x + \hat{\sigma}_y}{\sqrt{2}} \tag{2.3}$$

From the results $Pr(\hat{\sigma}_x = +1) = 1$ and $Pr(\hat{\sigma}_y = +1) = 1$, one might wonder why we couldn't insert both values, $\hat{\sigma}_{\rm x} = +1$ **and** $\hat{\sigma}_{\rm y} = +1$ into Eq. (2.3) and obtain $\hat{\sigma}_\xi = \frac{1+1}{\sqrt{2}} = \frac{2}{\sqrt{2}} = \sqrt{2}$ (see figure 2). Such a result is incorrect for an ideal-measurement because the eigenvalues of any spin operator, including $\hat{\sigma}_\xi$, must be ± 1. Performing this step of replacing $\hat{\sigma}_{\rm x} = +1$ *and* $\hat{\sigma}_{\rm y} = +1$ in Eq. (2.3) can only be done if $\hat{\sigma}_{\rm x}$ and $\hat{\sigma}_{\rm y}$ commute, which would allow both values simultaneously to be definite. The ABL statements, namely $Pr(\hat{\sigma}_x = +1) = 1$ and $Pr(\hat{\sigma}_y = +1) = 1$, are said to be "counterfactuals." Although it appears we have reached the end-of-the-line with this argument, nevertheless, it still seems that there should be some sense in which both $Pr(\hat{\sigma}_x = +1) = 1$ and $Pr(\hat{\sigma}_y = +1) = 1$ manifest themselves simultaneously to produce $\hat{\sigma}_\xi = \sqrt{2}$.

2.1.2.1. **Counterfactuals:** There is a widespread tendency to "resolve" many quantum mechanical paradoxes by pointing out that there is an element of counter-factual reasoning, just as we pointed out in the previous example. I.e., the contradictions arise only because inferences are made that do not refer to actual experiments. Had the experiment actually been performed, then standard measurement theory predicts that the system would have been disrupted so that no paradoxical implications arises.

We have proven [12, 19] that one shouldn't be so quick in throwing away counter-factual reasoning; though indeed counter-factual statements have no observational meaning, such reasoning is actually a very good pointer towards interesting physical situations. *Without invoking counter-factual reasoning*, we have shown that the apparently paradoxical reality implied counter-factually has new, *experimentally accessible* consequences. These observable consequences become evident in terms of *weak measurements*, which allow us to test – to some extent – assertions that have been otherwise regarded as counter-factual.

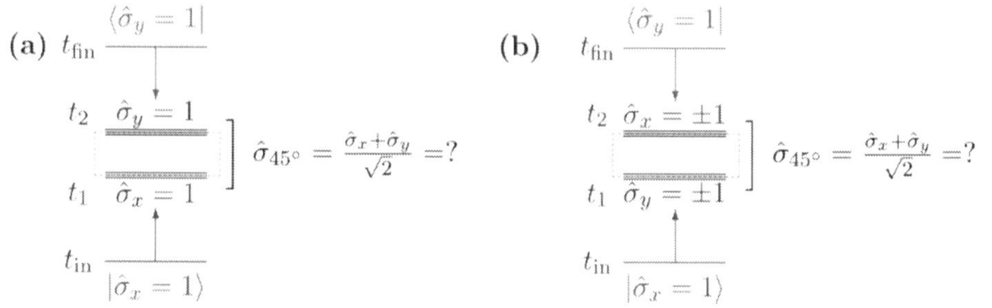

Figure 3. Measurement of $\hat{\sigma}_{45^\circ}$ is effectively a simultaneous measurement of both $\hat{\sigma}_x$ and $\hat{\sigma}_y$. In fig 2, we argued that if we measured either one separately, then we would obtain +1 with certainty. It is more complicated if we try to measure ***both*** of them together as required for a measurement of $\hat{\sigma}_{45^\circ}$. (a) E.g., if the first measurement performed during the intermediate time $t \in [t_{\rm in}, t_{\rm fin}]$ is in the same direction as the pre-selection (i.e. $\hat{\sigma}_x$) and the second measurement during the intermediate time $t \in [t_{\rm in}, t_{\rm fin}]$ is in the same direction as the post-selection (i.e. $\hat{\sigma}_y$) then both measurements will be determined with certainty (i.e. +1) as a result of the pre- and post-selection; (b) However, if we reverse the order of the measurements performed during the intermediate time $t \in [t_{\rm in}, t_{\rm fin}]$ then that certainty is destroyed. I.e. the first measurement, that of $\hat{\sigma}_y$ can result in either +1 or −1. The second measurement, that of $\hat{\sigma}_x$ can also result in either +1 or −1. In other words, $\hat{\sigma}_x$ is not determined by the pre-selection because the previous measurement performed during the intermediate time $t \in [t_{\rm in}, t_{\rm fin}]$ ***disturbs*** it.

For our spin-1/2 example, the counter-factual argument works as follows: if we verify $\hat{\sigma}_{\rm x}$ at $t = t_1$ and $\hat{\sigma}_{\rm y}$ at $t = t_2$, $t_{\rm in} < t_1 < t_2 < t_{\rm fin}$, then $Pr(\hat{\sigma}_x = +1) = 1$ and $Pr(\hat{\sigma}_y = +1) = 1$ are simultaneously true. But if we switch the order and perform $\hat{\sigma}_y$ before $\hat{\sigma}_x$, then $Pr(\hat{\sigma}_x = +1) = 1$ and $Pr(\hat{\sigma}_y = +1) = 1$ are not simultaneously true, since measuring $\hat{\sigma}_y$ at time $t = t_1$ would not allow the information from the earlier ($t_{\rm in} < t$) pre-selection of $\hat{\sigma}_{\rm x} = +1$ to propagate to the later time ($t_2 > t_1 > t_{\rm in}$) of the $\hat{\sigma}_{\rm x}$ measurement. As a consequence, the $\hat{\sigma}_{\rm x}$ measurement at time t_2 would yield both outcomes $\hat{\sigma}_x = \pm 1$ So, in general, the finding that $\hat{\sigma}_{\rm x} = +1$ with certainty or $\hat{\sigma}_{\rm y} = +1$ with certainty in the pre- and post-selected ensemble only held when *one* of these two measurements was performed in the intermediate time, not both. Therefore, we should not expect both $\hat{\sigma}_{\rm y} = +1$ and $\hat{\sigma}_{\rm x} = +1$ when measured simultaneously through $\hat{\sigma}_{\xi=45^\circ}$.

Since we have understood the reason why both statements are not simultaneously true as a result of disturbance, we can now see the "sense" in which the definite ABL assignments can be simultaneously relevant. Our main argument is that if one doesn't perform absolutely precise (ideal) measurements but is willing to accept some finite accuracy, then one can bound the disturbance on the system. For example, according to Heisenberg's uncertainty relations, a precise measurement of position reduces the uncertainty in the position of the measuring device to zero $\Delta Q_{\rm md} = 0$ but produces an infinite uncertainty in momentum $\Delta P_{\rm md} = \infty$. On the other hand, if we measure the position only up to some finite precision $\Delta Q_{\rm md} = \Delta$ we can limit the disturbance of momentum to a finite amount $\Delta P_{\rm md} \geq \hbar/\Delta$. By replacing precise measurements with a bounded-measurement paradigm, we produce limited-disturbance measurements. With this, there **is** a sense in which both $Pr(\hat{\sigma}_x = +1) = 1$ *and* $Pr(\hat{\sigma}_y = +1) = 1$ are simultaneously relevant because measurement of one does *not* disturb the other. With limited-disturbance measurements, we can simultaneously use both $\hat{\sigma}_x = +1$ and $\hat{\sigma}_y = +1$ to obtain $(\hat{\sigma}_{\xi=45^\circ})_{\rm w} = \frac{\langle\uparrow_y|\frac{\hat{\sigma}_y+\hat{\sigma}_x}{\sqrt{2}}|\uparrow_x\rangle}{\langle\uparrow_y|\uparrow_x\rangle} = \frac{\{\langle\uparrow_y|\hat{\sigma}_y\}+\{\hat{\sigma}_x|\uparrow_x\rangle\}}{\sqrt{2}\langle\uparrow_y|\uparrow_x\rangle} = \frac{\langle\uparrow_y|1+1|\uparrow_x\rangle}{\sqrt{2}\langle\uparrow_y|\uparrow_x\rangle} = \sqrt{2}$.

2.1.2.2. **Quantum Measurements:** Weak-measurements [11] originally grew out of the quantum measurement theory developed by von Neumann though weak measurements and their outcome, weak values, can be derived in all approaches to quantum measurement theory. E.g. the usual projective measurement typically utilized in quantum experiments is a special case of these weak-measurements [21]. Using an interaction Hamiltonian $H_{\rm int}$ of the form $H_{\rm int} = -g(t)\hat{Q}_{\rm md}\hat{A}$ with $\int_0^T g(t)dt = g_o$, we note that to make a more precise determination of $\hat{A}$, that is, to make an ideal measurement, requires that the shift in $P_{\rm md}$, i.e. $\delta P_{\rm md} = P_{\rm md}(T) - P_{\rm md}(0)$, be distinguishable from its uncertainty, $\Delta P_{\rm md}$. This occurs, e.g., if $P_{\rm md}(0)$ and $P_{\rm md}(T)$ are more precisely defined and/or if g_0 is sufficiently large (see figure 4.a). However, under these conditions (e.g. if the measuring device approaches a delta function in $P_{\rm md}$), then the disturbance or back-reaction on the system is increased due to a larger $H_{\rm int}$, the result of the larger $\Delta Q_{\rm md}$ ($\Delta Q_{\rm md} \geq \frac{1}{\Delta P_{\rm md}}$). When $\hat{A}$ is measured in this way, then any operator $\hat{O}$ ($[\hat{A}, \hat{O}] \neq 0$) is disturbed because it evolved according to $\frac{d}{dt}\hat{O} = ig(t)[\hat{A}, \hat{O}]\hat{Q}_{\rm md}$, and since $g_o\Delta Q_{\rm md}$ is not zero, $\hat{O}$ changes in an uncertain way proportional to $g_o\Delta Q_{\rm md}$. In the spin-1/2 example, the conditions for an ideal-measurement $\delta P^{\xi}_{\rm md} = g_o\hat{\sigma}_\xi \gg \Delta P^{\xi}_{\rm md}$ will also necessitate $\Delta Q^{\xi}_{\rm md} \gg \frac{1}{g_o\hat{\sigma}_\xi}$ which will thereby create a back-reaction causing a precession in the spin such that $\Delta\Theta \gg 1$ (i.e. more than one revolution), thereby destroying (i.e. making completely uncertain) the information that in the past we had $\hat{\sigma}_x = +1$, and in the future we will have $\hat{\sigma}_y = +1$. In the Schrödinger picture, the time evolution operator for the complete system from $t = t_0 - \varepsilon$ to $t = t_0 + \varepsilon$ is $\exp\{-i\int_{t_0-\varepsilon}^{t_0+\varepsilon} H(t)dt\} = \exp\{-ig_o\hat{Q}_{\rm md}\hat{A}\}$. This shifts $P_{\rm md}$ (see figure 4.a). If before the measurement the system was in a superposition of eigenstates of $\hat{A}$, then the measuring device will also be superposed in direct correspondence with the system. This leads to the "quantum measurement problem." A conventional solution to this problem is to argue that because the measuring device is macroscopic, it cannot be in a superposition, and so it will "collapse" into one of these states and the system will collapse with it.

2.1.2.3. **Weakening the interaction between system and measuring device:** Following the intuition gained in Fig. 3, we now perform measurements which do not disturb either the pre- or post-selections. The interaction $H_{\rm int} = -g(t)\hat{Q}_{\rm md}\hat{A}$ is weakened by minimizing $g_o\Delta Q_{\rm md}$. For simplicity, we consider $g_o \ll 1$ (assuming without lack of generality that the state of the measuring device is a Gaussian with spreads $\Delta P_{\rm md} = \Delta Q_{\rm md} = 1$). We may then set

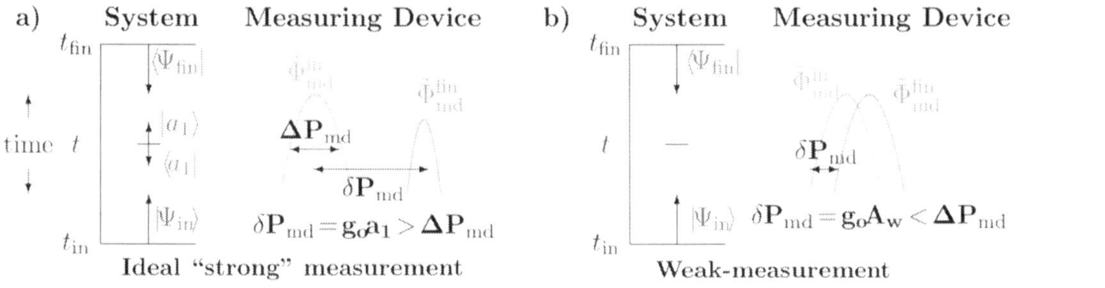

Figure 4. a) with an ideal or "strong" measurement at t (characterized e.g. by $\delta P_{\rm md} = g_o a_1 \gg \Delta P_{\rm md}$), then ABL gives the probability to obtain a collapse onto eigenstate a_1 by propagating $\langle\Psi_{\rm fin}|$ backwards in time from $t_{\rm fin}$ to t and $|\Psi_{\rm in}\rangle$ forwards in time from $t_{\rm in}$ to t; in addition, the collapse caused by ideal-measurement at t creates a new boundary condition $|a_1\rangle\langle a_1|$ at time $t \in [t_{\rm in}, t_{\rm fin}]$; b) if a weak-measurement is performed at t (characterized e.g. by $\delta P_{\rm md} = g_o A_w \ll \Delta P_{\rm md}$), then the outcome of the weak-measurement, the weak-value, can be calculated by propagating the state $\langle\Psi_{\rm fin}|$ backwards in time from $t_{\rm fin}$ to t and the state $|\Psi_{\rm in}\rangle$ forwards in time from $t_{\rm in}$ to t; the weak-measurement does not cause a collapse and thus no new boundary condition is created at time t.

$e^{-ig_o\hat{Q}_{\rm md}\hat{A}} \approx 1 - ig_o\hat{Q}_{\rm md}\hat{A}$ and use a theorem[2]:

$$\hat{A}|\Psi\rangle = \langle\hat{A}\rangle|\Psi\rangle + \Delta A|\Psi_\perp\rangle\,, \tag{2.4}$$

to show that before the post-selection, the system state is:

$$e^{-ig_o\hat{Q}_{\rm md}\hat{A}}|\Psi_{\rm in}\rangle = (1 - ig_o\hat{Q}_{\rm md}\hat{A})|\Psi_{\rm in}\rangle = (1 - ig_o\hat{Q}_{\rm md}\langle\hat{A}\rangle)|\Psi_{\rm in}\rangle - ig_o\hat{Q}_{\rm md}\Delta\hat{A}|\Psi_{\rm in\perp}\rangle. \tag{2.5}$$

Using the norm of this state $\| (1 - ig_o\hat{Q}_{\rm md}\hat{A})|\Psi_{\rm in}\rangle \|^2 = 1 + g_o^2\hat{Q}_{\rm md}^2\langle\hat{A}^2\rangle$, the probability to leave $|\Psi_{\rm in}\rangle$ un-changed after the measurement is:

$$\frac{1 + g_o^2\hat{Q}_{\rm md}^2\langle\hat{A}\rangle^2}{1 + g_o^2\hat{Q}_{\rm md}^2\langle\hat{A}^2\rangle} \longrightarrow 1 \quad (g_o \to 0) \tag{2.6}$$

while the probability to disturb the state (i.e. to obtain $|\Psi_{in\perp}\rangle$) is:

$$\frac{g_o^2\hat{Q}_{\rm md}^2\Delta\hat{A}^2}{1 + g_o^2\hat{Q}_{\rm md}^2\langle\hat{A}^2\rangle} \longrightarrow 0 \quad (g_o \to 0). \tag{2.7}$$

The final state of the measuring device is now a superposition of many substantially overlapping Gaussians with probability distribution given by $Pr(P_{\rm md}) = \sum_i |\langle a_i|\Psi_{\rm in}\rangle|^2 \exp\left\{-\frac{(P_{\rm md} - g_o a_i)^2}{2\Delta P_{\rm md}^2}\right\}$. This sum is a Gaussian mixture, so it can be approximated by a single Gaussian $\tilde{\Phi}_{\rm md}^{\rm fin}(P_{\rm md}) \approx \langle P_{\rm md}|e^{-ig_o\hat{Q}_{\rm md}\langle\hat{A}\rangle}|\Phi_{\rm md}^{\rm in}\rangle \approx \exp\left\{-\frac{(P_{\rm md} - g_o\langle\hat{A}\rangle)^2}{\Delta P_{\rm md}^2}\right\}$ centered on $g_o\langle\hat{A}\rangle$.

[2] where $\langle\hat{A}\rangle = \langle\Psi|\hat{A}|\Psi\rangle$, $|\Psi\rangle$ is any vector in Hilbert space, $\Delta A^2 = \langle\Psi|(\hat{A} - -\langle\hat{A}\rangle)^2|\Psi\rangle$, and $|\Psi_\perp\rangle$ is a state such that $\langle\Psi|\Psi_\perp\rangle = 0$.

2.1.2.4. **Information gain without disturbance – safety in numbers:** It follows from Eq. (2.7) that the probability for a collapse decreases as $O(g_o^2)$, but the measuring device's shift grows linearly $O(g_o)$, so $\delta P_{\rm md} = g_o a_i$ [22]. For a sufficiently weak interaction (e.g. $g_o \ll 1$), the probability for a collapse can be made arbitrarily small, while the measurement still yields information but becomes less precise because the shift in the measuring device is much smaller than its uncertainty $\delta P_{\rm md} \ll \Delta P_{\rm md}$ (figure 4.b). Nevertheless, if a large ($N \geq \frac{N'}{g_o}$) ensemble of particles is used, then the shift of all the measuring devices ($\delta P^{tot}_{\rm md} \approx g_o\langle \hat{A}\rangle \frac{N'}{g_o} = N'\langle \hat{A}\rangle$) becomes distinguishable because of repeated integrations, while the collapse probability still goes to zero.

Using these observations, we now emphasize that the average of any operator $\hat{A}$, i.e. $\langle \hat{A}\rangle \equiv \langle\Psi|\hat{A}|\Psi\rangle$, can be obtained in three distinct cases [18, 22]:

(i) **Statistical method with disturbance:** the traditional approach is to perform ideal-measurements of $\hat{A}$ on each particle, obtaining a variety of different eigenvalues, and then manually calculate the usual statistical average to obtain $\langle \hat{A}\rangle$.

(ii) **Statistical method without disturbance** as demonstrated by using $\hat{A}|\Psi\rangle = \langle \hat{A}\rangle|\Psi\rangle + \Delta A|\Psi_\perp\rangle$. Using an ensemble, we can verify there was no disturbance.

(iii) **Non-statistical method without disturbance** is the case where $\langle\Psi|\hat{A}|\Psi\rangle$ is the "eigenvalue" of a single "collective operator," $\hat{A}^{(N)} \equiv \frac{1}{N}\sum_{i=1}^{N}\hat{A}_i$ (with $\hat{A}_i$ the same operator $\hat{A}$ acting on the i-th particle). Using this, we are able to obtain information about $\langle\Psi|\hat{A}|\Psi\rangle$ without causing disturbance (or a collapse) and without using a statistical approach because any product state $|\Psi^{(N)}\rangle$ becomes an eigenstate of the operator $\hat{A}^{(N)}$. To see this, we apply the theorem $\hat{A}|\Psi\rangle = \langle \hat{A}\rangle|\Psi\rangle + \Delta A|\Psi_\perp\rangle$ to $\hat{A}^{(N)}|\Psi^{(N)}\rangle$, i.e.:

$$\hat{A}^{(N)}|\Psi^{(N)}\rangle = \frac{1}{N}\left[N\langle \hat{A}\rangle|\Psi^{(N)}\rangle + \Delta A\sum_i |\Psi_\perp^{(N)}(i)\rangle\right]. \tag{2.8}$$

where $\langle \hat{A}\rangle$ is the average for any one particle and the states $|\Psi_\perp^{(N)}(i)\rangle$ are mutually orthogonal and are given by $|\Psi_\perp^{(N)}(i)\rangle = |\Psi\rangle_1|\Psi\rangle_2...|\Psi_\perp\rangle_i...|\Psi\rangle_N$. That is, the ith state has particle i changed to an orthogonal state and all the other particles remain in the same state. If we further define a normalized state $|\Psi_\perp^{(N)}\rangle = \sum_i \frac{1}{\sqrt{N}}|\Psi_\perp^{(N)}(i)\rangle$ then the last term of Eq. (2.8) is $\frac{\Delta A}{\sqrt{N}}|\Psi_\perp^{(N)}\rangle$ and it's size is $|\frac{\Delta A}{\sqrt{N}}|\Psi_\perp^{(N)}\rangle|^2 \propto \frac{1}{N} \to 0$. Therefore, $|\Psi^{(N)}\rangle$ becomes an eigenstate of $\hat{A}^{(N)}$, with the value $\langle \hat{A}\rangle$ and not even a single particle has been disturbed (as $\hat{N} \to \infty$).

In the last case, the average for a single particle becomes a robust property over the entire ensemble, so a single experiment is sufficient to determine the average with great precision [22]. There is no longer any need to average over results obtained in multiple experiments. Tradition has dictated that when measurement interactions are limited so there is no disturbance on the system, then no information can be gained. However, we have now shown that when considered as a limiting process, the disturbance goes to zero more quickly than the shift in the measuring device, which means for a large enough ensemble, information (e.g. the expectation value) can be obtained even though not even a single particle is disturbed. This viewpoint thereby shifts the standard perspective on two fundamental postulates of QM.

2.1.2.5. **Adding a post-selection to the weakened interaction – Weak Values and Weak Measurements:** Having established a new measurement paradigm -information gain without disturbance – it is fruitful to inquire whether this type of measurement reveals new values or properties. With weak-measurements (which involve adding a post-selection to this ordinary – but weakened – von Neumann measurement), the measuring device registers a new value, the weak-value. As an indication of this, we insert a complete set of states $\{|\Psi_{\rm fin}\rangle_j\}$ into the outcome of the weak interaction of §2.1.2.3 (i.e. the expectation value $\langle\hat{A}\rangle$):

$$\langle\hat{A}\rangle = \langle\Psi_{\rm in}|\left[\sum_j |\Psi_{\rm fin}\rangle_j\langle\Psi_{\rm fin}|_j\right]\hat{A}|\Psi_{\rm in}\rangle = \sum_j |\langle\Psi_{\rm fin}|_j\Psi_{\rm in}\rangle|^2\ \frac{\langle\Psi_{\rm fin}|_j\ \hat{A}\ |\Psi_{\rm in}\rangle}{\langle\Psi_{\rm fin}|_j\Psi_{\rm in}\rangle} \tag{2.9}$$

If we interpret the states $|\Psi_{\rm fin}\rangle_j$ as the outcomes of a final ideal-measurement on the system (i.e. a post-selection) then performing a weak-measurement (e.g. with $g_o\Delta Q_{\rm md} \to 0$) during the intermediate time $t \in [t_{\rm in}, t_{\rm fin}]$, provides the coefficients for $|\langle\Psi_{\rm fin}|_j\Psi_{\rm in}\rangle|^2$ which gives the probabilities $Pr(j)$ for obtaining a pre-selection of $\langle\Psi_{\rm in}|$ and a post-selection of $|\Psi_{\rm fin}\rangle_j$. The intermediate weak-measurement does not disturb these states and the quantity $A_{\rm w}(j) \equiv \frac{\langle\Psi_{\rm fin}|_j\hat{A}|\Psi_{\rm in}\rangle}{\langle\Psi_{\rm fin}|_j\Psi_{\rm in}\rangle}$ is the weak-value of $\hat{A}$ given a particular final post-selection $\langle\Psi_{\rm fin}|_j$. From $\langle\hat{A}\rangle = \sum_j Pr(j)\,A_{\rm w}(j)$, one can think of $\langle\hat{A}\rangle$ for the whole ensemble as being constructed out of sub-ensembles of pre- and post-selected-states in which the weak-value is multiplied by a probability for a post-selected-state.

The weak-value arises naturally from a weakened measurement with post-selection: the final state of measuring device in the momentum representation becomes:

$$\begin{aligned}
\langle P_{\rm md}|\langle\Psi_{\rm fin}|e^{-ig_o\hat{Q}_{\rm md}\hat{A}}|\Psi_{\rm in}\rangle|\Phi^{\rm MD}_{\rm in}\rangle &\approx \langle P_{\rm md}|\langle\Psi_{\rm fin}|1 + ig_o\hat{Q}_{\rm md}\hat{A}|\Psi_{\rm in}\rangle|\Phi^{\rm MD}_{\rm in}\rangle\\
&\approx \langle P_{\rm md}|\langle\Psi_{\rm fin}|\ \Psi_{\rm in}\rangle\{1 + ig_o\hat{Q}\frac{\langle\Psi_{\rm fin}|\hat{A}|\Psi_{\rm in}\rangle}{\langle\Psi_{\rm fin}|\Psi_{\rm in}\rangle}\}|\Phi^{\rm MD}_{\rm in}\rangle\\
&\approx \langle\Psi_{\rm fin}|\Psi_{\rm in}\rangle\langle P_{\rm md}|e^{-ig_o\hat{Q}A_{\rm w}}|\Phi^{\rm MD}_{\rm in}\rangle\\
&\to \langle\Psi_{\rm fin}|\Psi_{\rm in}\rangle\exp\{-(P_{\rm md} - g_o\,A_{\rm w})^2\} \qquad (2.10)\\
\textit{where}\ A_{\rm w} &= \frac{\langle\Psi_{\rm fin}|\hat{A}|\Psi_{\rm in}\rangle}{\langle\Psi_{\rm fin}|\Psi_{\rm in}\rangle}
\end{aligned}$$

The final state of the measuring device is almost un-entangled with the system; it is shifted by a very unusual quantity, the weak-value, $A_{\rm w}$, which is not in general an eigenvalue of $\hat{A}$.

2.1.2.6. **How the weak-value of a spin-1/2 can be 100:** The weak-value for the spin-1/2 considered in §2.1.2 (which was confirmed experimentally for an analogous observable, the polarization [6]) is:

$$(\hat{\sigma}_{\xi=45^\circ})_{\rm w} = \frac{\langle\uparrow_y|\frac{\hat{\sigma}_y+\hat{\sigma}_x}{\sqrt{2}}|\uparrow_x\rangle}{\langle\uparrow_y|\uparrow_x\rangle} = \frac{\{\langle\uparrow_y|\hat{\sigma}_y\} + \{\hat{\sigma}_x|\uparrow_x\rangle\}}{\sqrt{2}\langle\uparrow_y|\uparrow_x\rangle} = \frac{\langle\uparrow_y|1+1|\uparrow_x\rangle}{\sqrt{2}\langle\uparrow_y|\uparrow_x\rangle} = \sqrt{2} \tag{2.11}$$

With a strong measurement, the component of spin $\hat{\sigma}_\xi$ is an eigenvalue, ±1, but the weak-value $(\hat{\sigma}_\xi)_w = \sqrt{2}$ (outside the range of eigenvalues of $\hat{\sigma}\cdot\mathbf{n}$) Weak values further outside the spectrum can be obtained by post-selecting states which are more anti-parallel to the pre-selection: e.g. if we post-select the +1 eigenstate of $(\cos\alpha)\sigma_x + (\sin\alpha)\sigma_z$, then $(\hat{\sigma}_z)_{\rm w} = g_o\tan\frac{\alpha}{2}$, yielding arbitrarily large values such as spin-100.

We have used such limited disturbance measurements to explore many paradoxes (see, e.g. [12, 19, 20]). A number of experiments have been performed to test the predictions made by weak-measurements and results have proven to be in very good agreement with theoretical predictions [4–8]. Since eigenvalues or expectation values can be **derived** from weak-values [3], we believe that the weak-value is indeed of fundamental importance in QM. In addition, the weak-value is the relevant quantity for all generalized weak interactions with an environment, not just measurement interactions. The only requirement being that the 2-vectors, i.e. the pre- and post-selection, are not significantly disturbed by the environment.

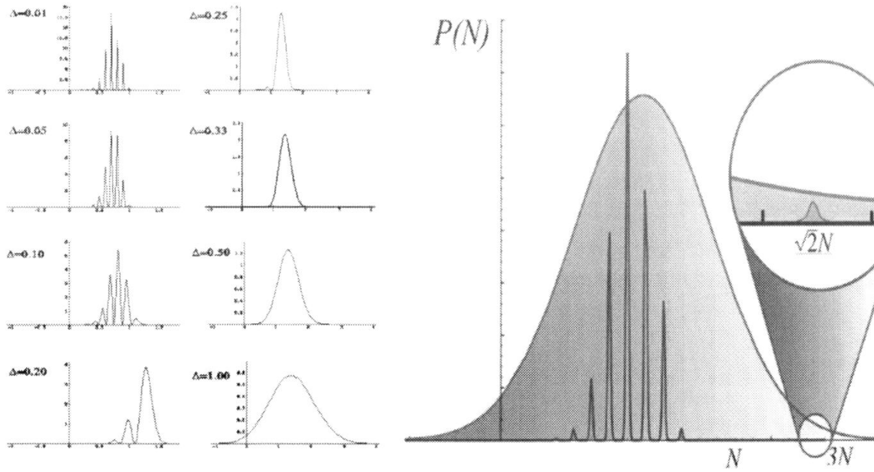

Figure 5. Left figure: "Measurement on a single system. Probability distribution of the pointer variable for the measurement of $A = (\sum_{i=1}^{20}(\sigma_i)_\xi)/20$ when the system of 20 spin-$\frac{1}{2}$ particles is pre-selected in the state $|\Psi_1\rangle = \prod_{i=1}^{20}|\uparrow_x\rangle_i$ and post-selected in the state $|\Psi_2\rangle = \prod_{i=1}^{20}|\uparrow_y\rangle_i$. While in the very strong measurements, $\Delta = 0.01 - 0.05$, the peaks of the distribution located at the eigenvalues, starting from $\Delta = 0.25$ there is essentially a single peak at the location of the weak value, $A_w = \sqrt{2}$." from [17]; **Right figure:** Combining the core results from the left figure, we draw here the probability distributions for strong and weak measurements. Before the post-selection is performed, the spikes in the distribution (colored in green) represent the possible measurement outcomes (which are eigenvalues) for an ideal measurement. The wider curve (colored in blue) represents the probabilities for a weak measurement. After the post-selection is performed, a single peak is left (colored in red) way out in the tails at the "impossible" location of the weak value, $A_w = N\sqrt{2}$.

2.1.2.7. **New approach to axiomatic structure of QM:** A central theme is that the future can only be relevant to the present if there is room to write-off its "influence" as a mistake. As can be seen in figure 5, the probability to obtain the weak value as an error of the measuring device is greater than the probability to actually obtain the weak value. This is essential to preserve both causality and free will: following a measurement of $\hat{A}$, we can choose to either post-select the system, or measure $\hat{A}$ again. If we post-select, we may interpret the result of the measurement of $\hat{A}$ as a weak-value A_w; if we re-measure $\hat{A}$, we may interpret the same result as an error. These two interpretations are consistent, for they apply to different ensembles. How we interpret a measured value depends on what else we choose to measure. We emphasize however that the weak value is not a random error. It is given by a simple heuristic expression; it is a highly predictable property; it always occurs whenenver we obtain a given post-selection and any weakened interaction, and not just measurements, are sensitive to the weak value.

Traditionally, the uncertainty of quantum mechanics was interpreted to mean that nature is "capricious." Using weak measurement we can derive the uncertainty from two axioms: 1) the future can be relevant to the present and 2) causality is maintained. That is, uncertainty is derived as a consequence of the consistency between causality and weak values: the amount of uncertainty is exactly what is needed in order that the future can be relevant for the present, without violating causality, thus providing a new answer to the question "Why does God play dice?" In order to enrich nature with temporal non-locality, and yet preserve cause-effect relations, we must have indeterminacy.

2.1.3. **TSQM led to new mathematics: Superoscillations** We might ask the question "How is it that the measuring device indicates this answer?" As we mentioned previously, during the process of the measurement, the pointer and the system get entangled. Upon performing the post-selection on the system, the entanglement is eliminated. This leaves the measuring device pointer in a superposition, i.e. shifted by eigenvalues of the measured operator. There is constructive interference around the weak value $\sqrt{2}N$ and destructive interference everywhere else.

TSQM is a **re-formulation** of QM, and therefore it must be possible to view the novel effects from the traditional single-vector perspective. This is precisely what super-oscillations teach us. In summary, there are 2 ways to understand weak values:

- the measuring device is registering the weak value as a property of the **system** as characterized by TSQM
- the weak value is a result of a complex interference effect in the **measuring device**, i.e. a superoscillation; the system continues to be described with a single-vector pursuant to the standard approach to QM

A universally accepted truth in spectral analysis is that signals, be they space dependent, as in optical imaging, or time dependent, cannot have details on a scale shorter than the shortest wavelength or shortest time period of their Fourier components. The new subject called *superoscillations* seems to violate this principle [23–27]. This is one of many examples in which TSQM has led to new mathematics, simplifications in calculations, and stimulated discoveries in other fields (see, e.g. [30]).

2.2. Emergence and weak values

In §2.2.1, we shall see that when properly analyzed, what seemed to be a "whole inseparable system" can in fact be physically separated into distinct parts residing at different locations. This opens a new door on the relationship between wholes and parts. In §2.2.2, we review a new category of Gedanken-experiment which uncovers a loophole in the correspondence principle and provides a new perspective on the transition from quantum to classical. With respect to item 4, i.e. "origin of laws", we show the transition is in fact dramatically more involved and it requires a complete revision of all our intuitions. Finally, in §2.2.3, because weak values follow a very different logical structure (e.g. the weak value of a product of observables is not equal to the product of their weak-values), we introduce even more dramatic categories of emergence.

2.2.1. **Quantum Cheshire cat** Another surprising effect originating from weak values is the ability to separate a system from it's properties [12, 40, 41], as suggested by the Cheshire cat story: "Well! I've often seen a cat without a grin," thought Alice; "but a grin without a cat! It's the most curious thing I ever saw in all my life!"

The essential property of a quantum Cheshire Cat is that the cat itself is located in one beam path, while its grin is located in the other one: i.e. the object and its property are spatially separated (see figure 6). In our experiment, the neutron plays the role of the cat and the cat's grin is represented by the neutron's spin, and, we have indeed successfully performed the first experimental confirmation of the quantum Cheshire Cat. [41]

To be explicit, we will first calculate a variety of operators such as $\langle\hat{\Pi}_{\mathrm{j}}\rangle \equiv |\mathrm{j}\rangle\langle\mathrm{j}|$, with $\mathrm{j} = I, II$, which are the projections on the path which the neutron takes, namely $|I\rangle$, stands for path I

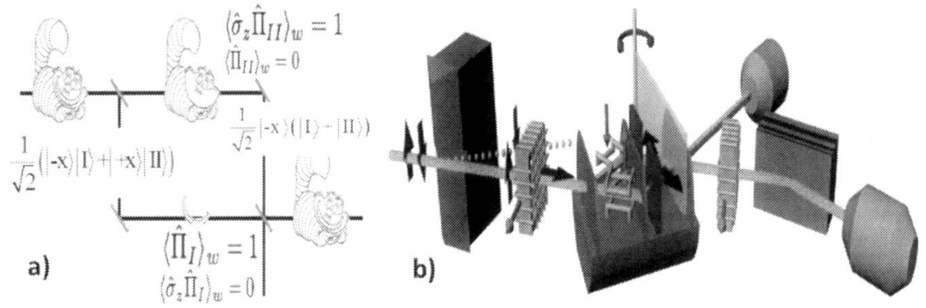

Figure 6. a) Artistic depiction of the quantum cheshire cat paradox: Inside the interferometer the cat travels along the upper beam path (path I), while its grin is located in the lower beampath (path II); b) Illustration of the experimental setup for the observation of a quantum Cheshire Cat in a neutron interferometer. [41]

and $|II\rangle$ for path II. We next consider operators corresponding to the cat's grin states which are given by the spin in the z-direction, namely $\hat{\Pi}_{\uparrow}$ the projector corresponding to $|\sigma_z = +1\rangle$ (grinning) and $\hat{\Pi}_{\downarrow}$, the projector corresponding to $|\sigma_z = -1\rangle$ (frowning). The pre-selected wave function for an individual particle is a superposition of the particle being in path I with a spin $+1$ in the x-direction superposed with the particle being in path II with a spin -1 in the x-direction:

$$|\Psi_{\text{in}}\rangle = \frac{1}{\sqrt{2}}|\sigma_x = +1\rangle|I\rangle + \frac{1}{\sqrt{2}}|\sigma_x = -1\rangle|II\rangle. \tag{2.12}$$

To make calculations easier and more intuitive, we re-write the pre-selection Eq. (2.12) in the z-basis as Eq. (2.13):

$$|\Psi_{\text{in}}\rangle = \frac{1}{\sqrt{2}}\Big[|\sigma_z = +1\rangle + |\sigma_z = -1\rangle\Big]|I\rangle + \frac{1}{\sqrt{2}}\Big[|\sigma_z = +1\rangle - -|\sigma_z = -1\rangle\Big]|II\rangle. \tag{2.13}$$

In order to measure the Cheshire Cat, after we pre-select the ensemble, we will next perform weak measurements on both the number of neutrons in a given path along with the value of the spin in a given path. After these weak measurements, the ensemble is then post-selected in the final state:

$$|\Psi_{\text{fin}}\rangle = \frac{1}{\sqrt{2}}|\sigma_x = -1\rangle\Big[|I\rangle + |II\rangle\Big]. \tag{2.14}$$

That is, it is post-selected in a superposition of both paths with a -1 spin in the x-direction. Again, it will be convenient to re-express this in the z-basis:

$$\begin{aligned} |\Psi_{\text{fin}}\rangle &= \frac{1}{\sqrt{2}}\Big[|\sigma_z = +1\rangle - -|\sigma_z = -1\rangle\Big]\Big[|I\rangle + |II\rangle\Big] \\ &= \frac{1}{\sqrt{2}}\Big[|\sigma_z = +1\rangle - -|\sigma_z = -1\rangle\Big]|I\rangle + \frac{1}{\sqrt{2}}\Big[|\sigma_z = +1\rangle - -|\sigma_z = -1\rangle\Big]|II\rangle. \end{aligned} \tag{2.15}$$

The ABL formula (2.2) tells us that if an ideal measurement is performed on Path I during the intermediate time, then, with certainty, the particle will not be found there. This can be seen intuitively: suppose we find the particle in Path I. The state is then $|\sigma_x = +1\rangle|I\rangle$ which is orthogonal to the post-selection. Because this is a contradiction, we conclude that the particle must be found in Path II. Calculating with ABL (ignoring unitary evolutions) confirms this:

$$|\langle\Psi_{\text{fin}}|\hat{\Pi}_I|\Psi_{\text{in}}\rangle|^2 = \frac{1}{2}\Big[\langle\sigma_z=+1| + \langle\sigma_z=-1|\Big]\frac{1}{2}\Big[|\sigma_z=+1\rangle - |\sigma_z=-1\rangle\Big] = 0 \tag{2.16}$$

and

$$|\langle\Psi_{\text{fin}}|\hat{\Pi}_{II}|\Psi_{\text{in}}\rangle|^2 = \frac{1}{2}\Big[\langle\sigma_z=+1| - \langle\sigma_z=-1|\Big]\frac{1}{2}\Big[|\sigma_z=+1\rangle - |\sigma_z=-1\rangle\Big] = \frac{1}{2}. \tag{2.17}$$

Therefore, the probability that the particle will be found in path II is:

$$\frac{|\langle\Psi_{\text{fin}}|\hat{\Pi}_{II}|\Psi_{\text{in}}\rangle|^2}{|\langle\Psi_{\text{fin}}|\hat{\Pi}_I|\Psi_{\text{in}}\rangle|^2 + |\langle\Psi_{\text{fin}}|\hat{\Pi}_{II}|\Psi_{\text{in}}\rangle|^2} = \frac{\frac{1}{2}}{(0+\frac{1}{2})} = 1 \tag{2.18}$$

and the probability that the particle will be found in path I is:

$$\frac{|\langle\Psi_{\text{fin}}|\hat{\Pi}_I|\Psi_{\text{in}}\rangle|^2}{|\langle\Psi_{\text{fin}}|\hat{\Pi}_I|\Psi_{\text{in}}\rangle|^2 + |\langle\Psi_{\text{fin}}|\hat{\Pi}_{II}|\Psi_{\text{in}}\rangle|^2} = \frac{0}{(0+\frac{1}{2})} = 0. \tag{2.19}$$

For the spin, we can again use ABL to ascertain its state:

$$|\langle\Psi_{\text{fin}}|\hat{\Pi}_\uparrow|\Psi_{\text{in}}\rangle|^2 = \frac{1}{2}\Big[\langle I| + \langle II|\Big]\frac{1}{2}\Big[|I\rangle + |II\rangle\Big] = \frac{1}{2} \tag{2.20}$$

and

$$|\langle\Psi_{\text{fin}}|\hat{\Pi}_\downarrow|\Psi_{\text{in}}\rangle|^2 = \frac{1}{2}\Big[\langle I| + \langle II|\Big]\frac{1}{2}\Big[|I\rangle - |II\rangle\Big] = 0. \tag{2.21}$$

Therefore, the probability that the spin is $\sigma_z = +1$ is:

$$\frac{|\langle\Psi_{\text{fin}}|\hat{\Pi}_\uparrow|\Psi_{\text{in}}\rangle|^2}{|\langle\Psi_{\text{fin}}|\hat{\Pi}_\uparrow|\Psi_{\text{in}}\rangle|^2 + |\langle\Psi_{\text{fin}}|\hat{\Pi}_\downarrow|\Psi_{\text{in}}\rangle|^2} = \frac{\frac{1}{2}}{(\frac{1}{2}+0)} = 1. \tag{2.22}$$

A theorem [9, 19, 32][3] states that if a single ideal measurement of an observable is performed between the pre- and post-selection, then if ABL asserts that the outcome is definite, then the weak value will be equal to this eigenvalue. Furthermore and more importantly, we may perform many weak measurements simultaneously on the same particle since they do not disturb each other. Therefore, all the counterfactual ABL statements can be measured (weakly) on the same particle and thereby simultaneously confirmed (to a certain extent). Using this theorem, we can ***simultaneously*** confirm Eq. (2.19), i.e. $\langle\hat{\Pi}_I\rangle_w = 0$ and Eq. (2.29), i.e. $\langle\hat{\Pi}_{II}\rangle_w = 1$. We can also simultaneously confirm Eq. (2.22), i.e. $\langle\sigma_z\rangle_w = \langle\hat{\Pi}_\uparrow - \hat{\Pi}_\downarrow\rangle_w = 1$ for all t. However, it is not the case that $\langle\sigma_z\rangle_w = 1$ along path II, i.e. where the particle is located. That is, the weak value of the spin in path II is

$$\langle\hat{\sigma}_z\hat{\Pi}_{II}\rangle_w = \langle\hat{\Pi}_\uparrow\hat{\Pi}_{II}\rangle_w - \langle\hat{\Pi}_\downarrow\hat{\Pi}_{II}\rangle_w = 0. \tag{2.23}$$

[3] This also provides a direct link to the counterfactual statements (§2.1.2.1) because all counterfactual statements which claim that something occurs with certainty, and which can actually be experimentally verified by **separate** ideal-measurements, continue to remain true when tested by weak-measurements. However, given that weak-measurements do not disturb each other, all these statements can be measured **simultaneously**.

The reason for this is that in general, the weak value of a product of observables is not equal to the product of their weak values (even when the observables commute). Therefore, to more fully understand what is going on, we will be interested in the weak values of all products of projection operators. Using Fig. 7, we demonstrate how these weak values, Eq. (2.9), can be easily ascertained. First, we can neglect normalization since the normalization of the numerator and denominator of Eq. (2.9) cancel. The denominator of Eq. (2.9), i.e. the scalar product $\langle\Psi_{\text{fin}}|\Psi_{\text{in}}\rangle$, is 2. The numerator of Eq. (2.9), $\langle\Psi_{\text{fin}}|\hat{A}|\Psi_{\text{in}}\rangle$, is then calculated by referring to figure 7, wherein each "column" yields ± 1:

$$\langle\hat{\Pi}_\uparrow\hat{\Pi}_I\rangle_w = \frac{1}{2} \quad (column\ 1) \tag{2.24}$$

$$\langle\hat{\Pi}_\downarrow\hat{\Pi}_I\rangle_w = -\frac{1}{2} \quad (column\ 2) \tag{2.25}$$

$$\langle\hat{\Pi}_\uparrow\hat{\Pi}_{II}\rangle_w = \frac{1}{2} \quad (column\ 3) \tag{2.26}$$

$$\langle\hat{\Pi}_\downarrow\hat{\Pi}_{II}\rangle_w = \frac{1}{2} \quad (column\ 4) \tag{2.27}$$

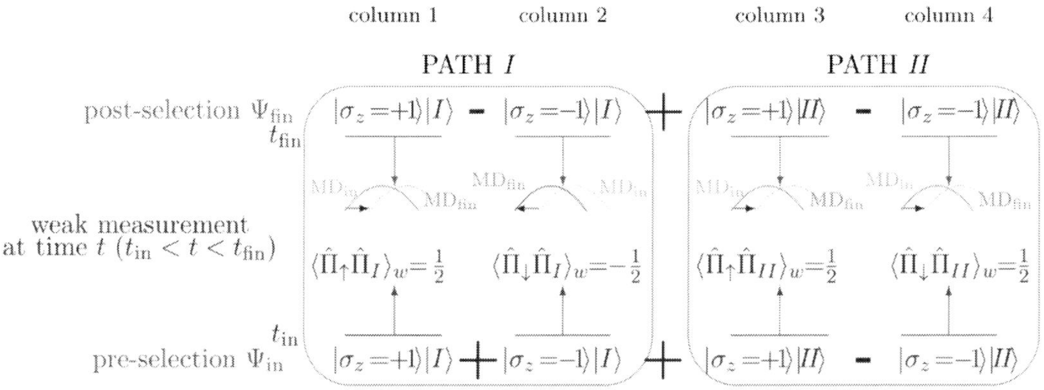

Figure 7. Graphical depiction of the Cheshire Cat calculations

We can now prove our main point, i.e. that the weak value of the spin in path I is

$$\langle\hat{\sigma}_z\hat{\Pi}_I\rangle_w = \langle\hat{\Pi}_\uparrow\hat{\Pi}_I\rangle_w - \langle\hat{\Pi}_\downarrow\hat{\Pi}_I\rangle_w = 1. \tag{2.28}$$

And if we now use an ensemble of N particles, we get the same results, only now with high precision: all N particles are (weakly) in path II, but there is no field there; there are no particles in path I, but there is a field equal to N spins there. Alice would say "Curiouser and curiouser."

Lastly, we see that $\langle\hat{\Pi}_\downarrow\hat{\Pi}_I\rangle_w = -\frac{1}{2}$ is not only outside the spectrum, it is even a seemingly impossible negative number. Such predictions have been made before [12, 19, 32] and confirmed experimentally [41]. More importantly, though strange, these weak values form a logical and consistent pattern, and can provide intuition for the discovery of new phenomena. For example, instead of performing the full calculation using Eq. (2.9) to give eqs. 2.24-2.27, let us derive the results using this "weak" logic. The weak value of the occupation in path II, $\langle\hat{\Pi}_{II}\rangle_w$, i.e. not caring what is the value of the spin, is just the sum of Eq. (2.26) and Eq. (2.27),

$$\langle\hat{\Pi}_\uparrow\hat{\Pi}_{II}\rangle_w + \langle\hat{\Pi}_\downarrow\hat{\Pi}_{II}\rangle_w = \langle\hat{\Pi}_{II}\rangle_w = 1 \tag{2.29}$$

Using Eq. (2.23) and Eq. (2.29) we easily derive Eq. (2.26) and Eq. (2.27). Furthermore, the weak value of the occupation in path I, $\langle \hat{\Pi}_I \rangle_w$, i.e. not caring what is the value of the spin, is just the sum of Eq. (2.24) and Eq. (2.25):

$$\langle \hat{\Pi}_\uparrow \hat{\Pi}_I \rangle_w + \langle \hat{\Pi}_\downarrow \hat{\Pi}_I \rangle_w = \langle \hat{\Pi}_I \rangle_w = 0. \tag{2.30}$$

and therefore, we can deduce that $\langle \hat{\Pi}_\uparrow \hat{\Pi}_I \rangle_w = -\langle \hat{\Pi}_\downarrow \hat{\Pi}_I \rangle_w$. Finally, since there is just one particle, we know:

$$1 = \langle \sigma_z \rangle_w = \langle \hat{\Pi}_\uparrow - \hat{\Pi}_\downarrow \rangle_w = \langle \hat{\Pi}_\uparrow \hat{\Pi}_I \rangle_w + \langle \hat{\Pi}_\uparrow \hat{\Pi}_{II} \rangle_w - \langle \hat{\Pi}_\downarrow \hat{\Pi}_I \rangle_w - \langle \hat{\Pi}_\downarrow \hat{\Pi}_{II} \rangle \tag{2.31}$$

from which we can easily derive Eq. (2.24) and Eq. (2.25).

2.2.2. **New perspectives on emergence and the correspondence principle** It is generally believed that, similar to the way in which relativistic mechanics reduces to Newtonian mechanics when c becomes infinite, also QM "reduces" to classical mechanics in a limit specified by the correspondence principle (i.e. in the classical limit of $\hbar \to 0$). Similarly, the reduction of quantum optics to wave optics has been considered to be relatively simple. It is not so. As we show in [35], the classical limit of quantum optics is dramatically more involved and requires a fundamental revision of our intuitions. The revised intuitions can serve as a guide to finding novel quantum effects.

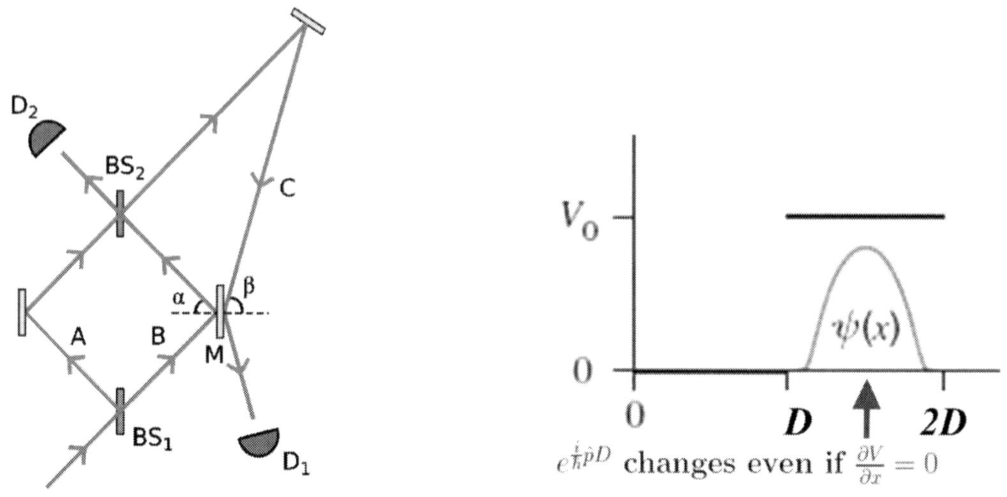

Figure 8. Left figure: Mach-Zehnder interferometer with one output beam reflected back onto the exterior side of mirror M. [35]; **Right figure:** A potential with 2 values and a wave-packet with support only in the interval $D < x < DL$.

Consider the interferometric experiment in Fig. 8. The three mirrors used in the experiment are perfectly reflecting, with mirror M being silvered on both sides. One of the (macroscopic) beams of light emerging from the interferometer is first reflected by a supplementary mirror onto the mirror M which then reflects it towards the detector. Mirror M receives therefore two momentum kicks, one from the light inside the interferometer and one from the beam that reflects on it from the outside. The quantum and classical calculations of course lead to the same result. The issue however is with the story each theory has to tell. Although the external beam has a shallower incidence angle than the inside beam, its intensity is much higher and the momentum kick given by it is larger, hence the mirror is pushed inwards. Classically, the external beam plays the central role – one would be tempted to assume that quantum mechanically the photons that constitute this beam are the ones responsible for the inward push. Remarkably, this is not so! According to QM, it is the photons that end up in the other detector, D_2, that kick M inside. Astonishingly, although they collide with the mirror only from the inside of the interferometer, they do not push the mirror outwards; rather they pull it in!

Since the story told by quantum mechanics is dramatically different from the classical one, this will have implications for our understanding of the relationship between different physical scales. It also serves as guide for intuition. E.g., suppose that, by a quantum fluctuation, we receive more than the average number of photons at detector D_1. The classical intuition will lead us to expect that now the mirror will receive an even larger momentum inwards. Our QM analysis tells us differently since the effect is due to photons going towards D_2 and now there are fewer of them, the inward momentum will be smaller.

2.2.3. **New aspects of non-locality and weak values** We recently introduced [14] a family of new weak value effects which have a number of implications for EmQM. First of all, up until now, it was generally thought that any unusual connection between particles must be due to the fact that the particles are in an entangled state. In order to create the entangled state, originally the particles needed to be near each other. However, we have proposed situations in which the particles could have been prepared at space-like separated intervals, i.e. they simply knew nothing about each other. Nevertheless, they all exhibit non-classical anti-correlation, even though the particles are in a product state, i.e. there is no entanglement involved. This is the first such example that we know of.

2.3. Weak values and hidden variables

In [32, 33], we analyzed contextuality in terms of pre- and post-selection, and showed that it is possible to assign definite values to observables in a new and surprising way. Novel experimental aspects of contextuality can be demonstrated with weak measurements. We also proved that every PPS-paradox with definite predictions directly implies 'quantum contextuality' which is introduced as the analogue of contextuality at the level of quantum mechanics rather than at the level of hidden variable theories. Finally, we argued that certain results of these measurements (e.g. eccentric weak values outside the eigenvalue spectrum) cannot be explained by a 'classical-like' hidden variable theory and introduce new restrictions on hidden variable approaches.

2.4. Weak information

We recently introduced [15] a novel approach to information called "weak information." We believe it is quintessentially unique to quantum mechanics and therefore a potential precursor and/or resource for quantum applications. Weak information has aided us in exploring the implications for our conventional measures and concepts of randomness. We applied weak information to quantum communication systems and were able to prove that it can be used to significantly enhance the security of practical quantum key distribution systems.

2.5. Dynamical nonlocality and the whole-part dialogue

Dynamical nonlocality [37] impacts the dialogue concerning the relationship between parts and wholes. Motivated by the AB non-locality and by weak measurements, we look for new manifestations of the dynamics of QM which are not predicted by the dynamics of classical mechanics. The key difference is that the equation of motion of QM exhibits a new kind of non-locality, which is best described by using modular variables. This is in contrast with Bell-inequality violations which follow from the Hilbert-space structure of QM (i.e. they are purely kinematic).

In [31] we applied this to the double-slit using single localized particles. From the perspective of a single particle, the central mystery is how does the localized particle passing through one of the slits sense whether or not the distant slit is open (closed), causing it to scatter (or not scatter) into a region of destructive interference? Feynman said "Nobody knows how it can be like that." We showed that the Heisenberg picture leads us to a **physical** explanation for the different behavior of a *single* particle when the distant slit is open or closed. Instead of having a quantum wave that passes through all slits, we have a localized particle with *non-local* interactions with the other slits.

Interference of course depends on the relative phase α between different lumps: $\Psi_\alpha = \psi_{\rm L} + e^{i\alpha}\psi_{\rm R}$. Until $\psi_{\rm L}(x,t)$ later overlaps with $\psi_{\rm R}(x,t)$ (during interference), neither the current nor the probability densities depend on α. Observables which are simple functions of position and momentum are not sensitive to the relative phase between the "lumps," suggesting that these dynamical variables are not the most appropriate to describe quantum interference phenomena. Operators that *are* sensitive to the relative phase are exponentials of the position and momentum. We call these "modular variables." These operators translate the different "lumps" so that they overlap. Using $H = \frac{p^2}{2m} + V(x)$ and $e^{\frac{i}{\hbar}\hat{p}D}V(x)e^{-\frac{i}{\hbar}\hat{p}D} = V(x+D)$, we find **non-local** [2, 10] Heisenberg equations of motion for modular variables:

$$\frac{d}{dt}e^{\frac{i}{\hbar}\hat{p}D} = \frac{i}{\hbar}[H, e^{\frac{i}{\hbar}\hat{p}D}] = \frac{i}{\hbar}[V(x) - V(x+D)]e^{\frac{i}{\hbar}\hat{p}D} \tag{2.32}$$

with $e^{\frac{i}{\hbar}\hat{p}D}$ changing even when $\frac{\partial V}{\partial x} = 0$. This, essentially quantum phenomenon, has no classical counterpart (see Fig. 8). The classical equations of motion for any function $f(p)$ derives from the Poisson bracket:

$$\frac{df(p)}{dt} = \{f(p), H\}_{PB} = -\frac{\partial f}{\partial p}\frac{\partial H}{\partial x} + \underbrace{\frac{\partial f}{\partial x}}_{=0}\frac{\partial H}{\partial p} = 0 \tag{2.33}$$

i.e. $f(p)$ changes only if $\frac{\partial V}{\partial x} \neq 0$ at the particle's location. We first proposed a way to experimentally observe dynamical non-locality (a method which requires weak measurements) [31] and this experiment has been successfully carried out [45].

2.6. Internal and external quantum frames-of-reference

Relevant to EmQM is our recent introduction of a new "internal" quantum frame-of-reference. E.g. ordinarily we consider a charged particle moving in the presence of fixed EM potentials. If however the charge is strong enough, it can influence these potentials. It will then see an effective, "private" potential, which rather than being determined by external charges only, is critically influenced by the particle under consideration- whereas other infinitesimal test charges, even if put at the very same location, see only the original, predetermined external "public" potential. The public potential can even vanish while the private potential remains non-zero. [36]

3. Generalizations of QM

While the alternative formulations of QM discussed in the previous section (§2) are equivalent to standard QM (i.e. they yield identical predictions), the underlying concepts are dramatically different. If new physics someday requires us to change QM, then these alternative formulations may provide a jumping-off point. We now present several generalizations of QM based on these fundamental conceptual shifts and discuss their implications for EmQM.

3.1. Reformulation of dynamics: Each moment a new universe

We review a generalization of QM suggested by TSQM [12,13,44] which addresses the "artificial" separation in all areas of theoretical physics between the kinematic and dynamical descriptions. We note [12,13,44] that the description of the time evolution given by QM does not appropriately represent multi-time-correlations (similar to EPR entanglement but instead of being between two particles in space, they are correlations for a single particle between two different *times*). Multi-time-correlations, however, can be represented by using TSQM. As a consequence, the general notion of time in QM is changed from the current conceptual framework which was inherited from classical mechanics, i.e.:

> **1)** the universe is viewed as unique, and the objects which inhabit it just change their state in time. In this view, time is "empty," it just propagates a state forward; the operators of the theory create the time evolution;

to a new conceptual framework in which:

> **2)** each instant corresponds to a new pair of Hilbert spaces, (i.e., each instant is a new degree of freedom; in a sense, a new universe); instead of the operators creating the time evolution as in the previous approach, an entangled state (in time) "creates" the propagation: a whole new set of structures within time is able to "propagate" a quantum state forward in time.

This new approach has a number of useful qualities, e.g.: 1) the dynamics and kinematics can both be represented simultaneously in the same language, a single entangled vector (in many Hilbert spaces), and 2) a new, more fundamental complementarity between dynamics and kinematics is naturally introduced. This approach also leads to a new solution to the measurement problem which we model by uncertain Hamiltonians.

While we leave all details to other publications [12, 13, 44]), in brief, consider a spin-1/2 particle, initially polarized "up" along the z axis, and having the Hamiltonian $H = 0$. In this case the time evolution of the particle is trivial,

$$|\Psi(t)\rangle = constant = |\sigma_z = 1\rangle. \tag{3.34}$$

To see the deficiency in representing multi-time-correlations, we will consider an isomorphism between the correlations for a **single** particle at **multiple** instants of time and the correlations between **multiple** particles at a **single** instant of time. Therefore, we ask if we could prepare N spin-1/2 particles such that if we perform measurements on them at some time t_0 we would obtain the same information as we would obtain by measuring the state of the original particle at N different time moments, t_1, $t_2 ... t_N$? Since the state of the original particle at all these moments is $|\sigma_z = 1\rangle$, one would suppose that this task can be accomplished by preparing the N particles each polarized "up" along the z axis, that is:

$$|\sigma_z = 1\rangle_1 |\sigma_z = 1\rangle_2 ... |\sigma_z = 1\rangle_N. \tag{3.35}$$

But this mapping is not appropriate for many reasons. One reason is that the time evolution (3.34) contains subtle correlations (i.e. multi-time-correlations) and which do not appear in the state (3.35) but which can actually be measured. It is generally believed that since the particle is at every moment in a definite state of the z-spin component, the z-spin component is the only thing we know with certainty about the particle – all other spin components do not commute with σ_z and cannot thus be well-defined. However, there are *multi-time* variables whose values are known with certainty, given the evolution (3.34). For example, although the x spin component is not well defined when the spin is in the $|\sigma_z = 1\rangle$ state, we know that it

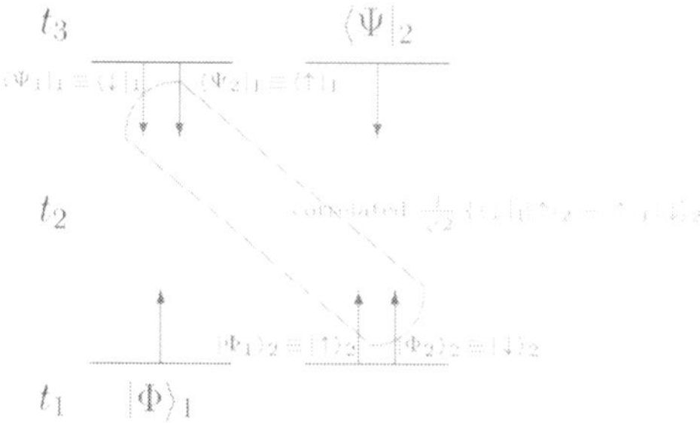

Figure 9. Particle 1 is correlated to the pre-selected state of particle 2.

is constant in time, since the Hamiltonian is zero. Thus, for example, the two-time observable $\sigma_x(t_4) - \sigma_x(t_2) = 0$ is definite ($t_2 < t_4$). However, there is no state of N spins such that

$$\hat{\sigma}^1_{\hat{n}} = \hat{\sigma}^2_{\hat{n}} = ... = \hat{\sigma}^N_{\hat{n}} \tag{3.36}$$

for every direction $\hat{n}$ as would be required for all the multi-time-correlations. At best, one may find a **two-particle state** (the EPR or Bell state) for which the spins are *anti-correlated* instead of correlated i.e. $\hat{\sigma}^1_{\hat{n}} = -\hat{\sigma}^2_{\hat{n}}$. However, e.g., for 3 particles, only 2 of them can be completely anti-correlated, thus it cannot be extended to N particles.

Although a state of N spin 1/2 particles with complete correlations among all their spin components as required by Eq. (3.36) doesn't exist in the usual sense, there are pre-and-post-selected states with this property given by TSQM. By way of example (see figure 9), the post-selected state of particle 1 can be completely correlated with the pre-selected state of particle 2 as described by the state $\Phi = \frac{1}{\sqrt{2}}\{\langle\downarrow|_1|\uparrow\rangle_2 - \langle\uparrow|_1|\downarrow\rangle_2\}$. We are now able to preserve the single particle's multi-time-correlations by simply "stacking" the N spin-1/2 particles "one on top of the other" along the time axis (Fig. 10). As a result of the correlations between the pre-and-post-selected states, a verification measurement of $\hat{\sigma}_x(t_4) - \hat{\sigma}_x(t_2)$ (see left part of Fig. 10), will yield 0, i.e. perfect multi-time correlations because $\hat{\sigma}_x(t_2, particle\,2) - \hat{\sigma}_x(t_2, particle\,1) = 0$ (see right part of Fig. 10). When "stacked" onto the time axis, these correlations act like the identity operator and thus evolve the state forward, handing-off or effectively propagating a state from one moment to the next (although nothing is "really" propagating in this picture).

3.2. Destiny states: New solution to measurement problem

Up until now we have limited ourselves to the possibility of 2 boundary conditions which obtain their assignment due to selections made before and after a measurement. It is feasible and even suggestive to consider an extension of QM to include both a wavefunction arriving from the past and a second "destiny" wavefunction coming from the future which are determined by 2 boundary conditions, rather than a measurement and selection. This proposal could solve the issue of the "collapse" of the wavefunction in a new and more natural way: every time a measurement takes place and the possible measurement outcomes decohere, then the future boundary condition simply selects one out of many possible outcomes [12, 34].

3.3. Emergence and origin of laws

Usually, whole-part interaction is considered only to be more *methodologically* efficient. However, the participants of the EmQM conference inquired about the controversial notion of the *ontological* status of top-down or whole-part interactions. Unlike the more familiar bottom-up form of causality, the idea of top-down causality is in general unstable with respect

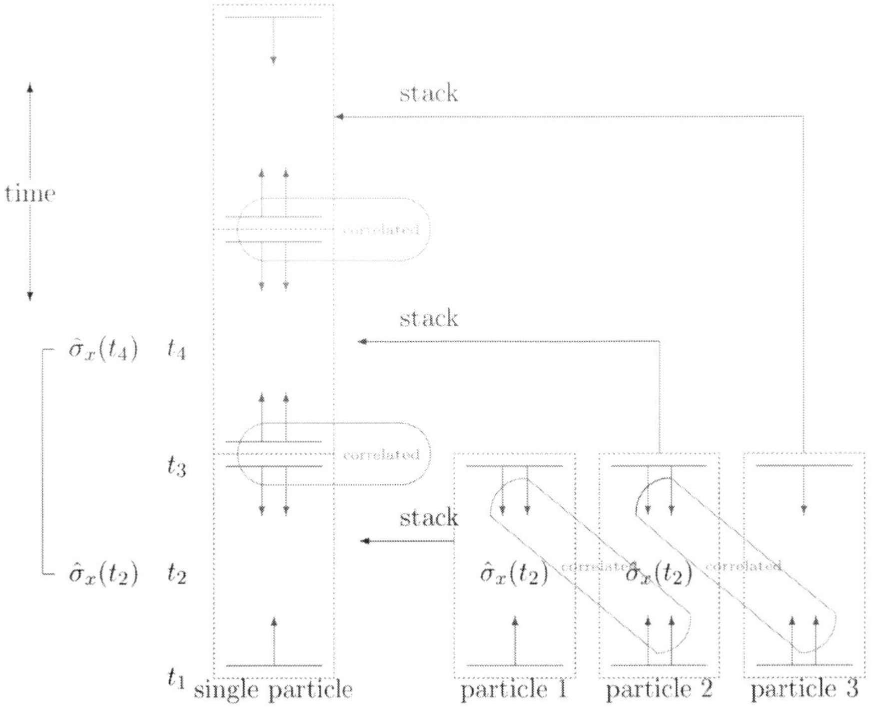

Figure 10. Measuring $\hat{\sigma}_x(t_4) - \hat{\sigma}_x(t_2)$ for the single spin-1/2 particle on the left ensures perfect multi-time-correlations because $\hat{\sigma}_x(t_2, particle\, 2) - \hat{\sigma}_x(t_2, particle\, 1) = 0$.

to fluctuations. Suggestions that quantum entanglement implies whole-part (top-down) interactions in atomic physics have been criticized by Shimony [42]:

> Atoms thus exhibit form in Aristotle's sense, and even have the tendency to maintain this form, which phenomenologically is like his final cause. But the Aristotelian form is achieved by Democritean means – by interactions among the electrons and the nucleus, which leave these building blocks intact.

To conclude this article, we mention some very speculative ways to address this. First, a "destiny-state" suggests a form of top-down causality which is stable to fluctuations because post-selections are performed on the entire Universe and by definition no fluctuation exists outside the Universe.

Secondly, the dynamics-kinematics generalization (§3.1) [12, 13, 44] suggests a novel way to think about dynamical laws. One implication is the fact that although we may know the dynamics on a particular time-scale T, this doesn't mean that we know anything about the dynamics on a smaller time-scale: consider a superposition of unitary evolutions (using $e^{-iHT} = \{e^{\frac{-iHT}{N}}\}^N$):

$$\int g(\nu)e^{-iH(\nu)t}d\nu \rightarrow \int g(\nu)\{1 + iH(\nu)t\}d\nu \underbrace{\longrightarrow}_{if\ \int g(\nu)d\nu=1} 1 + i\int g(\nu)H(\nu)td\nu. \tag{3.37}$$

This theory is the same as the usual theory but with an effective Hamiltonian $H_{eff} = \int g(\nu)H(\nu)td\nu$. The finer grained Hamiltonian can be expressed as a superposition of evolutions $e^{\frac{-iHT}{N}} = \sum_n \alpha_n e^{\frac{-i\beta_n HT}{N}}$, i.e. the Hamiltonian can be represented as a superposition of different laws given by pre- and post-selection [12,34]. This would allow new types of non-locality in which patterns at higher levels can set boundary conditions for lower levels. E.g. new potentials which become non-zero only when a certain number of particles are in a particular configuration so that new forms of non-locality come into being at different levels of complexity. Such new phenomena would not be inconsistent with existing experiments because at physical scales previously tested, per Eq. (3.37), these non-localities would destructively interfere with each other.

References

[1] Aharonov Y, Bergmann P G and Lebowitz J L 1964 *Phys. Rev.* **134** B1410
[2] Aharonov Y and Rohrlich D 2005 *Quantum paradoxes*, Weinheim; Cambridge: Wiley-VCH
[3] Aharonov Y and Bohm D 1961 *Phys. Rev.* **122** 1649
[4] Parks A D, Cullin D W and Stoudt D C 1979 *Proc. R. Soc. London, Ser. A* **454** 2997-3008
[5] Pryde G J, O'Brien J L, White A G, Ralph T C and Wiseman H M 2005 *Phys. Rev. Lett.* **94** (22) 220405
[6] Ritchie N W M, Story J G and Hulet R G 1991 *Phys. Rev. Lett.* **66** 1107
[7] Ahnert S E and Payne M C 2004 *Phys. Rev.* **A 70** (4) 042102
[8] Wiseman H M 2002 *Phys. Rev.* **A 65** (3) 032111 Part A
[9] Aharonov Y and Vaidman L 1991 *J. Phys.* **A 24** 2315
[10] Aharonov Y, Pendelton H and Petersen A 1969 *Int. J. Theor. Phys.* **2** 213
[11] Aharonov Y and Vaidman L 1990 *Phys. Rev.* **A 41** 11
[12] Tollaksen J 2001 *Quantum Reality and Nonlocal Aspects of Time* PhD thesis, Boston University
[13] Aharonov Y, Popescu S, Tollaksen J and Vaidman L 2009 *Phys. Rev.* **A 79** 052110
[14] Aharonov Y, Colombo F, Popescu S, Sabadini I, Struppa D C and Tollaksen J, *Quantum violation of the pigeonhole principle*, forthcoming
[15] Aharonov Y, Colombo F, Popescu S, Sabadini I, Struppa D C and Tollaksen J, *Weak Information*, forthcoming.
[16] Nussinov S and Tollaksen J 2012 *Found. Phys.* **42** (9) 1186-1199
[17] Aharonov Y and Vaidman L 2001 *Preprint* arXiv:quant-ph/0105101
[18] Tollaksen J 2007 *J. Phys.: Conf. Series* **70** 012015 *Preprint* quant-ph/0703038
[19] Aharonov Y, Botero A, Popescu S, Reznik B and Tollaksen J 2002 *Phys. Lett.* **A 301** (3-4) 130-138
[20] Brout R, Massar S, Parentani R, Popescu S and Spindel Ph 1995 *Phys. Rev.* **D 52** 1119; Nussinov S and Tollaksen J 2008 *Phys. Rev.* **D 78** 036007
[21] Oreshkov and Brun 2005 *Phys. Rev. Lett.* **95** 110409
[22] Tollaksen J 2007 *Non-statistical weak measurements* in: Quantum Information and Computation V, Ed E Donkor, A Pirich and H Brandt, *Proc of SPIE* **6573** (SPIE, Bellingham, WA), CID 6573-33
[23] Aharonov Y, Colombo F, Sabadini I, Struppa D C and Tollaksen J 2011 *J. Phys.* **A 44** 365304
[24] Aharonov Y, Colombo F, Sabadini I, Struppa D C and Tollaksen J 2013 *Complex Anal. Oper. Theory* **7** 1299–1310
[25] Aharonov Y, Colombo F, Sabadini I, Struppa D C and Tollaksen J 2013 *J. Math.: Pure Appl.* **99** (9) 165–173
[26] Aharonov Y, Colombo F, Nussinov S, Sabadini I, Struppa D C and Tollaksen J 2012 *Proc. R. Soc. London, Ser. A* **468** 3587–3600
[27] Aharonov Y, Colombo F, Sabadini I, Struppa D C and Tollaksen J 2013 *On superoscillations longevity: a windowed Fourier transform approach* in: Quantum Theory: A Two-Time Success Story, Springer, 313–325
[28] Aharonov Y, Colombo F, Sabadini I, Struppa D C and Tollaksen J 2013 *Evolution of superoscillatory data, Preprint*
[29] Aharonov Y, Popescu S and Tollaksen J 2010 *Physics Today* November, p.27
[30] Nussinov S and Tollaksen J 2008 *Phys. Rev.* **D 78** 036007
[31] Tollaksen J, Aharonov Y, Casher A, Kaufherr T and Nussinov S 2010 *New J. Phys.* **12** 013023
[32] Tollaksen J 2007 *J. Phys. A: Math. Gen.* **40** 9033-9066; *Preprint* quant-ph/0602226
[33] Tollaksen J 2007 *J. Phys.: Conf. Ser.* **70** 012014
[34] Aharonov Y and Tollaksen J 2009 *New insights on Time-Symmetry in Quantum Mechanics* in: Visions of Discovery, Ed. R Y Chiao, M L Cohen, A J Leggett, W D Phillips and C L Harper Jr, Cambridge: Cambridge University Press
[35] Aharonov Y, Botero A, Nussinov S, Popescu S and Tollaksen J 2013 *New J. Phys.* **15** 093006
[36] Kaufherr T, Aharonov Y, Nussinov S, Popescu S and Tollaksen J 2011 *Phys. Rev.* **A 83** 052127
[37] Tollaksen J and Aharonov Y 2009 *J. Phys.: Conf. Ser.* **196** 012006
[38] Aharonov Y, Botero A, Popescu S, Reznik B and Tollaksen J 2002 *Phys. Lett.* **A 301** (3-4) 130-138
[39] Aharonov Y, Massar S, Popescu S, Tollaksen J and Vaidman L 1996 *Phys. Rev. Lett.* **77** 983-987
[40] Aharonov Y, Popescu S, Rohrlich D and Skrzypczyk P 2013 *New J. Phys.* **15** 113015; *Preprint* arXiv:1202.0631

[41] Denkmayr T, Geppert H, Sponar S, Lemmel H, Matzkin A, Tollaksen J and Hasegawa Y 2013 *Preprint* arXiv:1312.3775
[42] Shimony A 1993 *The methodology of synthesis: parts and wholes in low-energy physics* in: Kelvin's Baltimore Lectures and Modern Theoretical Physics, Eds R Kargon and P Achinstein, 1987, M.I.T. Press, re-printed in *Search for a Naturalistic Worldview* by A Shimony, vol 2, Cambridge Univ. Press
[43] Cho A 2011 *Science* **333** 690-693
[44] Aharonov Y, Popescu S and Tollaksen J 2013 *Each instant of time a new universe* in: Quantum Theory: a Two-Time Success Story. Yakir Aharonov Festschrift. Eds D Struppa and J Tollaksen, Springer, Milan
[45] Spence S E and Parks A D 2012 *Found. Phys.* **42** (6) 803-815; *Preprint* arXiv:1010.3289

AUTHOR INDEX

Adler, S L 11
Bassi, A 239
Caticha, A 88
Cetto, A M 66
De Gosson, M A 289
De La Peña, L 66
De Raedt, H 274
Diósi, L 203
Elze, H-T 31
Faber, M 100
Flack, R 167
Fussy, S 51
Ghirardi, G C 225
Grössing, G 1, 51
Hasegawa, Y 261
Hiley, B J 167, 289
Hofer, W A 140
Hooft, G 'T 17
Hossenfelder, S 189
Hu, B L 210
Jizba, P 127
Khrennikov, A 193
Kupczynski, M 156
Michielsen, K 274
Nelson, E 136
Nieuwenhuizen, T M 78
Ord, G N 115
Ozawa, M 249
Pascasio, J Mesa 51
Rauch, H 180
Romano, R 225
Sanz, Á S 302
Scardigli, F 127
Schuch, D 40
Schwabl, H 51
Tollaksen, J 316
Ulbricht, H 239
Valdés-Hernández, A 66
Walleczek, J 1

Institute of Physics
Dirac House, Temple Back
Bristol BS1 6BE UK

proceedings.com